Mathematical Methods for Physicists
A concise introduction

This text is designed for an intermediate-level, two-semester undergraduate course in mathematical physics. It provides an accessible account of most of the current, important mathematical tools required in physics these days. It is assumed that the reader has an adequate preparation in general physics and calculus.

The book bridges the gap between an introductory physics course and more advanced courses in classical mechanics, electricity and magnetism, quantum mechanics, and thermal and statistical physics. The text contains a large number of worked examples to illustrate the mathematical techniques developed and to show their relevance to physics.

The book is designed primarily for undergraduate physics majors, but could also be used by students in other subjects, such as engineering, astronomy and mathematics.

TAI L. CHOW was born and raised in China. He received a BS degree in physics from the National Taiwan University, a Masters degree in physics from Case Western Reserve University, and a PhD in physics from the University of Rochester. Since 1969, Dr Chow has been in the Department of Physics at California State University, Stanislaus, and served as department chairman for 17 years, until 1992. He served as Visiting Professor of Physics at University of California (at Davis and Berkeley) during his sabbatical years. He also worked as Summer Faculty Research Fellow at Stanford University and at NASA. Dr Chow has published more than 35 articles in physics journals and is the author of two textbooks and a solutions manual.

Mathematical Methods for Physicists

A concise introduction

TAI L. CHOW

California State University

PUBLISHED BY THE PRESS SYNDICATE OF THE UNIVERSITY OF CAMBRIDGE
The Pitt Building, Trumpington Street, Cambridge, United Kingdom

CAMBRIDGE UNIVERSITY PRESS
The Edinburgh Building, Cambridge CB2 2RU, UK http://www.cup.cam.ac.uk
40 West 20th Street, New York, NY 10011-4211, USA http://www.cup.org
10 Stamford Road, Oakleigh, Melbourne 3166, Australia
Ruiz de Alarcón 13, 28014 Madrid, Spain

First published 2000

Printed in the United Kingdom at the University Press, Cambridge

Typeface Times New Roman 10/13pt

A catalogue record for this book is available from the British Library

Library of Congress Cataloguing in Publication data

Chow, Tai L.
Mathematical physics : mathematical methods for scientists and engineers / Tai L. Chow.
p. cm.
Includes bibliographical references and index.
ISBN 0-521-65227-8 (hc.) – ISBN 0-521-65544-7 (pbk.)
1. Mathematical physics. I. Title

QC20 .C57 2000
530.15 21–dc21   99–044592

ISBN 0 521 65227 8 hardback
ISBN 0 521 65544 7 paperback

Contents

Preface

This book evolved from a set of lecture notes for a course on 'Introduction to Mathematical Physics', that I have given at California State University, Stanislaus (CSUS) for many years. Physics majors at CSUS take introductory mathematical physics before the physics core courses, so that they may acquire the expected level of mathematical competency for the core course. It is assumed that the student has an adequate preparation in general physics and a good understanding of the mathematical manipulations of calculus. For the student who is in need of a review of calculus, however, Appendix 1 and Appendix 2 are included.

This book is not encyclopedic in character, nor does it give in a highly mathematical rigorous account. Our emphasis in the text is to provide an accessible working knowledge of some of the current important mathematical tools required in physics.

The student will find that a generous amount of detail has been given mathematical manipulations, and that 'it-may-be-shown-thats' have been kept to a minimum. However, to ensure that the student does not lose sight of the development underway, some of the more lengthy and tedious algebraic manipulations have been omitted when possible.

Each chapter contains a number of physics examples to illustrate the mathematical techniques just developed and to show their relevance to physics. They supplement or amplify the material in the text, and are arranged in the order in which the material is covered in the chapter. No effort has been made to trace the origins of the homework problems and examples in the book. A solution manual for instructors is available from the publishers upon adoption.

Many individuals have been very helpful in the preparation of this text. I wish to thank my colleagues in the physics department at CSUS.

Any suggestions for improvement of this text will be greatly appreciated.

Turlock, California   TAI L. CHOW
2000

1

Vector and tensor analysis

Vectors and scalars

Vector methods have become standard tools for the physicists. In this chapter we discuss the properties of the vectors and vector fields that occur in classical physics. We will do so in a way, and in a notation, that leads to the formation of abstract linear vector spaces in Chapter 5.

A physical quantity that is completely specified, in appropriate units, by a single number (called its magnitude) such as volume, mass, and temperature is called a scalar. Scalar quantities are treated as ordinary real numbers. They obey all the regular rules of algebraic addition, subtraction, multiplication, division, and so on.

There are also physical quantities which require a magnitude and a direction for their complete specification. These are called vectors *if* their combination with each other is commutative (that is the order of addition may be changed without affecting the result). Thus not all quantities possessing magnitude and direction are vectors. Angular displacement, for example, may be characterised by magnitude and direction but is not a vector, for the addition of two or more angular displacements is not, in general, commutative (Fig. 1.1).

In print, we shall denote vectors by boldface letters (such as **A**) and use ordinary italic letters (such as A) for their magnitudes; in writing, vectors are usually represented by a letter with an arrow above it such as $\vec{A}$. A given vector **A** (or $\vec{A}$) can be written as

$$\mathbf{A} = A\hat{A}, \tag{1.1}$$

where A is the magnitude of vector **A** and so it has unit and dimension, and $\hat{A}$ is a dimensionless unit vector with a unity magnitude having the direction of **A**. Thus $\hat{A} = \mathbf{A}/A$.

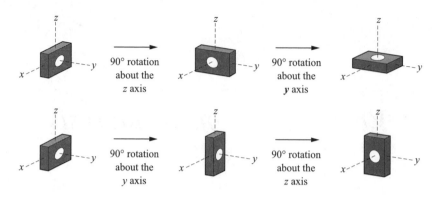

Figure 1.1.   Rotation of a parallelpiped about coordinate axes.

A vector quantity may be represented graphically by an arrow-tipped line segment. The length of the arrow represents the magnitude of the vector, and the direction of the arrow is that of the vector, as shown in Fig. 1.2. Alternatively, a vector can be specified by its components (projections along the coordinate axes) and the unit vectors along the coordinate axes (Fig. 1.3):

$$\mathbf{A} = A_1\hat{e}_1 + A_2\hat{e}_2 + A\hat{e}_3 = \sum_{i=1}^{3} A_i\hat{e}_i, \tag{1.2}$$

where $\hat{e}_i$ $(i = 1, 2, 3)$ are unit vectors along the rectangular axes x_i $(x_1 = x, x_2 = y, x_3 = z)$; they are normally written as $\hat{i}, \hat{j}, \hat{k}$ in general physics textbooks. The component triplet (A_1, A_2, A_3) is also often used as an alternate designation for vector $\mathbf{A}$:

$$\mathbf{A} = (A_1, A_2, A_3). \tag{1.2a}$$

This algebraic notation of a vector can be extended (or generalized) to spaces of dimension greater than three, where an ordered n-tuple of real numbers, $(A_1, A_2, \ldots, A_n)$, represents a vector. Even though we cannot construct physical vectors for $n > 3$, we can retain the geometrical language for these n-dimensional generalizations. Such abstract "vectors" will be the subject of Chapter 5.

Figure 1.2.   Graphical representation of vector $\mathbf{A}$.

2

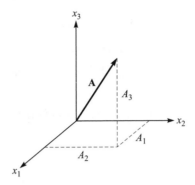

Figure 1.3. A vector **A** in Cartesian coordinates.

Direction angles and direction cosines

We can express the unit vector $\hat{A}$ in terms of the unit coordinate vectors $\hat{e}_i$. From Eq. (1.2), $\mathbf{A} = A_1\hat{e}_1 + A_2\hat{e}_2 + A\hat{e}_3$, we have

$$\mathbf{A} = A\left(\frac{A_1}{A}\hat{e}_1 + \frac{A_2}{A}\hat{e}_2 + \frac{A_3}{A}\hat{e}_3\right) = A\hat{A}.$$

Now $A_1/A = \cos\alpha$, $A_2/A = \cos\beta$, and $A_3/A = \cos\gamma$ are the direction cosines of the vector **A**, and α, β, and γ are the direction angles (Fig. 1.4). Thus we can write

$$\mathbf{A} = A(\cos\alpha\,\hat{e}_1 + \cos\beta\,\hat{e}_2 + \cos\gamma\,\hat{e}_3) = A\hat{A};$$

it follows that

$$\hat{A} = (\cos\alpha\,\hat{e}_1 + \cos\beta\,\hat{e}_2 + \cos\gamma\,\hat{e}_3) = (\cos\alpha, \cos\beta, \cos\gamma). \tag{1.3}$$

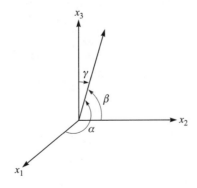

Figure 1.4. Direction angles of vector **A**.

3

Vector algebra

Equality of vectors

Two vectors, say **A** and **B**, are equal if, and only if, their respective components are equal:

$$\mathbf{A} = \mathbf{B} \quad \text{or} \quad (A_1, A_2, A_3) = (B_1, B_2, B_3)$$

is equivalent to the three equations

$$A_1 = B_1, A_2 = B_2, A_3 = B_3.$$

Geometrically, equal vectors are parallel and have the same length, but do not necessarily have the same position.

Vector addition

The addition of two vectors is defined by the equation

$$\mathbf{A} + \mathbf{B} = (A_1, A_2, A_3) + (B_1, B_2, B_3) = (A_1 + B_1, A_2 + B_2, A_3 + B_3).$$

That is, the sum of two vectors is a vector whose components are sums of the components of the two given vectors.

We can add two non-parallel vectors by graphical method as shown in Fig. 1.5. To add vector **B** to vector **A**, shift **B** parallel to itself until its tail is at the head of **A**. The vector sum **A** + **B** is a vector **C** drawn from the tail of **A** to the head of **B**. The order in which the vectors are added does not affect the result.

Multiplication by a scalar

If c is scalar then

$$c\mathbf{A} = (cA_1, cA_2, cA_3).$$

Geometrically, the vector $c\mathbf{A}$ is parallel to **A** and is c times the length of **A**. When $c = -1$, the vector $-\mathbf{A}$ is one whose direction is the reverse of that of **A**, but both

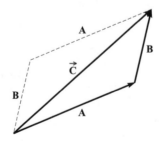

Figure 1.5. Addition of two vectors.

have the same length. Thus, subtraction of vector **B** from vector **A** is equivalent to adding −**B** to **A**:

$$\mathbf{A} - \mathbf{B} = \mathbf{A} + (-\mathbf{B}).$$

We see that vector addition has the following properties:

(a) $\mathbf{A} + \mathbf{B} = \mathbf{B} + \mathbf{A}$ (commutativity);
(b) $(\mathbf{A} + \mathbf{B}) + \mathbf{C} = \mathbf{A} + (\mathbf{B} + \mathbf{C})$ (associativity);
(c) $\mathbf{A} + \mathbf{0} = \mathbf{0} + \mathbf{A} = \mathbf{A};$
(d) $\mathbf{A} + (-\mathbf{A}) = \mathbf{0}.$

We now turn to vector multiplication. Note that division by a vector is not defined: expressions such as $k/\mathbf{A}$ or $\mathbf{B}/\mathbf{A}$ are meaningless.

There are several ways of multiplying two vectors, each of which has a special meaning; two types are defined.

The scalar product

The scalar (dot or inner) product of two vectors **A** and **B** is a real number defined (in geometrical language) as the product of their magnitude and the cosine of the (smaller) angle between them (Figure 1.6):

$$\mathbf{A} \cdot \mathbf{B} \equiv AB\cos\theta \qquad (0 \le \theta \le \pi). \tag{1.4}$$

It is clear from the definition (1.4) that the scalar product is commutative:

$$\mathbf{A} \cdot \mathbf{B} = \mathbf{B} \cdot \mathbf{A}, \tag{1.5}$$

and the product of a vector with itself gives the square of the dot product of the vector:

$$\mathbf{A} \cdot \mathbf{A} = A^2. \tag{1.6}$$

If $\mathbf{A} \cdot \mathbf{B} = 0$ and neither **A** nor **B** is a null (zero) vector, then **A** is perpendicular to **B**.

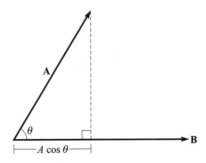

Figure 1.6. The scalar product of two vectors.

We can get a simple geometric interpretation of the dot product from an inspection of Fig. 1.6:

$(B\cos\theta)A$ = projection of **B** onto **A** multiplied by the magnitude of **A**,

$(A\cos\theta)B$ = projection of **A** onto **B** multiplied by the magnitude of **B**.

If only the components of **A** and **B** are known, then it would not be practical to calculate **A** · **B** from definition (1.4). But, in this case, we can calculate **A** · **B** in terms of the components:

$$\mathbf{A}\cdot\mathbf{B} = (A_1\hat{e}_1 + A_2\hat{e}_2 + A_3\hat{e}_3)\cdot(B_1\hat{e}_1 + B_2\hat{e}_2 + B_3\hat{e}_3);\qquad(1.7)$$

the right hand side has nine terms, all involving the product $\hat{e}_i \cdot \hat{e}_j$. Fortunately, the angle between each pair of unit vectors is $90°$, and from (1.4) and (1.6) we find that

$$\hat{e}_i \cdot \hat{e}_j = \delta_{ij}, \qquad i, j = 1, 2, 3,\qquad(1.8)$$

where δ_{ij} is the Kronecker delta symbol

$$\delta_{ij} = \begin{cases} 0, & \text{if } i \neq j, \\ 1, & \text{if } i = j. \end{cases}\qquad(1.9)$$

After we use (1.8) to simplify the resulting nine terms on the right-side of (7), we obtain

$$\mathbf{A}\cdot\mathbf{B} = A_1B_1 + A_2B_2 + A_3B_3 = \sum_{i=1}^{3} A_iB_i.\qquad(1.10)$$

The law of cosines for plane triangles can be easily proved with the application of the scalar product: refer to Fig. 1.7, where **C** is the resultant vector of **A** and **B**. Taking the dot product of **C** with itself, we obtain

$$C^2 = \mathbf{C}\cdot\mathbf{C} = (\mathbf{A}+\mathbf{B})\cdot(\mathbf{A}+\mathbf{B})$$

$$= A^2 + B^2 + 2\mathbf{A}\cdot\mathbf{B} = A^2 + B^2 + 2AB\cos\theta,$$

which is the law of cosines.

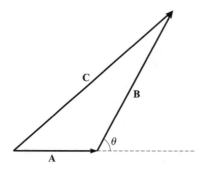

Figure 1.7.   Law of cosines.

A simple application of the scalar product in physics is the work W done by a constant force $\mathbf{F}$: $W = \mathbf{F} \cdot \mathbf{r}$, where $\mathbf{r}$ is the displacement vector of the object moved by $\mathbf{F}$.

The vector (cross or outer) product

The vector product of two vectors $\mathbf{A}$ and $\mathbf{B}$ is a vector and is written as

$$\mathbf{C} = \mathbf{A} \times \mathbf{B}. \tag{1.11}$$

As shown in Fig. 1.8, the two vectors $\mathbf{A}$ and $\mathbf{B}$ form two sides of a parallelogram. We define $\mathbf{C}$ to be perpendicular to the plane of this parallelogram with its magnitude equal to the area of the parallelogram. And we choose the direction of $\mathbf{C}$ along the thumb of the right hand when the fingers rotate from $\mathbf{A}$ to $\mathbf{B}$ (angle of rotation less than 180°).

$$\mathbf{C} = \mathbf{A} \times \mathbf{B} = AB \sin\theta\, \hat{e}_C \qquad (0 \le \theta \le \pi). \tag{1.12}$$

From the definition of the vector product and following the right hand rule, we can see immediately that

$$\mathbf{A} \times \mathbf{B} = -\mathbf{B} \times \mathbf{A}. \tag{1.13}$$

Hence the vector product is not commutative. If $\mathbf{A}$ and $\mathbf{B}$ are parallel, then it follows from Eq. (1.12) that

$$\mathbf{A} \times \mathbf{B} = 0. \tag{1.14}$$

In particular

$$\mathbf{A} \times \mathbf{A} = 0. \tag{1.14a}$$

In vector components, we have

$$\mathbf{A} \times \mathbf{B} = (A_1\hat{e}_1 + A_2\hat{e}_2 + A_3\hat{e}_3) \times (B_1\hat{e}_1 + B_2\hat{e}_2 + B_3\hat{e}_3). \tag{1.15}$$

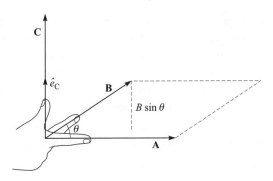

Figure 1.8.   The right hand rule for vector product.

Using the following relations

$$\hat{e}_i \times \hat{e}_i = 0, \; i = 1, 2, 3,$$

$$\hat{e}_1 \times \hat{e}_2 = \hat{e}_3, \; \hat{e}_2 \times \hat{e}_3 = \hat{e}_1, \; \hat{e}_3 \times \hat{e}_1 = \hat{e}_2,$$

(1.16)

Eq. (1.15) becomes

$$\mathbf{A} \times \mathbf{B} = (A_2 B_3 - A_3 B_2)\hat{e}_1 + (A_3 B_1 - A_1 B_3)\hat{e}_2 + (A_1 B_2 - A_2 B_1)\hat{e}_3.$$

(1.15a)

This can be written as an easily remembered determinant of third order:

$$\mathbf{A} \times \mathbf{B} = \begin{vmatrix} \hat{e}_1 & \hat{e}_2 & \hat{e}_3 \\ A_1 & A_2 & A_3 \\ B_1 & B_2 & B_3 \end{vmatrix}.$$

(1.17)

The expansion of a determinant of third order can be obtained by diagonal multiplication by repeating on the right the first two columns of the determinant and adding the signed products of the elements on the various diagonals in the resulting array:

$$\begin{bmatrix} a_1 & a_2 & a_3 \\ b_1 & b_2 & b_3 \\ c_1 & c_2 & c_c \end{bmatrix} \begin{matrix} a_1 & a_2 \\ b_1 & b_2 \\ c_1 & c_2 \end{matrix}$$

The non-commutativity of the vector product of two vectors now appears as a consequence of the fact that interchanging two rows of a determinant changes its sign, and the vanishing of the vector product of two vectors in the same direction appears as a consequence of the fact that a determinant vanishes if one of its rows is a multiple of another.

The determinant is a basic tool used in physics and engineering. The reader is assumed to be familiar with this subject. Those who are in need of review should read Appendix II.

The vector resulting from the vector product of two vectors is called an axial vector, while ordinary vectors are sometimes called polar vectors. Thus, in Eq. (1.11), $\mathbf{C}$ is a pseudovector, while $\mathbf{A}$ and $\mathbf{B}$ are axial vectors. On an inversion of coordinates, polar vectors change sign but an axial vector does not change sign.

A simple application of the vector product in physics is the torque $\boldsymbol{\tau}$ of a force $\mathbf{F}$ about a point O: $\boldsymbol{\tau} = \mathbf{F} \times \mathbf{r}$, where $\mathbf{r}$ is the vector from O to the initial point of the force $\mathbf{F}$ (Fig. 1.9).

We can write the nine equations implied by Eq. (1.16) in terms of permutation symbols ε_{ijk}:

$$\hat{e}_i \times \hat{e}_j = \varepsilon_{ijk}\hat{e}_k,$$

(1.16a)

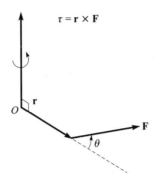

$$\tau = \mathbf{r} \times \mathbf{F}$$

Figure 1.9.   The torque of a force about a point O.

where ε_{ijk} is defined by

$$\varepsilon_{ijk} = \begin{cases} +1 & \text{if } (i,j,k) \text{ is an even permutation of } (1,2,3), \\ -1 & \text{if } (i,j,k) \text{ is an odd permutation of } (1,2,3), \\ 0 & \text{otherwise (for example, if 2 or more indices are equal).} \end{cases} \qquad (1.18)$$

It follows immediately that

$$\varepsilon_{ijk} = \varepsilon_{kij} = \varepsilon_{jki} = -\varepsilon_{jik} = -\varepsilon_{kji} = -\varepsilon_{ikj}.$$

There is a very useful identity relating the ε_{ijk} and the Kronecker delta symbol:

$$\sum_{k=1}^{3} \varepsilon_{mnk}\varepsilon_{ijk} = \delta_{mi}\delta_{nj} - \delta_{mj}\delta_{ni}, \qquad (1.19)$$

$$\sum_{j,k} \varepsilon_{mjk}\varepsilon_{njk} = 2\delta_{mn}, \qquad \sum_{i,j,k} \varepsilon_{ijk}^{2} = 6. \qquad (1.19a)$$

Using permutation symbols, we can now write the vector product $\mathbf{A} \times \mathbf{B}$ as

$$\mathbf{A} \times \mathbf{B} = \left(\sum_{i=1}^{3} A_i \hat{e}_i\right) \times \left(\sum_{j=1}^{3} B_j \hat{e}_j\right) = \sum_{i,j} A_i B_j \left(\hat{e}_i \times \hat{e}_j\right) = \sum_{i,j,k} \left(A_i B_j \varepsilon_{ijk}\right)\hat{e}_k.$$

Thus the kth component of $\mathbf{A} \times \mathbf{B}$ is

$$(\mathbf{A} \times \mathbf{B})_k = \sum_{i,j} A_i B_j \varepsilon_{ijk} = \sum_{i,j} \varepsilon_{kij} A_i B_j.$$

If $k = 1$, we obtain the usual geometrical result:

$$(\mathbf{A} \times \mathbf{B})_1 = \sum_{i,j} \varepsilon_{1ij} A_i B_j = \varepsilon_{123} A_2 B_3 + \varepsilon_{132} A_3 B_2 = A_2 B_3 - A_3 B_2.$$

The triple scalar product $\mathbf{A} \cdot (\mathbf{B} \times \mathbf{C})$

We now briefly discuss the scalar $\mathbf{A} \cdot (\mathbf{B} \times \mathbf{C})$. This scalar represents the volume of the parallelepiped formed by the coterminous sides $\mathbf{A}$, $\mathbf{B}$, $\mathbf{C}$, since

$$\mathbf{A} \cdot (\mathbf{B} \times \mathbf{C}) = ABC \sin \theta \cos \alpha = hS = \text{volume},$$

S being the area of the parallelogram with sides $\mathbf{B}$ and $\mathbf{C}$, and h the height of the parallelogram (Fig. 1.10).

Now

$$\mathbf{A} \cdot (\mathbf{B} \times \mathbf{C}) = (A_1\hat{e}_1 + A_2\hat{e}_2 + A_3\hat{e}_3) \cdot \begin{vmatrix} \hat{e}_1 & \hat{e}_2 & \hat{e}_3 \\ B_1 & B_2 & B_3 \\ C_1 & C_2 & C_3 \end{vmatrix}$$

$$= A_1(B_2C_3 - B_3C_2) + A_2(B_3C_1 - B_1C_3) + A_3(B_1C_2 - B_2C_1)$$

so that

$$\mathbf{A} \cdot (\mathbf{B} \times \mathbf{C}) = \begin{vmatrix} A_1 & A_2 & A_3 \\ B_1 & B_2 & B_3 \\ C_1 & C_2 & C_3 \end{vmatrix}. \tag{1.20}$$

The exchange of two rows (or two columns) changes the sign of the determinant but does not change its absolute value. Using this property, we find

$$\mathbf{A} \cdot (\mathbf{B} \times \mathbf{C}) = \begin{vmatrix} A_1 & A_2 & A_3 \\ B_1 & B_2 & B_3 \\ C_1 & C_2 & C_3 \end{vmatrix} = - \begin{vmatrix} C_1 & C_2 & C_3 \\ B_1 & B_2 & B_3 \\ A_1 & A_2 & A_3 \end{vmatrix} = \mathbf{C} \cdot (\mathbf{A} \times \mathbf{B}),$$

that is, the dot and the cross may be interchanged in the triple scalar product.

$$\mathbf{A} \cdot (\mathbf{B} \times \mathbf{C}) = (\mathbf{A} \times \mathbf{B}) \cdot \mathbf{C} \tag{1.21}$$

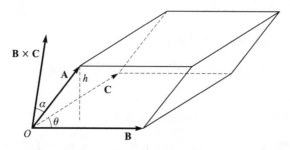

Figure 1.10.   The triple scalar product of three vectors $\mathbf{A}$, $\mathbf{B}$, $\mathbf{C}$.

In fact, as long as the three vectors appear in cyclic order, $\mathbf{A} \to \mathbf{B} \to \mathbf{C} \to \mathbf{A}$, then the dot and cross may be inserted between any pairs:

$$\mathbf{A} \cdot (\mathbf{B} \times \mathbf{C}) = \mathbf{B} \cdot (\mathbf{C} \times \mathbf{A}) = \mathbf{C} \cdot (\mathbf{A} \times \mathbf{B}).$$

It should be noted that the scalar resulting from the triple scalar product changes sign on an inversion of coordinates. For this reason, the triple scalar product is sometimes called a pseudoscalar.

The triple vector product

The triple product $\mathbf{A} \times (\mathbf{B} \times \mathbf{C})$ is a vector, since it is the vector product of two vectors: $\mathbf{A}$ and $\mathbf{B} \times \mathbf{C}$. This vector is perpendicular to $\mathbf{B} \times \mathbf{C}$ and so it lies in the plane of $\mathbf{B}$ and $\mathbf{C}$. If $\mathbf{B}$ is not parallel to $\mathbf{C}$, $\mathbf{A} \times (\mathbf{B} \times \mathbf{C}) = x\mathbf{B} + y\mathbf{C}$. Now dot both sides with $\mathbf{A}$ and we obtain $x(\mathbf{A} \cdot \mathbf{B}) + y(\mathbf{A} \cdot \mathbf{C}) = 0$, since $\mathbf{A} \cdot [\mathbf{A} \times (\mathbf{B} \times \mathbf{C})] = 0$. Thus

$$x/(\mathbf{A} \cdot \mathbf{C}) = -y/(\mathbf{A} \cdot \mathbf{B}) \equiv \lambda \quad (\lambda \text{ is a scalar})$$

and so

$$\mathbf{A} \times (\mathbf{B} \times \mathbf{C}) = x\mathbf{B} + y\mathbf{C} = \lambda[\mathbf{B}(\mathbf{A} \cdot \mathbf{C}) - \mathbf{C}(\mathbf{A} \cdot \mathbf{B})].$$

We now show that $\lambda = 1$. To do this, let us consider the special case when $\mathbf{B} = \mathbf{A}$. Dot the last equation with $\mathbf{C}$:

$$\mathbf{C} \times [\mathbf{A} \times (\mathbf{A} \times \mathbf{C})] = \lambda[(\mathbf{A} \cdot \mathbf{C})^2 - \mathbf{A}^2 \mathbf{C}^2],$$

or, by an interchange of dot and cross

$$-(\mathbf{A} \cdot \mathbf{C})^2 = \lambda[(\mathbf{A} \cdot \mathbf{C})^2 - \mathbf{A}^2 \mathbf{C}^2].$$

In terms of the angles between the vectors and their magnitudes the last equation becomes

$$-A^2 C^2 \sin^2 \theta = \lambda(A^2 C^2 \cos^2 \theta - A^2 C^2) = -\lambda A^2 C^2 \sin^2 \theta;$$

hence $\lambda = 1$. And so

$$\mathbf{A} \times (\mathbf{B} \times \mathbf{C}) = \mathbf{B}(\mathbf{A} \cdot \mathbf{C}) - \mathbf{C}(\mathbf{A} \cdot \mathbf{B}). \tag{1.22}$$

Change of coordinate system

Vector equations are independent of the coordinate system we happen to use. But the components of a vector quantity are different in different coordinate systems. We now make a brief study of how to represent a vector in different coordinate systems. As the rectangular Cartesian coordinate system is the basic type of coordinate system, we shall limit our discussion to it. Other coordinate systems will be introduced later. Consider the vector **A** expressed in terms of the unit coordinate vectors $(\hat{e}_1, \hat{e}_2, \hat{e}_3)$:

$$\mathbf{A} = A_1\hat{e}_1 + A_2\hat{e}_2 + A\hat{e}_3 = \sum_{i=1}^{3} A_i\hat{e}_i.$$

Relative to a new system $(\hat{e}_1', \hat{e}_2', \hat{e}_3')$ that has a different orientation from that of the old system $(\hat{e}_1, \hat{e}_2, \hat{e}_3)$, vector **A** is expressed as

$$\mathbf{A} = A_1'\hat{e}_1' + A_2'\hat{e}_2' + A'\hat{e}_3' = \sum_{i=1}^{3} A_i'\hat{e}_i'.$$

Note that the dot product $\mathbf{A} \cdot \hat{e}_1'$ is equal to A_1', the projection of **A** on the direction of $\hat{e}_1'$; $\mathbf{A} \cdot \hat{e}_2'$ is equal to A_2', and $\mathbf{A} \cdot \hat{e}_3'$ is equal to A_3'. Thus we may write

$$\left.\begin{aligned}
A_1' &= (\hat{e}_1 \cdot \hat{e}_1')A_1 + (\hat{e}_2 \cdot \hat{e}_1')A_2 + (\hat{e}_3 \cdot \hat{e}_1')A_3, \\
A_2' &= (\hat{e}_1 \cdot \hat{e}_2')A_1 + (\hat{e}_2 \cdot \hat{e}_2')A_2 + (\hat{e}_3 \cdot \hat{e}_2')A_3, \\
A_3' &= (\hat{e}_1 \cdot \hat{e}_3')A_1 + (\hat{e}_2 \cdot \hat{e}_3')A_2 + (\hat{e}_3 \cdot \hat{e}_3')A_3.
\end{aligned}\right\} \qquad (1.23)$$

The dot products $(\hat{e}_i \cdot \hat{e}_j')$ are the direction cosines of the axes of the new coordinate system relative to the old system: $\hat{e}_i' \cdot \hat{e}_j = \cos(x_i', x_j)$; they are often called the coefficients of transformation. In matrix notation, we can write the above system of equations as

$$\begin{pmatrix} A_1' \\ A_2' \\ A_3' \end{pmatrix} = \begin{pmatrix} \hat{e}_1 \cdot \hat{e}_1' & \hat{e}_2 \cdot \hat{e}_1' & \hat{e}_3 \cdot \hat{e}_1' \\ \hat{e}_1 \cdot \hat{e}_2' & \hat{e}_2 \cdot \hat{e}_2' & \hat{e}_3 \cdot \hat{e}_2' \\ \hat{e}_1 \cdot \hat{e}_3' & \hat{e}_2 \cdot \hat{e}_3' & \hat{e}_3 \cdot \hat{e}_3' \end{pmatrix} \begin{pmatrix} A_1 \\ A_2 \\ A_3 \end{pmatrix}.$$

The 3×3 matrix in the above equation is called the rotation (or transformation) matrix, and is an orthogonal matrix. One advantage of using a matrix is that successive transformations can be handled easily by means of matrix multiplication. Let us digress for a quick review of some basic matrix algebra. A full account of matrix method is given in Chapter 3.

A matrix is an ordered array of scalars that obeys prescribed rules of addition and multiplication. A particular matrix element is specified by its row number followed by its column number. Thus a_{ij} is the matrix element in the ith row and jth column. Alternative ways of representing matrix $\tilde{A}$ are $[a_{ij}]$ or the entire array

$$\tilde{A} = \begin{pmatrix} a_{11} & a_{12} & \cdots & a_{1n} \\ a_{21} & a_{22} & \cdots & a_{2n} \\ \cdots & \cdots & \cdots & \cdots \\ a_{m1} & a_{m2} & \cdots & a_{mn} \end{pmatrix}.$$

12

$\tilde{A}$ is an $n \times m$ matrix. A vector is represented in matrix form by writing its components as either a row or column array, such as

$$\tilde{B} = (b_{11} \ b_{12} \ b_{13}) \quad \text{or} \quad \tilde{C} = \begin{pmatrix} c_{11} \\ c_{21} \\ c_{31} \end{pmatrix},$$

where $b_{11} = b_x, b_{12} = b_y, b_{13} = b_z$, and $c_{11} = c_x, c_{21} = c_y, c_{31} = c_z$.

The multiplication of a matrix $\tilde{A}$ and a matrix $\tilde{B}$ is defined only when the number of columns of $\tilde{A}$ is equal to the number of rows of $\tilde{B}$, and is performed in the same way as the multiplication of two determinants: if $\tilde{C} = \tilde{A}\tilde{B}$, then

$$c_{ij} = \sum_k a_{ik} b_{kl}.$$

We illustrate the multiplication rule for the case of the 3×3 matrix $\tilde{A}$ multiplied by the 3×3 matrix $\tilde{B}$:

$$a_{11}b_{12} + a_{12}b_{22} + a_{13}b_{32} = c_{12}$$

$$\begin{pmatrix} a_{11} & a_{12} & a_{13} \\ a_{21} & a_{22} & a_{23} \\ a_{31} & a_{32} & a_{33} \end{pmatrix} \begin{pmatrix} b_{11} & b_{12} & b_{13} \\ b_{21} & b_{22} & b_{23} \\ b_{31} & b_{32} & b_{33} \end{pmatrix} = \begin{pmatrix} c_{11} & c_{12} & c_{13} \\ c_{21} & c_{22} & c_{23} \\ c_{31} & c_{32} & c_{33} \end{pmatrix}.$$

If we denote the direction cosines $\hat{e}_i' \cdot \hat{e}_j$ by λ_{ij}, then Eq. (1.23) can be written as

$$A_i' = \sum_{j=1}^{3} \hat{e}_i' \cdot \hat{e}_j A_j = \sum_{j=1}^{3} \lambda_{ij} A_j. \tag{1.23a}$$

It can be shown (Problem 1.9) that the quantities λ_{ij} satisfy the following relations

$$\sum_{i=1}^{3} \lambda_{ij} \lambda_{ik} = \delta_{jk} \quad (j, k = 1, 2, 3). \tag{1.24}$$

Any linear transformation, such as Eq. (1.23a), that has the properties required by Eq. (1.24) is called an orthogonal transformation, and Eq. (1.24) is known as the orthogonal condition.

The linear vector space V_n

We have found that it is very convenient to use vector components, in particular, the unit coordinate vectors $\hat{e}_i$ ($i = 1, 2, 3$). The three unit vectors $\hat{e}_i$ are orthogonal and normal, or, as we shall say, orthonormal. This orthonormal property is conveniently written as Eq. (1.8). But there is nothing special about these orthonormal unit vectors $\hat{e}_i$. If we refer the components of the vectors to a different system of rectangular coordinates, we need to introduce another set of three orthonormal unit vectors $\hat{f}_1, \hat{f}_2$, and $\hat{f}_3$:

$$\hat{f}_i \hat{f}_j = \delta_{ij} \quad (i, j = 1, 2, 3). \tag{1.8a}$$

For any vector **A** we now write

$$\mathbf{A} = \sum_{i=1}^{3} c_i \hat{f}_i, \quad \text{and} \quad c_i = \hat{f}_i \cdot \mathbf{A}.$$

We see that we can define a large number of different coordinate systems. But the physically significant quantities are the vectors themselves and certain functions of these, which are independent of the coordinate system used. The orthonormal condition (1.8) or (1.8a) is convenient in practice. If we also admit oblique Cartesian coordinates then the $\hat{f}_i$ need neither be normal nor orthogonal; they could be any three non-coplanar vectors, and any vector **A** can still be written as a linear superposition of the $\hat{f}_i$

$$\mathbf{A} = c_1 \hat{f}_1 + c_2 \hat{f}_2 + c_3 \hat{f}_3. \tag{1.25}$$

Starting with the vectors $\hat{f}_i$, we can find linear combinations of them by the algebraic operations of vector addition and multiplication of vectors by scalars, and then the collection of all such vectors makes up the three-dimensional linear space often called V_3 (V for vector) or R_3 (R for real) or E_3 (E for Euclidean). The vectors $\hat{f}_1, \hat{f}_2, \hat{f}_3$ are called the base vectors or bases of the vector space V_3. Any set of vectors, such as the $\hat{f}_i$, which can serve as the bases or base vectors of V_3 is called complete, and we say it spans the linear vector space. The base vectors are also linearly independent because no relation of the form

$$c_1 \hat{f}_1 + c_2 \hat{f}_2 + c_3 \hat{f}_3 = 0 \tag{1.26}$$

exists between them, unless $c_1 = c_2 = c_3 = 0$.

The notion of a vector space is much more general than the real vector space V_3. Extending the concept of V_3, it is convenient to call an ordered set of n matrices, or functions, or operators, a 'vector' (or an n-vector) in the n-dimensional space V_n. Chapter 5 will provide justification for doing this. Taking a cue from V_3, vector addition in V_n is defined to be

$$(x_1, \ldots, x_n) + (y_1, \ldots, y_n) = (x_1 + y_1, \ldots, x_n + y_n) \tag{1.27}$$

and multiplication by scalars is defined by

$$\alpha(x_1, \ldots, x_n) = (\alpha x_1, \ldots, \alpha x_n), \tag{1.28}$$

where α is real. With these two algebraic operations of vector addition and multiplication by scalars, we call V_n a vector space. In addition to this algebraic structure, V_n has geometric structure derived from the length defined to be

$$\left(\sum_{j=1}^{n} x_j^2\right)^{1/2} = \sqrt{x_1^2 + \cdots + x_n^2} \tag{1.29}$$

The dot product of two n-vectors can be defined by

$$(x_1, \ldots, x_n) \cdot (y_1, \ldots, y_n) = \sum_{j=1}^{n} x_j y_j. \tag{1.30}$$

In V_n, vectors are not directed line segments as in V_3; they may be an ordered set of n operators, matrices, or functions. We do not want to become sidetracked from our main goal of this chapter, so we end our discussion of vector space here.

Vector differentiation

Up to this point we have been concerned mainly with vector algebra. A vector may be a function of one or more scalars and vectors. We have encountered, for example, many important vectors in mechanics that are functions of time and position variables. We now turn to the study of the calculus of vectors.

Physicists like the concept of field and use it to represent a physical quantity that is a function of position in a given region. Temperature is a scalar field, because its value depends upon location: to each point (x, y, z) is associated a temperature $T(x, y, z)$. The function $T(x, y, z)$ is a scalar field, whose value is a real number depending only on the point in space but not on the particular choice of the coordinate system. A vector field, on the other hand, associates with each point a vector (that is, we associate three numbers at each point), such as the wind velocity or the strength of the electric or magnetic field. When described in a rotated system, for example, the three components of the vector associated with one and the same point will change in numerical value. Physically and geometrically important concepts in connection with scalar and vector fields are the gradient, divergence, curl, and the corresponding integral theorems.

The basic concepts of calculus, such as continuity and differentiability, can be naturally extended to vector calculus. Consider a vector $\mathbf{A}$, whose components are functions of a single variable u. If the vector $\mathbf{A}$ represents position or velocity, for example, then the parameter u is usually time t, but it can be any quantity that determines the components of $\mathbf{A}$. If we introduce a Cartesian coordinate system, the vector function $\mathbf{A}(u)$ may be written as

$$\mathbf{A}(u) = A_1(u)\hat{e}_1 + A_2(u)\hat{e}_2 + A_3(u)\hat{e}_3. \tag{1.31}$$

15

$\mathbf{A}(u)$ is said to be continuous at $u = u_0$ if it is defined in some neighborhood of u_0 and

$$\lim_{u \to u_0} A(u) = A(u_0). \tag{1.32}$$

Note that $\mathbf{A}(u)$ is continuous at u_0 if and only if its three components are continuous at u_0.

$\mathbf{A}(u)$ is said to be differentiable at a point u if the limit

$$\frac{d\mathbf{A}(u)}{du} = \lim_{\Delta u \to 0} \frac{\mathbf{A}(u + \Delta u) - \mathbf{A}(u)}{\Delta u} \tag{1.33}$$

exists. The vector $\mathbf{A}'(u) = d\mathbf{A}(u)/du$ is called the derivative of $\mathbf{A}(u)$; and to differentiate a vector function we differentiate each component separately:

$$\mathbf{A}'(u) = A_1'(u)\hat{e}_1 + A_2'(u)\hat{e}_2 + A_3'(u)\hat{e}_3. \tag{1.33a}$$

Note that the unit coordinate vectors are fixed in space. Higher derivatives of $\mathbf{A}(u)$ can be similarly defined.

If $\mathbf{A}$ is a vector depending on more than one scalar variable, say u, v for example, we write $\mathbf{A} = \mathbf{A}(u, v)$. Then

$$d\mathbf{A} = (\partial \mathbf{A}/\partial u)du + (\partial \mathbf{A}/\partial v)dv \tag{1.34}$$

is the differential of $\mathbf{A}$, and

$$\frac{\partial \mathbf{A}}{\partial u} = \lim_{\Delta u \to 0} \frac{\mathbf{A}(u + \Delta u, v) - \mathbf{A}(u, v)}{\partial u} \tag{1.34a}$$

and similarly for $\partial \mathbf{A}/\partial v$.

Derivatives of products obey rules similar to those for scalar functions. However, when cross products are involved the order may be important.

Space curves

As an application of vector differentiation, let us consider some basic facts about curves in space. If $\mathbf{A}(u)$ is the position vector $\mathbf{r}(u)$ joining the origin of a coordinate system and any point $P(x_1, x_2, x_3)$ in space as shown in Fig. 1.11, then Eq. (1.31) becomes

$$\mathbf{r}(u) = x_1(u)\hat{e}_1 + x_2(u)\hat{e}_2 + x_3(u)\hat{e}_3. \tag{1.35}$$

As u changes, the terminal point P of $\mathbf{r}$ describes a curve C in space. Eq. (1.35) is called a parametric representation of the curve C, and u is the parameter of this representation. Then

$$\frac{\Delta \mathbf{r}}{\Delta u} \left(= \frac{\mathbf{r}(u + \Delta u) - \mathbf{r}(u)}{\Delta u} \right)$$

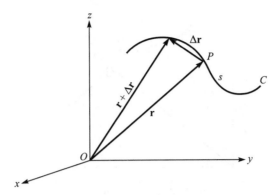

Figure 1.11.   Parametric representation of a curve.

is a vector in the direction of $\Delta\mathbf{r}$, and its limit (if it exists) $d\mathbf{r}/du$ is a vector in the direction of the tangent to the curve at (x_1, x_2, x_3). If u is the arc length s measured from some fixed point on the curve C, then $d\mathbf{r}/ds = \hat{T}$ is a unit tangent vector to the curve C. The rate at which $\hat{T}$ changes with respect to s is a measure of the curvature of C and is given by $d\hat{T}/ds$. The direction of $d\hat{T}/ds$ at any given point on C is normal to the curve at that point: $\hat{T} \cdot \hat{T} = 1$, $d(\hat{T} \cdot \hat{T})/ds = 0$, from this we get $\hat{T} \cdot d\hat{T}/ds = 0$, so they are normal to each other. If $\hat{N}$ is a unit vector in this normal direction (called the principal normal to the curve), then $d\hat{T}/ds = \kappa\hat{N}$, and κ is called the curvature of C at the specified point. The quantity $\rho = 1/\kappa$ is called the radius of curvature. In physics, we often study the motion of particles along curves, so the above results may be of value.

In mechanics, the parameter u is time t, then $d\mathbf{r}/dt = \mathbf{v}$ is the velocity of the particle which is tangent to the curve at the specific point. Now we can write

$$\mathbf{v} = \frac{d\mathbf{r}}{dt} = \frac{d\mathbf{r}}{ds}\frac{ds}{dt} = v\hat{T}$$

where v is the magnitude of $\mathbf{v}$, called the speed. Similarly, $\mathbf{a} = d\mathbf{v}/dt$ is the acceleration of the particle.

Motion in a plane

Consider a particle P moving in a plane along a curve C (Fig. 1.12). Now $\mathbf{r} = r\hat{e}_r$, where $\hat{e}_r$ is a unit vector in the direction of $\mathbf{r}$. Hence

$$\mathbf{v} = \frac{d\mathbf{r}}{dt} = \frac{dr}{dt}\hat{e}_r + r\frac{d\hat{e}_r}{dt}.$$

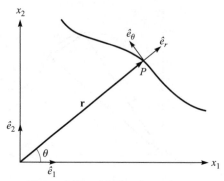

Figure 1.12.   Motion in a plane.

Now $d\hat{e}_r/dt$ is perpendicular to $\hat{e}_r$. Also $|d\hat{e}_r/dt| = d\theta/dt$; we can easily verify this by differentiating $\hat{e}_r = \cos\theta\hat{e}_1 + \sin\theta\hat{e}_2$. Hence

$$\mathbf{v} = \frac{d\mathbf{r}}{dt} = \frac{dr}{dt}\hat{e}_r + r\frac{d\theta}{dt}\hat{e}_\theta;$$

$\hat{e}_\theta$ is a unit vector perpendicular to $\hat{e}_r$.

Differentiating again we obtain

$$\mathbf{a} = \frac{d\mathbf{v}}{dt} = \frac{d^2r}{dt^2}\hat{e}_r + \frac{dr}{dt}\frac{d\hat{e}_r}{dt} + \frac{dr}{dt}\frac{d\theta}{dt}\hat{e}_\theta + r\frac{d^2\theta}{dt^2}\hat{e}_\theta + r\frac{d\theta}{dt}\hat{e}_\theta$$

$$= \frac{d^2r}{dt^2}\hat{e}_r + 2\frac{dr}{dt}\frac{d\theta}{dt}\hat{e}_\theta + r\frac{d^2\theta}{dt^2}\hat{e}_\theta - r\left(\frac{d\theta}{dt}\right)^2\hat{e}_r \quad \left(\because \frac{d\hat{e}_\theta}{dt} = -\frac{d\theta}{dt}\hat{e}_r\right).$$

Thus

$$\mathbf{a} = \left[\frac{d^2r}{dt^2} - r\left(\frac{d\theta}{dt}\right)^2\right]\hat{e}_r + \frac{1}{r}\frac{d}{dt}\left(r^2\frac{d\theta}{dt}\right)\hat{e}_\theta.$$

A vector treatment of classical orbit theory

To illustrate the power and use of vector methods, we now employ them to work out the Keplerian orbits. We first prove Kepler's second law which can be stated as: angular momentum is constant in a central force field. A central force is a force whose line of action passes through a single point or center and whose magnitude depends only on the distance from the center. Gravity and electrostatic forces are central forces. A general discussion on central force can be found in, for example, Chapter 6 of *Classical Mechanics*, Tai L. Chow, John Wiley, New York, 1995.

Differentiating the angular momentum $\mathbf{L} = \mathbf{r} \times \mathbf{p}$ with respect to time, we obtain

$$d\mathbf{L}/dt = d\mathbf{r}/dt \times \mathbf{p} + \mathbf{r} \times d\mathbf{p}/dt.$$

The first vector product vanishes because $\mathbf{p} = md\mathbf{r}/dt$ so $d\mathbf{r}/dt$ and $\mathbf{p}$ are parallel. The second vector product is simply $\mathbf{r} \times \mathbf{F}$ by Newton's second law, and hence vanishes for all forces directed along the position vector $\mathbf{r}$, that is, for all central forces. Thus the angular momentum $\mathbf{L}$ is a constant vector in central force motion. This implies that the position vector $\mathbf{r}$, and therefore the entire orbit, lies in a fixed plane in three-dimensional space. This result is essentially Kepler's second law, which is often stated in terms of the conservation of area velocity, $|\mathbf{L}|/2m$.

We now consider the inverse-square central force of gravitational and electrostatics. Newton's second law then gives

$$md\mathbf{v}/dt = -(k/r^2)\hat{n}, \tag{1.36}$$

where $\hat{n} = \mathbf{r}/r$ is a unit vector in the $\mathbf{r}$-direction, and $k = Gm_1m_2$ for the gravitational force, and $k = q_1q_2$ for the electrostatic force in cgs units. First we note that

$$\mathbf{v} = d\mathbf{r}/dt = dr/dt\hat{n} + rd\hat{n}/dt.$$

Then $\mathbf{L}$ becomes

$$\mathbf{L} = \mathbf{r} \times (m\mathbf{v}) = mr^2[\hat{n} \times (d\hat{n}/dt)]. \tag{1.37}$$

Now consider

$$\frac{d}{dt}(\mathbf{v} \times \mathbf{L}) = \frac{d\mathbf{v}}{dt} \times \mathbf{L} = -\frac{k}{mr^2}(\hat{n} \times \mathbf{L}) = -\frac{k}{mr^2}[\hat{n} \times mr^2(\hat{n} \times d\hat{n}/dt)]$$

$$= -k[\hat{n}(d\hat{n}/dt \cdot \hat{n}) - (d\hat{n}/dt)(\hat{n} \cdot \hat{n})].$$

Since $\hat{n} \cdot \hat{n} = 1$, it follows by differentiation that $\hat{n} \cdot d\hat{n}/dt = 0$. Thus we obtain

$$\frac{d}{dt}(\mathbf{v} \times \mathbf{L}) = kd\hat{n}/dt;$$

integration gives

$$\mathbf{v} \times \mathbf{L} = k\hat{n} + \mathbf{C}, \tag{1.38}$$

where $\mathbf{C}$ is a constant vector. It lies along, and fixes the position of, the major axis of the orbit as we shall see after we complete the derivation of the orbit. To find the orbit, we form the scalar quantity

$$L^2 = \mathbf{L} \cdot (\mathbf{r} \times m\mathbf{v}) = mr \cdot (\mathbf{v} \times \mathbf{L}) = mr(k + C\cos\theta), \tag{1.39}$$

where θ is the angle measured from $\mathbf{C}$ (which we may take to be the x-axis) to $\mathbf{r}$. Solving for r, we obtain

$$r = \frac{L^2/km}{1 + C/(k\cos\theta)} = \frac{A}{1 + \varepsilon\cos\theta}. \tag{1.40}$$

Eq. (1.40) is a conic section with one focus at the origin, where ε represents the eccentricity of the conic section; depending on its values, the conic section may be a circle, an ellipse, a parabola, or a hyperbola. The eccentricity can be easily determined in terms of the constants of motion:

$$\varepsilon = \frac{C}{k} = \frac{1}{k} |(\mathbf{v} \times \mathbf{L}) - k\hat{n}|$$

$$= \frac{1}{k} [|\mathbf{v} \times \mathbf{L}|^2 + k^2 - 2k\hat{n} \cdot (\mathbf{v} \times \mathbf{L})]^{1/2}$$

Now $|\mathbf{v} \times \mathbf{L}|^2 = v^2 L^2$ because $\mathbf{v}$ is perpendicular to $\mathbf{L}$. Using Eq. (1.39), we obtain

$$\varepsilon = \frac{1}{k} \left[v^2 L^2 + k^2 - \frac{2kL^2}{mr} \right]^{1/2} = \left[1 + \frac{2L^2}{mk^2} \left(\frac{1}{2} mv^2 - \frac{k}{r} \right) \right]^{1/2} = \left[1 + \frac{2L^2 E}{mk^2} \right]^{1/2},$$

where E is the constant energy of the system.

Vector differentiation of a scalar field and the gradient

Given a scalar field in a certain region of space given by a scalar function $\phi(x_1, x_2, x_3)$ that is defined and differentiable at each point with respect to the position coordinates (x_1, x_2, x_3), the total differential corresponding to an infinitesimal change $d\mathbf{r} = (dx_1, dx_2, dx_3)$ is

$$d\phi = \frac{\partial \phi}{\partial x_1} dx_1 + \frac{\partial \phi}{\partial x_2} dx_2 + \frac{\partial \phi}{\partial x_3} dx_3. \tag{1.41}$$

We can express $d\phi$ as a scalar product of two vectors:

$$d\phi = \frac{\partial \phi}{\partial x_1} dx_1 + \frac{\partial \phi}{\partial x_2} dx_2 + \frac{\partial \phi}{\partial x_3} dx_3 = (\nabla \phi) \cdot d\mathbf{r}, \tag{1.42}$$

where

$$\nabla \phi \equiv \frac{\partial \phi}{\partial x_1} \hat{e}_1 + \frac{\partial \phi}{\partial x_2} \hat{e}_2 + \frac{\partial \phi}{\partial x_3} \hat{e}_3 \tag{1.43}$$

is a vector field (or a vector point function). By this we mean to each point $\mathbf{r} = (x_1, x_2, x_3)$ in space we associate a vector $\nabla \phi$ as specified by its three components $(\partial \phi / \partial x_1, \partial \phi / \partial x_2, \partial \phi / \partial x_3)$: $\nabla \phi$ is called the *gradient* of ϕ and is often written as grad ϕ.

There is a simple geometric interpretation of $\nabla \phi$. Note that $\phi(x_1, x_2, x_3) = c$, where c is a constant, represents a surface. Let $\mathbf{r} = x_1 \hat{e}_1 + x_2 \hat{e}_2 + x_3 \hat{e}_3$ be the position vector to a point $P(x_1, x_2, x_3)$ on the surface. If we move along the surface to a nearby point $Q(\mathbf{r} + d\mathbf{r})$, then $d\mathbf{r} = dx_1 \hat{e}_1 + dx_2 \hat{e}_2 + dx_3 \hat{e}_3$ lies in the tangent plane to the surface at P. But as long as we move along the surface ϕ has a constant value and $d\phi = 0$. Consequently from (1.41),

$$d\mathbf{r} \cdot \nabla \phi = 0. \tag{1.44}$$

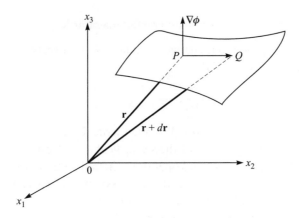

Figure 1.13. Gradient of a scalar.

Eq. (1.44) states that $\nabla\phi$ is perpendicular to $d\mathbf{r}$ and therefore to the surface (Fig. 1.13). Let us return to

$$d\phi = (\nabla\phi) \cdot d\mathbf{r}.$$

The vector $\nabla\phi$ is fixed at any point P, so that $d\phi$, the change in ϕ, will depend to a great extent on $d\mathbf{r}$. Consequently $d\phi$ will be a maximum when $d\mathbf{r}$ is parallel to $\nabla\phi$, since $d\mathbf{r} \cdot \nabla\phi = |d\mathbf{r}||\nabla\phi|\cos\theta$, and $\cos\theta$ is a maximum for $\theta = 0$. Thus $\nabla\phi$ is in the direction of maximum increase of $\phi(x_1, x_2, x_3)$. The component of $\nabla\phi$ in the direction of a unit vector $\hat{u}$ is given by $\nabla\phi \cdot \hat{u}$ and is called the directional derivative of ϕ in the direction $\hat{u}$. Physically, this is the rate of change of ϕ at (x_1, x_2, x_3) in the direction $\hat{u}$.

Conservative vector field

By definition, a vector field is said to be conservative if the line integral of the vector along any closed path vanishes. Thus, if $\mathbf{F}$ is a conservative vector field (say, a conservative force field in mechanics), then

$$\oint \mathbf{F} \cdot d\mathbf{s} = 0, \tag{1.45}$$

where $d\mathbf{s}$ is an element of the path. A (necessary and sufficient) condition for $\mathbf{F}$ to be conservative is that $\mathbf{F}$ can be expressed as the gradient of a scalar, say ϕ: $\mathbf{F} = -\text{grad } \phi$:

$$\int_a^b \mathbf{F} \cdot d\mathbf{s} = -\int_a^b \text{grad } \phi \cdot d\mathbf{s} = -\int_a^b d\phi = \phi(a) - \phi(b):$$

it is obvious that the line integral depends solely on the value of the scalar ϕ at the initial and final points, and $\oint \mathbf{F} \cdot d\mathbf{s} = -\oint \text{grad } \phi \cdot d\mathbf{s} = 0$.

The vector differential operator ∇

We denoted the operation that changes a scalar field to a vector field in Eq. (1.43) by the symbol ∇ (del or nabla):

$$\nabla \equiv \frac{\partial}{\partial x_1}\hat{e}_1 + \frac{\partial}{\partial x_2}\hat{e}_2 + \frac{\partial}{\partial x_3}\hat{e}_3, \tag{1.46}$$

which is called a gradient operator. We often write $\nabla\phi$ as grad ϕ, and the vector field $\nabla\phi(\mathbf{r})$ is called the gradient of the scalar field $\phi(\mathbf{r})$. Notice that the operator ∇ contains both partial differential operators and a direction: it is a vector differential operator. This important operator possesses properties analogous to those of ordinary vectors. It will help us in the future to keep in mind that ∇ acts both as a differential operator and as a vector.

Vector differentiation of a vector field

Vector differential operations on vector fields are more complicated because of the vector nature of both the operator and the field on which it operates. As we know there are two types of products involving two vectors, namely the scalar and vector products; vector differential operations on vector fields can also be separated into two types called the curl and the divergence.

The divergence of a vector

If $\mathbf{V}(x_1, x_2, x_3) = V_1\hat{e}_1 + V_2\hat{e}_2 + V_3\hat{e}_3$ is a differentiable vector field (that is, it is defined and differentiable at each point (x_1, x_2, x_3) in a certain region of space), the divergence of $\mathbf{V}$, written $\nabla \cdot \mathbf{V}$ or div $\mathbf{V}$, is defined by the scalar product

$$\nabla \cdot \mathbf{V} = \left(\frac{\partial}{\partial x_1}\hat{e}_1 + \frac{\partial}{\partial x_2}\hat{e}_2 + \frac{\partial}{\partial x_3}\hat{e}_3 \right) \cdot (V_1\hat{e}_1 + V_2\hat{e}_2 + V_3\hat{e}_3)$$

$$= \frac{\partial V_1}{\partial x_1} + \frac{\partial V_2}{\partial x_2} + \frac{\partial V_3}{\partial x_3}. \tag{1.47}$$

The result is a scalar field. Note the analogy with $\mathbf{A} \cdot \mathbf{B} = A_1B_1 + A_2B_2 + A_3B_3$, but also note that $\nabla \cdot \mathbf{V} \neq \mathbf{V} \cdot \nabla$ (bear in mind that ∇ is an operator). $\mathbf{V} \cdot \nabla$ is a scalar differential operator:

$$\mathbf{V} \cdot \nabla = V_1\frac{\partial}{\partial x_1} + V_2\frac{\partial}{\partial x_2} + V_3\frac{\partial}{\partial x_3}.$$

What is the physical significance of the divergence? Or why do we call the scalar product $\nabla \cdot \mathbf{V}$ the divergence of $\mathbf{V}$? To answer these questions, we consider, as an example, the steady motion of a fluid of density $\rho(x_1, x_2, x_3)$, and the velocity field is given by $\mathbf{v}(x_1, x_2, x_3) = v_1(x_1, x_2, x_3)e_1 + v_2(x_1, x_2, x_3)e_2 + v_3(x_1, x_2, x_3)e_3$. We now concentrate on the flow passing through a small parallelepiped $ABCDEFGH$ of dimensions $dx_1 dx_2 dx_3$ (Fig. 1.14). The x_1 and x_3 components of the velocity $\mathbf{v}$ contribute nothing to the flow through the face $ABCD$. The mass of fluid entering $ABCD$ per unit time is given by $\rho v_2 dx_1 dx_3$ and the amount leaving the face $EFGH$ per unit time is

$$\left[\rho v_2 + \frac{\partial(\rho v_2)}{\partial x_2} dx_2\right] dx_1 dx_3.$$

So the loss of mass per unit time is $[\partial(\rho v_2)/\partial x_2] dx_1 dx_2 dx_3$. Adding the net rate of flow out all three pairs of surfaces of our parallelepiped, the total mass loss per unit time is

$$\left[\frac{\partial}{\partial x_1}(\rho v_1) + \frac{\partial}{\partial x_2}(\rho v_2) + \frac{\partial}{\partial x_3}(\rho v_3)\right] dx_1 dx_2 dx_3 = \nabla \cdot (\rho \mathbf{v}) dx_1 dx_2 dx_3.$$

So the mass loss per unit time per unit volume is $\nabla \cdot (\rho \mathbf{v})$. Hence the name divergence.

The divergence of any vector $\mathbf{V}$ is defined as $\nabla \cdot \mathbf{V}$. We now calculate $\nabla \cdot (f\mathbf{V})$, where f is a scalar:

$$\nabla \cdot (f\,\mathbf{V}) = \frac{\partial}{\partial x_1}(fV_1) + \frac{\partial}{\partial x_2}(fV_2) + \frac{\partial}{\partial x_3}(fV_3)$$

$$= f\left(\frac{\partial V_1}{\partial x_1} + \frac{\partial V_2}{\partial x_2} + \frac{\partial V_3}{\partial x_3}\right) + \left(V_1 \frac{\partial f}{\partial x_1} + V_2 \frac{\partial f}{\partial x_2} + V_3 \frac{\partial f}{\partial x_3}\right)$$

or

$$\nabla \cdot (f\mathbf{V}) = f\nabla \cdot \mathbf{V} + \mathbf{V} \cdot \nabla f. \tag{1.48}$$

It is easy to remember this result if we remember that ∇ acts both as a differential operator and a vector. Thus, when operating on $f\,\mathbf{V}$, we first keep f fixed and let ∇

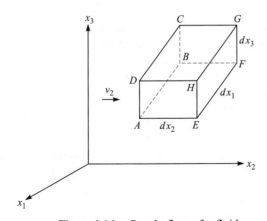

Figure 1.14. Steady flow of a fluid.

operate on $\mathbf{V}$, and then we keep $\mathbf{V}$ fixed and let ∇ operate on f ($\nabla \cdot f$ is nonsense), and as ∇f and $\mathbf{V}$ are vectors we complete their multiplication by taking their dot product.

A vector $\mathbf{V}$ is said to be solenoidal if its divergence is zero: $\nabla \cdot \mathbf{V} = 0$.

The operator ∇^2, the Laplacian

The divergence of a vector field is defined by the scalar product of the operator ∇ with the vector field. What is the scalar product of ∇ with itself?

$$\nabla^2 = \nabla \cdot \nabla = \left(\frac{\partial}{\partial x_1} \hat{e}_1 + \frac{\partial}{\partial x_2} \hat{e}_2 + \frac{\partial}{\partial x_3} \hat{e}_3 \right) \cdot \left(\frac{\partial}{\partial x_1} \hat{e}_1 + \frac{\partial}{\partial x_2} \hat{e}_2 + \frac{\partial}{\partial x_3} \hat{e}_3 \right)$$

$$= \frac{\partial^2}{\partial x_1^2} + \frac{\partial^2}{\partial x_2^2} + \frac{\partial^2}{\partial x_3^2}.$$

This important quantity

$$\nabla^2 = \frac{\partial^2}{\partial x_1^2} + \frac{\partial^2}{\partial x_2^2} + \frac{\partial^2}{\partial x_3^2} \tag{1.49}$$

is a scalar differential operator which is called the Laplacian, after a French mathematician of the eighteenth century named Laplace. Now, what is the divergence of a gradient?

Since the Laplacian is a scalar differential operator, it does not change the vector character of the field on which it operates. Thus $\nabla^2 \phi(\mathbf{r})$ is a scalar field if $\phi(\mathbf{r})$ is a scalar field, and $\nabla^2 [\nabla \phi(\mathbf{r})]$ is a vector field because the gradient $\nabla \phi(\mathbf{r})$ is a vector field.

The equation $\nabla^2 \phi = 0$ is called Laplace's equation.

The curl of a vector

If $\mathbf{V}(x_1, x_2, x_3)$ is a differentiable vector field, then the curl or rotation of $\mathbf{V}$, written $\nabla \times \mathbf{V}$ (or curl $\mathbf{V}$ or rot $\mathbf{V}$), is defined by the vector product

$$\text{curl } \mathbf{V} = \nabla \times \mathbf{V} = \begin{vmatrix} \hat{e}_1 & \hat{e}_2 & \hat{e}_3 \\ \dfrac{\partial}{\partial x_1} & \dfrac{\partial}{\partial x_2} & \dfrac{\partial}{\partial x_3} \\ V_1 & V_2 & V_3 \end{vmatrix}$$

$$= \hat{e}_1 \left(\frac{\partial V_3}{\partial x_2} - \frac{\partial V_2}{\partial x_3} \right) + \hat{e}_2 \left(\frac{\partial V_1}{\partial x_3} - \frac{\partial V_3}{\partial x_1} \right) + \hat{e}_3 \left(\frac{\partial V_2}{\partial x_1} - \frac{\partial V_1}{\partial x_2} \right)$$

$$= \sum_{i,j,k} \varepsilon_{ijk} \hat{e}_i \frac{\partial V_k}{\partial x_j}. \tag{1.50}$$

The result is a vector field. In the expansion of the determinant the operators $\partial/\partial x_i$ must precede V_i; $\sum_{ijk}$ stands for $\sum_i \sum_j \sum_k$; and ε_{ijk} are the permutation symbols: an even permutation of ijk will not change the value of the resulting permutation symbol, but an odd permutation gives an opposite sign. That is,

$$\varepsilon_{ijk} = \varepsilon_{jki} = \varepsilon_{kij} = -\varepsilon_{jik} = -\varepsilon_{kji} = -\varepsilon_{ikj}, \quad \text{and}$$

$$\varepsilon_{ijk} = 0 \text{ if two or more indices are equal.}$$

A vector $\mathbf{V}$ is said to be irrotational if its curl is zero: $\nabla \times \mathbf{V}(\mathbf{r}) = 0$. From this definition we see that the gradient of any scalar field $\phi(\mathbf{r})$ is irrotational. The proof is simple:

$$\nabla \times (\nabla \phi) = \begin{vmatrix} \hat{e}_1 & \hat{e}_2 & \hat{e}_3 \\ \dfrac{\partial}{\partial x_1} & \dfrac{\partial}{\partial x_2} & \dfrac{\partial}{\partial x_3} \\ \dfrac{\partial}{\partial x_1} & \dfrac{\partial}{\partial x_2} & \dfrac{\partial}{\partial x_3} \end{vmatrix} \phi(x_1, x_2, x_3) = 0 \tag{1.51}$$

because there are two identical rows in the determinant. Or, in terms of the permutation symbols, we can write $\nabla \times (\nabla \phi)$ as

$$\nabla \times (\nabla \phi) = \sum_{ijk} \varepsilon_{ijk} \hat{e}_i \frac{\partial}{\partial x_j} \frac{\partial}{\partial x_k} \phi(x_1, x_2, x_3).$$

Now ε_{ijk} is antisymmetric in j, k, but $\partial^2/\partial x_j \partial x_k$ is symmetric, hence each term in the sum is always cancelled by another term:

$$\varepsilon_{ijk} \frac{\partial}{\partial x_j} \frac{\partial}{\partial x_k} + \varepsilon_{ikj} \frac{\partial}{\partial x_k} \frac{\partial}{\partial x_j} = 0,$$

and consequently $\nabla \times (\nabla \phi) = 0$. Thus, for a conservative vector field $\mathbf{F}$, we have curl $\mathbf{F} = $ curl (grad ϕ) $= 0$.

We learned above that a vector $\mathbf{V}$ is solenoidal (or divergence-free) if its divergence is zero. From this we see that the curl of any vector field $\mathbf{V}(\mathbf{r})$ must be solenoidal:

$$\nabla \cdot (\nabla \times \mathbf{V}) = \sum_i \frac{\partial}{\partial x_i} (\nabla \times \mathbf{V})_i = \sum_i \frac{\partial}{\partial x_i} \left(\sum_{j,k} \varepsilon_{ijk} \frac{\partial}{\partial x_j} V_k \right) = 0, \tag{1.52}$$

because ε_{ijk} is antisymmetric in i, j.

If $\phi(\mathbf{r})$ is a scalar field and $\mathbf{V}(\mathbf{r})$ is a vector field, then

$$\nabla \times (\phi \mathbf{V}) = \phi(\nabla \times \mathbf{V}) + (\nabla \phi) \times \mathbf{V}. \tag{1.53}$$

25

We first write

$$\nabla \times (\phi \mathbf{V}) = \begin{vmatrix} \hat{e}_1 & \hat{e}_2 & \hat{e}_3 \\ \dfrac{\partial}{\partial x_1} & \dfrac{\partial}{\partial x_2} & \dfrac{\partial}{\partial x_3} \\ \phi V_1 & \phi V_2 & \phi V_3 \end{vmatrix},$$

then notice that

$$\frac{\partial}{\partial x_1}(\phi V_2) = \phi \frac{\partial V_2}{\partial x_1} + \frac{\partial \phi}{\partial x_1} V_2,$$

so we can expand the determinant in the above equation as a sum of two determinants:

$$\nabla \times (\phi \mathbf{V}) = \phi \begin{vmatrix} \hat{e}_1 & \hat{e}_2 & \hat{e}_3 \\ \dfrac{\partial}{\partial x_1} & \dfrac{\partial}{\partial x_2} & \dfrac{\partial}{\partial x_3} \\ V_1 & V_2 & V_3 \end{vmatrix} + \begin{vmatrix} \hat{e}_1 & \hat{e}_2 & \hat{e}_3 \\ \dfrac{\partial \phi}{\partial x_1} & \dfrac{\partial \phi}{\partial x_2} & \dfrac{\partial \phi}{\partial x_3} \\ V_1 & V_2 & V_3 \end{vmatrix}$$

$$= \phi(\nabla \times \mathbf{V}) + (\nabla \phi) \times \mathbf{V}.$$

Alternatively, we can simplify the proof with the help of the permutation symbols ε_{ijk}:

$$\nabla \times (\phi \mathbf{V}) = \sum_{i,j,k} \varepsilon_{ijk} \hat{e}_i \frac{\partial}{\partial x_j} (\phi V_k)$$

$$= \phi \sum_{i,j,k} \varepsilon_{ijk} \hat{e}_i \frac{\partial V_k}{\partial x_j} + \sum_{i,j,k} \varepsilon_{ijk} \hat{e}_i \frac{\partial \phi}{\partial x_j} V_k$$

$$= \phi(\nabla \times \mathbf{V}) + (\nabla \phi) \times \mathbf{V}.$$

A vector field that has non-vanishing curl is called a vortex field, and the curl of the field vector is a measure of the vorticity of the vector field.

The physical significance of the curl of a vector is not quite as transparent as that of the divergence. The following example from fluid flow will help us to develop a better feeling. Fig. 1.15 shows that as the component v_2 of the velocity $\mathbf{v}$ of the fluid increases with x_3, the fluid curls about the x_1-axis in a negative sense (rule of the right-hand screw), where $\partial v_2 / \partial x_3$ is considered positive. Similarly, a positive curling about the x_1-axis would result from v_3 if $\partial v_3 / \partial x_2$ were positive. Therefore, the total x_1 component of the curl of $\mathbf{v}$ is

$$[\text{curl } \mathbf{v}]_1 = \partial v_3 / (\partial x_2 - \partial v_2 / \partial x_3,$$

which is the same as the x_1 component of Eq. (1.50).

26

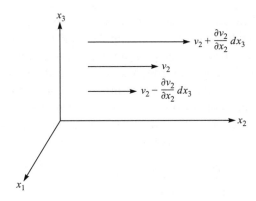

Figure 1.15.   Curl of a fluid flow.

Formulas involving ∇

We now list some important formulas involving the vector differential operator ∇, some of which are recapitulation. In these formulas, $\mathbf{A}$ and $\mathbf{B}$ are differentiable vector field functions, and f and g are differentiable scalar field functions of position (x_1, x_2, x_3):

(1) $\nabla(fg) = f\nabla g + g\nabla f$;

(2) $\nabla \cdot (f\mathbf{A}) = f\nabla \cdot \mathbf{A} + \nabla f \cdot \mathbf{A}$;

(3) $\nabla \times (f\mathbf{A}) = f\nabla \times \mathbf{A} + \nabla f \times \mathbf{A}$;

(4) $\nabla \times (\nabla f) = 0$;

(5) $\nabla \cdot (\nabla \times \mathbf{A}) = 0$;

(6) $\nabla \cdot (\mathbf{A} \times \mathbf{B}) = (\nabla \times \mathbf{A}) \cdot \mathbf{B} - (\nabla \times \mathbf{B}) \times \mathbf{A}$;

(7) $\nabla \times (\mathbf{A} \times \mathbf{B}) = (\mathbf{B} \cdot \nabla)\mathbf{A} - \mathbf{B}(\nabla \cdot \mathbf{A}) + \mathbf{A}(\nabla \cdot \mathbf{B}) - (\mathbf{A} \cdot \nabla)\mathbf{B}$;

(8) $\nabla \times (\nabla \times \mathbf{A}) = \nabla(\nabla \cdot \mathbf{A}) - \nabla^2\mathbf{A}$;

(9) $\nabla(\mathbf{A} \cdot \mathbf{B}) = \mathbf{A} \times (\nabla \times \mathbf{B}) + \mathbf{B} \times (\nabla \times \mathbf{A}) + (\mathbf{A} \cdot \nabla)\mathbf{B} + (\mathbf{B} \cdot \nabla)\mathbf{A}$;

(10) $(\mathbf{A} \cdot \nabla)\mathbf{r} = \mathbf{A}$;

(11) $\nabla \cdot \mathbf{r} = 3$;

(12) $\nabla \times \mathbf{r} = 0$;

(13) $\nabla \cdot (r^{-3}\mathbf{r}) = 0$;

(14) $d\mathbf{F} = (d\mathbf{r} \cdot \nabla)\mathbf{F} + \dfrac{\partial \mathbf{F}}{\partial t}dt$ ($\mathbf{F}$ a differentiable vector field quantity);

(15) $d\varphi = d\mathbf{r} \cdot \nabla\varphi + \dfrac{\partial \varphi}{\partial t}dt$ (φ a differentiable scalar field quantity).

Orthogonal curvilinear coordinates

Up to this point all calculations have been performed in rectangular Cartesian coordinates. Many calculations in physics can be greatly simplified by using, instead of the familiar rectangular Cartesian coordinate system, another kind of

27

system which takes advantage of the relations of symmetry involved in the particular problem under consideration. For example, if we are dealing with sphere, we will find it expedient to describe the position of a point in sphere by the spherical coordinates (r, θ, ϕ). Spherical coordinates are a special case of the orthogonal curvilinear coordinate system. Let us now proceed to discuss these more general coordinate systems in order to obtain expressions for the gradient, divergence, curl, and Laplacian. Let the new coordinates u_1, u_2, u_3 be defined by specifying the Cartesian coordinates (x_1, x_2, x_3) as functions of (u_1, u_2, u_3):

$$x_1 = f(u_1, u_2, u_3), \quad x_2 = g(u_1, u_2, u_3), \quad x_3 = h(u_1, u_2, u_3), \qquad (1.54)$$

where f, g, h are assumed to be continuous, differentiable. A point P (Fig. 1.16) in space can then be defined not only by the rectangular coordinates (x_1, x_2, x_3) but also by curvilinear coordinates (u_1, u_2, u_3).

If u_2 and u_3 are constant as u_1 varies, P (or its position vector $\mathbf{r}$) describes a curve which we call the u_1 coordinate curve. Similarly, we can define the u_2 and u_3 coordinate curves through P. We adopt the convention that the new coordinate system is a right handed system, like the old one. In the new system $d\mathbf{r}$ takes the form:

$$d\mathbf{r} = \frac{\partial \mathbf{r}}{\partial u_1} du_1 + \frac{\partial \mathbf{r}}{\partial u_2} du_2 + \frac{\partial \mathbf{r}}{\partial u_3} du_3.$$

The vector $\partial \mathbf{r}/\partial u_1$ is tangent to the u_1 coordinate curve at P. If $\hat{u}_1$ is a unit vector at P in this direction, then $\hat{u}_1 = \partial \mathbf{r}/\partial u_1 / |\partial \mathbf{r}/\partial u_1|$, so we can write $\partial \mathbf{r}/\partial u_1 = h_1 \hat{u}_1$, where $h_1 = |\partial \mathbf{r}/\partial u_1|$. Similarly we can write $\partial \mathbf{r}/\partial u_2 = h_2 \hat{u}_2$ and $\partial \mathbf{r}/\partial u_3 = h_3 \hat{u}_3$, where $h_2 = |\partial \mathbf{r}/\partial u_2|$ and $h_3 = |\partial \mathbf{r}/\partial u_3|$, respectively. Then $d\mathbf{r}$ can be written

$$d\mathbf{r} = h_1 du_1 \hat{u}_1 + h_2 du_2 \hat{u}_2 + h_3 du_3 \hat{u}_3. \qquad (1.55)$$

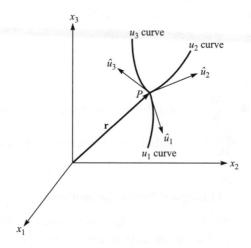

Figure 1.16.   Curvilinear coordinates.

The quantities h_1, h_2, h_3 are sometimes called scale factors. The unit vectors $\hat{u}_1, \hat{u}_2, \hat{u}_3$ are in the direction of increasing u_1, u_2, u_3, respectively.

If $\hat{u}_1, \hat{u}_2, \hat{u}_3$ are mutually perpendicular at any point P, the curvilinear coordinates are called orthogonal. In such a case the element of arc length ds is given by

$$ds^2 = d\mathbf{r} \cdot d\mathbf{r} = h_1^2 du_1^2 + h_2^2 du_2^2 + h_3^2 du_3^2. \tag{1.56}$$

Along a u_1 curve, u_2 and u_3 are constants so that $d\mathbf{r} = h_1 du_1 \hat{u}_1$. Then the differential of arc length ds_1 along u_1 at P is $h_1 du_1$. Similarly the differential arc lengths along u_2 and u_3 at P are $ds_2 = h_2 du_2$, $ds_3 = h_3 du_3$ respectively.

The volume of the parallelepiped is given by

$$dV = |(h_1 du_1 \hat{u}_1) \cdot (h_2 du_2 \hat{u}_2) \times (h_3 du_3 \hat{u}_3)| = h_1 h_2 h_3 du_1 du_2 du_3$$

since $|\hat{u}_1 \cdot \hat{u}_2 \times \hat{u}_3| = 1$. Alternatively dV can be written as

$$dV = \left| \frac{\partial \mathbf{r}}{\partial u_1} \cdot \frac{\partial \mathbf{r}}{\partial u_2} \times \frac{\partial \mathbf{r}}{\partial u_3} \right| du_1 du_2 du_3 = \left| \frac{\partial(x_1, x_2, x_3)}{\partial(u_1, u_2, u_3)} \right| du_1 du_2 du_3, \tag{1.57}$$

where

$$J = \frac{\partial(x_1, x_2, x_3)}{\partial(u_1, u_2, u_3)} = \begin{vmatrix} \dfrac{\partial x_1}{\partial u_1} & \dfrac{\partial x_1}{\partial u_2} & \dfrac{\partial x_1}{\partial u_3} \\[2mm] \dfrac{\partial x_2}{\partial u_1} & \dfrac{\partial x_2}{\partial u_2} & \dfrac{\partial x_2}{\partial u_3} \\[2mm] \dfrac{\partial x_3}{\partial u_1} & \dfrac{\partial x_3}{\partial u_2} & \dfrac{\partial x_3}{\partial u_3} \end{vmatrix}$$

is called the Jacobian of the transformation.

We assume that the Jacobian $J \neq 0$ so that the transformation (1.54) is one to one in the neighborhood of a point.

We are now ready to express the gradient, divergence, and curl in terms of $u_1, u_2,$ and u_3. If ϕ is a scalar function of $u_1, u_2,$ and u_3, then the gradient takes the form

$$\nabla \phi = \operatorname{grad} \phi = \frac{1}{h_1} \frac{\partial \phi}{\partial u_1} \hat{u}_1 + \frac{1}{h_2} \frac{\partial \phi}{\partial u_2} \hat{u}_2 + \frac{1}{h_3} \frac{\partial \phi}{\partial u_3} \hat{u}_3. \tag{1.58}$$

To derive this, let

$$\nabla \phi = f_1 \hat{u}_1 + f_2 \hat{u}_2 + f_3 \hat{u}_3, \tag{1.59}$$

where f_1, f_2, f_3 are to be determined. Since

$$d\mathbf{r} = \frac{\partial \mathbf{r}}{\partial u_1} du_1 + \frac{\partial \mathbf{r}}{\partial u_2} du_2 + \frac{\partial \mathbf{r}}{\partial u_3} du_3$$

$$= h_1 du_1 \hat{u}_1 + h_2 du_2 \hat{u}_2 + h_3 du_3 \hat{u}_3,$$

29

we have

$$d\phi = \nabla\phi \cdot d\mathbf{r} = h_1 f_1 du_1 + h_2 f_2 du_2 + h_3 f_3 du_3.$$

But

$$d\phi = \frac{\partial\phi}{\partial u_1} du_1 + \frac{\partial\phi}{\partial u_2} du_2 + \frac{\partial\phi}{\partial u_3} du_3,$$

and on equating the two equations, we find

$$f_i = \frac{1}{h_i}\frac{\partial\phi}{\partial u_i}, \quad i = 1, 2, 3.$$

Substituting these into Eq. (1.57), we obtain the result Eq. (1.58).

From Eq. (1.58) we see that the operator ∇ takes the form

$$\nabla = \frac{\hat{u}_1}{h_1}\frac{\partial}{\partial u_1} + \frac{\hat{u}_2}{h_2}\frac{\partial}{\partial u_2} + \frac{\hat{u}_3}{h_3}\frac{\partial}{\partial u_3}. \tag{1.60}$$

Because we will need them later, we now proceed to prove the following two relations:

(a) $|\nabla u_i| = h_i^{-1}, i = 1, 2, 3$.

(b) $\hat{u}_1 = h_2 h_3 \nabla u_2 \times \nabla u_3$ with similar equations for $\hat{u}_2$ and $\hat{u}_3$. $\tag{1.61}$

Proof: (a) Let $\phi = u_1$ in Eq. (1.51), we then obtain $\nabla u_1 = \hat{u}_1/h_1$ and so

$$|\nabla u_1| = |\hat{u}_1| h_1^{-1} = h_1^{-1}, \text{ since } |\hat{u}_1| = 1.$$

Similarly by letting $\phi = u_2$ and u_3, we obtain the relations for $i = 2$ and 3.

(b) From (a) we have

$$\nabla u_1 = \hat{u}_1/h_1, \quad \nabla u_2 = \hat{u}_2/h_2, \quad \text{and} \quad \nabla u_3 = \hat{u}_3/h_3.$$

Then

$$\nabla u_2 \times \nabla u_3 = \frac{\hat{u}_2 \times \hat{u}_3}{h_2 h_3} = \frac{\hat{u}_1}{h_2 h_3} \quad \text{and} \quad \hat{u}_1 = h_2 h_3 \nabla u_2 \times \nabla u_3.$$

Similarly

$$\hat{u}_2 = h_3 h_1 \nabla u_3 \times \nabla u_1 \quad \text{and} \quad \hat{u}_3 = h_1 h_2 \nabla u_1 \times \nabla u_2.$$

We are now ready to express the divergence in terms of curvilinear coordinates. If $\mathbf{A} = A_1\hat{u}_1 + A_2\hat{u}_2 + A_3\hat{u}_3$ is a vector function of orthogonal curvilinear coordinates u_1, u_2, and u_3, the divergence will take the form

$$\nabla \cdot \mathbf{A} = \operatorname{div}\mathbf{A} = \frac{1}{h_1 h_2 h_3}\left[\frac{\partial}{\partial u_1}(h_2 h_3 A_1) + \frac{\partial}{\partial u_2}(h_3 h_1 A_2) + \frac{\partial}{\partial u_3}(h_1 h_2 A_3)\right]. \tag{1.62}$$

To derive (1.62), we first write $\nabla \cdot \mathbf{A}$ as

$$\nabla \cdot \mathbf{A} = \nabla \cdot (A_1\hat{u}_1) + \nabla \cdot (A_2\hat{u}_2) + \nabla \cdot (A_3\hat{u}_3), \tag{1.63}$$

then, because $\hat{u}_1 = h_1 h_2 \nabla u_2 \times \nabla u_3$, we express $\nabla \cdot (A_1 \hat{u}_1)$ as

$$\nabla \cdot (A_1 \hat{u}_1) = \nabla \cdot (A_1 h_2 h_3 \nabla u_2 \times \nabla u_3) \quad (\hat{u}_1 = h_2 h_3 \nabla u_2 \times \nabla u_3)$$

$$= \nabla(A_1 h_2 h_3) \cdot \nabla u_2 \times \nabla u_3 + A_1 h_2 h_3 \nabla \cdot (\nabla u_2 \times \nabla u_3),$$

where in the last step we have used the vector identity: $\nabla \cdot (\phi \mathbf{A}) = (\nabla \phi) \cdot \mathbf{A} + \phi(\nabla \times \mathbf{A})$. Now $\nabla u_i = \hat{u}_i / h_i, i = 1, 2, 3$, so $\nabla \cdot (A_1 \hat{u}_1)$ can be rewritten as

$$\nabla \cdot (A_1 \hat{u}_1) = \nabla(A_1 h_2 h_3) \cdot \frac{\hat{u}_2}{h_2} \times \frac{\hat{u}_3}{h_3} + 0 = \nabla(A_1 h_2 h_3) \cdot \frac{\hat{u}_1}{h_2 h_3}.$$

The gradient $\nabla(A_1 h_2 h_3)$ is given by Eq. (1.58), and we have

$$\nabla \cdot (A_1 \hat{u}_1) = \left[\frac{\hat{u}_1}{h_1} \frac{\partial}{\partial u_1} (A_1 h_2 h_3) + \frac{\hat{u}_2}{h_2} \frac{\partial}{\partial u_2} (A_1 h_2 h_3) + \frac{\hat{u}_3}{h_3} \frac{\partial}{\partial u_3} (A_1 h_2 h_3) \right] \cdot \frac{\hat{u}_1}{h_2 h_3}$$

$$= \frac{1}{h_1 h_2 h_3} \frac{\partial}{\partial u_1} (A_1 h_2 h_3).$$

Similarly, we have

$$\nabla \cdot (A_2 \hat{u}_2) = \frac{1}{h_1 h_2 h_3} \frac{\partial}{\partial u_2} (A_2 h_3 h_1), \quad \text{and} \quad \nabla \cdot (A_3 \hat{u}_3) = \frac{1}{h_1 h_2 h_3} \frac{\partial}{\partial u_3} (A_3 h_2 h_1).$$

Substituting these into Eq. (1.63), we obtain the result, Eq. (1.62).

In the same manner we can derive a formula for curl $\mathbf{A}$. We first write it as

$$\nabla \times \mathbf{A} = \nabla \times (A_1 \hat{u}_1 + A_2 \hat{u}_2 + A_3 \hat{u}_3)$$

and then evaluate $\nabla \times A_i \hat{u}_i$.

Now $\hat{u}_i = h_i \nabla u_i, i = 1, 2, 3$, and we express $\nabla \times (A_1 \hat{u}_1)$ as

$$\nabla \times (A_1 \hat{u}_1) = \nabla \times (A_1 h_1 \nabla u_1)$$

$$= \nabla(A_1 h_1) \times \nabla u_1 + A_1 h_1 \nabla \times \nabla u_1$$

$$= \nabla(A_1 h_1) \times \frac{\hat{u}_1}{h_1} + 0$$

$$= \left[\frac{\hat{u}_1}{h_1} \frac{\partial}{\partial u_1} (A_1 h_1) + \frac{\hat{u}_2}{h_2} \frac{\partial}{\partial u_2} (A_2 h_2) + \frac{\hat{u}_3}{h_3} \frac{\partial}{\partial u_3} (A_3 h_3) \right] \times \frac{\hat{u}_1}{h_1}$$

$$= \frac{\hat{u}_2}{h_3 h_1} \frac{\partial}{\partial u_3} (A_1 h_1) - \frac{\hat{u}_3}{h_1 h_2} \frac{\partial}{\partial u_2} (A_1 h_1),$$

with similar expressions for $\nabla \times (A_2 \hat{u}_2)$ and $\nabla \times (A_3 \hat{u}_3)$. Adding these together, we get $\nabla \times \mathbf{A}$ in orthogonal curvilinear coordinates:

$$\nabla \times \mathbf{A} = \frac{\hat{u}_1}{h_2 h_3} \left[\frac{\partial}{\partial u_2} (A_3 h_3) - \frac{\partial}{\partial u_3} (A_2 h_2) \right] + \frac{\hat{u}_2}{h_3 h_1} \left[\frac{\partial}{\partial u_3} (A_1 h_1) - \frac{\partial}{\partial u_1} (A_3 h_3) \right]$$
$$+ \frac{\hat{u}_3}{h_1 h_2} \left[\frac{\partial}{\partial u_1} (A_2 h_2) - \frac{\partial}{\partial u_2} (A_1 h_1) \right]. \tag{1.64}$$

This can be written in determinant form:

$$\nabla \times \mathbf{A} = \frac{1}{h_1 h_2 h_3} \begin{vmatrix} h_1 \hat{u}_1 & h_2 \hat{u}_2 & h_3 \hat{u}_3 \\ \dfrac{\partial}{\partial u_1} & \dfrac{\partial}{\partial u_2} & \dfrac{\partial}{\partial u_3} \\ A_1 h_1 & A_2 h_2 & A_3 h_3 \end{vmatrix}. \tag{1.65}$$

We now express the Laplacian in orthogonal curvilinear coordinates. From Eqs. (1.58) and (1.62) we have

$$\nabla \phi = \operatorname{grad} \phi = \frac{1}{h_1} \frac{\partial \phi}{\partial u_1} \hat{u}_1 + \frac{1}{h_2} \frac{\partial \phi}{\partial u_2} \hat{u} + \frac{1}{h_3} \frac{\partial \phi}{\partial u_3} \hat{u}_3,$$

$$\nabla \cdot \mathbf{A} = \operatorname{div} \mathbf{A} = \frac{1}{h_1 h_2 h_3} \left[\frac{\partial}{\partial u_1} (h_2 h_3 A_1) + \frac{\partial}{\partial u_2} (h_3 h_1 A_2) + \frac{\partial}{\partial u_3} (h_1 h_2 A_3) \right].$$

If $\mathbf{A} = \nabla \phi$, then $A_i = (1/h_i) \partial \phi / \partial u_i$, $i = 1, 2, 3$; and

$$\nabla \cdot \mathbf{A} = \nabla \cdot \nabla \phi = \nabla^2 \phi$$

$$= \frac{1}{h_1 h_2 h_3} \left[\frac{\partial}{\partial u_1} \left(\frac{h_2 h_3}{h_1} \frac{\partial \phi}{\partial u_1} \right) + \frac{\partial}{\partial u_2} \left(\frac{h_3 h_1}{h_2} \frac{\partial \phi}{\partial u_2} \right) + \frac{\partial}{\partial u_3} \left(\frac{h_1 h_2}{h_3} \frac{\partial \phi}{\partial u_3} \right) \right]. \tag{1.66}$$

Special orthogonal coordinate systems

There are at least nine special orthogonal coordinates systems, the most common and useful ones are the cylindrical and spherical coordinates; we introduce these two coordinates in this section.

Cylindrical coordinates (ρ, ϕ, z)

$$u_1 = \rho, u_2 = \phi, u_3 = z; \quad \text{and} \quad \hat{u}_1 = e_\rho, \hat{u}_2 = e_\phi \hat{u}_3 = e_z.$$

From Fig. 1.17 we see that

$$x_1 = \rho \cos \phi, x_2 = \rho \sin \phi, x_3 = z$$

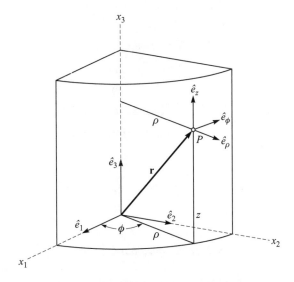

Figure 1.17.   Cylindrical coordinates.

where

$$\rho \geq 0, 0 \leq \phi \leq 2\pi, -\infty < z < \infty.$$

The square of the element of arc length is given by

$$ds^2 = h_1^2(d\rho)^2 + h_2^2(d\phi)^2 + h_3^2(dz)^2.$$

To find the scale factors h_i, we notice that $ds^2 = d\mathbf{r} \cdot d\mathbf{r}$ where

$$\mathbf{r} = \rho\cos\phi e_1 + \rho\sin\phi e_2 + ze_3.$$

Thus

$$ds^2 = d\mathbf{r} \cdot d\mathbf{r} = (d\rho)^2 + \rho^2(d\phi)^2 + (dz)^2.$$

Equating the two ds^2, we find the scale factors:

$$h_1 = h_\rho = 1, h_2 = h_\phi = \rho, h_3 = h_z = 1. \tag{1.67}$$

From Eqs. (1.58), (1.62), (1.64), and (1.66) we find the gradient, divergence, curl, and Laplacian in cylindrical coordinates:

$$\nabla\Phi = \frac{\partial\Phi}{\partial\rho}e_\rho + \frac{1}{\rho}\frac{\partial\Phi}{\partial\phi}e_\phi + \frac{\partial\Phi}{\partial z}e_z, \tag{1.68}$$

where $\Phi = \Phi(\rho, \phi, z)$ is a scalar function;

$$\nabla \cdot \mathbf{A} = \frac{1}{\rho}\left[\frac{\partial}{\partial\rho}(\rho A_\rho) + \frac{\partial A_\phi}{\partial\phi} + \frac{\partial}{\partial z}(\rho A_z)\right]; \tag{1.69}$$

where

$$\mathbf{A} = A_\rho e_\rho + A_\phi e_\phi + A_z e_z;$$

$$\nabla \times \mathbf{A} = \frac{1}{\rho} \begin{vmatrix} e_\rho & \rho e_\phi & e_z \\ \dfrac{\partial}{\partial \rho} & \dfrac{\partial}{\partial \phi} & \dfrac{\partial}{\partial z} \\ A_\rho & \rho A_\phi & A_z \end{vmatrix};$$

$$(1.70)$$

and

$$\nabla^2 \Phi = \frac{1}{\rho} \frac{\partial}{\partial \rho} \left(\rho \frac{\partial \Phi}{\partial \rho} \right) + \frac{1}{\rho^2} \frac{\partial^2 \Phi}{\partial \phi^2} + \frac{\partial^2 \Phi}{\partial z^2}.$$

$$(1.71)$$

Spherical coordinates (r, θ, ϕ)

$$u_1 = r, u_2 = \theta, u_3 = \phi; \hat{u}_1 = e_r, \hat{u}_2 = e_\theta, \hat{u}_3 = e_\phi$$

From Fig. 1.18 we see that

$$x_1 = r \sin \theta \cos \phi, x_2 = r \sin \theta \sin \phi, x_3 = r \cos \theta.$$

Now

$$ds^2 = h_1^2 (dr)^2 + h_2^2 (d\theta)^2 + h_3^2 (d\phi)^2$$

but

$$\mathbf{r} = r \sin \theta \cos \phi \hat{e}_1 + r \sin \theta \sin \phi \hat{e}_2 + r \cos \theta \hat{e}_3,$$

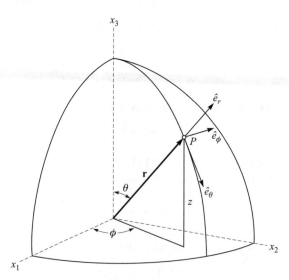

Figure 1.18.   Spherical coordinates.

34

so

$$ds^2 = d\mathbf{r} \cdot d\mathbf{r} = (dr)^2 + r^2(d\theta)^2 + r^2 \sin^2 \theta (d\phi)^2.$$

Equating the two ds^2, we find the scale factors: $h_1 = h_r = 1$, $h_2 = h_\theta = r$, $h_3 = h_\phi = r \sin \theta$. We then find, from Eqs. (1.58), (1.62), (1.64), and (1.66), the gradient, divergence, curl, and the Laplacian in spherical coordinates:

$$\nabla \Phi = \hat{e}_r \frac{\partial \Phi}{\partial r} + \hat{e}_\theta \frac{1}{r} \frac{\partial \Phi}{\partial \theta} + \hat{e}_\phi \frac{1}{r \sin \theta} \frac{\partial \Phi}{\partial \phi}; \tag{1.72}$$

$$\nabla \cdot \mathbf{A} = \frac{1}{r^2 \sin \theta} \left[\sin \theta \frac{\partial}{\partial r} (r^2 A_r) + r \frac{\partial}{\partial \theta} (\sin \theta A_\theta) + r \frac{\partial A_\phi}{\partial \phi} \right]; \tag{1.73}$$

$$\nabla \times \mathbf{A} = \frac{1}{r^2 \sin \theta} \begin{vmatrix} \hat{e}_r & r\hat{e}_\theta & r \sin \theta \hat{e}_\phi \\ \dfrac{\partial}{\partial r} & \dfrac{\partial}{\partial \theta} & \dfrac{\partial}{\partial \phi} \\ A_r & rA_r & r \sin \theta A_\phi \end{vmatrix}; \tag{1.74}$$

$$\nabla^2 \Phi = \frac{1}{r^2 \sin \theta} \left[\sin \theta \frac{\partial}{\partial r} \left(r^2 \frac{\partial \Phi}{\partial r} \right) + \frac{\partial}{\partial \theta} \left(\sin \theta \frac{\partial \Phi}{\partial \theta} \right) + \frac{1}{\sin \theta} \frac{\partial^2 \Phi}{\partial \phi^2} \right]. \tag{1.75}$$

Vector integration and integral theorems

Having discussed vector differentiation, we now turn to a discussion of vector integration. After defining the concepts of line, surface, and volume integrals of vector fields, we then proceed to the important integral theorems of Gauss, Stokes, and Green.

The integration of a vector, which is a function of a single scalar u, can proceed as ordinary scalar integration. Given a vector

$$\mathbf{A}(u) = A_1(u)\hat{e}_1 + A_2(u)\hat{e}_2 + A_3(u)\hat{e}_3,$$

then

$$\int \mathbf{A}(u)du = \hat{e}_1 \int A_1(u)du + \hat{e}_2 \int A_2(u)du + \hat{e}_3 \int A_3(u)du + \mathbf{B},$$

where $\mathbf{B}$ is a constant of integration, a constant vector. Now consider the integral of the scalar product of a vector $\mathbf{A}(x_1, x_2, x_3)$ and $d\mathbf{r}$ between the limit $P_1(x_1, x_2, x_3)$ and $P_2(x_1, x_2, x_3)$:

$$\int_{P_1}^{P_2} \mathbf{A} \cdot d\mathbf{r} = \int_{P_1}^{P_2} (A_1\hat{e}_1 + A_2\hat{e}_2 + A_3\hat{e}_3) \cdot (dx_1\hat{e}_1 + dx_2\hat{e}_2 + dx_3\hat{e}_3)$$

$$= \int_{P_1}^{P_2} A_1(x_1, x_2, x_3)dx_1 + \int_{P_1}^{P_2} A_2(x_1, x_2, x_3)dx_2$$

$$+ \int_{P_1}^{P_2} A_3(x_1, x_2, x_3)dx_3.$$

Each integral on the right hand side requires for its execution more than a knowledge of the limits. In fact, the three integrals on the right hand side are not completely defined because in the first integral, for example, we do not the know value of x_2 and x_3 in A_1:

$$I_1 = \int_{P_1}^{P_2} A_1(x_1, x_2, x_3)dx_1. \tag{1.76}$$

What is needed is a statement such as

$$x_2 = f(x_1), x_3 = g(x_1) \tag{1.77}$$

that specifies x_2, x_3 for each value of x_1. The integrand now reduces to $A_1(x_1, x_2, x_3) = A_1(x_1, f(x_1), g(x_1)) = B_1(x_1)$ so that the integral I_1 becomes well defined. But its value depends on the constraints in Eq. (1.77). The constraints specify paths on the $x_1 x_2$ and $x_3 x_1$ planes connecting the starting point P_1 to the end point P_2. The x_1 integration in (1.76) is carried out along these paths. It is a path-dependent integral and is called a line integral (or a path integral). It is very helpful to keep in mind that: *when the number of integration variables is less than the number of variables in the integrand, the integral is not yet completely defined and it is path-dependent*. However, if the scalar product $\mathbf{A} \cdot d\mathbf{r}$ is equal to an exact differential, $\mathbf{A} \cdot d\mathbf{r} = d\varphi = \nabla\varphi \cdot d\mathbf{r}$, the integration depends only upon the limits and is therefore path-independent:

$$\int_{P_1}^{P_2} \mathbf{A} \cdot d\mathbf{r} = \int_{P_1}^{P_2} d\varphi = \varphi_2 - \varphi_1.$$

A vector field $\mathbf{A}$ which has above (path-independent) property is termed conservative. It is clear that the line integral above is zero along any close path, and the curl of a conservative vector field is zero ($\nabla \times \mathbf{A} = \nabla \times (\nabla\varphi) = 0$). A typical example of a conservative vector field in mechanics is a conservative force.

The surface integral of a vector function $\mathbf{A}(x_1, x_2, x_3)$ over the surface S is an important quantity; it is defined to be

$$\int_S \mathbf{A} \cdot d\mathbf{a},$$

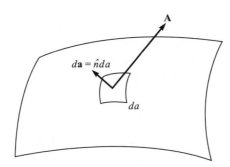

Figure 1.19.   Surface integral over a surface S.

where the surface integral symbol $\int_s$ stands for a double integral over a certain surface S, and $d\mathbf{a}$ is an element of area of the surface (Fig. 1.19), a vector quantity. We attribute to $d\mathbf{a}$ a magnitude da and also a direction corresponding the normal, $\hat{n}$, to the surface at the point in question, thus

$$d\mathbf{a} = \hat{n}da.$$

The normal $\hat{n}$ to a surface may be taken to lie in either of two possible directions. But if da is part of a closed surface, the sign of $\hat{n}$ relative to da is so chosen that it points outward away from the interior. In rectangular coordinates we may write

$$d\mathbf{a} = \hat{e}_1 da_1 + \hat{e}_2 da_2 + \hat{e}_3 da_3 = \hat{e}_1 dx_2 dx_3 + \hat{e}_2 dx_3 dx_1 + \hat{e}_3 dx_1 dx_2.$$

If a surface integral is to be evaluated over a closed surface S, the integral is written as

$$\oint_S \mathbf{A} \cdot d\mathbf{a}.$$

Note that this is different from a closed-path line integral. When the path of integration is closed, the line integral is write it as

$$\oint_\Gamma \mathbf{A} \cdot d\mathbf{s},$$

where Γ specifies the closed path, and $d\mathbf{s}$ is an element of length along the given path. By convention, $d\mathbf{s}$ is taken positive along the direction in which the path is traversed. Here we are only considering simple closed curves. A simple closed curve does not intersect itself anywhere.

Gauss' theorem (the divergence theorem)

This theorem relates the surface integral of a given vector function and the volume integral of the divergence of that vector. It was introduced by Joseph Louis Lagrange and was first used in the modern sense by George Green. Gauss' name is associated with this theorem because of his extensive work on general problems of double and triple integrals.

If a continuous, differentiable vector field **A** is defined in a simply connected region of volume V bounded by a closed surface S, then the theorem states that

$$\int_V \nabla \cdot \mathbf{A} \, dV = \oint_S \mathbf{A} \cdot d\mathbf{a}, \tag{1.78}$$

where $dV = dx_1 dx_2 dx_3$. A simple connected region V has the property that every simple closed curve within it can be continuously shrunk to a point without leaving the region. To prove this, we first write

$$\int_V \nabla \cdot \mathbf{A} \, dV = \int_V \sum_{i=1}^{3} \frac{\partial A_i}{\partial x_i} \, dV,$$

then integrate the right hand side with respect to x_1 while keeping $x_2 x_3$ constant, thus summing up the contribution from a rod of cross section $dx_2 dx_3$ (Fig. 1.20). The rod intersects the surface S at the points P and Q and thus defines two elements of area $d\mathbf{a}_P$ and $d\mathbf{a}_Q$:

$$\int_V \frac{\partial A_1}{\partial x_1} \, dV = \oint_S dx_2 dx_3 \int_P^Q \frac{\partial A_1}{\partial x_1} \, dx_1 = \oint_S dx_2 dx_3 \int_P^Q dA_1,$$

where we have used the relation $dA_1 = (\partial A_1/\partial x_1) dx_1$ along the rod. The last integration on the right hand side can be performed at once and we have

$$\int_V \frac{\partial A_1}{\partial x_1} \, dV = \oint_S [A_1(Q) - A_1(P)] dx_2 dx_3,$$

where $A_1(Q)$ denotes the value of A_1 evaluated at the coordinates of the point Q, and similarly for $A_1(P)$.

The component of the surface element $d\mathbf{a}$ which lies in the x_1-direction is $da_1 = dx_2 dx_3$ at the point Q, and $da_1 = -dx_2 dx_3$ at the point P. The minus sign

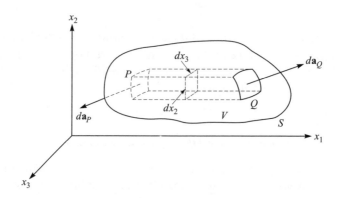

Figure 1.20.   A square tube of cross section $dx_2 dx_3$.

arises since the x_1 component of $d\mathbf{a}$ at P is in the direction of negative x_1. We can now rewrite the above integral as

$$\int_V \frac{\partial A_1}{\partial x_1} dV = \int_{S_Q} A_1(Q) da_1 + \int_{S_P} A_1(P) da_1,$$

where S_Q denotes that portion of the surface for which the x_1 component of the outward normal to the surface element da_1 is in the positive x_1-direction, and S_P denotes that portion of the surface for which da_1 is in the negative direction. The two surface integrals then combine to yield the surface integral over the entire surface S (if the surface is sufficiently concave, there may be several such as right hand and left hand portions of the surfaces):

$$\int_V \frac{\partial A_1}{\partial x_1} dV = \oint_S A_1 da_1.$$

Similarly we can evaluate the x_2 and x_3 components. Summing all these together, we have Gauss' theorem:

$$\int_V \sum_i \frac{\partial A_i}{\partial x_i} dV = \oint_S \sum_i A_i da_i \quad \text{or} \quad \int_V \nabla \cdot \mathbf{A} dV = \oint_S \mathbf{A} \cdot d\mathbf{a}.$$

We have proved Gauss' theorem for a simply connected region (a volume bounded by a single surface), but we can extend the proof to a multiply connected region (a region bounded by several surfaces, such as a hollow ball). For interested readers, we recommend the book *Electromagnetic Fields*, Roald K. Wangsness, John Wiley, New York, 1986.

Continuity equation

Consider a fluid of density $\rho(\mathbf{r})$ which moves with velocity $\mathbf{v}(\mathbf{r})$ in a certain region. If there are no sources or sinks, the following continuity equation must be satisfied:

$$\partial \rho(\mathbf{r})/\partial t + \nabla \cdot \mathbf{j}(\mathbf{r}) = 0, \tag{1.79}$$

where $\mathbf{j}$ is the current

$$\mathbf{j}(\mathbf{r}) = \rho(\mathbf{r})\mathbf{v}(\mathbf{r}) \tag{1.79a}$$

and Eq. (1.79) is called the continuity equation for a conserved current.

To derive this important equation, let us consider an arbitrary surface S enclosing a volume V of the fluid. At any time the mass of fluid within V is $M = \int_V \rho dV$ and the time rate of mass increase (due to mass flowing into V) is

$$\frac{\partial M}{\partial t} = \frac{\partial}{\partial t} \int_V \rho dV = \int_V \frac{\partial \rho}{\partial t} dV,$$

while the mass of fluid leaving V per unit time is

$$\int_S \rho\mathbf{v} \cdot \hat{n}\,ds = \int_V \nabla \cdot (\rho\mathbf{v})dV,$$

where Gauss' theorem is used in changing the surface integral to volume integral. Since there is neither a source nor a sink, mass conservation requires an exact balance between these effects:

$$\int_V \frac{\partial \rho}{\partial t}dV = -\int_V \nabla \cdot (\rho\mathbf{v})dV, \quad \text{or} \quad \int_V \left(\frac{\partial \rho}{\partial t} + \nabla \cdot (\rho\mathbf{v})\right)dV = 0.$$

Also since V is arbitrary, mass conservation requires that the continuity equation

$$\frac{\partial \rho}{\partial t} + \nabla \cdot (\rho\mathbf{v}) = \frac{\partial \rho}{\partial t}\nabla \cdot \mathbf{j} = 0$$

must be satisfied everywhere in the region.

Stokes' theorem

This theorem relates the line integral of a vector function and the surface integral of the curl of that vector. It was first discovered by Lord Kelvin in 1850 and rediscovered by George Gabriel Stokes four years later.

If a continuous, differentiable vector field $\mathbf{A}$ is defined a three-dimensional region V, and S is a regular open surface embedded in V bounded by a simple closed curve Γ, the theorem states that

$$\int_S \nabla \times \mathbf{A} \cdot d\mathbf{a} = \oint_\Gamma \mathbf{A} \cdot d\mathbf{l}; \tag{1.80}$$

where the line integral is to be taken completely around the curve Γ and $d\mathbf{l}$ is an element of line (Fig. 1.21).

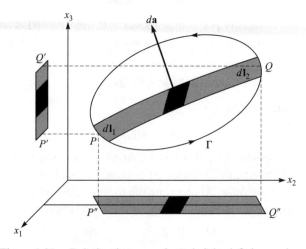

Figure 1.21.   Relation between $d\mathbf{a}$ and $d\mathbf{l}$ in defining curl.

40

The surface S, bounded by a simple closed curve, is an open surface; and the normal to an open surface can point in two opposite directions. We adopt the usual convention, namely the right hand rule: when the fingers of the right hand follow the direction of $d\mathbf{l}$, the thumb points in the $d\mathbf{a}$ direction, as shown in Fig. 1.21.

Note that Eq. (1.80) does not specify the shape of the surface S other than that it be bounded by Γ; thus there are many possibilities in choosing the surface. But Stokes' theorem enables us to reduce the evaluation of surface integrals which depend upon the shape of the surface to the calculation of a line integral which depends only on the values of $\mathbf{A}$ along the common perimeter.

To prove the theorem, we first expand the left hand side of Eq. (1.80); with the aid of Eq. (1.50), it becomes

$$\int_S \nabla \times \mathbf{A} \cdot d\mathbf{a} = \int_S \left(\frac{\partial A_1}{\partial x_3} da_2 - \frac{\partial A_1}{\partial x_2} da_3 \right) + \int_S \left(\frac{\partial A_2}{\partial x_1} da_3 - \frac{\partial A_2}{\partial x_3} da_1 \right)$$
$$+ \int_S \left(\frac{\partial A_3}{\partial x_2} da_1 - \frac{\partial A_3}{\partial x_1} da_2 \right), \tag{1.81}$$

where we have grouped the terms by components of $\mathbf{A}$. We next subdivide the surface S into a large number of small strips, and integrate the first integral on the right hand side of Eq. (1.81), denoted by I_1, over one such a strip of width dx_1, which is parallel to the $x_2 x_3$ plane and a distance x_1 from it, as shown in Fig. 1.21. Then, by integrating over x_1, we sum up the contributions from all of the strips.

Fig. 1.21 also shows the projections of the strip on the $x_1 x_3$ and $x_1 x_2$ planes that will help us to visualize the orientation of the surface. The element area $d\mathbf{a}$ is shown at an intermediate stage of the integration, when the direction angles have values such that α and γ are less than $90°$ and β is greater than $90°$. Thus, $da_2 = -dx_1 dx_3$ and $da_3 = dx_1 dx_2$ and we can write

$$I_1 = - \int_{\text{strips}} dx_1 \int_P^Q \left(\frac{\partial A_1}{\partial x_2} dx_2 + \frac{\partial A_1}{\partial x_3} dx_3 \right). \tag{1.82}$$

Note that dx_2 and dx_3 in the parentheses are not independent because x_2 and x_3 are related by the equation for the surface S and the value of x_1 involved. Since the second integral in Eq. (1.82) is being evaluated on the strip from P to Q for which $x_1 = \text{const.}$, $dx_1 = 0$ and we can add $(\partial A_1/\partial x_1)dx_1 = 0$ to the integrand to make it dA_1:

$$\frac{\partial A_1}{\partial x_1} dx_1 + \frac{\partial A_1}{\partial x_2} dx_2 + \frac{\partial A_1}{\partial x_3} dx_3 = dA_1.$$

And Eq. (1.82) becomes

$$I_1 = - \int_{\text{strips}} dx_1 \int_P^Q dA_1 = \int_{\text{strips}} [A_1(P) - A_1(Q)]dx_1.$$

Next we consider the line integral of $\mathbf{A}$ around the lines bounding each of the small strips. If we trace each one of these lines in the same sense as we trace the path Γ, then we will trace all of the interior lines twice (once in each direction) and all of the contributions to the line integral from the interior lines will cancel, leaving only the result from the boundary line Γ. Thus, the sum of all of the line integrals around the small strips will equal the line integral Γ of A_1:

$$\int_S \left(\frac{\partial A_1}{\partial x_3} da_2 - \frac{\partial A_1}{\partial x_2} da_3 \right) = \oint_\Gamma A_1 dl_1. \tag{1.83}$$

Similarly, the last two integrals of Eq. (1.81) can be shown to have the respective values

$$\oint_\Gamma A_2 dl_2 \quad \text{and} \quad \oint_\Gamma A_3 dl_3.$$

Substituting these results and Eq. (1.83) into Eq. (1.81) we obtain Stokes' theorem:

$$\int_S \nabla \times \mathbf{A} \cdot d\mathbf{a} = \oint_\Gamma (A_1 dl_1 + A_2 dl_2 + A_3 dl_3) = \oint_\Gamma \mathbf{A} \cdot d\mathbf{l}.$$

Stokes' theorem in Eq. (1.80) is valid whether or not the closed curve Γ lies in a plane, because in general the surface S is not a planar surface. Stokes' theorem holds for any surface bounded by Γ.

In fluid dynamics, the curl of the velocity field $\mathbf{v}(\mathbf{r})$ is called its *vorticity* (for example, the whirls that one creates in a cup of coffee on stirring it). If the velocity field is derivable from a potential

$$\mathbf{v}(\mathbf{r}) = -\nabla \phi(\mathbf{r})$$

it must be irrotational (see Eq. (1.51)). For this reason, an irrotational flow is also called a potential flow, which describes a steady flow of the fluid, free of vortices and eddies.

One of Maxwell's equations of electromagnetism (Ampère's law) states that

$$\nabla \times \mathbf{B} = \mu_0 \mathbf{j},$$

where $\mathbf{B}$ is the magnetic induction, $\mathbf{j}$ is the current density (per unit area), and μ_0 is the permeability of free space. From this equation, current densities may be visualized as vortices of $\mathbf{B}$. Applying Stokes' theorem, we can rewrite Ampère's law as

$$\oint_\Gamma \mathbf{B} \cdot d\mathbf{r} = \mu_0 \int_S \mathbf{j} \cdot d\mathbf{a} = \mu_0 I;$$

it states that the circulation of the magnetic induction is proportional to the total current I passing through the surface S enclosed by Γ.

Green's theorem

Green's theorem is an important corollary of the divergence theorem, and it has many applications in many branches of physics. Recall that the divergence theorem Eq. (1.78) states that

$$\int_V \nabla \cdot \mathbf{A} dV = \oint_S \mathbf{A} \cdot d\mathbf{a}.$$

Let $\mathbf{A} = \psi \mathbf{B}$, where ψ is a scalar function and $\mathbf{B}$ a vector function, then $\nabla \cdot \mathbf{A}$ becomes

$$\nabla \cdot \mathbf{A} = \nabla \cdot (\psi \mathbf{B}) = \psi \nabla \cdot \mathbf{B} + \mathbf{B} \cdot \nabla \psi.$$

Substituting these into the divergence theorem, we have

$$\oint_S \psi \mathbf{B} \cdot d\mathbf{a} = \int_V (\psi \nabla \cdot \mathbf{B} + \mathbf{B} \cdot \nabla \psi) dV. \tag{1.84}$$

If $\mathbf{B}$ represents an irrotational vector field, we can express it as a gradient of a scalar function, say, φ:

$$\mathbf{B} \equiv \nabla \varphi.$$

Then Eq. (1.84) becomes

$$\oint_S \psi \mathbf{B} \cdot d\mathbf{a} = \int_V [\psi \nabla \cdot (\nabla \varphi) + (\nabla \varphi) \cdot (\nabla \psi)] dV. \tag{1.85}$$

Now

$$\mathbf{B} \cdot d\mathbf{a} = (\nabla \varphi) \cdot \hat{n} da.$$

The quantity $(\nabla \varphi) \cdot \hat{n}$ represents the rate of change of ϕ in the direction of the outward normal; it is called the normal derivative and is written as

$$(\nabla \varphi) \cdot \hat{n} \equiv \partial \varphi / \partial n.$$

Substituting this and the identity $\nabla \cdot (\nabla \varphi) = \nabla^2 \varphi$ into Eq. (1.85), we have

$$\oint_S \psi \frac{\partial \varphi}{\partial n} da = \int_V [\psi \nabla^2 \varphi + \nabla \varphi \cdot \nabla \psi] dV. \tag{1.86}$$

Eq. (1.86) is known as Green's theorem in the first form.

Now let us interchange φ and ψ, then Eq. (1.86) becomes

$$\oint_S \varphi \frac{\partial \psi}{\partial n} da = \int_V [\varphi \nabla^2 \psi + \nabla \varphi \cdot \nabla \psi] dV.$$

Subtracting this from Eq. (1.85):

$$\oint_S \left(\psi \frac{\partial \varphi}{\partial n} - \varphi \frac{\partial \psi}{\partial n} \right) da = \int_V (\psi \nabla^2 \varphi - \varphi \nabla^2 \psi) dV. \tag{1.87}$$

This important result is known as the second form of Green's theorem, and has many applications.

Green's theorem in the plane

Consider the two-dimensional vector field $\mathbf{A} = M(x_1, x_2)\hat{e}_1 + N(x_1, x_2)\hat{e}_2$. From Stokes' theorem

$$\oint_\Gamma \mathbf{A} \cdot d\mathbf{r} = \int_S \nabla \times \mathbf{A} \cdot d\mathbf{a} = \int_S \left(\frac{\partial N}{\partial x_1} - \frac{\partial M}{\partial x_2} \right) dx_1 dx_2, \tag{1.88}$$

which is often called Green's theorem in the plane.

Since $\oint_\Gamma \mathbf{A} \cdot d\mathbf{r} = \oint_\Gamma (M dx_1 + N dx_2)$, Green's theorem in the plane can be written as

$$\oint_\Gamma M dx_1 + N dx_2 = \int_S \left(\frac{\partial N}{\partial x_1} - \frac{\partial M}{\partial x_2} \right) dx_1 dx_2. \tag{1.88a}$$

As an illustrative example, let us apply Green's theorem in the plane to show that the area bounded by a simple closed curve Γ is given by

$$\frac{1}{2} \oint_\Gamma x_1 dx_2 - x_2 dx_1.$$

Into Green's theorem in the plane, let us put $M = -x_2, N = x_1$, giving

$$\oint_\Gamma x_1 dx_2 - x_2 dx_1 = \int_S \left(\frac{\partial}{\partial x_1} x_1 - \frac{\partial}{\partial x_2} (-x_2) \right) dx_1 dx_2 = 2 \int_S dx_1 dx_2 = 2A,$$

where A is the required area. Thus $A = \frac{1}{2} \oint_\Gamma x_1 dx_2 - x_2 dx_1$.

Helmholtz's theorem

The divergence and curl of a vector field play very important roles in physics. We learned in previous sections that a divergence-free field is solenoidal and a curl-free field is irrotational. We may classify vector fields in accordance with their being solenoidal and/or irrotational. A vector field $\mathbf{V}$ is:

(1) Solenoidal and irrotational if $\nabla \cdot \mathbf{V} = 0$ and $\nabla \times \mathbf{V} = 0$. A static electric field in a charge-free region is a good example.
(2) Solenoidal if $\nabla \cdot \mathbf{V} = 0$ but $\nabla \times \mathbf{V} \neq 0$. A steady magnetic field in a current-carrying conductor meets these conditions.
(3) Irrotational if $\nabla \times \mathbf{V} = 0$ but $\nabla \cdot \mathbf{V} = 0$. A static electric field in a charged region is an irrotational field.

The most general vector field, such as an electric field in a charged medium with a time-varying magnetic field, is neither solenoidal nor irrotational, but can be considered as the sum of a solenoidal field and an irrotational field. This is made clear by Helmholtz's theorem, which can be stated as (C. W. Wong: *Introduction to Mathematical Physics*, Oxford University Press, Oxford 1991; p. 53):

> A vector field is uniquely determined by its divergence and curl in a region of space, and its normal component over the boundary of the region. In particular, if both divergence and curl are specified everywhere and if they both disappear at infinity sufficiently rapidly, then the vector field can be written as a unique sum of an irrotational part and a solenoidal part.

In other words, we may write

$$\mathbf{V}(\mathbf{r}) = -\nabla\phi(\mathbf{r}) + \nabla \times \mathbf{A}(\mathbf{r}), \tag{1.89}$$

where $-\nabla\phi$ is the irrotational part and $\nabla \times \mathbf{A}$ is the solenoidal part, and $\phi(\mathbf{r})$ and $\mathbf{A}(\mathbf{r})$ are called the scalar and the vector potential, respectively, of $\mathbf{V}(\mathbf{r})$. If both $\mathbf{A}$ and ϕ can be determined, the theorem is verified. How, then, can we determine $\mathbf{A}$ and ϕ? If the vector field $\mathbf{V}(\mathbf{r})$ is such that

$$\nabla \cdot \mathbf{V}(\mathbf{r}) = \rho, \quad \text{and} \quad \nabla \times \mathbf{V}(\mathbf{r}) = \mathbf{v},$$

then we have

$$\nabla \cdot \mathbf{V}(\mathbf{r}) = \rho = -\nabla \cdot (\nabla\phi) + \nabla \cdot (\nabla \times \mathbf{A})$$

or

$$\nabla^2 \phi = -\rho,$$

which is known as Poisson's equation. Next, we have

$$\nabla \times \mathbf{V}(\mathbf{r}) = \mathbf{v} = \nabla \times [-\nabla\phi + \nabla \times \mathbf{A}(\mathbf{r})]$$

or

$$\nabla^2 \mathbf{A} = \mathbf{v};$$

or in component, we have

$$\nabla^2 A_i = v_i, i = 1, 2, 3$$

where these are also Poisson's equations. Thus, both $\mathbf{A}$ and ϕ can be determined by solving Poisson's equations.

Some useful integral relations

These relations are closely related to the general integral theorems that we have proved in preceding sections.

(1) The line integral along a curve C between two points a and b is given by

$$\int_a^b (\nabla\phi) \cdot d\mathbf{l} = \phi(b) - \phi(a). \tag{1.90}$$

Proof:

$$\int_a^b (\nabla \phi) \cdot d\mathbf{l} = \int_a^b \left(\frac{\partial \phi}{\partial x} \hat{i} + \frac{\partial \phi}{\partial y} \hat{j} + \frac{\partial \phi}{\partial z} \hat{k} \right) \cdot (dx\hat{i} + dy\hat{j} + dz\hat{k})$$

$$= \int_a^b \left(\frac{\partial \phi}{\partial x} dx + \frac{\partial \phi}{\partial y} dy + \frac{\partial \phi}{\partial z} dz \right)$$

$$= \int_a^b \left(\frac{\partial \phi}{\partial x} \frac{dx}{dt} + \frac{\partial \phi}{\partial y} \frac{dy}{dt} + \frac{\partial \phi}{\partial z} \frac{dz}{dt} \right) dt$$

$$= \int_a^b \left(\frac{d\phi}{dt} \right) dt = \phi(b) - \phi(a).$$

(2)
$$\oint_S \frac{\partial \varphi}{\partial n} da = \int_V \nabla^2 \varphi dV. \tag{1.91}$$

Proof: Set $\psi = 1$ in Eq. (1.87), then $\partial \psi / \partial n = 0 = \nabla^2 \psi$ and Eq. (1.87) reduces to Eq. (1.91).

(3)
$$\int_V \nabla \varphi dV = \oint_S \varphi \hat{n} da. \tag{1.92}$$

Proof: In Gauss' theorem (1.78), let $\mathbf{A} = \varphi \mathbf{C}$, where $\mathbf{C}$ is constant vector. Then we have

$$\int_V \nabla \cdot (\varphi \mathbf{C}) dV = \int_S \varphi \mathbf{C} \cdot \hat{n} da.$$

Since

$$\nabla \cdot (\varphi \mathbf{C}) = \nabla \varphi \cdot \mathbf{C} = \mathbf{C} \cdot \nabla \varphi \quad \text{and} \quad \varphi \mathbf{C} \cdot \hat{n} = \mathbf{C} \cdot (\varphi \hat{n}),$$

we have

$$\int_V \mathbf{C} \cdot \nabla \varphi dV = \int_S \mathbf{C} \cdot (\varphi \hat{n}) da.$$

Taking $\mathbf{C}$ outside the integrals,

$$\mathbf{C} \cdot \int_V \nabla \varphi dV = \mathbf{C} \cdot \int_S (\varphi \hat{n}) da$$

and since $\mathbf{C}$ is an arbitrary constant vector, we have

$$\int_V \nabla \varphi dV = \oint_S \varphi \hat{n} da.$$

(4)
$$\int_V \nabla \times \mathbf{B} dV = \int_S \hat{n} \times \mathbf{B} da \tag{1.93}$$

Proof: In Gauss' theorem (1.78), let $\mathbf{A} = \mathbf{B} \times \mathbf{C}$ where $\mathbf{C}$ is a constant vector. We then have

$$\int_V \nabla \cdot (\mathbf{B} \times \mathbf{C}) dV = \int_S (\mathbf{B} \times \mathbf{C}) \cdot \hat{n} da.$$

Since $\nabla \cdot (\mathbf{B} \times \mathbf{C}) = \mathbf{C} \cdot (\nabla \times \mathbf{B})$ and $(\mathbf{B} \times \mathbf{C}) \cdot \hat{n} = \mathbf{B} \cdot (\mathbf{C} \times \hat{n}) = (\mathbf{C} \times \hat{n}) \cdot \mathbf{B} = \mathbf{C} \cdot (\hat{n} \times \mathbf{B})$,

$$\int_V \mathbf{C} \cdot (\nabla \times \mathbf{B}) dV = \int_S \mathbf{C} \cdot (\hat{n} \times \mathbf{B}) da.$$

Taking $\mathbf{C}$ outside the integrals

$$\mathbf{C} \cdot \int_V (\nabla \times \mathbf{B}) dV = \mathbf{C} \cdot \int_S (\hat{n} \times \mathbf{B}) da$$

and since $\mathbf{C}$ is an arbitrary constant vector, we have

$$\int_V \nabla \times \mathbf{B} dV = \int_S \hat{n} \times \mathbf{B} da.$$

Tensor analysis

Tensors are a natural generalization of vectors. The beginnings of tensor analysis can be traced back more than a century to Gauss' works on curved surfaces. Today tensor analysis finds applications in theoretical physics (for example, general theory of relativity, mechanics, and electromagnetic theory) and to certain areas of engineering (for example, aerodynamics and fluid mechanics). The general theory of relativity uses tensor calculus of curved space-time, and engineers mainly use tensor calculus of Euclidean space. Only general tensors are considered in this section. The general definition of a tensor is given, followed by a concise discussion of tensor algebra and tensor calculus (covariant differentiation).

Tensors are defined by means of their properties of transformation under coordinate transformation. Let us consider the transformation from one coordinate system $(x^1, x^2, \ldots, x^N)$ to another $(x'^1, x'^2, \ldots, x'^N)$ in an N-dimensional space V_N. Note that in writing x^μ, the index μ is a superscript and should not be mistaken for an exponent. In three-dimensional space we use subscripts. We now use superscripts in order that we may maintain a 'balancing' of the indices in all the general equations. The meaning of 'balancing' will become clear a little later. When we transform the coordinates, their differentials transform according to the relation

$$dx^\mu = \frac{\partial x^\mu}{\partial x'^\nu} dx'^\nu. \tag{1.94}$$

Here we have used Einstein's summation convention: repeated indexes which appear once in the lower and once in the upper position are automatically summed over. Thus,

$$\sum_{\mu=1}^{N} A_{\mu} A^{\mu} = A_{\mu} A^{\mu}.$$

It is important to remember that indexes repeated in the lower part or upper part alone are not summed over. An index which is repeated and over which summation is implied is called a dummy index. Clearly, a dummy index can be replaced by any other index that does not appear in the same term.

Contravariant and covariant vectors

A set of N quantities $A^{\mu}(\mu = 1, 2, \ldots, N)$ which, under a coordinate change, transform like the coordinate differentials, are called the components of a contravariant vector or a contravariant tensor of the first rank or first order:

$$A^{\mu} = \frac{\partial x^{\mu}}{\partial x'^{\nu}} A'^{\nu}. \tag{1.95}$$

This relation can easily be inverted to express A'^{ν} in terms of A^{μ}. We shall leave this as homework for the reader (Problem 1.32).

If N quantities $A^{\mu}(\mu = 1, 2, \ldots, N)$ in a coordinate system $(x^1, x^2, \ldots, x^N)$ are related to N other quantities $A'_{\nu}(\nu = 1, 2, \ldots, N)$ in another coordinate system $(x'^1, x'^2, \ldots, x'^N)$ by the transformation equations

$$A_{\mu} = \frac{\partial x'^{\nu}}{\partial x^{\mu}} A_{\nu} \tag{1.96}$$

they are called components of a covariant vector or covariant tensor of the first rank or first order.

One can show easily that velocity and acceleration are contravariant vectors and that the gradient of a scalar field is a covariant vector (Problem 1.33).

Instead of speaking of a tensor whose components are A^{μ} or A_{μ} we shall simply refer to the tensor A^{μ} or A_{μ}.

Tensors of second rank

From two contravariant vectors A^{μ} and B^{ν} we may form the N^2 quantities $A^{\mu} B^{\nu}$. This is known as the outer product of tensors. These N^2 quantities form the components of a contravariant tensor of the second rank: any aggregate of N^2 quantities $T^{\mu\nu}$ which, under a coordinate change, transform like the product of two contravariant vectors

$$T^{\mu\nu} = \frac{\partial x^\mu}{\partial x'^\alpha}\frac{\partial x^\nu}{\partial x'^\beta} T'_{\alpha\beta}, \tag{1.97}$$

is a contravariant tensor of rank two. We may also form a covariant tensor of rank two from two covariant vectors, which transforms according to the formula

$$T_{\mu\nu} = \frac{\partial x'\alpha}{\partial x^\mu}\frac{\partial x'\beta}{\partial x^\nu} T'_{\alpha\beta}. \tag{1.98}$$

Similarly, we can form a mixed tensor $T^\mu{}_\nu$ of order two that transforms as follows:

$$T^\mu{}_\nu = \frac{\partial x^\mu}{\partial x'\alpha}\frac{\partial x'\beta}{\partial x^\nu} T'^\alpha{}_\beta. \tag{1.99}$$

We may continue this process and multiply more than two vectors together, taking care that their indexes are all different. In this way we can construct tensors of higher rank. The total number of free indexes of a tensor is its rank (or order).

In a Cartesian coordinate system, the distinction between the contravariant and the covariant tensors vanishes. This can be illustrated with the velocity and gradient vectors. Velocity and acceleration are contravariant vectors, they are represented in terms of components in the directions of coordinate increase; the gradient vector is a covariant vector and it is represented in terms of components in the directions orthogonal to the constant coordinate surfaces. In a Cartesian coordinate system, the coordinate direction x^μ coincides with the direction orthogonal to the constant-x^μ surface, hence the distinction between the covariant and the contravariant vectors vanishes. In fact, this is the essential difference between contravariant and covariant tensors: a covariant tensor is represented by components in directions orthogonal to like constant coordinate surface, and a contravariant tensor is represented by components in the directions of coordinate increase.

If two tensors have the same contravariant rank and the same covariant rank, we say that they are of the same type.

Basic operations with tensors

(1) Equality: Two tensors are said to be equal if and only if they have the same covariant rank and the same contravariant rank, and every component of one is equal to the corresponding component of the other:

$$A^{\alpha\beta}{}_\mu = B^{\alpha\beta}{}_\mu.$$

(2) Addition (subtraction): The sum (difference) of two or more tensors of the same type and rank is also a tensor of the same type and rank. Addition of tensors is commutative and associative.

(3) Outer product of tensors: The product of two tensors is a tensor whose rank is the sum of the ranks of the given two tensors. This product involves ordinary multiplication of the components of the tensor and it is called the outer product. For example, $A_\mu{}^{\nu\alpha} B^\beta{}_\lambda = C_{\mu\lambda}{}^{\nu\alpha\beta}$ is the outer product of $A_\mu{}^{\nu\alpha}$ and $B^\beta{}_\lambda$.

(4) Contraction: If a covariant and a contravariant index of a mixed tensor are set equal, a summation over the equal indices is to be taken according to the summation convention. The resulting tensor is a tensor of rank two less than that of the original tensor. This process is called contraction. For example, if we start with a fourth-order tensor $T^\mu{}_{\nu\rho}{}^\delta$, one way of contracting it is to set $\delta = \rho$, which gives the second rank tensor $T^\mu{}_{\nu\rho}{}^\rho$. We could contract it again to get the scalar $T^\mu{}_{\mu\rho}{}^\rho$.

(5) Inner product of tensors: The inner product of two tensors is produced by contracting the outer product of the tensors. For example, given two tensors $A^{\alpha\beta}{}_\delta$ and $B^\mu{}_\nu$, the outer product is $A^{\alpha\beta}{}_\delta B^\mu{}_\nu$. Setting $\delta = \mu$, we obtain the inner product $A^{\alpha\beta}{}_\mu B^\mu{}_\nu$.

(6) Symmetric and antisymmetric tensors: A tensor is called symmetric with respect to two contravariant or two covariant indices if its components remain unchanged upon interchange of the indices:

$$A^{\alpha\beta} = A^{\beta\alpha}, A_{\alpha\beta} = A_{\beta\alpha}.$$

A tensor is called anti-symmetric with respect to two contravariant or two covariant indices if its components change sign upon interchange of the indices:

$$A^{\alpha\beta} = -A^{\beta\alpha}, A_{\alpha\beta} = -A_{\beta\alpha}.$$

Symmetry and anti-symmetry can be defined only for similar indices, not when one index is up and the other is down.

Quotient law

A quantity $Q^{\alpha\cdots}{}_{\mu\cdots}$ with various up and down indexes may or may not be a tensor. We can test whether it is a tensor or not by using the quotient law, which can be stated as follows:

Suppose it is not known whether a quantity X is a tensor or not.
If an inner product of X with an arbitrary tensor is a tensor, then X is also a tensor.

As an example, let $X = P_{\lambda\mu\nu}$, A^λ be an arbitrary contravariant vector, and $A^\lambda P_{\lambda\mu\nu}$ be a tensor, say $Q_{\mu\nu}$: $A^\lambda P_{\lambda\mu\nu} = Q_{\mu\nu}$, then

$$A^\lambda P_{\lambda\mu\nu} = \frac{\partial x'^\alpha}{\partial x^\mu}\frac{\partial x'^\beta}{\partial x^\nu} A'^\gamma P'_{\gamma\alpha\beta}.$$

But

$$A'^\gamma = \frac{\partial x'^\gamma}{\partial x^\lambda} A^\lambda$$

and so

$$A^\lambda P_{\lambda\mu\nu} = \frac{\partial x'^\alpha}{\partial x^\mu}\frac{\partial x'^\beta}{\partial x^\nu}\frac{\partial x'^\gamma}{\partial x^\lambda} A'^\lambda P'_{\gamma\alpha\beta}.$$

This equation must hold for all values of A^λ, hence we have, after canceling the arbitrary A^λ,

$$P_{\lambda\mu\nu} = \frac{\partial x'^\alpha}{\partial x^\mu}\frac{\partial x'^\beta}{\partial x^\nu}\frac{\partial x'^\gamma}{\partial x^\lambda} P'_{\gamma\alpha\beta},$$

which shows that $P_{\lambda\mu\nu}$ is a tensor (contravariant tensor of rank 3).

The line element and metric tensor

So far covariant and contravariant tensors have nothing to do each other except that their product is an invariant:

$$A'_\mu B'^\mu = \frac{\partial x^\alpha}{\partial x'^\mu}\frac{\partial x'^\mu}{\partial x^\beta} A_\alpha A^\beta = \frac{\partial x^\alpha}{\partial x^\beta} A_\alpha A^\beta = \delta^\alpha_{\ \beta} A_\alpha A^\beta = A_\alpha A^\alpha.$$

A space in which covariant and contravariant tensors exist separately is called affine. Physical quantities are independent of the particular choice of the mode of description (that is, independent of the possible choice of contravariance or covariance). Such a space is called a metric space. In a metric space, contravariant and covariant tensors can be converted into each other with the help of the metric tensor $g_{\mu\nu}$. That is, in metric spaces there exists the concept of a tensor that may be described by covariant indices, or by contravariant indices. These two descriptions are now equivalent.

To introduce the metric tensor $g_{\mu\nu}$, let us consider the line element in V_N. In rectangular coordinates the line element (the differential of arc length) ds is given by

$$ds^2 = dx^2 + dy^2 + dz^2 = (dx^1)^2 + (dx^2)^2 + (dx^3)^2;$$

there are no cross terms $dx^i dx^j$. In curvilinear coordinates ds^2 cannot be represented as a sum of squares of the coordinate differentials. As an example, in spherical coordinates we have

$$ds^2 = dr^2 + r^2 d\theta^2 + r^2 \sin^2\theta d\phi^2$$

which can be in a quadratic form, with $x^1 = r, x^2 = \theta, x^3 = \phi$.

A generalization to V_N is immediate. We define the line element ds in V_N to be given by the following quadratic form, called the metric form, or metric

$$ds^2 = \sum_{\mu=1}^{3} \sum_{\nu=1}^{3} g_{\mu\nu} dx^\mu dx^\nu = g_{\mu\nu} dx^\mu dx^\nu. \tag{1.100}$$

For the special cases of rectangular coordinates and spherical coordinates, we have

$$\tilde{g} = (g_{\mu\nu}) = \begin{pmatrix} 1 & 0 & 0 \\ 0 & 1 & 0 \\ 0 & 0 & 1 \end{pmatrix}, \qquad \tilde{g} = (g_{\mu\nu}) = \begin{pmatrix} 1 & 0 & 0 \\ 0 & r^2 & 0 \\ 0 & 0 & r^2 \sin^2 \theta \end{pmatrix}. \tag{1.101}$$

In an N-dimensional orthogonal coordinate system $g_{\mu\nu} = 0$ for $\mu \neq \nu$. And in a Cartesian coordinate system $g_{\mu\mu} = 1$ and $g_{\mu\nu} = 0$ for $\mu \neq \nu$. In the general case of Riemannian space, the $g_{\mu\nu}$ are functions of the coordinates $x^\mu (\mu = 1, 2, \ldots, N)$.

Since the inner product of $g_{\mu\nu}$ and the contravariant tensor $dx^\mu dx^\nu$ is a scalar (ds^2, the square of line element), then according to the quotient law $g_{\mu\nu}$ is a covariant tensor. This can be demonstrated directly:

$$ds^2 = g_{\alpha\beta} dx^\alpha dx^\beta = g'_{\alpha\beta} dx'^\alpha dx'^\beta.$$

Now $dx'^\alpha = (\partial x'^\alpha / \partial x^\mu) dx^\mu$, so that

$$g'_{\alpha\beta} \frac{\partial x'^\alpha}{\partial x^\mu} \frac{\partial x'^\beta}{\partial x^\nu} dx^\mu dx^\nu = g_{\mu\nu} dx^\mu dx^\nu$$

or

$$\left(g'_{\alpha\beta} \frac{\partial x'^\alpha}{\partial x^\mu} \frac{\partial x'^\beta}{\partial x^\nu} - g_{\mu\nu} \right) dx^\mu dx^\nu = 0.$$

The above equation is identically zero for arbitrary dx^μ, so we have

$$g_{\mu\nu} = \frac{\partial x'^\alpha}{\partial x^\mu} \frac{\partial x'^\beta}{\partial x^\nu} g'_{\alpha\beta}, \tag{1.102}$$

which shows that $g_{\mu\nu}$ is a covariant tensor of rank two. It is called the metric tensor or the fundamental tensor.

Now contravariant and covariant tensors can be converted into each other with the help of the metric tensor. For example, we can get the covariant vector (tensor of rank one) A_μ from the contravariant vector A^ν:

$$A_\mu = g_{\mu\nu} A^\nu. \tag{1.103}$$

Since we expect that the determinant of $g_{\mu\nu}$ does not vanish, the above equations can be solved for A^ν in terms of the A_μ. Let the result be

$$A^\nu = g^{\nu\mu} A_\mu. \tag{1.104}$$

By combining Eqs. (1.103) and (1.104) we get

$$A_\mu = g_{\mu\nu}g^{\nu\alpha}A_\alpha.$$

Since the equation must hold for any arbitrary A_μ, we have

$$g_{\mu\nu}g^{\nu\alpha} = \delta_\mu{}^\alpha, \tag{1.105}$$

where $\delta_\mu{}^\alpha$ is Kronecker's delta symbol. Thus, $g^{\mu\nu}$ is the inverse of $g_{\mu\nu}$ and vice versa; $g^{\mu\nu}$ is often called the conjugate or reciprocal tensor of $g_{\mu\nu}$. But remember that $g^{\mu\nu}$ and $g_{\mu\nu}$ are the contravariant and covariant components of the same tensor, that is the metric tensor. Notice that the matrix $(g^{\mu\nu})$ is just the inverse of the matrix $(g_{\mu\nu})$.

We can use $g_{\mu\nu}$ to lower any upper index occurring in a tensor, and use $g^{\mu\nu}$ to raise any lower index. It is necessary to remember the position from which the index was lowered or raised, because when we bring the index back to its original site, we do not want to interchange the order of indexes, in general $T^{\mu\nu} \neq T^{\nu\mu}$. Thus, for example

$$A^p{}_q = g^{rp}A_{rq},\ A^{pq} = g^{rp}g^{sq}A_{rs},\ A^p{}_{rs} = g_{rq}A^{pq}{}_s.$$

Associated tensors

All tensors obtained from a given tensor by forming an inner product with the metric tensor are called associated tensors of the given tensor. For example, A^α and A_α are associated tensors:

$$A_\alpha = g_{\alpha\beta}A^\beta, \quad A^\alpha = g^{\alpha\beta}A_\beta.$$

Geodesics in a Riemannian space

In a Euclidean space, the shortest path between two points is a straight line joining the two points. In a Riemannian space, the shortest path between two points, called the geodesic, may be a curved path. To find the geodesic, let us consider a space curve in a Riemannian space given by $x^\mu = f^\mu(t)$ and compute the distance between two points of the curve, which is given by the formula

$$s = \int_P^Q \sqrt{g_{\lambda\mu}dx^\lambda dx^\mu} = \int_{t_1}^{t_2} \sqrt{g_{\lambda\mu}d\dot{x}^\lambda d\dot{x}^\mu}dt, \tag{1.106}$$

where $d\dot{x}^\lambda = dx^\lambda/dt$, and t (a parameter) varies from point to point of the geodesic curve described by the relations which we are seeking. A geodesic joining

53

two points P and Q has a stationary value compared with any other neighboring path that connects P and Q. Thus, to find the geodesic we extremalize (1.106), and this leads to the differential equation of the geodesic (Problem 1.37)

$$\frac{d}{dt}\left(\frac{\partial F}{\partial \dot{x}}\right) - \frac{\partial F}{\partial x} = 0, \tag{1.107}$$

where $F = \sqrt{g_{\alpha\beta}\dot{x}^\alpha \dot{x}^\beta}$, and $\dot{x} = dx/dt$. Now

$$\frac{\partial F}{\partial x^\gamma} = \frac{1}{2}\left(g_{\alpha\beta}\dot{x}^\alpha \dot{x}^\beta\right)^{-1/2}\frac{\partial g_{\alpha\beta}}{\partial x^\gamma}\dot{x}^\alpha \dot{x}^\beta, \quad \frac{\partial F}{\partial \dot{x}^\gamma} = \frac{1}{2}\left(g_{\alpha\beta}\dot{x}^\alpha \dot{x}^\beta\right)^{-1/2}2g_{\alpha\gamma}\dot{x}^\alpha$$

and

$$ds/dt = \sqrt{g_{\alpha\beta}\dot{x}^\alpha \dot{x}^\beta}.$$

Substituting these into (1.107) we obtain

$$\frac{d}{dt}\left(g_{\alpha\gamma}\dot{x}^\alpha \dot{s}^{-1}\right) - \frac{1}{2}\frac{\partial g_{\alpha\beta}}{\partial x^\gamma}\dot{x}^\alpha \dot{x}^\beta \dot{s}^{-1} = 0, \quad \dot{s} = \frac{ds}{dt}$$

or

$$g_{\alpha\gamma}\ddot{x}^\alpha + \frac{\partial g_{\alpha\gamma}}{\partial x^\beta}\dot{x}^\alpha \dot{x}^\beta - \frac{1}{2}\frac{\partial g_{\alpha\beta}}{\partial x^\gamma}\dot{x}^\alpha \dot{x}^\beta = g_{\alpha\gamma}\dot{x}^\alpha \ddot{s}\dot{s}^{-1}.$$

We can simplify this equation by writing

$$\frac{\partial g_{\alpha\gamma}}{\partial x^\beta}\dot{x}^\alpha \dot{x}^\beta = \frac{1}{2}\left(\frac{\partial g_{\alpha\gamma}}{\partial x^\beta} + \frac{\partial g_{\beta\gamma}}{\partial x^\alpha}\right)\dot{x}^\alpha \dot{x}^\beta,$$

then we have

$$g_{\alpha\gamma}\ddot{x}^\alpha + [\alpha\beta, \gamma]\dot{x}^\alpha \dot{x}^\beta = g_{\alpha\gamma}\dot{x}^\alpha \ddot{s}\dot{s}^{-1}.$$

We can further simplify this equation by taking arc length as the parameter t, then $\dot{s} = 1, \ddot{s} = 0$ and we have

$$g_{\alpha\gamma}\frac{d^2 x^\alpha}{ds^2} + [\alpha\beta, \gamma]\frac{dx^\alpha}{ds}\frac{dx^\beta}{ds} = 0. \tag{1.108}$$

where the functions

$$[\alpha\beta, \gamma] = \Gamma_{\alpha\beta,\gamma} = \frac{1}{2}\left(\frac{\partial g_{\alpha\gamma}}{\partial x^\beta} + \frac{\partial g_{\beta\gamma}}{\partial x^\alpha} - \frac{\partial g_{\alpha\beta}}{\partial x^\gamma}\right) \tag{1.109}$$

are called the Christoffel symbols of the first kind.

Multiplying (1.108) by $g^{\rho\gamma}$, we obtain

$$\frac{d^2 x^\rho}{ds^2} + \left\{\begin{matrix}\rho \\ \alpha\beta\end{matrix}\right\}\frac{dx^\alpha}{ds}\frac{dx^\beta}{ds} = 0, \tag{1.110}$$

where the functions

$$\left\{ {\rho \atop \alpha\beta} \right\} = \Gamma^{\rho}{}_{\alpha\beta} = g^{\rho\gamma}[\alpha\beta, \gamma] \qquad (1.111)$$

are the Christoffel symbol of the second kind.

Eq. (1.110) is, of course, a set of N coupled differential equations; they are the equations of the geodesic. In Euclidean spaces, geodesics are straight lines. In a Euclidean space, $g_{\alpha\beta}$ are independent of the coordinates x^{μ}, so that the Christoffel symbols identically vanish, and Eq. (1.110) reduces to

$$\frac{d^2 x^{\rho}}{ds^2} = 0$$

with the solution

$$x^{\rho} = a_{\rho}s + b_{\rho},$$

where a_{ρ} and b_{ρ} are constants independent of s. This solution is clearly a straight line.

The Christoffel symbols are not tensors. Using the defining Eqs. (1.109) and the transformation of the metric tensor, we can find the transformation laws of the Christoffel symbol. We now give the result, without the mathematical details:

$$\bar{\Gamma}_{\mu\nu,\lambda} = \Gamma_{\alpha\beta,\gamma} \frac{\partial x^{\alpha}}{\partial \bar{x}^{\mu}} \frac{\partial x^{\beta}}{\partial \bar{x}^{\nu}} \frac{\partial x^{\gamma}}{\partial \bar{x}^{\lambda}} + g_{\alpha\beta} \frac{\partial x^{\alpha}}{\partial x^{\lambda}} \frac{\partial^2 x^{\beta}}{\partial \bar{x}^{\mu} \partial \bar{x}^{\nu}}. \qquad (1.112)$$

The Christoffel symbols are not tensors because of the presence of the second term on the right hand side.

Covariant differentiation

We have seen that a covariant vector is transformed according to the formula

$$\bar{A}_{\mu} = \frac{\partial x^{\nu}}{\partial \bar{x}^{\mu}} A_{\nu},$$

where the coefficients are functions of the coordinates, and so vectors at different points transform differently. Because of this fact, dA_{μ} is not a vector, since it is the difference of vectors located at two (infinitesimally separated) points. We can verify this directly:

$$\frac{\partial \bar{A}_{\mu}}{\partial \bar{x}^{\gamma}} = \frac{\partial A_{\nu}}{\partial x^{\beta}} \frac{\partial x^{\nu}}{\partial \bar{x}^{\mu}} \frac{\partial x^{\beta}}{\partial \bar{x}^{\gamma}} + A_{\nu} \frac{\partial^2 x^{\nu}}{\partial \bar{x}^{\mu} \partial \bar{x}^{\gamma}}, \qquad (1.113)$$

which shows that $\partial A/\partial x^\beta$ are not the components of a tensor because of the second term on the right hand side. The same also applies to the differential of a contravariant vector. But we can construct a tensor by the following device.

From Eq. (1.111) we have

$$\bar{\Gamma}^\alpha_{\mu_\gamma} = \Gamma^\rho_{\sigma\tau} \frac{\partial x^\sigma}{\partial \bar{x}^\mu} \frac{\partial x^\tau}{\partial \bar{x}^\gamma} \frac{\partial \bar{x}^\alpha}{\partial x^\rho} + \frac{\partial^2 x^\sigma}{\partial \bar{x}^\mu \partial \bar{x}^\gamma} \frac{\partial \bar{x}^\alpha}{\partial x^\sigma}. \tag{1.114}$$

Multiplying (1.114) by $\bar{A}_\alpha$ and subtracting from (1.113), we obtain

$$\frac{\partial \bar{A}_\mu}{\partial \bar{x}^\gamma} - \bar{A}_\alpha \bar{\Gamma}^\alpha_{\mu\gamma} = \left(\frac{\partial A_\alpha}{\partial x^\beta} - A_\rho \Gamma^\rho_{\alpha\beta} \right) \frac{\partial x^\alpha}{\partial \bar{x}^\mu} \frac{\partial x^\beta}{\partial \bar{x}^\gamma}. \tag{1.115}$$

If we define

$$A_{\alpha;\beta} = \frac{\partial A_\alpha}{\partial x^\beta} - A_\rho \Gamma^\rho_{\alpha\beta}, \tag{1.116}$$

then (1.115) can be rewritten as

$$\bar{A}_{\mu;\gamma} = A_{\alpha;\beta} \frac{\partial x^\alpha}{\partial \bar{x}^\mu} \frac{\partial x^\beta}{\partial \bar{x}^\gamma},$$

which shows that $A_{\alpha;\beta}$ is a covariant tensor of rank 2. This tensor is called the covariant derivative of A_α with respect to x^β. The semicolon denotes covariant differentiation. In a Cartesian coordinate system, the Christoffel symbols vanish, and so covariant differentiation reduces to ordinary differentiation.

The contravariant derivative is found by raising the index which denotes differentiation:

$$A^{\mu;\sigma} = g^{\sigma\alpha} A^\mu_{;\alpha}. \tag{1.117}$$

We can similarly determine the covariant derivative of a tensor of arbitrary rank. In doing so we find the following simple rule helps greatly:

> *To obtain the covariant derivative of the tensor $T^{\cdots}_{\cdots}$ with respect to x^μ, we add to the ordinary derivative $\partial T^{\cdots}_{\cdots}/\partial x^\mu$ for each covariant index $\nu(T^{\cdots}_{\cdots\nu\cdot})$ a term $-\Gamma^\alpha_{\mu\nu}T^{\cdots}_{\cdots\alpha\cdot}$, and for each contravariant index $\nu(T^{\cdots\nu\cdots}_{\cdots})$ a term $+\Gamma^\alpha_{\nu\mu}T^{\cdots\alpha\cdots}_{\cdots}$.*

Thus,

$$T_{\mu\nu;\alpha} = \frac{\partial T_{\mu\nu}}{\partial x^\alpha} - \Gamma^\beta_{\mu\alpha} T_{\beta\nu} - \Gamma^\beta_{\nu\alpha} T_{\mu\beta},$$

$$T^\mu_{\nu;\alpha} = \frac{\partial T^\mu_{\nu}}{\partial x^\alpha} - \Gamma^\beta_{\nu\alpha} T^\mu_{\beta} + \Gamma^\mu_{\beta\alpha} T^\beta_{\nu}.$$

The covariant derivatives of both the metric tensor and the Kronnecker delta are identically zero (Problem 1.38).

Problems

1.1. Given the vector $\mathbf{A} = (2, 2, -1)$ and $\mathbf{B} = (6, -3, 2)$, determine:
(a) $6\mathbf{A} - 3\mathbf{B}$, (b) $A^2 + B^2$, (c) $\mathbf{A} \cdot \mathbf{B}$, (d) the angle between $\mathbf{A}$ and $\mathbf{B}$, (e) the direction cosines of $\mathbf{A}$, (f) the component of $\mathbf{B}$ in the direction of $\mathbf{A}$.

1.2. Find a unit vector perpendicular to the plane of $\mathbf{A} = (2, -6, -3)$ and $\mathbf{B} = (4, 3, -1)$.

1.3. Prove that:
(a) the median to the base of an isosceles triangle is perpendicular to the base; (b) an angle inscribed in a semicircle is a right angle.

1.4. Given two vectors $\mathbf{A} = (2, 1, -1)$, $\mathbf{B} = (1, -1, 2)$ find: (a) $\mathbf{A} \times \mathbf{B}$, and (b) a unit vector perpendicular to the plane containing vectors $\mathbf{A}$ and $\mathbf{B}$.

1.5. Prove: (a) the law of sines for plane triangles, and (b) Eq. (1.16a).

1.6. Evaluate $(2\hat{e}_1 - 3\hat{e}_2) \cdot [(\hat{e}_1 + \hat{e}_2 - \hat{e}_3) \times (3\hat{e}_1 - \hat{e}_3)]$.

1.7. (a) Prove that a necessary and sufficient condition for the vectors $\mathbf{A}$, $\mathbf{B}$ and $\mathbf{C}$ to be coplanar is that $\mathbf{A} \cdot (\mathbf{B} \times \mathbf{C}) = 0$.
(b) Find an equation for the plane determined by the three points $P_1(2, -1, 1)$, $P_2(3, 2, -1)$ and $P_3(-1, 3, 2)$.

1.8. (a) Find the transformation matrix for a rotation of new coordinate system through an angle ϕ about the $x_3 (= z)$-axis.
(b) Express the vector $\mathbf{A} = 3\hat{e}_1 + 2\hat{e}_2 + \hat{e}_3$ in terms of the triad $\hat{e}_1' \hat{e}_2' \hat{e}_3'$ where the $x_1' x_2'$ axes are rotated $45°$ about the x_3-axis (the x_3- and x_3'-axes coinciding).

1.9. Consider the linear transformation $A_i' = \sum_{j=1}^{3} \hat{e}_i' \cdot \hat{e}_j A_j = \sum_{j=1}^{3} \lambda_{ij} A_j$. Show, using the fact that the magnitude of the vector is the same in both systems, that

$$\sum_{i=1}^{3} \lambda_{ij} \lambda_{ik} = \delta_{jk} \quad (j, k = 1, 2, 3).$$

1.10. A curve C is defined by the parametric equation

$$\mathbf{r}(u) = x_1(u)\hat{e}_1 + x_2(u)\hat{e}_2 + x_3(u)\hat{e}_3,$$

where u is the arc length of C measured from a fixed point on C, and $\mathbf{r}$ is the position vector of any point on C; show that:
(a) $d\mathbf{r}/du$ is a unit vector tangent to C;
(b) the radius of curvature of the curve C is given by

$$\rho = \left[\left(\frac{d^2 x_1}{du^2}\right)^2 + \left(\frac{d^2 x_2}{du^2}\right)^2 + \left(\frac{d^2 x_3}{du^2}\right)^2 \right]^{-1/2}.$$

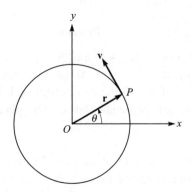

Figure 1.22.   Motion on a circle.

1.11. (a) Show that the acceleration $\mathbf{a}$ of a particle which travels along a space curve with velocity $\mathbf{v}$ is given by

$$\mathbf{a} = \frac{dv}{dt}\hat{T} + \frac{v^2}{\rho}\hat{N},$$

where $\hat{T}$, $\hat{N}$, and ρ are as defined in the text.

(b) Consider a particle P moving on a circular path of radius $\mathbf{r}$ with constant angular speed $\omega = d\theta/dt$ (Fig. 1.22). Show that the acceleration $\mathbf{a}$ of the particle is given by

$$\mathbf{a} = -\omega^2\mathbf{r}.$$

1.12. A particle moves along the curve $x_1 = 2t^2$, $x_2 = t^2 - 4t$, $x_3 = 3t - 5$, where t is the time. Find the components of the particle's velocity and acceleration at time $t = 1$ in the direction $\hat{e}_1 - 3\hat{e}_2 + 2\hat{e}_3$.

1.13. (a) Find a unit vector normal to the surface $x_1^2 + x_2^2 - x_3 = 1$ at the point $P(1,1,1)$.

(b) Find the directional derivative of $\phi = x_1^2 x_2 x_3 + 4x_1 x_3^2$ at $(1, -2, -1)$ in the direction $2\hat{e}_1 - \hat{e}_2 - 2\hat{e}_3$.

1.14. Consider the ellipse given by $r_1 + r_2 = $ const. (Fig. 1.23). Show that r_1 and r_2 make equal angles with the tangent to the ellipse.

1.15. Find the angle between the surfaces $x_1^2 + x_2^2 + x_3^2 = 9$ and $x_3 = x_1^2 + x_2^2 - 3$. at the point $(2, -1, 2)$.

1.16. (a) If f and g are differentiable scalar functions, show that

$$\nabla(fg) = f\nabla g + g\nabla f.$$

(b) Find ∇r if $r = (x_1^2 + x_2^2 + x_3^2)^{1/2}$.

(c) Show that $\nabla r^n = nr^{n-2}\mathbf{r}$.

1.17. Show that:

(a) $\nabla \cdot (\mathbf{r}/r^3) = 0$. Thus the divergence of an inverse-square force is zero.

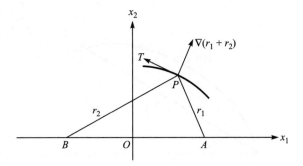

Figure 1.23.

(b) If f is a differentiable function and $\mathbf{A}$ is a differentiable vector function, then

$$\nabla \cdot (f\mathbf{A}) = (\nabla f) \cdot \mathbf{A} + f(\nabla \cdot \mathbf{A}).$$

1.18. (a) What is the divergence of a gradient?

(b) Show that $\nabla^2(1/\mathbf{r}) = 0$.

(c) Show that $\mathbf{r} \cdot (\nabla \cdot \mathbf{r}) \neq (\mathbf{r}\nabla)\mathbf{r}$.

1.19 Given $\nabla \cdot \mathbf{E} = 0, \nabla \cdot \mathbf{H} = 0, \nabla \times \mathbf{E} = -\partial H/\partial t, \nabla \times \mathbf{H} = \partial E/\partial t$, show that $\mathbf{E}$ and $\mathbf{H}$ satisfy the wave equation $\nabla^2 u = \partial^2 u/\partial t^2$.

The given equations are related to the source-free Maxwell's equations of electromagnetic theory, $\mathbf{E}$ and $\mathbf{H}$ are the electric field and magnetic field intensities.

1.20. (a) Find constants a, b, c such that

$$\mathbf{A} = (x_1 + 2x_2 + ax_3)\hat{e}_1 + (bx_1 - 3x_2 - x_3)\hat{e}_2 + (4x_1 + cx_2 + 2x_3)\hat{e}_3$$

is irrotational.

(b) Show that $\mathbf{A}$ can be expressed as the gradient of a scalar function.

1.21. Show that a cylindrical coordinate system is orthogonal.

1.22. Find the volume element dV in: (a) cylindrical and (b) spherical coordinates. Hint: The volume element in orthogonal curvilinear coordinates is

$$dV = h_1 h_2 h_3 du_1 du_2 du_3 = \left|\frac{\partial(x_1, x_2, x_3)}{\partial(u_1, u_2, u_3)}\right| du_1 du_2 du_3.$$

1.23. Evaluate the integral $\int_{(0,1)}^{(1,2)} (x^2 - y)dx + (y^2 + x)dy$ along

(a) a straight line from (0, 1) to (1, 2);

(b) the parabola $x = t, y = t^2 + 1$;

(c) straight lines from (0, 1) to (1, 1) and then from (1, 1) to (1, 2).

1.24. Evaluate the integral $\int_{(0,0)}^{(1,1)} (x^2 + y^2)dx$ along (see Fig. 1.24):

(a) the straight line $y = x$,

(b) the circle arc of radius 1 $(x - 1)^2 + y^2 = 1$.

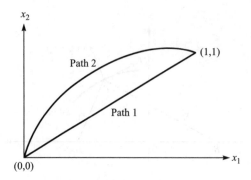

Figure 1.24. Paths for a path integral.

1.25. Evaluate the surface integral $\int_S \mathbf{A} \cdot d\mathbf{a} = \int_S \mathbf{A} \cdot \hat{n} da$, where $\mathbf{A} = x_1 x_2 \hat{e}_1 - x_1^2 \hat{e}_2 + (x_1 + x_2)\hat{e}_3$, S is that portion of the plane $2x_1 + 2x_2 + x_3 = 6$ included in the first octant.

1.26. Verify Gauss' theorem for $\mathbf{A} = (2x_1 - x_3)\hat{e}_1 + x_1^2 x_2 \hat{e}_2 - x_1 x_3^2 \hat{e}_3$ taken over the region bounded by $x_1 = 0, x_1 = 1, x_2 = 0, x_2 = 1, x_3 = 0, x_3 = 1$.

1.28 Show that the electrostatic field intensity $E(\mathbf{r})$ of a point charge Q at the origin has an inverse-square dependence on $\mathbf{r}$.

1.28. Show, by using Stokes' theorem, that the gradient of a scalar field is irrotational:

$$\nabla \times (\nabla \phi(\mathbf{r})) = 0.$$

1.29. Verify Stokes' theorem for $\mathbf{A} = (2x_1 - x_2)\hat{e}_1 - x_2 x_3^2 \hat{e}_2 - x_2^2 x_3 \hat{e}_3$, where S is the upper half surface of the sphere $x_1^2 + x_2^2 + x_3^2 = 1$ and Γ is its boundary (a circle in the $x_1 x_2$ plane of radius 1 with its center at the origin).

1.30. Find the area of the ellipse $x_1 = a\cos\theta, x_2 = b\sin\theta$.

1.31. Show that $\int_S \mathbf{r} \times \hat{n} da = 0$, where S is a closed surface which encloses a volume V.

1.33. Starting with Eq. (1.95), express A'^{ν} in terms of A^{μ}.

1.33. Show that velocity and acceleration are contravariant vectors and that the gradient of a scalar field is a covariant vector.

1.34. The Cartesian components of the acceleration vector are

$$a_x = \frac{d^2 x}{dt^2}, \qquad a_y = \frac{d^2 y}{dt^2}, \qquad a_z = \frac{d^2 z}{dt^2}.$$

Find the component of the acceleration vector in the spherical polar coordinates.

1.35. Show that the property of symmetry (or anti-symmetry) with respect to indexes of a tensor is invariant under coordinate transformation.

1.36. A covariant tensor has components $xy, 2y - z^2, xz$ in rectangular coordinates, find its covariant components in spherical coordinates.

1.37. Prove that a necessary condition that $I = \int_{t_P}^{t_Q} F(t, x, \dot{x})dt$ be an extremum (maximum or minimum) is that

$$\frac{d}{dt}\left(\frac{\partial F}{\partial \dot{x}}\right) - \frac{\partial F}{\partial x} = 0.$$

1.38. Show that the covariant derivatives of: (a) the metric tensor, and (b) the Kronecker delta are identically zero.

2

Ordinary differential equations

Physicists have a variety of reasons for studying differential equations: almost all the elementary and numerous of the advanced parts of theoretical physics are posed mathematically in terms of differential equations. We devote three chapters to differential equations. This chapter will be limited to ordinary differential equations that are reducible to a linear form. Partial differential equations and special functions of mathematical physics will be dealt with in Chapters 10 and 7.

A differential equation is an equation that contains derivatives of an unknown function which expresses the relationship we seek. If there is only one independent variable and, as a consequence, total derivatives like dx/dt, the equation is called an ordinary differential equation (ODE). A partial differential equation (PDE) contains several independent variables and hence partial derivatives.

The *order* of a differential equation is the order of the highest derivative appearing in the equation; its *degree* is the power of the derivative of highest order after the equation has been rationalized, that is, after fractional powers of all derivatives have been removed. Thus the equation

$$\frac{d^2y}{dx^2} + 3\frac{dy}{dx} + 2y = 0$$

is of second order and first degree, and

$$\frac{d^3y}{dx^3} = \sqrt{1 + (dy/dx)^3}$$

is of third order and second degree, since it contains the term $(d^3y/dx^3)^2$ after it is rationalized.

A differential equation is said to be *linear* if each term in it is such that the dependent variable or its derivatives occur only once, and only to the first power. Thus

$$\frac{d^3 y}{dx^3} + y \frac{dy}{dx} = 0$$

is not linear, but

$$x^3 \frac{d^3 y}{dx^3} + e^x \sin x \frac{dy}{dx} + y = \ln x$$

is linear. If in a linear differential equation there are no terms independent of y, the dependent variable, the equation is also said to be *homogeneous*; this would have been true for the last equation above if the 'ln x' term on the right hand side had been replaced by zero.

A very important property of linear homogeneous equations is that, if we know two solutions y_1 and y_2, we can construct others as linear combinations of them. This is known as the principle of superposition and will be proved later when we deal with such equations.

Sometimes differential equations look unfamiliar. A trivial change of variables can reduce a seemingly impossible equation into one whose type is readily recognizable.

Many differential equations are very difficult to solve. There are only a relatively small number of types of differential equation that can be solved in closed form. We start with equations of first order. A first-order differential equation can always be solved, although the solution may not always be expressible in terms of familiar functions. A solution (or integral) of a differential equation is the relation between the variables, not involving differential coefficients, which satisfies the differential equation. The solution of a differential equation of order n in general involves n arbitrary constants.

First-order differential equations

A differential equation of the general form

$$\frac{dy}{dx} = -\frac{f(x,y)}{g(x,y)}, \quad \text{or} \quad g(x,y)dy + f(x,y)dx = 0 \tag{2.1}$$

is clearly a first-order differential equation.

Separable variables

If $f(x,y)$ and $g(x,y)$ are reducible to $P(x)$ and $Q(y)$, respectively, then we have

$$Q(y)dy + P(x)dx = 0. \tag{2.2}$$

Its solution is found at once by integrating.

The reader may notice that dy/dx has been treated as if it were a ratio of dy and dx, that can be manipulated independently. Mathematicians may be unhappy about this treatment. But, if necessary, we can justify it by considering dy and dx to represent small finite changes δy and δx, before we have actually reached the limit where each becomes infinitesimal.

Example 2.1

Consider the differential equation

$$dy/dx = -y^2 e^x.$$

We can rewrite it in the following form $-dy/y^2 = e^x dx$ which can be integrated separately giving the solution

$$1/y = e^x + c,$$

where c is an integration constant.

Sometimes when the variables are not separable a differential equation may be reduced to one in which they are separable by a change of variable. The general form of differential equation amenable to this approach is

$$dy/dx = f(ax + by), \qquad (2.3)$$

where f is an arbitrary function and a and b are constants. If we let $w = ax + by$, then $b\,dy/dx = dw/dx - a$, and the differential equation becomes

$$dw/dx - a = bf(w)$$

from which we obtain

$$\frac{dw}{a + bf(w)} = dx$$

in which the variables are separated.

Example 2.2

Solve the equation

$$dy/dx = 8x + 4y + (2x + y - 1)^2.$$

Solution: Let $w = 2x + y$, then $dy/dx = dw/dx - 2$, and the differential equation becomes

$$dw/dx + 2 = 4w + (w - 1)^2$$

or

$$dw/[4w + (w - 1)^2 - 2] = dx.$$

The variables are separated and the equation can be solved.

A homogeneous differential equation which has the general form

$$dy/dx = f(y/x) \tag{2.4}$$

may also be reduced, by a change of variable, to one with separable variables. This can be illustrated by the following example:

Example 2.3
Solve the equation

$$\frac{dy}{dx} = \frac{y^2 + xy}{x^2}.$$

Solution: The right hand side can be rewritten as $(y/x)^2 + (y/x)$, and hence is a function of the single variable

$$v = y/x.$$

We thus use v both for simplifying the right hand side of our equation, and also for rewriting dy/dx in terms of v and x. Now

$$\frac{dy}{dx} = \frac{d}{dx}(xv) = v + x\frac{dv}{dx}$$

and our equation becomes

$$v + x\frac{dv}{dx} = v^2 + v$$

from which we have

$$\frac{dv}{v^2} = \frac{dx}{x}.$$

Integration gives

$$-\frac{1}{v} = \ln x + c \quad \text{or} \quad x = Ae^{-x/y},$$

where c and $A\ (= e^{-c})$ are constants.

Sometimes a nearly homogeneous differential equation can be reduced to homogeneous form which can then be solved by variable separation. This can be illustrated by the by the following:

Example 2.4
Solve the equation

$$dy/dx = (y + x - 5)/(y - 3x - 1).$$

Solution: Our equation would be homogeneous if it were not for the constants -5 and -1 in the numerator and denominator respectively. But we can eliminate them by a change of variable:

$$x' = x + \alpha, \quad y' = y + \beta,$$

where α and β are constants specially chosen in order to make our equation homogeneous:

$$dy'/dx' = (y' + x')/y' - 3x'.$$

Note that $dy'/dx' = dy/dx$. Trivial algebra yields $\alpha = -1, \beta = -4$. Now let $v = y'/x'$, then

$$\frac{dy'}{dx'} = \frac{d}{dx'}(x'v) = v + x'\frac{dv}{dx'}$$

and our equation becomes

$$v + x'\frac{dv}{dx'} = \frac{v+1}{v-3}, \quad \text{or} \quad \frac{v-3}{-v^2 + 4v + 1}dv = \frac{dx'}{x'}$$

in which the variables are separated and the equation can be solved by integration.

Example 2.5
Fall of a skydiver.

Solution: Assuming the parachute opens at the beginning of the fall, there are two forces acting on the parachute: the downward force of gravity mg, and the upward force of air resistance kv^2. If we choose a coordinate system that has $y = 0$ at the earth's surface and increases upward, then the equation of motion of the falling diver, according to Newton's second law, is

$$mdv/dt = -mg + kv^2,$$

where m is the mass, g the gravitational acceleration, and k a positive constant. In general the air resistance is very complicated, but the power-law approximation is useful in many instances in which the velocity does not vary appreciably. Experiments show that for a subsonic velocity up to 300 m/s, the air resistance is approximately proportional to v^2.

The equation of motion is separable:

$$\frac{mdv}{mg - kv^2} = dt$$

or, to make the integration easier

$$\frac{dv}{v^2 - (mg/k)} = -\frac{k}{m} dt.$$

Now

$$\frac{1}{v^2 - (mg/k)} = \frac{1}{(v + v_t)(v - v_t)} = \frac{1}{2v_t}\left(\frac{1}{v - v_t} - \frac{1}{v + v_t}\right),$$

where $v_t^2 = mg/k$. Thus

$$\frac{1}{2v_t}\left(\frac{dv}{v - v_t} - \frac{dv}{v - v_t}\right) = -\frac{k}{m} dt.$$

Integrating yields

$$\frac{1}{2v_t}\ln\left(\frac{v - v_t}{v + v_t}\right) = -\frac{k}{m}t + c,$$

where c is an integration constant.

Solving for v we finally obtain

$$v(t) = \frac{v_t[1 + B\exp(-2gt/v_t)]}{1 - B\exp(-2gt/v_t)},$$

where $B = \exp(2v_t C)$.

It is easy to see that as $t \to \infty$, $\exp(-2gt/v_t) \to 0$, and so $v \to v_t$; that is, if he falls from a sufficient height, the diver will eventually reach a constant velocity given by v_t, the terminal velocity. To determine the constants of integration, we need to know the value of k, which is about 30 kg/m for the earth's atmosphere and a standard parachute.

Exact equations

We may integrate Eq. (2.1) directly if its left hand side is the differential du of some function $u(x, y)$, in which case the solution is of the form

$$u(x, y) = C \tag{2.5}$$

and Eq. (2.1) is said to be exact. A convenient test to see if Eq. (2.1) is exact is does

$$\frac{\partial g(x, y)}{\partial x} = \frac{\partial f(x, y)}{\partial y}. \tag{2.6}$$

To see this, let us go back to Eq. (2.5) and we have

$$d[u(x, y)] = 0.$$

On performing the differentiation we obtain

$$\frac{\partial u}{\partial x} dx + \frac{\partial u}{\partial y} dy = 0. \tag{2.7}$$

It is a general property of partial derivatives of any well-behaved function that the order of differentiation is immaterial. Thus we have

$$\frac{\partial}{\partial y}\left(\frac{\partial u}{\partial x}\right) = \frac{\partial}{\partial x}\left(\frac{\partial u}{\partial y}\right). \tag{2.8}$$

Now if our differential equation (2.1) is of the form of Eq. (2.7), we must be able to identify

$$f(x, y) = \partial u/\partial x \quad \text{and} \quad g(x, y) = \partial u/\partial y. \tag{2.9}$$

Then it follows from Eq. (2.8) that

$$\frac{\partial g(x, y)}{\partial x} = \frac{\partial f(x, y)}{\partial y},$$

which is Eq. (2.6).

Example 2.6

Show that the equation $x \, dy/dx + (x + y) = 0$ is exact and find its general solution.

Solution: We first write the equation in standard form

$$(x + y)dx + x \, dy = 0.$$

Applying the test of Eq. (2.6) we notice that

$$\frac{\partial f}{\partial y} = \frac{\partial}{\partial y}(x + y) = 1 \quad \text{and} \quad \frac{\partial g}{\partial x} = \frac{\partial x}{\partial x} = 1.$$

Therefore the equation is exact, and the solution is of the form indicated by Eq. (2.7). From Eq. (2.9) we have

$$\partial u/\partial x = x + y, \quad \partial u/\partial y = x,$$

from which it follows that

$$u(x, y) = x^2/2 + xy + h(y), \quad u(x, y) = xy + k(x),$$

where $h(y)$ and $k(x)$ arise from integrating $u(x, y)$ with respect to x and y, respectively. For consistency, we require that

$$h(y) = 0 \quad \text{and} \quad k(x) = x^2/2.$$

Thus the required solution is

$$x^2/2 + xy = c.$$

It is interesting to consider a differential equation of the type

$$g(x, y)\frac{dy}{dx} + f(x, y) = k(x), \tag{2.10}$$

where the left hand side is an exact differential $(d/dx)[u(x,y)]$, and $k(x)$ on the right hand side is a function of x only. Then the solution of the differential equation can be written as

$$u(x,y) = \int k(x)dx. \tag{2.11}$$

Alternatively Eq. (2.10) can be rewritten as

$$g(x,y)\frac{dy}{dx} + [f(x,y) - k(x)] = 0. \tag{2.10a}$$

Since the left hand side of Eq. (2.10) is exact, we have

$$\partial g/\partial x = \partial f/\partial y.$$

Then Eq. (2.10a) is exact as well. To see why, let us apply the test for exactness for Eq. (2.10a) which requires

$$\frac{\partial}{\partial x}[g(x,y)] = \frac{\partial}{\partial y}[f(x,y) - k(x)] = \frac{\partial}{\partial y}[f(x,y)].$$

Thus Eq. (2.10a) satisfies the necessary requirement for being exact. We can thus write its solution as

$$U(x,y) = c,$$

where

$$\frac{\partial U}{\partial y} = g(x,y) \quad \text{and} \quad \frac{\partial U}{\partial x} = f(x,y) - k(x).$$

Of course, the solution $U(x,y) = c$ must agree with Eq. (2.11).

Integrating factors

If a differential equation in the form of Eq. (2.1) is not already exact, it sometimes can be made so by multiplying by a suitable factor, called an integrating factor. Although an integrating factor always exists for each equation in the form of Eq. (2.1), it may be difficult to find it. However, if the equation is linear, that is, if can be written

$$\frac{dy}{dx} + f(x)y = g(x) \tag{2.12}$$

an integrating factor of the form

$$\exp\left(\int f(x)dx\right) \tag{2.13}$$

is always available. It is easy to verify this. Suppose that $R(x)$ is the integrating factor we are looking for. Multiplying Eq. (2.12) by R, we have

$$R\frac{dy}{dx} + Rf(x)y = Rg(x), \quad \text{or} \quad Rdy + Rf(x)ydx = Rg(x)dx.$$

The right hand side is already integrable; the condition that the left hand side of Eq. (2.12) be exact gives

$$\frac{\partial}{\partial y}[Rf(x)y] = \frac{\partial R}{\partial x},$$

which yields

$$dR/dx = Rf(x), \quad \text{or} \quad dR/R = f(x)dx,$$

and integrating gives

$$\ln R = \int f(x)dx$$

from which we obtain the integrating factor R we were looking for

$$R = \exp\left(\int f(x)dx\right).$$

It is now possible to write the general solution of Eq. (2.12). On applying the integrating factor, Eq. (2.12) becomes

$$\frac{d(ye^F)}{dx} = g(x)e^F,$$

where $F(x) = \int f(x)dx$. The solution is clearly given by

$$y = e^{-F}\left[\int e^F g(x)dx + C\right].$$

Example 2.7
Show that the equation $xdy/dx + 2y + x^2 = 0$ is not exact; then find a suitable integrating factor that makes the equation exact. What is the solution of this equation?

Solution: We first write the equation in the standard form

$$(2y + x^2)dx + xdy = 0;$$

then we notice that

$$\frac{\partial}{\partial y}(2y + x^2) = 2 \quad \text{and} \quad \frac{\partial}{\partial x}x = 1,$$

which indicates that our equation is not exact. To find the required integrating factor that makes our equation exact, we rewrite our equation in the form of Eq. (2.12):

$$\frac{dy}{dx} + \frac{2y}{x} = -x$$

from which we find $f(x) = 1/x$, and so the required integrating factor is

$$\exp\left(\int (1/x)dx\right) = \exp(\ln x) = x.$$

Applying this to our equation gives

$$x^2\frac{dy}{dx} + 2xy + x^3 = 0 \quad \text{or} \quad \frac{d}{dx}\left(x^2y + x^4/4\right) = 0$$

which integrates to

$$x^2y + x^4/4 = c,$$

or

$$y = \frac{c - x^4}{4x^2}.$$

Example 2.8
RL circuits: A typical *RL* circuit is shown in Fig. 2.1. Find the current $I(t)$ in the circuit as a function of time t.

Solution: We need first to establish the differential equation for the current flowing in the circuit. The resistance R and the inductance L are both constant. The voltage drop across the resistance is IR, and the voltage drop across the inductance is LdI/dt. Kirchhoff's second law for circuits then gives

$$L\frac{dI(t)}{dt} + RI(t) = E(t),$$

which is in the form of Eq. (2.12), but with t as the independent variable instead of x and I as the dependent variable instead of y. Thus we immediately have the general solution

$$I(t) = \frac{1}{L}e^{-Rt/L}\int e^{Rt/L}E(t)dt + ke^{-Rt/L},$$

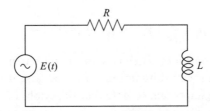

Figure 2.1. *RL* circuit.

where k is a constant of integration (in electric circuits, C is used for capacitance). Given E this equation can be solved for $I(t)$. If the voltage E is constant, we obtain

$$I(t) = \frac{1}{L}e^{-Rt/L}\left(E\frac{L}{R}e^{-Rt/L}\right) + ke^{-Rt/L} = \frac{E}{R} + ke^{-Rt/L}.$$

Regardless of the value of k, we see that

$$I(t) \rightarrow E/R \quad \text{as} \quad t \rightarrow \infty.$$

Setting $t = 0$ in the solution, we find

$$k = I(0) - E/R.$$

Bernoulli's equation

Bernoulli's equation is a non-linear first-order equation that occurs occasionally in physical problems:

$$\frac{dy}{dx} + f(x)y = g(x)y^n, \tag{2.14}$$

where n is not necessarily integer.

This equation can be made linear by the substitution $w = y^a$ with α suitably chosen. We find this can be achieved if $\alpha = 1 - n$:

$$w = y^{1-n} \quad \text{or} \quad y = w^{1/(1-n)}.$$

This converts Bernoulli's equation into

$$\frac{dw}{dx} + (1 - n)f(x)w = (1 - n)g(x),$$

which can be made exact using the integrating factor $\exp(\int(1 - n)f(x)dx)$.

Second-order equations with constant coefficients

The general form of the nth-order linear differential equation with constant coefficients is

$$\frac{d^n y}{dx^n} + p_1\frac{d^{n-1}y}{dx^{n-1}} + \cdots + p_{n-1}\frac{dy}{dx} + p_n y = (D^n + p_1 D^{n-1} + \cdots + p_{n-1}D + p_n)y = f(x),$$

where $p_1, p_2, \ldots$ are constants, $f(x)$ is some function of x, and $D \equiv d/dx$. If $f(x) = 0$, the equation is called homogeneous; otherwise it is called a non-homogeneous equation. It is important to note that the symbol D is meaningless unless applied to a function of x and is therefore not a mathematical quantity in the usual sense. D is an operator.

Many of the differential equations of this type which arise in physical problems are of second order and we shall consider in detail the solution of the equation

$$\frac{d^2y}{dt^2} + a\frac{dy}{dt} + by = (D^2 + aD + b)y = f(t), \tag{2.15}$$

where a and b are constants, and t is the independent variable. As an example, the equation of motion for a mass on a spring is of the form Eq. (2.15), with a representing the friction, c being the constant of proportionality in Hooke's law for the spring, and $f(t)$ some time-dependent external force acting on the mass. Eq. (2.15) can also apply to an electric circuit consisting of an inductor, a resistor, a capacitor and a varying external voltage.

The solution of Eq. (2.15) involves first finding the solution of the equation with $f(t)$ replaced by zero, that is,

$$\frac{d^2y}{dt^2} + a\frac{dy}{dt} + by = (D^2 + aD + b)y = 0; \tag{2.16}$$

this is called the reduced or homogeneous equation corresponding to Eq. (2.15).

Nature of the solution of linear equations

We now establish some results for linear equations in general. For simplicity, we consider the second-order reduced equation (2.16). If y_1 and y_2 are independent solutions of (2.16) and A and B are any constants, then

$$D(Ay_1 + By_2) = ADy_1 + BDy_2, \quad D^2(Ay_1 + By_2) = AD^2y_1 + BD^2y_2$$

and hence

$$(D^2 + aD + b)(Ay_1 + By_2) = A(D^2 + aD + b)y_1 + B(D^2 + aD + b)y_2 = 0.$$

Thus $y = Ay_1 + By_2$ is a solution of Eq. (2.16), and since it contains two arbitrary constants, it is the general solution. A necessary and sufficient condition for two solutions y_1 and y_2 to be linearly independent is that the Wronskian determinant of these functions does not vanish:

$$\begin{vmatrix} y_1 & y_2 \\ \dfrac{dy_1}{dt} & \dfrac{dy_2}{dt} \end{vmatrix} \neq 0.$$

Similarly, if $y_1, y_2, \ldots, y_n$ are n linearly independent solutions of the nth-order linear equations, then the general solution is

$$y = A_1y_1 + A_2y_2 + \cdots + A_ny_n,$$

where $A_1, A_2, \ldots, A_n$ are arbitrary constants. This is known as the superposition principle.

73

General solutions of the second-order equations

Suppose that we can find one solution, $y_p(t)$ say, of Eq. (2.15):

$$(D^2 + aD + b)y_p(t) = f(t). \tag{2.15a}$$

Then on defining

$$y_c(t) = y(t) - y_p(t)$$

we find by subtracting Eq. (2.15a) from Eq. (2.15) that

$$(D^2 + aD + b)y_c(t) = 0.$$

That is, $y_c(t)$ satisfies the corresponding homogeneous equation (2.16), and it is known as the complementary function $y_c(t)$ of non-homogeneous equation (2.15). while the solution $y_p(t)$ is called a particular integral of Eq. (2.15). Thus, the general solution of Eq. (2.15) is given by

$$y(t) = Ay_c(t) + By_p(t). \tag{2.17}$$

Finding the complementary function

Clearly the complementary function is independent of $f(t)$, and hence has nothing to do with the behavior of the system in response to the external applied influence. What it does represent is the free motion of the system. Thus, for example, even without external forces applied, a spring can oscillate, because of any initial displacement and/or velocity. Similarly, had a capacitor already been charged at $t = 0$, the circuit would subsequently display current oscillations even if there is no applied voltage.

   In order to solve Eq. (2.16) for $y_c(t)$, we first consider the linear first-order equation

$$a\frac{dy}{dt} + by = 0.$$

Separating the variables and integrating, we obtain

$$y = Ae^{-bt/a},$$

where A is an arbitrary constant of integration. This solution suggests that Eq. (2.16) might be satisfied by an expression of the type

$$y = e^{pt},$$

where p is a constant. Putting this into Eq. (2.16), we have

$$e^{pt}(p^2 + ap + b) = 0.$$

Therefore $y = e^{pt}$ is a solution of Eq. (2.16) if

$$p^2 + ap + b = 0.$$

This is called the auxiliary (or characteristic) equation of Eq. (2.16). Solving it gives

$$p_1 = \frac{-a + \sqrt{a^2 - 4b}}{2}, \qquad p_2 = \frac{-a - \sqrt{a^2 - 4b}}{2}. \tag{2.18}$$

We now distinguish between the cases in which the roots are real and distinct, complex or coincident.

(*i*) Real and distinct roots $(a^2 - 4b > 0)$
In this case, we have two independent solutions $y_1 = e^{p_1 t}, y_2 = e^{p_2 t}$ and the general solution of Eq. (2.16) is a linear combination of these two:

$$y = Ae^{p_1 t} + Be^{p_2 t}, \tag{2.19}$$

where A and B are constants.

Example 2.9
Solve the equation $(D^2 - 2D - 3)y = 0$, given that $y = 1$ and $y' = dy/dx = 2$ when $t = 0$.

Solution: The auxiliary equation is $p^2 - 2p - 3 = 0$, from which we find $p = -1$ or $p = 3$. Hence the general solution is

$$y = Ae^{-t} + Be^{3t}.$$

The constants A and B can be determined by the boundary conditions at $t = 0$. Since $y = 1$ when $t = 0$, we have

$$1 = A + B.$$

Now

$$y' = -Ae^{-t} + 3Be^{3t}$$

and since $y' = 2$ when $t = 0$, we have $2 = -A + 3B$. Hence

$$A = 1/4, \quad B = 3/4$$

and the solution is

$$4y = e^{-t} + 3e^{3t}.$$

(*ii*) Complex roots $(a^2 - 4b < 0)$
If the roots p_1, p_2 of the auxiliary equation are imaginary, the solution given by Eq. (2.18) is still correct. In order to give the solutions in terms of real quantities, we can use the Euler relations to express the exponentials. If we let $r = -a/2, is = \sqrt{a^2 - 4b}/2$, then

$$e^{p_1 t} = e^{rt}e^{ist} = e^{rt}[\cos st + i \sin st],$$

$$e^{p_2 t} = e^{rt}e^{ist} = e^{rt}[\cos st - i \sin st]$$

and the general solution can be written as

$$y = Ae^{p_1 t} + Be^{p_2 t}$$

$$= e^{rt}[(A + B)\cos st + i(A - B)\sin st]$$

$$= e^{rt}[A_0 \cos st + B_0 \sin st] \qquad (2.20)$$

with $A_0 = A + B, B_0 = i(A - B)$.

The solution (2.20) may be expressed in a slightly different and often more useful form by writing $B_0/A_0 = \tan \delta$. Then

$$y = (A_0^2 + B_0^2)^{1/2}e^{rt}(\cos \delta \cos st + \sin \delta \sin st) = Ce^{rt}\cos(st - \delta), \qquad (2.20a)$$

where C and δ are arbitrary constants.

Example 2.10

Solve the equation $(D^2 + 4D + 13)y = 0$, given that $y = 1$ and $y' = 2$ when $t = 0$.

Solution: The auxiliary equation is $p^2 + 4p + 13 = 0$, and hence $p = -2 \pm 3i$. The general solution is therefore, from Eq. (2.20),

$$y = e^{-2t}(A_0 \cos 3t + B_0 \sin 3t).$$

Since $y = l$ when $t = 0$, we have $A_0 = 1$. Now

$$y' = -2e^{-2t}(A_0 \cos 3t + B_0 \sin 3t) + 3e^{-2t}(-A_0 \sin 3t + B_0 \cos 3t)$$

and since $y' = 2$ when $t = 0$, we have $2 = -2A_0 + 3B_0$. Hence $B_0 = 4/3$, and the solution is

$$3y = e^{-2t}(3\cos 3t + 4\sin 3t).$$

(*iii*) Coincident roots

When $a^2 = 4b$, the auxiliary equation yields only one value for p, namely $p = \alpha = -a/2$, and hence the solution $y = Ae^{\alpha t}$. This is not the general solution as it does not contain the necessary two arbitrary constants. In order to obtain the general solution we proceed as follows. Assume that $y = ve^{\alpha t}$, where v is a function of t to be determined. Then

$$y' = v'e^{\alpha t} + \alpha ve^{\alpha t}, y'' = v''e^{\alpha t} + 2\alpha v'e^{\alpha t} + \alpha^2 ve^{\alpha t}.$$

Substituting for y, y', and y'' in the differential equation we have

$$e^{\alpha t}[v'' + 2\alpha v' + \alpha^2 v + a(v' + \alpha v) + bv] = 0$$

and hence

$$v'' + v'(a + 2\alpha) + v(\alpha^2 + a\alpha + b) = 0.$$

76

Now

$$\alpha^2 + a\alpha + b = 0, \quad \text{and} \quad a + 2\alpha = 0$$

so that

$$v'' = 0.$$

Hence, integrating gives

$$v = At + B,$$

where A and B are arbitrary constants, and the general solution of Eq. (2.16) is

$$y = (At + B)e^{\alpha t} \tag{2.21}$$

Example 2.11
Solve the equation $(D^2 - 4D + 4)y = 0$ given that $y = 1$ and $Dy = 3$ when $t = 0$.

Solution: The auxiliary equation is $p^2 - 4p + 4 = (p - 2)^2 = 0$ which has one root $p = 2$. The general solution is therefore, from Eq. (2.21)

$$y = (At + B)e^{2t}.$$

Since $y = 1$ when $t = 0$, we have $B = 1$. Now

$$y' = 2(At + B)e^{2t} + Ae^{2t}$$

and since $Dy = 3$ when $t = 0$,

$$3 = 2B + A.$$

Hence $A = 1$ and the solution is

$$y = (t + 1)e^{2t}.$$

Finding the particular integral

The particular integral is a solution of Eq. (2.15) that takes the term $f(t)$ on the right hand side into account. The complementary function is transient in nature, so from a physical point of view, the particular integral will usually dominate the response of the system at large times.

The method of determining the particular integral is to guess a suitable functional form containing arbitrary constants, and then to choose the constants to ensure it is indeed the solution. If our guess is incorrect, then no values of these constants will satisfy the differential equation, and so we have to try a different form. Clearly this procedure could take a long time; fortunately, there are some guiding rules on what to try for the common examples of $f(t)$:

(1) $f(t) = $ a polynomial in t.

If $f(t)$ is a polynomial in t with highest power t^n, then the trial particular integral is also a polynomial in t, with terms up to the same power. Note that the trial particular integral is a power series in t, even if $f(t)$ contains only a single terms At^n.

(2) $f(t) = Ae^{kt}$.

The trial particular integral is $y = Be^{kt}$.

(3) $f(t) = A \sin kt$ or $A \cos kt$.

The trial particular integral is $y = A \sin kt + C \cos kt$. That is, even though $f(t)$ contains only a sine or cosine term, we need both sine and cosine terms for the particular integral.

(4) $f(t) = Ae^{\alpha t} \sin \beta t$ or $Ae^{\alpha t} \cos \beta t$.

The trial particular integral is $y = e^{\alpha t}(B \sin \beta t + C \cos \beta t)$.

(5) $f(t)$ is a polynomial of order n in t, multiplied by e^{kt}.

The trial particular integral is a polynomial in t with coefficients to be determined, multiplied by e^{kt}.

(6) $f(t)$ is a polynomial of order n in t, multiplied by $\sin kt$.

The trial particular integral is $y = \Sigma_{j=0}^{n}(B_j \sin kt + C_j \cos kt)t^j$. Can we try $y = (B \sin kt + C \cos kt) \Sigma_{j=0}^{n} D_j t^j$? The answer is no. Do you know why?

If the trial particular integral or part of it is identical to one of the terms of the complementary function, then the trial particular integral must be multiplied by an extra power of t. Therefore, we need to find the complementary function before we try to work out the particular integral. What do we mean by 'identical in form'? It means that the ratio of their t-dependences is a constant. Thus $-2e^{-t}$ and Ae^{-t} are identical in form, but e^{-t} and e^{-2t} are not.

Particular integral and the operator D $(= d/dx)$

We now describe an alternative method that can be used for finding particular integrals. As compared with the method described in previous section, it involves less guesswork as to what the form of the solution is, and the constants multiplying the functional forms of the answer are obtained automatically. It does, however, require a fair amount of practice to ensure that you are familiar with how to use it.

The technique involves using the differential operator $D \equiv d( )/dt$, which is an interesting and simple example of a linear operator without a matrix representation. It is obvious that D obeys the relevant laws of operator algebra: suppose f and g are functions of t, and a is a constant, then

(i) $D(f + g) = Df + Dg$ (distributive);

(ii) $Daf = aDf$ (commutative);

(iii) $D^n D^m f = D^{n+m} f$ (index law).

We can form a polynomial function of D and write

$$F(D) = a_0 D^n + a_1 D^{n-1} + \cdots + a_{n-1}D + a_n$$

so that

$$F(D)f(t) = a_0 D^n f + a_1 D^{n-1}f + \cdots + a_{n-1}Df + a_n f$$

and we can interpret D^{-1} as follows

$$D^{-1}Df(t) = f(t)$$

and

$$\int (Df)dt = f.$$

Hence D^{-1} indicates the operation of integration (the inverse of differentiation). Similarly $D^{-m}f$ means 'integrate $f(t)m$ times'.

These properties of the linear operator D can be used to find the particular integral of Eq. (2.15):

$$\frac{d^2 y}{dt^2} + a\frac{dy}{dt} + by = (D^2 + aD + b)y = f(t)$$

from which we obtain

$$y = \frac{1}{D^2 + aD + b}f(t) = \frac{1}{F(D)}f(t), \tag{2.22}$$

where

$$F(D) = D^2 + aD + b.$$

The trouble with Eq. (2.22) is that it contains an expression involving Ds in the denominator. It requires a fair amount of practice to use Eq. (2.22) to express y in terms of conventional functions. For this, there are several rules to help us.

Rules for D operators

Given a power series of D

$$G(D) = a_0 + a_1 D + \cdots + a_n D^n + \cdots$$

and since $D^n e^{\alpha t} = \alpha^n e^{\alpha t}$, it follows that

$$G(D)e^{\alpha t} = (a_0 + a_1 D + \cdots + a_n D^n + \cdots)e^{\alpha t} = G(\alpha)e^{\alpha t}.$$

Thus we have

Rule (a): $G(D)e^{\alpha t} = G(\alpha)e^{\alpha t}$ provided $G(\alpha)$ is convergent.

When $G(D)$ is the expansion of $1/F(D)$ this rule gives

$$\frac{1}{F(D)}e^{\alpha t} = \frac{1}{F(\alpha)}e^{\alpha t} \quad \text{provided } F(\alpha) \neq 0.$$

Now let us operate $G(D)$ on a product function $e^{\alpha t}V(t)$:

$$G(D)[e^{\alpha t}V(t)] = [G(D)e^{\alpha t}]V(t) + e^{\alpha t}[G(D)V(t)]$$

$$= e^{\alpha t}[G(\alpha) + G(D)]V(t) = e^{\alpha t}G(D+\alpha)[V(t)].$$

That is, we have

Rule (b): $G(D)[e^{\alpha t}V(t)] = e^{\alpha t}G(D+\alpha)[V(t)].$

Thus, for example

$$D^2[e^{\alpha t}t^2] = e^{\alpha t}(D+\alpha)^2[t^2].$$

Rule (c): $G(D^2)\sin kt = G(-k^2)\sin kt.$

Thus, for example

$$\frac{1}{D^2}(\sin 3t) = -\frac{1}{9}\sin 3t.$$

Example 2.12 Damped oscillations (Fig. 2.2)

Suppose we have a spring of natural length L (that is, in its unstretched state). If we hang a ball of mass m from it and leave the system in equilibrium, the spring stretches an amount d, so that the ball is now $L + d$ from the suspension point. We measure the vertical displacement of the ball from this static equilibrium point. Thus, $L + d$ is $y = 0$, and y is chosen to be positive in the downward direction, and negative upward. If we pull down on the ball and then release it, it oscillates up and down about the equilibrium position. To analyze the oscillation of the ball, we need to know the forces acting on it:

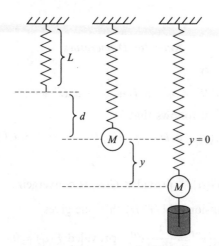

Figure 2.2. Damped spring system.

(1) the downward force of gravity, mg:
(2) the restoring force ky which always opposes the motion (Hooke's law), where k is the spring constant of the spring. If the ball is pulled down a distance y from its static equilibrium position, this force is $-k(d+y)$.

Thus, the total net force acting on the ball is

$$mg - k(d+y) = mg - kd - ky.$$

In static equilibrium, $y = 0$ and all forces balances. Hence

$$kd = mg$$

and the net force acting on the spring is just $-ky$; and the equation of motion of the ball is given by Newton's second law of motion:

$$m\frac{d^2y}{dt^2} = -ky,$$

which describes free oscillation of the ball. If the ball is connected to a dashpot (Fig. 2.2), a damping force will come into play. Experiment shows that the damping force is given by $-bdy/dt$, where the constant b is called the damping constant. The equation of motion of the ball now is

$$m\frac{d^2y}{dt^2} = -ky - b\frac{dy}{dt} \quad \text{or} \quad y'' + \frac{b}{m}y' + \frac{k}{m}y = 0.$$

The auxiliary equation is

$$p^2 + \frac{b}{m}p + \frac{k}{m} = 0$$

with roots

$$p_1 = -\frac{b}{2m} + \frac{1}{2m}\sqrt{b^2 - 4km}, \qquad p_2 = -\frac{b}{2m} - \frac{1}{2m}\sqrt{b^2 - 4km}.$$

We now have three cases, resulting in quite different motions of the oscillator.

Case 1 $b^2 - 4km > 0$ (overdamping)
The solution is of the form

$$y(t) = c_1 e^{p_1 t} + c_2 e^{p_2 t}.$$

Now, both b and k are positive, so

$$\frac{1}{2m}\sqrt{b^2 - 4km} < \frac{b}{2m}$$

and accordingly

$$p_1 = -\frac{b}{2m} + \frac{1}{2m}\sqrt{b^2 - 4km} < 0.$$

Obviously $p_2 < 0$ also. Thus, $y(t) \to 0$ as $t \to \infty$. This means that the oscillation dies out with time and eventually the mass will assume the static equilibrium position.

Case 2 $b^2 - 4km = 0$ (critical damping)
The solution is of the form

$$y(t) = e^{-bt/2m}(c_1 + c_2 t).$$

As both b and m are positive, $y(t) \to 0$ as $t \to \infty$ as in case 1. But c_1 and c_2 play a significant role here. Since $e^{-bt/2m} \neq 0$ for finite t, $y(t)$ can be zero only when $c_1 + c_2 t = 0$, and this happens when

$$t = -c_1/c_2.$$

If the number on the right is positive, the mass passes through the equilibrium position $y = 0$ at that time. If the number on the right is negative, the mass never passes through the equilibrium position.

It is interesting to note that $c_1 = y(0)$, that is, c_1 measures the initial position. Next, we note that

$$y'(0) = c_2 - bc_1/2m, \quad \text{or} \quad c_2 = y'(0) + by(0)/2m.$$

Case 3 $b^2 - 4km < 0$ (underdamping)
The auxiliary equation now has complex roots

$$p_1 = -\frac{b}{2m} + \frac{i}{2m}\sqrt{4km - b^2}, \quad p_2 = -\frac{b}{2m} - \frac{i}{2m}\sqrt{4km - b^2}$$

and the solution is of the form

$$y(t) = e^{-bt/2m}\left[c_1 \cos\left(\sqrt{4km - b^2}\,\frac{t}{2m}\right) + c_2 \sin\left(\sqrt{4km - b^2}\,\frac{t}{2m}\right)\right],$$

which can be rewritten as

$$y(t) = ce^{-bt/2m}\cos(\omega t - \alpha),$$

where

$$c = \sqrt{c_1^2 + c_2^2}, \alpha = \tan^{-1}\left(\frac{c_2}{c_1}\right), \quad \text{and} \quad \omega = \sqrt{4km - b^2}/2m.$$

As in case 2, $e^{-bt/2m} \to 0$ as $t \to \infty$, and the oscillation gradually dies down to zero with increasing time. As the oscillator dies down, it oscillates with a frequency $\omega/2\pi$. But the oscillation is not periodic.

The Euler linear equation

The linear equation with variable coefficients

$$x^n \frac{d^n y}{dx^n} + p_1 x^{n-1} \frac{d^{n-1} y}{dx^{n-1}} + \cdots + p_{n-1} x \frac{dy}{dx} + p_n y = f(x), \qquad (2.23)$$

in which the derivative of the jth order is multiplied by x^j and by a constant, is known as the Euler or Cauchy equation. It can be reduced, by the substitution $x = e^t$, to a linear equation with constant coefficients with t as the independent variable. Now if $x = e^t$, then $dx/dt = x$, and

$$\frac{dy}{dx} = \frac{dy}{dt}\frac{dt}{dx} = \frac{1}{x}\frac{dy}{dt}, \quad \text{or} \quad x\frac{dy}{dx} = \frac{dy}{dt}$$

and

$$\frac{d^2 y}{dx^2} = \frac{d}{dx}\left(\frac{dy}{dx}\right) = \frac{d}{dt}\left(\frac{1}{x}\frac{dy}{dt}\right)\frac{dt}{dx} = \frac{1}{x}\frac{d}{dt}\left(\frac{1}{x}\frac{dy}{dt}\right)$$

or

$$x\frac{d^2 y}{dx^2} = \frac{1}{x}\frac{d^2 y}{dt^2} + \frac{dy}{dt}\frac{d}{dt}\left(\frac{1}{x}\right) = \frac{1}{x}\frac{d^2 y}{dx^2} - \frac{1}{x}\frac{dy}{dt}$$

and hence

$$x^2 \frac{d^2 y}{dx^2} = \frac{d^2 y}{dt^2} - \frac{dy}{dt} = \frac{d}{dt}\left(\frac{dy}{dt} - 1\right)y.$$

Similarly

$$x^3 \frac{d^3 y}{dx^3} = \frac{d}{dt}\left(\frac{d}{dt} - 1\right)\left(\frac{d}{dt} - 2\right)y,$$

and

$$x^n \frac{d^n y}{dx^n} = \frac{d}{dt}\left(\frac{d}{dt} - 1\right)\left(\frac{d}{dt} - 2\right) \cdots \left(\frac{d}{dt} - n + 1\right)y.$$

Substituting for $x^j(d^j y/dx^j)$ in Eq. (2.23) the equation transforms into

$$\frac{d^n y}{dt^n} + q_1 \frac{d^{n-1} y}{dt^{n-1}} + \cdots + q_{n-1}\frac{dy}{dt} + q_n y = f(e^t)$$

in which $q_1, q_2, \ldots, q_n$ are constants.

Example 2.13
Solve the equation

$$x^2 \frac{d^2 y}{dx^2} + 6x\frac{dy}{dx} + 6y = \frac{1}{x^2}.$$

Solution: Put $x = e^t$, then

$$x\frac{dy}{dx} = \frac{dy}{dt}, \quad x^2\frac{d^2y}{dx^2} = \frac{d^2y}{dt^2} - \frac{dy}{dt}.$$

Substituting these in the equation gives

$$\frac{d^2y}{dt^2} + 5\frac{dy}{dt} + 6y = e^t.$$

The auxiliary equation $p^2 + 5p + 6 = (p+2)(p+3) = 0$ has two roots: $p_1 = -2$, $p_2 = 3$. So the complementary function is of the form $y_c = Ae^{-2t} + Be^{-3t}$ and the particular integral is

$$y_p = \frac{1}{(D+2)(D+3)}e^{-2t} = te^{-2t}.$$

The general solution is

$$y = Ae^{-2t} + Be^{-3t} + te^{-2t}.$$

The Euler equation is a special case of the general linear second-order equation

$$D^2y + p(x)Dy + q(x)y = f(x),$$

where $p(x)$, $q(x)$, and $f(x)$ are given functions of x. In general this type of equation can be solved by series approximation methods which will be introduced in next section, but in some instances we may solve it by means of a variable substitution, as shown by the following example:

$$D^2y + (4x - x^{-1})Dy + 4x^2y = 0,$$

where

$$p(x) = (4x - x^{-1}), \quad q(x) = 4x^2, \quad \text{and} \quad f(x) = 0.$$

If we let

$$x = z^{1/2}$$

the above equation is transformed into the following equation with constant coefficients:

$$D^2y + 2Dy + y = 0,$$

which has the solution

$$y = (A + Bz)e^{-z}.$$

Thus the general solution of the original equation is $y = (A + Bx^2)e^{-x^2}$.

Solutions in power series

In many problems in physics and engineering, the differential equations are of such a form that it is not possible to express the solution in terms of elementary functions such as exponential, sine, cosine, etc.; but solutions can be obtained as convergent infinite series. What is the basis of this method? To see it, let us consider the following simple second-order linear differential equation

$$\frac{d^2y}{dx^2} + y = 0.$$

Now assuming the solution is given by $y = a_0 + a_1x + a_2x^2 + \cdots$, we further assume the series is convergent and differentiable term by term for sufficiently small x. Then

$$dy/dx = a_1 + 2a_2x + 3a_3x^2 + \cdots$$

and

$$d^2y/dx^2 = 2a_2 + 2 \cdot 3a_3x + 3 \cdot 4a_4x^2 + \cdots.$$

Substituting the series for y and d^2y/dx^2 in the given differential equation and collecting like powers of x yields the identity

$$(2a_2 + a_0) + (2 \times 3a_3 + a_1)x + (3 \times 4a_4 + a_2)x^2 + \cdots = 0.$$

Since if a power series is identically zero all of its coefficients are zero, equating to zero the term independent of x and coefficients of x, $x^2, \ldots$, gives

$$2a_2 + a_0 = 0, \qquad 4 \times 5a_5 + a_3 = 0,$$

$$2 \times 3a_3 + a_1 = 0, \qquad 5 \times 6a_6 + a_4 = 0,$$

$$3 \times 4a_4 + a_2 = 0, \qquad \cdots$$

and it follows that

$$a_2 = -\frac{a_0}{2}, \quad a_3 = -\frac{a_1}{2 \times 3} = -\frac{a_1}{3!}, \quad a_4 = -\frac{a_2}{3 \times 4} - \frac{a_0}{4!}$$

$$a_5 = -\frac{a_3}{4 \times 5} = \frac{a_1}{5!}, \quad a_6 = -\frac{a_4}{5 \times 6} = -\frac{a_0}{6!}, \ldots.$$

The required solution is

$$y = a_0\left(1 - \frac{x^2}{2!} + \frac{x^4}{4!} - \frac{x^6}{6!} + - \cdots\right) + a_1\left(x - \frac{x^3}{3!} + \frac{x^5}{5!} - + \cdots\right);$$

you should recognize this as equivalent to the usual solution $y = a_0 \cos x + a_1 \sin x$, a_0 and a_1 being arbitrary constants.

Ordinary and singular points of a differential equation

We shall concentrate on the linear second-order differential equation of the form

$$\frac{d^2y}{dx^2} + P(x)\frac{dy}{dx} + Q(x)y = 0 \tag{2.24}$$

which plays a very important part in physical problems, and introduce certain definitions and state (without proofs) some important results applicable to equations of this type. With some small modifications, these are applicable to linear equation of any order. If both the functions P and Q can be expanded in Taylor series in the neighborhood of $x = \alpha$, then Eq. (2.24) is said to possess an ordinary point at $x = \alpha$. But when either of the functions P or Q does not possess a Taylor series in the neighborhood of $x = \alpha$, Eq. (2.24) is said to have a singular point at $x = \alpha$. If

$$P = \lambda(x)/(x - \alpha) \quad \text{and} \quad Q = \mu(x)/(x - \alpha)^2$$

and $\lambda(x)$ and $\mu(x)$ can be expanded in Taylor series near $x = \alpha$. In such cases, $x = \alpha$ is a singular point but the singularity is said to be regular.

Frobenius and Fuchs theorem

Frobenius and Fuchs showed that:

(1) If $P(x)$ and $Q(x)$ are regular at $x = \alpha$, then the differential equation (2.24) possesses two distinct solutions of the form

$$y = \sum_{\lambda=0}^{\infty} a_\lambda(x - \alpha)^\lambda \quad (a_0 \neq 0). \tag{2.25}$$

(2) If $P(x)$ and $Q(x)$ are singular at $x = \alpha$, but $(x - \alpha)P(x)$ and $(x - \alpha)^2 Q(x)$ are regular at $x = \alpha$, then there is at least one solution of the differential equation (2.24) of the form

$$y = \sum_{\lambda=0}^{\infty} a_\lambda(x - \alpha)^{\lambda+\rho} \quad (a_0 \neq 0), \tag{2.26}$$

where ρ is some constant, which is valid for $|x - \alpha| < \beta$ whenever the Taylor series for $\lambda(x)$ and $\mu(x)$ are valid for these values of x.

(3) If $P(x)$ and $Q(x)$ are irregular singular at $x = \alpha$ (that is, $\lambda(x)$ and $\mu(x)$ are singular at $x = \alpha$), then regular solutions of the differential equation (2.24) may not exist.

The proofs of these results are beyond the scope of the book, but they can be found, for example, in E. L. Ince's *Ordinary Differential Equations*, Dover Publications Inc., New York, 1944.

The first step in finding a solution of a second-order differential equation relative to a regular singular point $x = \alpha$ is to determine possible values for the index ρ in the solution (2.26). This is done by substituting series (2.26) and its appropriate differential coefficients into the differential equation and equating to zero the resulting coefficient of the lowest power of $x - \alpha$. This leads to a quadratic equation, called the indicial equation, from which suitable values of ρ can be found. In the simplest case, these values of ρ will give two different series solutions and the general solution of the differential equation is then given by a linear combination of the separate solutions. The complete procedure is shown in Example 2.14 below.

Example 2.14

Find the general solution of the equation

$$4x\frac{d^2y}{dx^2} + 2\frac{dy}{dx} + y = 0.$$

Solution: The origin is a regular singular point and, writing $y = \sum_{\lambda=0}^{\infty} a_\lambda x^{\lambda+\rho} (a_0 \neq 0)$ we have

$$dy/dx = \sum_{\lambda=0}^{\infty} a_\lambda(\lambda + \rho)x^{\lambda+\rho-1}, \quad d^2y/dx^2 = \sum_{\lambda=0}^{\infty} a_\lambda(\lambda + \rho)(\lambda + \rho - 1)x^{\lambda+\rho-2}.$$

Before substituting in the differential equation, it is convenient to rewrite it in the form

$$\left\{4x\frac{d^2y}{dx^2} + 2\frac{dy}{dx}\right\} + \{y\} = 0.$$

When $a_\lambda x^{\lambda+\rho}$ is substituted for y, each term in the first bracket yields a multiple of $x^{\lambda+\rho-1}$, while the second bracket gives a multiple of $x^{\lambda+\rho}$ and, in this form, the differential equation is said to be arranged according to weight, the weights of the bracketed terms differing by unity. When the assumed series and its differential coefficients are substituted in the differential equation, the term containing the lowest power of x is obtained by writing $y = a_0 x^\rho$ in the first bracket. Since the coefficient of the lowest power of x must be zero and, since $a_0 \neq 0$, this gives the indicial equation

$$4\rho(\rho - 1) + 2\rho = 2\rho(2\rho - 1) = 0;$$

its roots are $\rho = 0$, $\rho = 1/2$.

The term in $x^{\lambda+\rho}$ is obtained by writing $y = a_{\lambda+1}x^{\lambda+\rho+1}$ in first bracket and $y = a_{\lambda}x^{\lambda+\rho}$ in the second. Equating to zero the coefficient of the term obtained in this way we have

$$\{4(\lambda+\rho+1)(\lambda+\rho) + 2(\lambda+\rho+1)\}a_{\lambda+1} + a_{\lambda} = 0,$$

giving, with λ replaced by n,

$$a_{n+1} = -\frac{1}{2(\rho+n+1)(2\rho+2n+1)}a_n.$$

This relation is true for $n = 1, 2, 3, \ldots$ and is called the recurrence relation for the coefficients. Using the first root $\rho = 0$ of the indicial equation, the recurrence relation gives

$$a_{n+1} = \frac{1}{2(n+1)(2n+1)}a_n$$

and hence

$$a_1 = -\frac{a_0}{2}, \quad a_2 = -\frac{a_1}{12} = \frac{a_0}{4!}, \quad a_3 = -\frac{a_2}{30} = -\frac{a_0}{6!}, \ldots.$$

Thus one solution of the differential equation is the series

$$a_0\left(1 - \frac{x}{2!} + \frac{x^2}{4!} - \frac{x^3}{6!} + -\cdots\right).$$

With the second root $\rho = 1/2$, the recurrence relation becomes

$$a_{n+1} = -\frac{1}{(2n+3)(2n+2)}a_n.$$

Replacing a_0 (which is arbitrary) by b_0, this gives

$$a_1 = -\frac{b_0}{3\times 2} = -\frac{b_0}{3!}, \quad a_2 = -\frac{a_1}{5\times 4} = \frac{b_0}{5!}, \quad a_3 = -\frac{a_2}{7\times 6} = -\frac{b_0}{7!}, \ldots.$$

and a second solution is

$$b_0 x^{1/2}\left(1 - \frac{x}{3!} + \frac{x^2}{5!} - \frac{x^3}{7!} + -\cdots\right).$$

The general solution of the equation is a linear combination of these two solutions.

Many physical problems require solutions which are valid for large values of the independent variable x. By using the transformation $x = 1/t$, the differential equation can be transformed into a linear equation in the new variable t and the solutions required will be those valid for small t.

In Example 2.14 the indicial equation has two distinct roots. But there are two other possibilities: (a) the indicial equation has a double root; (b) the roots of the indicial equation differ by an integer. We now take a general look at these cases. For this purpose, let us consider the following differential equation which is highly important in mathematical physics:

$$x^2 y'' + x g(x) y' + h(x) y = 0, \qquad (2.27)$$

where the functions $g(x)$ and $h(x)$ are analytic at $x = 0$. Since the coefficients are not analyic at $x = 0$, the solution is of the form

$$y(x) = x^r \sum_{m=0}^{\infty} a_m x^m \quad (a_0 \neq 0). \qquad (2.28)$$

We first expand $g(x)$ and $h(x)$ in power series,

$$g(x) = g_0 + g_1 x + g_2 x^2 + \cdots \quad h(x) = h_0 + h_1 x + h_2 x^2 + \cdots.$$

Then differentiating Eq. (2.28) term by term, we find

$$y'(x) = \sum_{m=0}^{\infty} (m + r) a_m x^{m+r-1}, \quad y''(x) = \sum_{m=0}^{\infty} (m + r)(m + r - 1) a_m x^{m+r-2}.$$

By inserting all these into Eq. (2.27) we obtain

$$x^r [r(r - 1) a_0 + \cdots] + (g_0 + g_1 x + \cdots) x^r (r a_0 + \cdots)$$
$$+ (h_0 + h_1 x + \cdots) x^r (a_0 + a_1 x + \cdots) = 0.$$

Equating the sum of the coefficients of each power of x to zero, as before, yields a system of equations involving the unknown coefficients a_m. The smallest power is x^r, and the corresponding equation is

$$[r(r - 1) + g_0 r + h_0] a_0 = 0.$$

Since by assumption $a_0 \neq 0$, we obtain

$$r(r - 1) + g_0 r + h_0 = 0 \quad \text{or} \quad r^2 + (g_0 - 1) r + h_0 = 0. \qquad (2.29)$$

This is the indicial equation of the differential equation (2.27). We shall see that our series method will yield a fundamental system of solutions; one of the solutions will always be of the form (2.28), but for the form of other solution there will be three different possibilities corresponding to the following cases.

Case 1 The roots of the indicial equation are distinct and do not differ by an integer.

Case 2 The indicial equation has a double root.

Case 3 The roots of the indicial equation differ by an integer.

We now discuss these cases separately.

Case 1 Distinct roots not differing by an integer
This is the simplest case. Let r_1 and r_2 be the roots of the indicial equation (2.29). If we insert $r = r_1$ into the recurrence relation and determine the coefficients $a_1, a_2, \ldots$ successively, as before, then we obtain a solution

$$y_1(x) = x^{r_1}(a_0 + a_1 x + a_2 x^2 + \cdots).$$

Similarly, by inserting the second root $r = r_2$ into the recurrence relation, we will obtain a second solution

$$y_2(x) = x^{r_2}(a_0^* + a_1^* x + a_2^* x^2 + \cdots).$$

Linear independence of y_1 and y_2 follows from the fact that y_1/y_2 is not constant because $r_1 - r_2$ is not an integer.

Case 2 Double roots
The indicial equation (2.29) has a double root r if, and only if, $(g_0 - 1)^2 - 4h_0 = 0$, and then $r = (1 - g_0)/2$. We may determine a first solution

$$y_1(x) = x^r(a_0 + a_1 x + a_2 x^2 + \cdots)\left(r = \frac{1 - g_0}{2}\right) \tag{2.30}$$

as before. To find another solution we may apply the method of variation of parameters, that is, we replace constant c in the solution $cy_1(x)$ by a function $u(x)$ to be determined, such that

$$y_2(x) = u(x)y_1(x) \tag{2.31}$$

is a solution of Eq. (2.27). Inserting y_2 and the derivatives

$$y_2' = u'y_1 + uy_1' \qquad y_2'' = u''y_1 + 2u'y_1' + uy_1''$$

into the differential equation (2.27) we obtain

$$x^2(u''y_1 + 2u'y_1' + uy_1'') + xg(u'y_1 + uy_1') + huy_1 = 0$$

or

$$x^2 y_1 u'' + 2x^2 y_1' u' + xgy_1 u' + (x^2 y_1'' + xgy_1' + hy_1)u = 0.$$

Since y_1 is a solution of Eq. (2.27), the quantity inside the bracket vanishes; and the last equation reduces to

$$x^2 y_1 u'' + 2x^2 y_1' u' + xgy_1 u' = 0.$$

Dividing by $x^2 y_1$ and inserting the power series for g we obtain

$$u'' + \left(2\frac{y_1'}{y_1} + \frac{g_0}{x} + \cdots\right)u' = 0.$$

Here and in the following the dots designate terms which are constants or involve positive powers of x. Now from Eq. (2.30) it follows that

$$\frac{y_1'}{y_1} = \frac{x^{r-1}[ra_0 + (r+1)a_1 x + \cdots]}{x^r[a_0 + a_1 x + \cdots]} = \frac{1}{x}\frac{ra_0 + (r+1)a_1 x + \cdots}{a_0 + a_1 x + \cdots} = \frac{r}{x} + \cdots.$$

Hence the last equation can be written

$$u'' + \left(\frac{2r + g_0}{x} + \cdots\right)u' = 0. \tag{2.32}$$

Since $r = (1 - g_0)/2$ the term $(2r + g_0)/x$ equals $1/x$, and by dividing by u' we thus have

$$\frac{u''}{u'} = -\frac{1}{x} + \cdots.$$

By integration we obtain

$$\ln u' = -\ln x + \cdots \quad \text{or} \quad u' = \frac{1}{x}e^{(\cdots)}.$$

Expanding the exponential function in powers of x and integrating once more, we see that the expression for u will be of the form

$$u = \ln x + k_1 x + k_2 x^2 + \cdots.$$

By inserting this into Eq. (2.31) we find that the second solution is of the form

$$y_2(x) = y_1(x)\ln x + x^r \sum_{m=1}^{\infty} A_m x^m. \tag{2.33}$$

Case 3 Roots differing by an integer

If the roots r_1 and r_2 of the indicial equation (2.29) differ by an integer, say, $r_1 = r$ and $r_2 = r - p$, where p is a positive integer, then we may always determine one solution as before, namely, the solution corresponding to r_1:

$$y_1(x) = x^{r_1}(a_0 + a_1 x + a_2 x^2 + \cdots).$$

To determine a second solution y_2, we may proceed as in Case 2. The first steps are literally the same and yield Eq. (2.32). We determine $2r + g_0$ in Eq. (2.32). Then from the indicial equation (2.29), we find $-(r_1 + r_2) = g_0 - 1$. In our case, $r_1 = r$ and $r_2 = r - p$, therefore, $g_0 - 1 = p - 2r$. Hence in Eq. (2.32) we have $2r + g_0 = p + 1$, and we thus obtain

$$\frac{u''}{u'} = -\left(\frac{p+1}{x} + \cdots\right).$$

Integrating, we find

$$\ln u' = -(p+1)\ln x + \cdots \quad \text{or} \quad u' = x^{-(p+1)}e^{(\cdots)},$$

where the dots stand for some series of positive powers of x. By expanding the exponential function as before we obtain a series of the form

$$u' = \frac{1}{x^{p+1}} + \frac{k_1}{x^p} + \cdots + \frac{k_p}{x} + k_{p+1} + k_{p+2}x + \cdots.$$

Integrating, we have

$$u = -\frac{1}{px^p} - \cdots + k_p \ln x + k_{p+1} x + \cdots.$$ (2.34)

Multiplying this expression by the series

$$y_1(x) = x^{r_1}(a_0 + a_1 x + a_2 x^2 + \cdots)$$

and remembering that $r_1 - p = r_2$ we see that $y_2 = u y_1$ is of the form

$$y_2(x) = k_p y_1(x) \ln x + x^{r_2} \sum_{m=0}^{\infty} a_m x^m.$$ (2.35)

While for a double root of Eq. (2.29) the second solution always contains a logarithmic term, the coefficient k_p may be zero and so the logarithmic term may be missing, as shown by the following example.

Example 2.15
Solve the differential equation

$$x^2 y'' + xy' + (x^2 - \tfrac{1}{4}) y = 0.$$

Solution: Substituting Eq. (2.28) and its derivatives into this equation, we obtain

$$\sum_{m=0}^{\infty} [(m+r)(m+r-1) + (m+r) - \tfrac{1}{4}] a_m x^{m+r} + \sum_{m=0}^{\infty} a_m x^{m+r+2} = 0.$$

By equating the coefficient of x^r to zero we get the indicial equation

$$r(r-1) + r - \tfrac{1}{4} = 0 \quad \text{or} \quad r^2 = \tfrac{1}{4}.$$

The roots $r_1 = \tfrac{1}{2}$ and $r_2 = -\tfrac{1}{2}$ differ by an integer. By equating the sum of the coefficients of x^{s+r} to zero we find

$$[(r+1)r + (r-1) - \tfrac{1}{4}] a_1 = 0 \quad (s = 1).$$ (2.36a)

$$[(s+r)(s+r-1) + s + r - \tfrac{1}{4}] a_s + a_{s-2} = 0 \quad (s = 2, 3, \ldots).$$ (2.36b)

For $r = r_1 = \tfrac{1}{2}$, Eq. (2.36a) yields $a_1 = 0$, and the indicial equation (2.36b) becomes

$$(s+1)s a_s + a_{s-2} = 0.$$

From this and $a_1 = 0$ we obtain $a_3 = 0$, $a_5 = 0$, etc. Solving the indicial equation for a_s and setting $s = 2p$, we get

$$a_{2p} = -\frac{a_{2p-2}}{2p(2p+1)} \quad (p = 1, 2, \ldots).$$

Hence the non-zero coefficients are

$$a_2 = -\frac{a_0}{3!}, \quad a_4 = -\frac{a_2}{4 \times 5} = \frac{a_0}{5!}, \quad a_6 = -\frac{a_0}{7!}, \text{etc.},$$

and the solution y_1 is

$$y_1(x) = a_0\sqrt{x}\sum_{m=0}^{\infty}\frac{(-1)^m x^{2m}}{(2m+1)!} = a_0 x^{-1/2}\sum_{m=0}^{\infty}\frac{(-1)^m x^{2m+1}}{(2m+1)!} = a_0\frac{\sin x}{\sqrt{x}}. \tag{2.37}$$

From Eq. (2.35) we see that a second independent solution is of the form

$$y_2(x) = ky_1(x)\ln x + x^{-1/2}\sum_{m=0}^{\infty}a_m x^m.$$

Substituting this and the derivatives into the differential equation, we see that the three expressions involving $\ln x$ and the expressions ky_1 and $-ky_1$ drop out. Simplifying the remaining equation, we thus obtain

$$2kxy_1' + \sum_{m=0}^{\infty}m(m-1)a_m x^{m-1/2} + \sum_{m=0}^{\infty}a_m x^{m+3/2} = 0.$$

From Eq. (2.37) we find $2kxy' = -ka_0 x^{1/2} + \cdots$. Since there is no further term involving $x^{1/2}$ and $a_0 \neq 0$, we must have $k = 0$. The sum of the coefficients of the power $x^{s-1/2}$ is

$$s(s-1)a_s + a_{s-2} \quad (s = 2, 3, \ldots).$$

Equating this to zero and solving for a_s, we have

$$a_s = -a_{s-2}/[s(s-1)] \quad (s = 2, 3, \ldots),$$

from which we obtain

$$a_2 = -\frac{a_0}{2!}, a_4 = -\frac{a_2}{4 \times 3} = \frac{a_0}{4!}, a_6 = -\frac{a_0}{6!}, \text{etc.},$$

$$a_3 = -\frac{a_1}{3!}, a_5 = -\frac{a_3}{5 \times 4} = \frac{a_1}{5!}, a_7 = -\frac{a_1}{7!}, \text{etc.}$$

We may take $a_1 = 0$, because the odd powers would yield $a_1 y_1/a_0$. Then

$$y_2(x) = a_0 x^{-1/2}\sum_{m=0}^{\infty}\frac{(-1)^m x^{2m}}{(2m)!} = a_0\frac{\cos x}{\sqrt{x}}.$$

Simultaneous equations

In some physics and engineering problems we may face simultaneous differential equations in two or more dependent variables. The general solution

93

of simultaneous equations may be found by solving for each dependent variable separately, as shown by the following example

$$\left.\begin{array}{r} Dx + 2y + 3x = 0 \\ 3x + Dy - 2y = 0 \end{array}\right\} \quad (D = d/dt)$$

which can be rewritten as

$$\left.\begin{array}{r} (D+3)x + 2y = 0, \\ 3x + (D-2)y = 0. \end{array}\right\}$$

We then operate on the first equation with $(D-2)$ and multiply the second by a factor 2:

$$\left.\begin{array}{r} (D-2)(D+3)x + 2(D-2)y = 0, \\ 6x + 2(D-2)y = 0. \end{array}\right\}$$

Subtracting the first from the second leads to

$$(D^2 + D - 6)x - 6x = (D^2 + D - 12)x = 0,$$

which can easily be solved and its solution is of the form

$$x(t) = Ae^{3t} + Be^{-4t}.$$

Now inserting $x(t)$ back into the original equation to find y gives:

$$y(t) = -3Ae^{3t} + \tfrac{1}{2}Be^{-4t}.$$

The gamma and beta functions

The factorial notation $n! = n(n-1)(n-2)\cdots 3 \times 2 \times 1$ has proved useful in writing down the coefficients in some of the series solutions of the differential equations. However, this notation is meaningless when n is not a positive integer. A useful extension is provided by the gamma (or Euler) function, which is defined by the integral

$$\Gamma(\alpha) = \int_0^\infty e^{-x} x^{\alpha-1} dx \quad (\alpha > 0) \tag{2.38}$$

and it follows immediately that

$$\Gamma(1) = \int_0^\infty e^{-x} dx = [-e^{-x}]_0^\infty = 1. \tag{2.39}$$

Integration by parts gives

$$\Gamma(\alpha + 1) = \int_0^\infty e^{-x} x^\alpha dx = [-e^{-x} x^\alpha]_0^\infty + \alpha \int_0^\infty e^{-x} x^{\alpha-1} dx = \alpha\Gamma(\alpha). \tag{2.40}$$

When $\alpha = n$, a positive integer, repeated application of Eq. (2.40) and use of Eq. (2.39) gives

$$\Gamma(n+1) = n\Gamma(n) = n(n-1)\Gamma(n-1) = \ldots = n(n-1)\cdots 3 \times 2 \times \Gamma(1)$$

$$= n(n-1)\cdots 3 \times 2 \times 1 = n!.$$

Thus the gamma function is a generalization of the factorial function. Eq. (2.40) enables the values of the gamma function for any positive value of α to be calculated: thus

$$\Gamma(\tfrac{7}{2}) = (\tfrac{5}{2})\Gamma(\tfrac{5}{2}) = (\tfrac{5}{2})(\tfrac{3}{2})\Gamma(\tfrac{3}{2}) = (\tfrac{5}{2})(\tfrac{3}{2})(\tfrac{1}{2})\Gamma(\tfrac{1}{2}).$$

Write $u = +\sqrt{x}$ in Eq. (2.38) and we then obtain

$$\Gamma(\alpha) = 2\int_0^\infty u^{2\alpha-1}e^{-u^2}\,du,$$

so that

$$\Gamma(\tfrac{1}{2}) = 2\int_0^\infty e^{-u^2}\,du = \sqrt{\pi}.$$

The function $\Gamma(\alpha)$ has been tabulated for values of α between 0 and 1.

When $\alpha < 0$ we can define $\Gamma(\alpha)$ with the help of Eq. (2.40) and write

$$\Gamma(\alpha) = \Gamma(\alpha+1)/\alpha.$$

Thus

$$\Gamma(-\tfrac{3}{2}) = -\tfrac{2}{3}\Gamma(-\tfrac{1}{2}) = -\tfrac{2}{3}(-\tfrac{2}{1})\Gamma(\tfrac{1}{2}) = \tfrac{4}{3}\sqrt{\pi}.$$

When $\alpha \to 0$, $\int_0^\infty e^{-x}x^{\alpha-1}\,dx$ diverges so that $\Gamma(0)$ is not defined.

Another function which will be useful later is the beta function which is defined by

$$B(p,q) = \int_0^1 t^{p-1}(1-t)^{q-1}\,dt \quad (p, q > 0). \tag{2.41}$$

Substituting $t = v/(1+v)$, this can be written in the alternative form

$$B(p,q) = \int_0^\infty v^{p-1}(1+v)^{-p-q}\,dv. \tag{2.42}$$

By writing $t' = 1 - t$ we deduce that $B(p,q) = B(q,p)$.

The beta function can be expressed in terms of gamma functions as follows:

$$B(p,q) = \frac{\Gamma(p)\Gamma(q)}{\Gamma(p+q)}. \tag{2.43}$$

To prove this, write $x = at$ $(a > 0)$ in the integral (2.38) defining $\Gamma(\alpha)$, and it is straightforward to show that

$$\frac{\Gamma(\alpha)}{a^\alpha} = \int_0^\infty e^{-at}t^{\alpha-1}\,dt \tag{2.44}$$

and, with $\alpha = p + q$, $a = 1 + v$, this can be written

$$\Gamma(p+q)(1+v)^{-p-q} = \int_0^\infty e^{-(1+v)t} t^{p+q-1} dt.$$

Multiplying by v^{p-1} and integrating with respect to v between 0 and ∞,

$$\Gamma(p+q) \int_0^\infty v^{p-1}(1+v)^{-p-q} dv = \int_0^\infty v^{p-1} dv \int_0^\infty e^{-(1+v)t} t^{p+q+1} dt.$$

Then interchanging the order of integration in the double integral on the right and using Eq. (2.42),

$$\Gamma(p+q)B(p,q) = \int_0^\infty e^{-t} t^{p+q-1} dt \int_0^\infty e^{-vt} v^{p-1} dv$$

$$= \int_0^\infty e^{-t} t^{p+q-1} \frac{\Gamma(p)}{t^p} dt, \quad \text{using Eq. (2.44)}$$

$$= \Gamma(p) \int_0^\infty e^{-t} t^{q-1} dt = \Gamma(p)\Gamma(q).$$

Example 2.15

Evaluate the integral $\int_0^\infty 3^{-4x^2} dx$.

Solution: We first notice that $3 = e^{\ln 3}$, so we can rewrite the integral as

$$\int_0^\infty 3^{-4x^2} dx = \int_0^\infty (e^{\ln 3})^{(-4x^2)} dx = \int_0^\infty e^{-(4\ln 3)x^2} dx.$$

Now let $(4 \ln 3)x^2 = z$, then the integral becomes

$$\int_0^\infty e^{-z} d\left(\frac{z^{1/2}}{\sqrt{4\ln 3}}\right) = \frac{1}{2\sqrt{4\ln 3}} \int_0^\infty z^{-1/2} e^{-z} dz = \frac{\Gamma(\frac{1}{2})}{2\sqrt{4\ln 3}} = \frac{\sqrt{\pi}}{2\sqrt{4\ln 3}}.$$

Problems

2.1 Solve the following equations:
(a) $xdy/dx + y^2 = 1$;
(b) $dy/dx = (x+y)^2$.

2.2 Melting of a sphere of ice: Assume that a sphere of ice melts at a rate proportional to its surface area. Find an expression for the volume at any time t.

2.3 Show that $(3x^2 + y\cos x)dx + (\sin x - 4y^3)dy = 0$ is an exact differential equation and find its general solution.

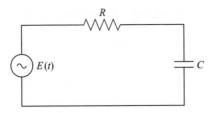

Figure 2.3.   *RC* circuit.

2.4   *RC* circuits: A typical *RC* circuit is shown in Fig. 2.3. Find current flow $I(t)$ in the circuit, assuming $E(t) = E_0$.

Hint: the voltage drop across the capacitor is given Q/C, with $Q(t)$ the charge on the capacitor at time t.

2.5   Find a constant α such that $(x + y)^\alpha$ is an integrating factor of the equation

$$(4x^2 + 2xy + 6y)dx + (2x^2 + 9y + 3x)dy = 0.$$

What is the solution of this equation?

2.6   Solve $dy/dx + y = y^3 x$.

2.7   Solve:

(a) the equation $(D^2 - D - 12)y = 0$ with the boundary conditions $y = 0$, $Dy = 3$ when $t = 0$;

(b) the equation $(D^2 + 2D + 3)y = 0$ with the boundary conditions $y = 2$, $Dy = 0$ when $t = 0$;

(c) the equation $(D^2 - 2D + 1)y = 0$ with the boundary conditions $y = 5$, $Dy = 3$ when $t = 0$.

2.8   Find the particular integral of $(D^2 + 2D - 1)y = 3 + t^3$.

2.9   Find the particular integral of $(2D^2 + 5D + 7) = 3e^{2t}$.

2.10   Find the particular integral of $(3D^2 + D - 5)y = \cos 3t$.

2.11   Simple harmonic motion of a pendulum (Fig. 2.4): Suspend a ball of mass m at the end of a massless rod of length L and set it in motion swinging back and forth in a vertical plane. Show that the equation of motion of the ball is

$$\frac{d^2\theta}{dt^2} + \frac{g}{L}\sin\theta = 0,$$

where g is the local gravitational acceleration. Solve this pendulum equation for small displacements by replacing $\sin\theta$ by θ.

2.12   Forced oscillations with damping: If we allow an external driving force $F(t)$ in addition to damping (Example 2.12), the motion of the oscillator is governed by

$$y'' + \frac{b}{m}y' + \frac{k}{m}y = F(t),$$

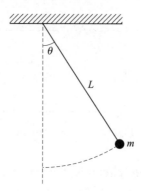

Figure 2.4. Simple pendulum.

a constant coefficient non-homogeneous equation. Solve this equation for
$F(t) = A\cos(\omega t)$.

2.13 Solve the equation

$$r^2 \frac{d^2 R}{dr^2} + 2r\frac{dR}{dr} - n(n+1)R = 0 \quad (n \text{ constant}).$$

2.14 The first-order non-linear equation

$$\frac{dy}{dx} + y^2 + Q(x)y + R(x) = 0$$

is known as Riccati's equation. Show that, by use of a change of dependent
variable

$$y = \frac{1}{z}\frac{dz}{dx};$$

Riccati's equation transforms into a second-order linear differential equation

$$\frac{d^2 z}{dx^2} + Q(x)\frac{dz}{dx} + R(x)z = 0.$$

Sometimes Riccati's equation is written as

$$\frac{dy}{dx} + P(x)y^2 + Q(x)y + R(x) = 0.$$

Then the transformation becomes

$$y = -\frac{1}{P(x)}\frac{dz}{dx}$$

and the second-order equation takes the form

$$\frac{d^2 z}{dx^2} + \left(Q + \frac{1}{P}\frac{dP}{dx}\right)\frac{dz}{dx} + PRz = 0.$$

2.15 Solve the equation $4x^2y'' + 4xy' + (x^2 - 1)y = 0$ by using Frobenius' method, where $y' = dy/dx$, and $y'' = d^2y/dx^2$.

2.16 Find a series solution, valid for large values of x, of the equation

$$(1 - x^2)y'' - 2xy' + 2y = 0.$$

2.17 Show that a series solution of Airy's equation $y'' - xy = 0$ is

$$y = a_0 \left(1 + \frac{x^3}{2 \times 3} + \frac{x^6}{2 \times 3 \times 5 \times 6} + \cdots \right)$$

$$+ b_0 \left(x + \frac{x^4}{3 \times 4} + \frac{x^7}{3 \times 4 \times 6 \times 7} + \cdots \right).$$

2.18 Show that Weber's equation $y'' + (n + \frac{1}{2} - \frac{1}{4}x^2)y = 0$ is reduced by the substitution $y = e^{-x^2/4}v$ to the equation $d^2v/dx^2 - x(dv/dx) + nv = 0$. Show that two solutions of this latter equation are

$$v_1 = 1 - \frac{n}{2!}x^2 + \frac{n(n-2)}{4!}x^4 - \frac{n(n-2)(n-4)}{6!}x^6 + - \cdots,$$

$$v_2 = x - \frac{(n-1)}{3!}x^3 + \frac{(n-1)(n-3)}{5!}x^5 - \frac{(n-1)(n-3)(n-5)}{7!}x^7 + - \cdots.$$

2.19 Solve the following simultaneous equations

$$\left. \begin{array}{l} Dx + y = t^3 \\ Dy - x = t \end{array} \right\} \quad (D = d/dt).$$

2.20 Evaluate the integrals:

(a) $\displaystyle\int_0^\infty x^3 e^{-x} dx.$

(b) $\displaystyle\int_0^\infty x^6 e^{-2x} dx$ (hint: let $y = 2x$).

(c) $\displaystyle\int_0^\infty \sqrt{y} e^{-y^2} dy$ (hint: let $y^2 = x$).

(d) $\displaystyle\int_0^1 \frac{dx}{\sqrt{-\ln x}}$ (hint : let $-\ln x = u$).

2.21 (a) Prove that $\displaystyle B(p, q) = 2 \int_0^{\pi/2} \sin^{2p-1}\theta \cos^{2q-1}\theta \, d\theta.$

(b) Evaluate the integral $\displaystyle\int_0^1 x^4(1 - x)^3 dx.$

2.22 Show that $n! \approx \sqrt{2\pi n} \, n^n e^{-n}$. This is known as Stirling's factorial approximation or asymptotic formula for $n!$.

3

Matrix algebra

As vector methods have become standard tools for physicists, so too matrix methods are becoming very useful tools in sciences and engineering. Matrices occur in physics in at least two ways: in handling the eigenvalue problems in classical and quantum mechanics, and in the solutions of systems of linear equations. In this chapter, we introduce matrices and related concepts, and define some basic matrix algebra. In Chapter 5 we will discuss various operations with matrices in dealing with transformations of vectors in vector spaces and the operation of linear operators on vector spaces.

Definition of a matrix

A matrix consists of a rectangular block or ordered array of numbers that obeys prescribed rules of addition and multiplication. The numbers may be real or complex. The array is usually enclosed within curved brackets. Thus

$$\begin{pmatrix} 1 & 2 & 4 \\ 2 & -1 & 7 \end{pmatrix}$$

is a matrix consisting of 2 rows and 3 columns, and it is called a 2×3 (2 by 3) matrix. An $m \times n$ matrix consists of m rows and n columns, which is usually expressed in a double suffix notation:

$$\tilde{A} = \begin{pmatrix} a_{11} & a_{12} & a_{13} & \cdots & a_{1n} \\ a_{21} & a_{22} & a_{23} & \cdots & a_{2n} \\ \vdots & \vdots & \vdots & & \vdots \\ a_{m1} & a_{m2} & a_{m3} & \cdots & a_{mn} \end{pmatrix}. \tag{3.1}$$

Each number a_{ij} is called an element of the matrix, where the first subscript i denotes the row, while the second subscript j indicates the column. Thus, a_{23}

100

refers to the element in the second row and third column. The element a_{ij} should be distinguished from the element a_{ji}.

It should be pointed out that a matrix has no single numerical value; therefore it must be carefully distinguished from a determinant.

We will denote a matrix by a letter with a tilde over it, such as $\tilde{A}$ in (3.1). Sometimes we write (a_{ij}) or $(a_{ij})_{mn}$, if we wish to express explicitly the particular form of element contained in $\tilde{A}$.

Although we have defined a matrix here with reference to numbers, it is easy to extend the definition to a matrix whose elements are functions $f_i(x)$; for a 2×3 matrix, for example, we have

$$\begin{pmatrix} f_1(x) & f_2(x) & f_3(x) \\ f_4(x) & f_5(x) & f_6(x) \end{pmatrix}.$$

A matrix having only one row is called a row matrix or a row vector, while a matrix having only one column is called a column matrix or a column vector. An ordinary vector $\mathbf{A} = A_1\hat{e}_1 + A_2\hat{e}_2 + A_3\hat{e}_3$ can be represented either by a row matrix or by a column matrix.

If the numbers of rows m and columns n are equal, the matrix is called a square matrix of order n.

In a square matrix of order n, the elements $a_{11}, a_{22}, \ldots, a_{nn}$ form what is called the principal (or leading) diagonal, that is, the diagonal from the top left hand corner to the bottom right hand corner. The diagonal from the top right hand corner to the bottom left hand corner is sometimes termed the trailing diagonal. Only a square matrix possesses a principal diagonal and a trailing diagonal.

The sum of all elements down the principal diagonal is called the trace, or spur, of the matrix. We write

$$\operatorname{Tr} \tilde{A} = \sum_{i=1}^{n} a_{ii}.$$

If all elements of the principal diagonal of a square matrix are unity while all other elements are zero, then it is called a unit matrix (for a reason to be explained later) and is denoted by $\tilde{I}$. Thus the unit matrix of order 3 is

$$\tilde{I} = \begin{pmatrix} 1 & 0 & 0 \\ 0 & 1 & 0 \\ 0 & 0 & 1 \end{pmatrix}.$$

A square matrix in which all elements other than those along the principal diagonal are zero is called a diagonal matrix.

A matrix with all elements zero is known as the null (or zero) matrix and is denoted by the symbol $\tilde{0}$, since it is not an ordinary number, but an array of zeros.

Four basic algebra operations for matrices

Equality of matrices

Two matrices $\tilde{A} = (a_{jk})$ and $\tilde{B} = (b_{jk})$ are equal if and only if $\tilde{A}$ and $\tilde{B}$ have the same order (equal numbers of rows and columns) and corresponding elements are equal, that is

$$a_{jk} = b_{jk} \quad \text{for all } j \text{ and } k.$$

Then we write

$$\tilde{A} = \tilde{B}.$$

Addition of matrices

Addition of matrices is defined only for matrices of the same order. If $\tilde{A} = (a_{jk})$ and $\tilde{B} = (b_{jk})$ have the same order, the sum of $\tilde{A}$ and $\tilde{B}$ is a matrix of the same order

$$\tilde{C} = \tilde{A} + \tilde{B}$$

with elements

$$c_{jk} = a_{jk} + b_{jk}. \tag{3.2}$$

We see that $\tilde{C}$ is obtained by adding corresponding elements of $\tilde{A}$ and $\tilde{B}$.

Example 3.1

If

$$\tilde{A} = \begin{pmatrix} 2 & 1 & 4 \\ 3 & 0 & 2 \end{pmatrix}, \quad \tilde{B} = \begin{pmatrix} 3 & 5 & 1 \\ 2 & 1 & -3 \end{pmatrix}$$

hen

$$\tilde{C} = \tilde{A} + \tilde{B} = \begin{pmatrix} 2 & 1 & 4 \\ 3 & 0 & 2 \end{pmatrix} + \begin{pmatrix} 3 & 5 & 1 \\ 2 & 1 & -3 \end{pmatrix} = \begin{pmatrix} 2+3 & 1+5 & 4+1 \\ 3+2 & 0+1 & 2-3 \end{pmatrix}$$

$$= \begin{pmatrix} 5 & 6 & 5 \\ 5 & 1 & -1 \end{pmatrix}.$$

From the definitions we see that matrix addition obeys the commutative and associative laws, that is, for any matrices $\tilde{A}$, $\tilde{B}$, $\tilde{C}$ of the same order

$$\tilde{A} + \tilde{B} = \tilde{B} + \tilde{A}, \quad \tilde{A} + (\tilde{B} + \tilde{C}) = (\tilde{A} + \tilde{B}) + \tilde{C}. \tag{3.3}$$

Similarly, if $\tilde{A} = (a_{jk})$ and $\tilde{B} = (b_{jk})$ have the same order, we define the difference of $\tilde{A}$ and $\tilde{B}$ as

$$\tilde{D} = \tilde{A} - \tilde{B}$$

with elements

$$d_{jk} = a_{jk} - b_{jk}. \tag{3.4}$$

Multiplication of a matrix by a number

If $\tilde{A} = (a_{jk})$ and c is a number (or scalar), then we define the product of $\tilde{A}$ and c as

$$c\tilde{A} = \tilde{A}c = (ca_{jk}); \tag{3.5}$$

we see that $c\tilde{A}$ is the matrix obtained by multiplying each element of $\tilde{A}$ by c.

We see from the definition that for any matrices and any numbers,

$$c(\tilde{A} + \tilde{B}) = c\tilde{A} + c\tilde{B}, \quad (c + k)\tilde{A} = c\tilde{A} + k\tilde{A}, \quad c(k\tilde{A}) = ck\tilde{A}. \tag{3.6}$$

Example 3.2

$$7\begin{pmatrix} a & b & c \\ d & e & f \end{pmatrix} = \begin{pmatrix} 7a & 7b & 7c \\ 7d & 7e & 7f \end{pmatrix}.$$

Formulas (3.3) and (3.6) express the properties which are characteristic for a vector space. This gives vector spaces of matrices. We will discuss this further in Chapter 5.

Matrix multiplication

The matrix product $\tilde{A}\tilde{B}$ of the matrices $\tilde{A}$ and $\tilde{B}$ is defined *if and only if* the number of columns in $\tilde{A}$ is equal to the number of rows in $\tilde{B}$. Such matrices are sometimes called 'conformable'. If $\tilde{A} = (a_{jk})$ is an $n \times s$ matrix and $\tilde{B} = (b_{jk})$ is an $s \times m$ matrix, then $\tilde{A}$ and $\tilde{B}$ are conformable and their matrix product, written $\tilde{C} = \tilde{A}\tilde{B}$, is an $n \times m$ matrix formed according to the rule

$$c_{ik} = \sum_{j=1}^{s} a_{ij}b_{jk}, \quad i = 1, 2, \ldots, n \quad k = 1, 2, \ldots, m. \tag{3.7}$$

Consequently, to determine the ijth element of matrix $\tilde{C}$, the corresponding terms of the ith row of $\tilde{A}$ and jth column of $\tilde{B}$ are multiplied and the resulting products added to form c_{ij}.

Example 3.3
Let

$$\tilde{A} = \begin{pmatrix} 2 & 1 & 4 \\ -3 & 0 & 2 \end{pmatrix}, \quad \tilde{B} = \begin{pmatrix} 3 & 5 \\ 2 & -1 \\ 4 & 2 \end{pmatrix}$$

then

$$\tilde{A}\tilde{B} = \begin{pmatrix} 2 \times 3 + 1 \times 2 + 4 \times 4 & 2 \times 5 + 1 \times (-1) + 4 \times 2 \\ (-3) \times 3 + 0 \times 2 + 2 \times 4 & (-3) \times 5 + 0 \times (-1) + 2 \times 2 \end{pmatrix}$$

$$= \begin{pmatrix} 24 & 17 \\ -1 & -11 \end{pmatrix}.$$

The reader should master matrix multiplication, since it is used throughout the rest of the book.

In general, matrix multiplication is not commutative: $\tilde{A}\tilde{B} \neq \tilde{B}\tilde{A}$. In fact, $\tilde{B}\tilde{A}$ is often not defined for non-square matrices, as shown in the following example.

Example 3.4
If

$$\tilde{A} = \begin{pmatrix} 1 & 2 \\ 3 & 4 \end{pmatrix}, \quad \tilde{B} = \begin{pmatrix} 3 \\ 7 \end{pmatrix}$$

then

$$\tilde{A}\tilde{B} = \begin{pmatrix} 1 & 2 \\ 3 & 4 \end{pmatrix}\begin{pmatrix} 3 \\ 7 \end{pmatrix} = \begin{pmatrix} 1 \times 3 + 2 \times 7 \\ 3 \times 3 + 4 \times 7 \end{pmatrix} = \begin{pmatrix} 17 \\ 37 \end{pmatrix}.$$

But

$$\tilde{B}\tilde{A} = \begin{pmatrix} 3 \\ 7 \end{pmatrix}\begin{pmatrix} 1 & 2 \\ 3 & 4 \end{pmatrix}$$

is not defined.

Matrix multiplication is associative and distributive:

$$(\tilde{A}\tilde{B})\tilde{C} = \tilde{A}(\tilde{B}\tilde{C}), \quad (\tilde{A} + \tilde{B})\tilde{C} = \tilde{A}\tilde{C} + \tilde{B}\tilde{C}.$$

To prove the associative law, we start with the matrix product $\tilde{A}\tilde{B}$, then multiply this product from the right by $\tilde{C}$:

$$\tilde{A}\tilde{B} = \sum_k a_{ik} b_{kj},$$

$$(\tilde{A}\tilde{B})\tilde{C} = \sum_j \left[\left(\sum_k a_{ik} b_{kj} \right) c_{js} \right] = \sum_k a_{ik} \left(\sum_j b_{kj} c_{js} \right) = \tilde{A}(\tilde{B}\tilde{C}).$$

Products of matrices differ from products of ordinary numbers in many remarkable ways. For example, $\tilde{A}\tilde{B} = 0$ does not imply $\tilde{A} = 0$ or $\tilde{B} = 0$. Even more bizarre is the case where $\tilde{A}^2 = 0$, $\tilde{A} \neq 0$; an example of which is

$$\tilde{A} = \begin{pmatrix} 0 & 1 \\ 0 & 0 \end{pmatrix}.$$

When you first run into Eq. (3.7), the rule for matrix multiplication, you might ask how anyone would arrive at it. It is suggested by the use of matrices in connection with linear transformations. For simplicity, we consider a very simple case: three coordinates systems in the plane denoted by the $x_1 x_2$-system, the $y_1 y_2$-system, and the $z_1 z_2$-system. We assume that these systems are related by the following linear transformations

$$x_1 = a_{11} y_1 + a_{12} y_2, \quad x_2 = a_{21} y_1 + a_{22} y_2, \tag{3.8}$$

$$y_1 = b_{11} z_1 + b_{12} z_2, \quad y_2 = b_{21} z_1 + b_{22} z_2. \tag{3.9}$$

Clearly, the $x_1 x_2$-coordinates can be obtained directly from the $z_1 z_2$-coordinates by a single linear transformation

$$x_1 = c_{11} z_1 + c_{12} z_2, \quad x_2 = c_{21} z_1 + c_{22} z_2, \tag{3.10}$$

whose coefficients can be found by inserting (3.9) into (3.8),

$$x_1 = a_{11}(b_{11} z_1 + b_{12} z_2) + a_{12}(b_{21} z_1 + b_{22} z_2),$$

$$x_2 = a_{21}(b_{11} z_1 + b_{12} z_2) + a_{22}(b_{21} z_1 + b_{22} z_2).$$

Comparing this with (3.10), we find

$$c_{11} = a_{11} b_{11} + a_{12} b_{21}, \quad c_{12} = a_{11} b_{12} + a_{12} b_{22},$$

$$c_{21} = a_{21} b_{11} + a_{22} b_{21}, \quad c_{22} = a_{21} b_{12} + a_{22} b_{22},$$

or briefly

$$c_{jk} = \sum_{i=1}^{2} a_{ji} b_{ik}, \quad j, k = 1, 2, \tag{3.11}$$

which is in the form of (3.7).

Now we rewrite the transformations (3.8), (3.9) and (3.10) in matrix form:

$$\tilde{X} = \tilde{A}\tilde{Y}, \quad \tilde{Y} = \tilde{B}\tilde{Z}, \quad \text{and} \quad \tilde{X} = \tilde{C}\tilde{Z},$$

where

$$\tilde{X} = \begin{pmatrix} x_1 \\ x_2 \end{pmatrix}, \quad \tilde{Y} = \begin{pmatrix} y_1 \\ y_2 \end{pmatrix}, \quad \tilde{Z} = \begin{pmatrix} z_1 \\ z_2 \end{pmatrix},$$

$$\tilde{A} = \begin{pmatrix} a_{11} & a_{12} \\ a_{21} & a_{22} \end{pmatrix}, \quad \tilde{B} = \begin{pmatrix} b_{11} & b_{12} \\ b_{21} & b_{22} \end{pmatrix}, \quad \tilde{C} = \begin{pmatrix} c_{11} & c_{12} \\ c_{21} & c_{22} \end{pmatrix}.$$

We then see that $\tilde{C} = \tilde{A}\tilde{B}$, and the elements of $\tilde{C}$ are given by (3.11).

Example 3.5
Rotations in three-dimensional space: An example of the use of matrix multiplication is provided by the representation of rotations in three-dimensional

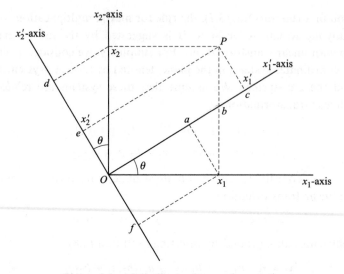

Figure 3.1. Coordinate changes by rotation.

space. In Fig. 3.1, the primed coordinates are obtained from the unprimed coordinates by a rotation through an angle θ about the x_3-axis. We see that x_1' is the sum of the projection of x_1 onto the x_1'-axis and the projection of x_2 onto the x_1'-axis:

$$x_1' = x_1 \cos\theta + x_2 \cos(\pi/2 - \theta) = x_1 \cos\theta + x_2 \sin\theta;$$

similarly

$$x_2' = x_1 \cos(\pi/2 + \theta) + x_2 \cos\theta = -x_1 \sin\theta + x_2 \cos\theta$$

and

$$x_3' = x_3.$$

We can put these in matrix form

$$X' = R_\theta X,$$

where

$$X' = \begin{pmatrix} x_1' \\ x_2' \\ x_3' \end{pmatrix}, \quad X = \begin{pmatrix} x_1 \\ x_2 \\ x_3 \end{pmatrix}, \quad R_\theta = \begin{pmatrix} \cos\theta & \sin\theta & 0 \\ -\sin\theta & \cos\theta & 0 \\ 0 & 0 & 1 \end{pmatrix}.$$

The commutator

Even if matrices $\tilde{A}$ and $\tilde{B}$ are both square matrices of order n, the products $\tilde{A}\tilde{B}$ and $\tilde{B}\tilde{A}$, although both square matrices of order n, are in general quite different, since their individual elements are formed differently. For example,

$$\begin{pmatrix} 1 & 2 \\ 1 & 3 \end{pmatrix}\begin{pmatrix} 1 & 0 \\ 1 & 2 \end{pmatrix} = \begin{pmatrix} 3 & 4 \\ 4 & 6 \end{pmatrix} \quad \text{but} \quad \begin{pmatrix} 1 & 0 \\ 1 & 2 \end{pmatrix}\begin{pmatrix} 1 & 2 \\ 1 & 3 \end{pmatrix} = \begin{pmatrix} 1 & 2 \\ 3 & 8 \end{pmatrix}.$$

The difference between the two products $\tilde{A}\tilde{B}$ and $\tilde{B}\tilde{A}$ is known as the commutator of $\tilde{A}$ and $\tilde{B}$ and is denoted by

$$[\tilde{A}, \tilde{B}] = \tilde{A}\tilde{B} - \tilde{B}\tilde{A}. \tag{3.12}$$

It is obvious that

$$[\tilde{B}, \tilde{A}] = -[\tilde{A}, \tilde{B}]. \tag{3.13}$$

If two square matrices $\tilde{A}$ and $\tilde{B}$ are very carefully chosen, it is possible to make the product identical. That is $\tilde{A}\tilde{B} = \tilde{B}\tilde{A}$. Two such matrices are said to commute with each other. Commuting matrices play an important role in quantum mechanics.

If $\tilde{A}$ commutes with $\tilde{B}$ and $\tilde{B}$ commutes with $\tilde{C}$, it does not necessarily follow that $\tilde{A}$ commutes with $\tilde{C}$.

Powers of a matrix

If n is a positive integer and $\tilde{A}$ is a square matrix, then $\tilde{A}^2 = \tilde{A}\tilde{A}$, $\tilde{A}^3 = \tilde{A}\tilde{A}\tilde{A}$, and in general, $\tilde{A}^n = \tilde{A}\tilde{A}\cdots\tilde{A}$ (n times). In particular, $\tilde{A}^0 = \tilde{I}$.

Functions of matrices

As we define and study various functions of a variable in algebra, it is possible to define and evaluate functions of matrices. We shall briefly discuss the following functions of matrices in this section: integral powers and exponential.

A simple example of integral powers of a matrix is polynomials such as

$$f(\tilde{A}) = \tilde{A}^2 + 3\tilde{A}^5.$$

Note that a matrix can be multiplied by itself if and only if it is a square matrix. Thus $\tilde{A}$ here is a square matrix and we denote the product $\tilde{A} \cdot \tilde{A}$ as $\tilde{A}^2$. More fancy examples can be obtained by taking series, such as

$$\tilde{S} = \sum_{k=0}^{\infty} a_k \tilde{A}^k,$$

where a_k are scalar coefficients. Of course, the sum has no meaning if it does not converge. The convergence of the matrix series means every matrix element of the infinite sum of matrices converges to a limit. We will not discuss the general theory of convergence of matrix functions. Another very common series is defined by

$$e^{\tilde{A}} = \sum_{n=0}^{\infty} \frac{\tilde{A}^n}{n!}.$$

Transpose of a matrix

Consider an $m \times n$ matrix $\tilde{A}$, if the rows and columns are systematically changed to columns to rows, without changing the order in which they occur, the new matrix is called the transpose of matrix $\tilde{A}$. It is denoted by $\tilde{A}^{T}$:

$$\tilde{A} = \begin{pmatrix} a_{11} & a_{12} & a_{13} & \cdots & a_{1n} \\ a_{21} & a_{22} & a_{23} & \cdots & a_{2n} \\ \vdots & \vdots & \vdots & & \vdots \\ a_{m1} & a_{m2} & a_{m3} & \cdots & a_{mn} \end{pmatrix}, \quad \tilde{A}^{T} = \begin{pmatrix} a_{11} & a_{21} & a_{31} & \cdots & a_{m1} \\ a_{12} & a_{22} & a_{32} & \cdots & a_{m2} \\ \vdots & \vdots & \vdots & & \vdots \\ a_{n1} & a_{2n} & a_{3n} & \cdots & a_{mn} \end{pmatrix}.$$

Thus the transpose matrix has n rows and m columns. If $\tilde{A}$ is written as (a_{jk}), then $\tilde{A}^{T}$ may be written as (a_{kj}).

$$\tilde{A} = (a_{jk}), \quad \tilde{A}^{T} = (a_{kj}). \tag{3.14}$$

The transpose of a row matrix is a column matrix, and vice versa.

Example 3.6

$$\tilde{A} = \begin{pmatrix} 1 & 2 & 3 \\ 4 & 5 & 6 \end{pmatrix}, \quad \tilde{A}^{T} = \begin{pmatrix} 1 & 4 \\ 2 & 5 \\ 3 & 6 \end{pmatrix}; \quad \tilde{B} = (1 \quad 2 \quad 3), \quad \tilde{B}^{T} = \begin{pmatrix} 1 \\ 2 \\ 3 \end{pmatrix}.$$

It is obvious that $(\tilde{A}^{T})^{T} = \tilde{A}$, and $(\tilde{A} + \tilde{B})^{T} = \tilde{A}^{T} + \tilde{B}^{T}$. It is also easy to prove that the transpose of the product is the product of the transposes in reverse:

$$(\tilde{A}\tilde{B})^{T} = \tilde{B}^{T}\tilde{A}^{T}. \tag{3.15}$$

Proof:

$$(\tilde{A}\tilde{B})_{ij}^{T} = (\tilde{A}\tilde{B})_{ji} \quad \text{by definition}$$

$$= \sum_{k} A_{jk} B_{ki}$$

$$= \sum_{k} B_{ik}^{T} A_{kj}^{T}$$

$$= (\tilde{B}^{T}\tilde{A}^{T})_{ij}$$

so that

$$(AB)^T = B^T A^T \quad \text{q.e.d.}$$

Because of (3.15), even if $\tilde{A} = \tilde{A}^T$ and $\tilde{B} = \tilde{B}^T$, $(\tilde{A}\tilde{B})^T \neq \tilde{A}\tilde{B}$ unless the matrices commute.

Symmetric and skew-symmetric matrices

A square matrix $\tilde{A} = (a_{jk})$ is said to be symmetric if all its elements satisfy the equations

$$a_{kj} = a_{jk}, \tag{3.16}$$

that is, $\tilde{A}$ and its transpose are equal $\tilde{A} = \tilde{A}^T$. For example,

$$\tilde{A} = \begin{pmatrix} 1 & 5 & 7 \\ 5 & 3 & -4 \\ 7 & -4 & 0 \end{pmatrix}$$

is a third-order symmetric matrix: the elements of the ith row equal the elements of ith column, for all i.

On the other hand, if the elements of $\tilde{A}$ satisfy the equations

$$a_{kj} = -a_{jk}, \tag{3.17}$$

then $\tilde{A}$ is said to be skew-symmetric, or antisymmetric. Thus, for a skew-symmetric $\tilde{A}$, its transpose equals minus $-\tilde{A}$: $\tilde{A}^T = -\tilde{A}$.

Since the elements a_{jj} along the principal diagonal satisfy the equations $a_{jj} = -a_{jj}$, it is evident that they must all vanish. For example,

$$\tilde{A} = \begin{pmatrix} 0 & -2 & 5 \\ 2 & 0 & 1 \\ -5 & -1 & 0 \end{pmatrix}$$

is a skew-symmetric matrix.

Any real square matrix $\tilde{A}$ may be expressed as the sum of a symmetric matrix $\tilde{R}$ and a skew-symmetric matrix $\tilde{S}$, where

$$\tilde{R} = \tfrac{1}{2}(\tilde{A} + \tilde{A}^T) \quad \text{and} \quad \tilde{S} = \tfrac{1}{2}(\tilde{A} - \tilde{A}^T). \tag{3.18}$$

Example 3.7
The matrix

$$\tilde{A} = \begin{pmatrix} 2 & 3 \\ 5 & -1 \end{pmatrix}$$

may be written in the form $\tilde{A} = \tilde{R} + \tilde{S}$, where

$$\tilde{R} = \frac{1}{2}(\tilde{A} + \tilde{A}^T) = \begin{pmatrix} 2 & 4 \\ 4 & -1 \end{pmatrix} \quad \tilde{S} = \frac{1}{2}(\tilde{A} - \tilde{A}^T) = \begin{pmatrix} 0 & -1 \\ 1 & 0 \end{pmatrix}.$$

The product of two symmetric matrices need not be symmetric. This is because of (3.15): even if $\tilde{A} = \tilde{A}^T$ and $\tilde{B} = \tilde{B}^T$, $(\tilde{A}\tilde{B})^T \neq \tilde{A}\tilde{B}$ unless the matrices commute.

A square matrix whose elements above or below the principal diagonal are all zero is called a triangular matrix. The following two matrices are triangular matrices:

$$\begin{pmatrix} 1 & 0 & 0 \\ 2 & 3 & 0 \\ 5 & 0 & 2 \end{pmatrix}, \quad \begin{pmatrix} 1 & 6 & -1 \\ 0 & 2 & 3 \\ 0 & 0 & 4 \end{pmatrix}.$$

A square matrix $\tilde{A}$ is said to be singular if $\det \tilde{A} = 0$, and non-singular if $\det \tilde{A} \neq 0$, where $\det \tilde{A}$ is the determinant of the matrix $\tilde{A}$.

The matrix representation of a vector product

The scalar product defined in ordinary vector theory has its counterpart in matrix theory. Consider two vectors $\mathbf{A} = (A_1, A_2, A_3)$ and $\mathbf{B} = (B_1, B_2, B_3)$ the counterpart of the scalar product is given by

$$\tilde{A}\tilde{B}^T = (A_1 \quad A_2 \quad A_3) \begin{pmatrix} B_1 \\ B_2 \\ B_3 \end{pmatrix} = A_1B_1 + A_2B_2 + A_3B_3.$$

Note that $\tilde{B}\tilde{A}^T$ is the transpose of $\tilde{A}\tilde{B}^T$, and, being a 1×1 matrix, the transpose equals itself. Thus a scalar product may be written in these two equivalent forms.

Similarly, the vector product used in ordinary vector theory must be replaced by something more in keeping with the definition of matrix multiplication. Note that the vector product

$$\mathbf{A} \times \mathbf{B} = (A_2B_3 - A_3B_2)\hat{e}_1 + (A_3B_1 - A_1B_3)\hat{e}_2 + (A_1B_2 - A_2B_1)\hat{e}_3$$

can be represented by the column matrix

$$\begin{pmatrix} A_2B_3 - A_3B_2 \\ A_3B_1 - A_1B_3 \\ A_1B_2 - A_2B_1 \end{pmatrix}.$$

This can be split into the product of two matrices

$$\begin{pmatrix} A_2B_3 - A_3B_2 \\ A_3B_1 - A_1B_3 \\ A_1B_2 - A_2B_1 \end{pmatrix} = \begin{pmatrix} 0 & -A_2 & A_2 \\ A_3 & 0 & -A_1 \\ -A_2 & A_1 & 0 \end{pmatrix} \begin{pmatrix} B_1 \\ B_2 \\ B_3 \end{pmatrix}$$

or

$$\begin{pmatrix} A_2B_3 - A_3B_2 \\ A_3B_1 - A_1B_3 \\ A_1B_2 - A_2B_1 \end{pmatrix} = \begin{pmatrix} 0 & -B_2 & B_2 \\ B_3 & 0 & -B_1 \\ -B_2 & B_1 & 0 \end{pmatrix} \begin{pmatrix} A_1 \\ A_2 \\ A_3 \end{pmatrix}.$$

Thus the vector product may be represented as the product of a skew-symmetric matrix and a column matrix. However, this definition only holds for 3×3 matrices.

Similarly, curl A may be represented in terms of a skew-symmetric matrix operator, given in Cartesian coordinates by

$$\nabla \times A = \begin{pmatrix} 0 & -\partial/\partial x_3 & \partial/\partial x_2 \\ \partial/\partial x_3 & 0 & -\partial/\partial x_1 \\ -\partial/\partial x_2 & \partial/\partial x_1 & 0 \end{pmatrix} \begin{pmatrix} A_1 \\ A_2 \\ A_3 \end{pmatrix}.$$

In a similar way, we can investigate the triple scalar product and the triple vector product.

The inverse of a matrix

If for a given square matrix $\tilde{A}$ there exists a matrix $\tilde{B}$ such that $\tilde{A}\tilde{B} = \tilde{B}\tilde{A} = \tilde{I}$, where $\tilde{I}$ is a unit matrix, then $\tilde{B}$ is called an inverse of matrix $\tilde{A}$.

Example 3.8
The matrix

$$\tilde{B} = \begin{pmatrix} 3 & 5 \\ 1 & 2 \end{pmatrix}$$

is an inverse of

$$\tilde{A} = \begin{pmatrix} 2 & -5 \\ -1 & 3 \end{pmatrix},$$

since

$$\tilde{A}\tilde{B} = \begin{pmatrix} 2 & -5 \\ -1 & 3 \end{pmatrix} \begin{pmatrix} 3 & 5 \\ 1 & 2 \end{pmatrix} = \begin{pmatrix} 1 & 0 \\ 0 & 1 \end{pmatrix} = \tilde{I}$$

and

$$\tilde{B}\tilde{A} = \begin{pmatrix} 3 & 5 \\ 1 & 2 \end{pmatrix} \begin{pmatrix} 2 & -5 \\ -1 & 3 \end{pmatrix} = \begin{pmatrix} 1 & 0 \\ 0 & 1 \end{pmatrix} = \tilde{I}.$$

An invertible matrix has a unique inverse. That is, if $\tilde{B}$ and $\tilde{C}$ are both inverses of the matrix $\tilde{A}$, then $\tilde{B} = \tilde{C}$. The proof is simple. Since $\tilde{B}$ is an inverse of $\tilde{A}$,

$\tilde{B}\tilde{A} = \tilde{I}$. Multiplying both sides on the right by $\tilde{C}$ gives $(\mathbf{B}\tilde{A})\tilde{C} = \tilde{I}\tilde{C} = \tilde{C}$. On the other hand, $(\tilde{B}\tilde{A})\tilde{C} = \tilde{B}(\tilde{A}\tilde{C}) = \tilde{B}\tilde{I} = \tilde{B}$, so that $\tilde{B} = \tilde{C}$. As a consequence of this result, we can now speak of *the* inverse of an invertible matrix. If $\tilde{A}$ is invertible, then its inverse will be denoted by $\tilde{A}^{-1}$. Thus

$$\tilde{A}\tilde{A}^{-1} = \tilde{A}^{-1}\tilde{A} = \tilde{I}. \tag{3.19}$$

It is obvious that the inverse of the inverse is the given matrix, that is,

$$(\tilde{A}^{-1})^{-1} = \tilde{A}. \tag{3.20}$$

It is easy to prove that the inverse of the product is the product of the inverse in reverse order, that is,

$$(\tilde{A}\tilde{B})^{-1} = \tilde{B}^{-1}\tilde{A}^{-1}. \tag{3.21}$$

To prove (3.21), we start with $\tilde{A}\tilde{A}^{-1} = \tilde{I}$, with $\tilde{A}$ replaced by $\tilde{A}\tilde{B}$, that is,

$$\tilde{A}\tilde{B}(\tilde{A}\tilde{B})^{-1} = \tilde{I}.$$

By premultiplying this by $\tilde{A}^{-1}$ we get

$$\tilde{B}(\tilde{A}\tilde{B})^{-1} = \tilde{A}^{-1}.$$

If we premultiply this by $\tilde{B}^{-1}$, the result follows.

A method for finding $\tilde{A}^{-1}$

The positive power for a square matrix $\tilde{A}$ is defined as $\tilde{A}^n = \tilde{A}\tilde{A} \cdots \tilde{A}$ (n factors) and $\tilde{A}^0 = \tilde{I}$, where n is a positive integer. If, in addition, $\tilde{A}$ is invertible, we define

$$\tilde{A}^{-n} = (\tilde{A}^{-1})^n = \tilde{A}^{-1}\tilde{A}^{-1} \cdots \tilde{A}^{-1}(n \text{ factors}).$$

We are now in position to construct the inverse of an invertible matrix $\tilde{A}$:

$$\tilde{A} = \begin{pmatrix} a_{11} & a_{12} & \cdots & a_{1n} \\ a_{21} & a_{22} & \cdots & a_{2n} \\ \vdots & \vdots & & \vdots \\ a_{n1} & a_{n2} & \cdots & a_{nn} \end{pmatrix}.$$

The a_{jk} are known. Now let

$$\tilde{A}^{-1} = \begin{pmatrix} a'_{11} & a'_{12} & \cdots & a'_{1n} \\ a'_{21} & a'_{22} & \cdots & a'_{2n} \\ \vdots & \vdots & & \vdots \\ a'_{n1} & a'_{n2} & \cdots & a'_{nn} \end{pmatrix}.$$

The a'_{jk} are required to construct $\tilde{A}^{-1}$. Since $\tilde{A}\tilde{A}^{-1} = \tilde{I}$, we have

$$a_{11}a'_{11} + a_{12}a'_{12} + \cdots + a_{1n}a'_{1n} = 1,$$
$$a_{21}a'_{21} + a_{22}a'_{22} + \cdots + a_{2n}a'_{2n} = 0,$$
$$\vdots \tag{3.22}$$
$$a_{n1}a'_{n1} + a_{n2}a'_{n2} + \cdots + a_{nn}a'_{nn} = 0.$$

The solution to the above set of linear algebraic equations (3.22) may be facilitated by applying Cramer's rule. Thus

$$a'_{jk} = \frac{\text{cofactor } a_{kj}}{\det \tilde{A}}. \tag{3.23}$$

From (3.23) it is clear that $\tilde{A}^{-1}$ exists if and only if matrix $\tilde{A}$ is non-singular (that is, $\det \tilde{A} \neq 0$).

Systems of linear equations and the inverse of a matrix

As an immediate application, let us apply the concept of an inverse matrix to a system of n linear equations in n unknowns $(x_1, \ldots, x_n)$:

$$a_{11}x_1 + a_{12}x_2 + \cdots + a_{1n}x_n = b_1;$$
$$a_{21}x_2 + a_{22}x_2 + \cdots + a_{2n}x_n = b_2;$$
$$\vdots$$
$$a_{n1}x_n + a_{n2}x_n + \cdots + a_{nn}x_n = b_n;$$

in matrix form we have

$$\tilde{A}\tilde{X} = \tilde{B}, \tag{3.24}$$

where

$$\tilde{A} = \begin{pmatrix} a_{11} & a_{12} & \cdots & a_{1n} \\ a_{21} & a_{22} & \cdots & a_{2n} \\ \vdots & \vdots & & \vdots \\ a_{n1} & a_{n2} & \cdots & a_{nn} \end{pmatrix}, \quad \tilde{X} = \begin{pmatrix} x_1 \\ x_2 \\ \vdots \\ x_n \end{pmatrix}, \quad \tilde{B} = \begin{pmatrix} b_1 \\ b_2 \\ \vdots \\ b_n \end{pmatrix}.$$

We can prove that the above linear system possesses a unique solution given by

$$\tilde{X} = \tilde{A}^{-1}\tilde{B}. \tag{3.25}$$

The proof is simple. If $\tilde{A}$ is non-singular it has a unique inverse $\tilde{A}^{-1}$. Now premultiplying (3.24) by $\tilde{A}^{-1}$ we obtain

$$\tilde{A}^{-1}(\tilde{A}\tilde{X}) = \tilde{A}^{-1}\tilde{B},$$

but

$$\tilde{A}^{-1}(\tilde{A}\tilde{X}) = (\tilde{A}^{-1}\tilde{A})\tilde{X} = \tilde{X}$$

so that

$$\tilde{X} = \tilde{A}^{-1}\tilde{B} \text{ is a solution to (3.24)}, \ \tilde{A}\tilde{X} = \tilde{B}.$$

Complex conjugate of a matrix

If $\tilde{A} = (a_{jk})$ is an arbitrary matrix whose elements may be complex numbers, the complex conjugate matrix, denoted by $\tilde{A}^*$, is also a matrix of the same order, every element of which is the complex conjugate of the corresponding element of $\tilde{A}$, that is,

$$(A^*)_{jk} = a_{jk}^*. \tag{3.26}$$

Hermitian conjugation

If $\tilde{A} = (a_{jk})$ is an arbitrary matrix whose elements may be complex numbers, when the two operations of transposition and complex conjugation are carried out on $\tilde{A}$, the resulting matrix is called the hermitian conjugate (or hermitian adjoint) of the original matrix $\tilde{A}$ and will be denoted by $\tilde{A}^\dagger$. We frequently call $\tilde{A}^\dagger$ A-dagger. The order of the two operations is immaterial:

$$\tilde{A}^\dagger = (\tilde{A}^{\mathrm{T}})^* = (\tilde{A}^*)^{\mathrm{T}}. \tag{3.27}$$

In terms of the elements, we have

$$(\tilde{A}^\dagger)_{jk} = a_{kj}^*. \tag{3.27a}$$

It is clear that if $\tilde{A}$ is a matrix of order $m \times n$, then $\tilde{A}^\dagger$ is a matrix of order $n \times m$. We can prove that, as in the case of the transpose of a product, the adjoint of the product is the product of the adjoints in reverse:

$$(\tilde{A}\tilde{B})^\dagger = \tilde{B}^\dagger \tilde{A}^\dagger. \tag{3.28}$$

Hermitian/anti-hermitian matrix

A matrix $\tilde{A}$ that obeys

$$\tilde{A}^\dagger = \tilde{A} \tag{3.29}$$

is called a hermitian matrix. It is very clear the following matrices are hermitian:

$$\begin{pmatrix} 1 & -i \\ i & 2 \end{pmatrix}, \quad \begin{pmatrix} 4 & 5+2i & 6+3i \\ 5-2i & 5 & -1-2i \\ 6-3i & -1+2i & 6 \end{pmatrix}, \quad \text{where} \quad i = \sqrt{-1}.$$

114

Table 3.1. Operations on matrices

Operation	Matrix element		$\tilde{A}$	$\tilde{B}$	If $\tilde{B} = \tilde{A}$
Transposition	$\tilde{B} = \tilde{A}^{\mathrm{T}}$	$b_{ij} = a_{ji}$	$m \times n$	$n \times m$	Symmetric[a]
Complex conjugation	$\tilde{B} = \tilde{A}^{*}$	$b_{ij} = a_{ij}^{*}$	$m \times n$	$m \times n$	Real
Hermitian conjugation	$\tilde{B} = \tilde{A}^{\mathrm{T}*}$	$b_{ij} = a_{ji}^{*}$	$m \times n$	$n \times m$	Hermitian

[a] For square matrices only.

Evidently all the elements along the principal diagonal of a hermitian matrix must be real.

A hermitian matrix is also defined as a matrix whose transpose equals its complex conjugate:

$$\tilde{A}^{\mathrm{T}} = \tilde{A}^{*} \quad (\text{that is}, a_{kj} = a_{jk}^{*}). \tag{3.29a}$$

These two definitions are the same. First note that the elements in the principal diagonal of a hermitian matrix are always real. Furthermore, any real symmetric matrix is hermitian, so a real hermitian matrix is a symmetric matrix.

The product of two hermitian matrices is not generally hermitian unless they commute. This is because of property (3.28): even if $\tilde{A}^{\dagger} = \tilde{A}$ and $\tilde{B}^{\dagger} = \tilde{B}$, $(\tilde{A}\tilde{B})^{\dagger} \neq \tilde{A}\tilde{B}$ unless the matrices commute.

A matrix $\tilde{A}$ that obeys

$$\tilde{A}^{\dagger} = -\tilde{A} \tag{3.30}$$

is called an anti-hermitian (or skew-hermitian) matrix. All the elements along the principal diagonal must be pure imaginary. An example is

$$\begin{pmatrix} 6i & 5+2i & 6+3i \\ -5+2i & -8i & -1-2i \\ -6+3i & 1-2i & 0 \end{pmatrix}.$$

We summarize the three operations on matrices discussed above in Table 3.1.

Orthogonal matrix (real)

A matrix $\tilde{A} = (a_{jk})_{mn}$ satisfying the relations

$$\tilde{A}\tilde{A}^{\mathrm{T}} = \tilde{I}_{n}, \tag{3.31a}$$

$$\tilde{A}^{\mathrm{T}}\tilde{A} = \tilde{I}_{m} \tag{3.31b}$$

is called an orthogonal matrix. It can be shown that if $\tilde{A}$ is a finite matrix satisfying both relations (3.31a) and (3.31b), then $\tilde{A}$ must be square, and we have

$$\tilde{A}\tilde{A}^{\mathrm{T}} = \tilde{A}^{\mathrm{T}}\tilde{A} = \tilde{I}. \tag{3.32}$$

But if $\tilde{A}$ is an infinite matrix, then $\tilde{A}$ is orthogonal if and only if both (3.31a) and (3.31b) are simultaneously satisfied.

Now taking the determinant of both sides of Eq. (3.32), we have $(\det \tilde{A})^2 = 1$, or $\det \tilde{A} = \pm 1$. This shows that $\tilde{A}$ is non-singular, and so $\tilde{A}^{-1}$ exists. Premultiplying (3.32) by $\tilde{A}^{-1}$ we have

$$\tilde{A}^{-1} = \tilde{A}^{\mathrm{T}}. \tag{3.33}$$

This is often used as an alternative way of defining an orthogonal matrix.

The elements of an orthogonal matrix are not all independent. To find the conditions between them, let us first equate the ijth element of both sides of $\tilde{A}\tilde{A}^{\mathrm{T}} = \tilde{I}$; we find that

$$\sum_{k=1}^{n} a_{ik} a_{jk} = \delta_{ij}. \tag{3.34a}$$

Similarly, equating the ijth element of both sides of $\tilde{A}^{\mathrm{T}}\tilde{A} = \tilde{I}$, we obtain

$$\sum_{k=1}^{n} a_{ki} a_{kj} = \delta_{ij}. \tag{3.34b}$$

Note that either (3.34a) and (3.34b) gives $2n(n+1)$ relations. Thus, for a real orthogonal matrix of order n, there are only $n^2 - n(n+1)/2 = n(n-1)/2$ different elements.

Unitary matrix

A matrix $\tilde{U} = (u_{jk})_{mn}$ satisfying the relations

$$\tilde{U}\tilde{U}^{\dagger} = \tilde{I}_n, \tag{3.35a}$$

$$\tilde{U}^{\dagger}\tilde{U} = \tilde{I}_m \tag{3.35b}$$

is called a unitary matrix. If $\tilde{U}$ is a finite matrix satisfying both (3.35a) and (3.35b), then $\tilde{U}$ must be a square matrix, and we have

$$\tilde{U}\tilde{U}^{\dagger} = \tilde{U}^{\dagger}\tilde{U} = \tilde{I}. \tag{3.36}$$

This is the complex generalization of the real orthogonal matrix. The elements of a unitary matrix may be complex, for example

$$\frac{1}{\sqrt{2}} \begin{pmatrix} 1 & i \\ i & 1 \end{pmatrix}$$

is unitary. From the definition (3.35), a real unitary matrix is orthogonal.

Taking the determinant of both sides of (3.36) and noting that $\det \tilde{U}^{\dagger} = (\det \tilde{U})^*$, we have

$$(\det \tilde{U})(\det \tilde{U})^* = 1 \quad \text{or} \quad |\det \tilde{U}| = 1. \tag{3.37}$$

This shows that the determinant of a unitary matrix can be a complex number of unit magnitude, that is, a number of the form $e^{i\alpha}$, where α is a real number. It also shows that a unitary matrix is non-singular and possesses an inverse. Premultiplying (3.35a) by $\tilde{U}^{-1}$, we get

$$\tilde{U}^\dagger = \tilde{U}^{-1}. \tag{3.38}$$

This is often used as an alternative way of defining a unitary matrix.

Just as in the case of an orthogonal matrix that is a special (real) case of a unitary matrix, the elements of a unitary matrix satisfy the following conditions:

$$\sum_{k=1}^n u_{ik} u_{jk}^* = \delta_{ij}, \quad \sum_{k=1}^n u_{ki} u_{kj}^* = \delta_{ij}. \tag{3.39}$$

The product of two unitary matrices is unitary. The reason is as follows. If $\tilde{U}_1$ and $\tilde{U}_2$ are two unitary matrices, then

$$\tilde{U}_1 \tilde{U}_2 (\tilde{U}_1 \tilde{U}_2)^\dagger = \tilde{U}_1 \tilde{U}_2 (\tilde{U}_2^\dagger \tilde{U}_1^\dagger) = \tilde{U}_1 \tilde{U}_1^\dagger = \tilde{I}, \tag{3.40}$$

which shows that $U_1 U_2$ is unitary.

Rotation matrices

Let us revisit Example 3.5. Our discussion will illustrate the power and usefulness of matrix methods. We will also see that rotation matrices are orthogonal matrices. Consider a point P with Cartesian coordinates (x_1, x_2, x_3) (see Fig. 3.2). We rotate the coordinate axes about the x_3-axis through an angle θ and create a new coordinate system, the primed system. The point P now has the coordinates (x_1', x_2', x_3') in the primed system. Thus the position vector $\mathbf{r}$ of point P can be written as

$$\mathbf{r} = \sum_{i=1}^3 x_i \hat{e}_i = \sum_{i=1}^3 x_i' \hat{e}_i'. \tag{3.41}$$

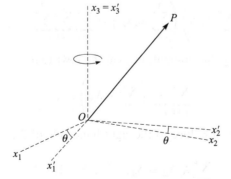

Figure 3.2.　Coordinate change by rotation.

Taking the dot product of Eq. (3.41) with $\hat{e}'_1$ and using the orthonormal relation $\hat{e}'_i \cdot \hat{e}'_j = \delta_{ij}$ (where δ_{ij} is the Kronecker delta symbol), we obtain $x'_1 = \mathbf{r} \cdot \hat{e}'_1$. Similarly, we have $x'_2 = \mathbf{r} \cdot \hat{e}'_2$ and $x'_3 = \mathbf{r} \cdot \hat{e}'_3$. Combining these results we have

$$x'_i = \sum_{j=1}^{3} \hat{e}'_i \cdot \hat{e}_j x_j = \sum_{j=1}^{3} \lambda_{ij} x_j, \qquad i = 1, 2, 3. \tag{3.42}$$

The quantities $\lambda_{ij} = \hat{e}'_i \cdot \hat{e}_j$ are called the coefficients of transformation. They are the direction cosines of the primed coordinate axes relative to the unprimed ones

$$\lambda_{ij} = \hat{e}'_i \cdot \hat{e}_j = \cos(x'_i, x_j), \qquad i, j = 1, 2, 3. \tag{3.42a}$$

Eq. (3.42) can be written conveniently in the following matrix form

$$\begin{pmatrix} x'_1 \\ x'_2 \\ x'_3 \end{pmatrix} = \begin{pmatrix} \lambda_{11} & \lambda_{12} & \lambda_{13} \\ \lambda_{21} & \lambda_{22} & \lambda_{23} \\ \lambda_{31} & \lambda_{32} & \lambda_{33} \end{pmatrix} \begin{pmatrix} x_1 \\ x_2 \\ x_3 \end{pmatrix} \tag{3.43a}$$

or

$$\tilde{X}' = \tilde{\lambda}(\theta)\tilde{X}, \tag{3.43b}$$

where $\tilde{X}'$ and $\tilde{X}$ are the column matrices, $\tilde{\lambda}(\theta)$ is called a transformation (or rotation) matrix; it acts as a linear operator which transforms the vector $\mathbf{X}$ into the vector $\mathbf{X}'$. Strictly speaking, we should describe the matrix $\tilde{\lambda}(\theta)$ as the matrix representation of the linear operator $\hat{\lambda}$. The concept of linear operator is more general than that of matrix.

Not all of the nine quantities λ_{ij} are independent; six relations exist among the λ_{ij}, hence only three of them are independent. These six relations are found by using the fact that the magnitude of the vector must be the same in both systems:

$$\sum_{i=1}^{3} (x'_i)^2 = \sum_{i=1}^{3} x_i^2. \tag{3.44}$$

With the help of Eq. (3.42), the left hand side of the last equation becomes

$$\sum_{i=1}^{3} \left(\sum_{j=1}^{3} \lambda_{ij} x_j \right) \left(\sum_{k=1}^{3} \lambda_{ik} x_k \right) = \sum_{i=1}^{3} \sum_{j=1}^{3} \sum_{k=1}^{3} \lambda_{ij} \lambda_{ik} x_j x_k,$$

which, by rearranging the summations, can be rewritten as

$$\sum_{k=1}^{3} \sum_{j=1}^{3} \left(\sum_{i=1}^{3} \lambda_{ij} \lambda_{ik} \right) x_j x_k.$$

This last expression will reduce to the right hand side of Eq. (3.43) if and only if

$$\sum_{i=1}^{3} \lambda_{ij} \lambda_{ik} = \delta_{jk}, \qquad j, k = 1, 2, 3. \tag{3.45}$$

Eq. (3.45) gives six relations among the λ_{ij}, and is known as the orthogonal condition.

If the primed coordinates system is generated by a rotation about the x_3-axis through an angle θ as shown in Fig. 3.2. Then from Example 3.5, we have

$$x_1' = x_1 \cos\theta + x_2 \sin\theta, \quad x_2' = -x_1 \sin\theta + x_2 \cos\theta, \quad x_3' = x_3. \tag{3.46}$$

Thus

$$\lambda_{11} = \cos\theta, \quad \lambda_{12} = \sin\theta, \quad \lambda_{13} = 0,$$

$$\lambda_{21} = -\sin\theta, \quad \lambda_{22} = \cos\theta, \quad \lambda_{23} = 0,$$

$$\lambda_{31} = 0, \quad \lambda_{32} = 0, \quad \lambda_{33} = 1.$$

We can also obtain these elements from Eq. (3.42a). It is obvious that only three of them are independent, and it is easy to check that they satisfy the condition given in Eq. (3.45). Now the rotation matrix takes the simple form

$$\tilde{\lambda}(\theta) = \begin{pmatrix} \cos\theta & \sin\theta & 0 \\ -\sin\theta & \cos\theta & 0 \\ 0 & 0 & 1 \end{pmatrix} \tag{3.47}$$

and its transpose is

$$\tilde{\lambda}^{T}(\theta) = \begin{pmatrix} \cos\theta & -\sin\theta & 0 \\ \sin\theta & \cos\theta & 0 \\ 0 & 0 & 1 \end{pmatrix}.$$

Now take the product

$$\tilde{\lambda}^{T}(\theta)\tilde{\lambda}(\theta) = \begin{pmatrix} \cos\theta & \sin\theta & 0 \\ -\sin\theta & \cos\theta & 0 \\ 0 & 0 & 1 \end{pmatrix} \begin{pmatrix} \cos\theta & -\sin\theta & 0 \\ \sin\theta & \cos\theta & 0 \\ 0 & 0 & 1 \end{pmatrix} = \begin{pmatrix} 1 & 0 & 0 \\ 0 & 1 & 0 \\ 0 & 0 & 1 \end{pmatrix} = \tilde{I},$$

which shows that the rotation matrix is an orthogonal matrix. In fact, rotation matrices are orthogonal matrices, not limited to $\tilde{\lambda}(\theta)$ of Eq. (3.47). The proof of this is easy. Since coordinate transformations are reversible by interchanging old and new indices, we must have

$$\left(\tilde{\lambda}^{-1}\right)_{ij} = \hat{e}_i^{old} \cdot \hat{e}_j^{new} = \hat{e}_j^{new} \cdot \hat{e}_i^{old} = \lambda_{ji} = \left(\tilde{\lambda}^{T}\right)_{ij}.$$

Hence rotation matrices are orthogonal matrices. It is obvious that the inverse of an orthogonal matrix is equal to its transpose.

A rotation matrix such as given in Eq. (3.47) is a continuous function of its argument θ. So its determinant is also a continuous function of θ and, in fact, it is equal to 1 for any θ. There are matrices of coordinate changes with a determinant of -1. These correspond to inversion of the coordinate axes about the origin and

119

change the handedness of the coordinate system. Examples of such parity transformations are

$$\tilde{P}_1 = \begin{pmatrix} -1 & 0 & 0 \\ 0 & 1 & 0 \\ 0 & 0 & 1 \end{pmatrix}, \qquad \tilde{P}_3 = \begin{pmatrix} -1 & 0 & 0 \\ 0 & -1 & 0 \\ 0 & 0 & -1 \end{pmatrix}, \qquad \tilde{P}_i^2 = I.$$

They change the signs of an odd number of coordinates of a fixed point $\mathbf{r}$ in space (Fig. 3.3).

What is the advantage of using matrices in describing rotation in space? One of the advantages is that successive transformations $1, 2, \ldots, m$ of the coordinate axes about the origin are described by successive matrix multiplications as far as their effects on the coordinates of a fixed point are concerned:

If $\tilde{X}^{(1)} = \tilde{\lambda}_1 \tilde{X}, \tilde{X}^{(2)} = \tilde{\lambda}_2 \tilde{X}^{(1)}, \ldots$, then

$$\tilde{X}^{(m)} = \tilde{\lambda}_m \tilde{X}^{(m-1)} = (\tilde{\lambda}_m \tilde{\lambda}_{m-1} \cdots \tilde{\lambda}_1)\tilde{X} = \tilde{R}\tilde{X}$$

where

$$\tilde{R} = \tilde{\lambda}_m \tilde{\lambda}_{m-1} \cdots \tilde{\lambda}_1$$

is the resultant (or net) rotation matrix for the m successive transformations taken place in the specified manner.

Example 3.9
Consider a rotation of the x_1-, x_2-axes about the x_3-axis by an angle θ. If this rotation is followed by a back-rotation of the same angle in the opposite direction,

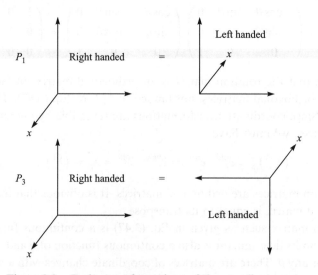

Figure 3.3. Parity transformations of the coordinate system.

that is, by $-\theta$, we recover the original coordinate system. Thus

$$\tilde{R}(-\theta)\tilde{R}(\theta) = \begin{pmatrix} 1 & 0 & 0 \\ 0 & 1 & 0 \\ 0 & 0 & 1 \end{pmatrix} = \tilde{R}^{-1}(\theta)\tilde{R}(\theta).$$

Hence

$$\tilde{R}^{-1}(\theta) = \tilde{R}(-\theta) = \begin{pmatrix} \cos\theta & -\sin\theta & 0 \\ \sin\theta & \cos\theta & 0 \\ 0 & 0 & 1 \end{pmatrix} = \tilde{R}^{T}(\theta),$$

which shows that a rotation matrix is an orthogonal matrix.

We would like to make one remark on rotation in space. In the above discussion, we have considered the vector to be fixed and rotated the coordinate axes. The rotation matrix can be thought of as an operator that, acting on the unprimed system, transforms it into the primed system. This view is often called the passive view of rotation. We could equally well keep the coordinate axes fixed and rotate the vector through an equal angle, but in the opposite direction. Then the rotation matrix would be thought of as an operator acting on the vector, say $\mathbf{X}$, and changing it into $\mathbf{X}'$. This procedure is called the active view of the rotation.

Trace of a matrix

Recall that the trace of a square matrix $\tilde{A}$ is defined as the sum of all the principal diagonal elements:

$$\mathrm{Tr}\,\tilde{A} = \sum_{k} a_{kk}.$$

It can be proved that the trace of the product of a finite number of matrices is invariant under any *cyclic* permutation of the matrices. We leave this as home work.

Orthogonal and unitary transformations

Eq. (3.42) is a linear transformation and it is called an orthogonal transformation, because the rotation matrix is an orthogonal matrix. One of the properties of an orthogonal transformation is that it preserves the length of a vector. A more useful linear transformation in physics is the unitary transformation:

$$\tilde{Y} = \tilde{U}\tilde{X} \qquad (3.48)$$

in which $\tilde{X}$ and $\tilde{Y}$ are column matrices (vectors) of order $n \times 1$ and $\tilde{U}$ is a unitary matrix of order $n \times n$. One of the properties of a unitary transformation is that it preserves the norm of a vector. To see this, premultiplying Eq. (3.48) by $\tilde{Y}^{\dagger}(= \tilde{X}^{\dagger}\tilde{U}^{\dagger})$ and using the condition $\tilde{U}^{\dagger}\tilde{U} = \tilde{I}$, we obtain

$$\tilde{Y}^{\dagger}\tilde{Y} = \tilde{X}^{\dagger}\tilde{U}^{\dagger}\tilde{U}\tilde{X} = \tilde{X}^{\dagger}\tilde{X} \tag{3.49a}$$

or

$$\sum_{k=1}^{n} y_k^* y_k = \sum_{k=1}^{n} x_k^* x_k. \tag{3.49b}$$

This shows that the norm of a vector remains invariant under a unitary transformation. If the matrix $\tilde{U}$ of transformation happens to be real, then $\tilde{U}$ is also an orthogonal matrix and the transformation (3.48) is an orthogonal transformation, and Eqs. (3.49) reduce to

$$\tilde{Y}^{T}\tilde{Y} = \tilde{X}^{T}\tilde{X}, \tag{3.50a}$$

$$\sum_{k=1}^{n} y_k^2 = \sum_{k=1}^{n} x_k^2, \tag{3.50b}$$

as we expected.

Similarity transformation

We now consider a different linear transformation, the similarity transformation that, we shall see later, is very useful in diagonalization of a matrix. To get the idea about similarity transformations, we consider vectors $\mathbf{r}$ and $\mathbf{R}$ in a particular basis, the coordinate system $Ox_1x_2x_3$, which are connected by a square matrix $\tilde{A}$:

$$\mathbf{R} = \tilde{A}\mathbf{r}. \tag{3.51a}$$

Now rotating the coordinate system about the origin O we obtain a new system $Ox_1'x_2'x_3'$ (a new basis). The vectors $\mathbf{r}$ and $\mathbf{R}$ have not been affected by this rotation. Their components, however, will have different values in the new system, and we now have

$$\mathbf{R}' = \tilde{A}'\mathbf{r}'. \tag{3.51b}$$

The matrix $\tilde{A}'$ in the new (primed) system is called similar to the matrix $\tilde{A}$ in the old (unprimed) system, since they perform same function. Then what is the relationship between matrices $\tilde{A}$ and $\tilde{A}'$? This information is given in the form of coordinate transformation. We learned in the previous section that the components of a vector in the primed and unprimed systems are connected by a matrix equation similar to Eq. (3.43). Thus we have

$$\mathbf{r} = \tilde{S}\mathbf{r}' \quad \text{and} \quad \mathbf{R} = \tilde{S}\mathbf{R}',$$

where $\tilde{S}$ is a non-singular matrix, the transition matrix from the new coordinate system to the old system. With these, Eq. (3.51a) becomes

$$\tilde{S}\mathbf{R}' = \tilde{A}\tilde{S}\mathbf{r}'$$

or

$$\mathbf{R}' = \tilde{S}^{-1}\tilde{A}\tilde{S}\mathbf{r}'.$$

Combining this with Eq. (3.51) gives

$$\tilde{A}' = \tilde{S}^{-1}\tilde{A}\tilde{S}, \tag{3.52}$$

where $\tilde{A}'$ and $\tilde{A}$ are similar matrices. Eq. (3.52) is called a similarity transformation.

Generalization of this idea to n-dimensional vectors is straightforward. In this case, we take $\mathbf{r}$ and $\mathbf{R}$ as two n-dimensional vectors in a particular basis, having their coordinates connected by the matrix $\tilde{A}$ (a $n \times n$ square matrix) through Eq. (3.51a). In another basis they are connected by Eq. (3.51b). The relationship between $\tilde{A}$ and $\tilde{A}'$ is given by Eq. (3.52). The transformation of $\tilde{A}$ into $\tilde{S}^{-1}\tilde{A}\tilde{S}$ is called a similarity transformation.

All identities involving vectors and matrices will remain invariant under a similarity transformation since this arises only in connection with a change in basis. That this is so can be seen in the following two simple examples.

Example 3.10
Given the matrix equation $\tilde{A}\tilde{B} = \tilde{C}$, and the matrices $\tilde{A}$, $\tilde{B}$, $\tilde{C}$ subjected to the same similarity transformation, show that the matrix equation is invariant.

Solution: Since the three matrices are all subjected to the same similarity transformation, we have

$$\tilde{A}' = \tilde{S}\tilde{A}\tilde{S}^{-1}, \quad \tilde{B}' = \tilde{S}\tilde{B}\tilde{S}^{-1}, \quad \tilde{C}' = \tilde{S}\tilde{C}\tilde{S}^{-1}$$

and it follows that

$$\tilde{A}'\tilde{B}' = (\tilde{S}\tilde{A}\tilde{S}^{-1})(\tilde{S}\tilde{B}\tilde{S}^{-1}) = \tilde{S}\tilde{A}\tilde{I}\tilde{B}\tilde{S}^{-1} = \tilde{S}\tilde{A}\tilde{B}\tilde{S}^{-1} = \tilde{S}\tilde{C}\tilde{S}^{-1} = \tilde{C}'.$$

Example 3.11
Show that the relation $\tilde{A}\mathbf{R} = \tilde{B}\mathbf{r}$ is invariant under a similarity transformation.

Solution: Since matrices $\tilde{A}$ and $\tilde{B}$ are subjected to the same similarity transformation, we have

$$\tilde{A}' = \tilde{S}\tilde{A}\tilde{S}^{-1}, \quad \tilde{B}' = \tilde{S}\tilde{B}\tilde{S}^{-1}$$

we also have

$$\mathbf{R}' = \tilde{S}\mathbf{R}, \quad \mathbf{r}' = \tilde{S}\mathbf{r}.$$

123

Then

$$\tilde{A}'\mathbf{R}' = (\tilde{S}\tilde{A}\tilde{S}^{-1})(S\mathbf{R}) = \tilde{S}\tilde{A}\mathbf{R} \quad \text{and} \quad \tilde{B}'\mathbf{r}' = (\tilde{S}\tilde{B}\tilde{S}^{-1})(S\mathbf{r}) = \tilde{S}\tilde{B}\mathbf{r}$$

thus

$$\tilde{A}'\mathbf{R}' = \tilde{B}'\mathbf{r}'.$$

We shall see in the following section that similarity transformations are very useful in diagonalization of a matrix, and that two similar matrices have the same eigenvalues.

The matrix eigenvalue problem

As we saw in preceding sections, a linear transformation generally carries a vector $\mathbf{X} = (x_1, x_2, \ldots, x_n)$ into a vector $\mathbf{Y} = (y_1, y_2, \ldots, y_n)$. However, there may exist certain non-zero vectors for which $\tilde{A}\mathbf{X}$ is just $\mathbf{X}$ multiplied by a constant λ

$$\tilde{A}\mathbf{X} = \lambda\mathbf{X}. \tag{3.53}$$

That is, the transformation represented by the matrix (operator) $\tilde{A}$ just multiplies the vector $\mathbf{X}$ by a number λ. Such a vector is called an eigenvector of the matrix $\tilde{A}$, and λ is called an eigenvalue (German: *eigenwert*) or characteristic value of the matrix $\tilde{A}$. The eigenvector is said to 'belong' (or correspond) to the eigenvalue. And the set of the eigenvalues of a matrix (an operator) is called its eigenvalue spectrum.

The problem of finding the eigenvalues and eigenvectors of a matrix is called an eigenvalue problem. We encounter problems of this type in all branches of physics, classical or quantum. Various methods for the approximate determination of eigenvalues have been developed, but here we only discuss the fundamental ideas and concepts that are important for the topics discussed in this book.

There are two parts to every eigenvalue problem. First, we compute the eigenvalue λ, given the matrix $\tilde{A}$. Then, we compute an eigenvector X for each previously computed eigenvalue λ.

Determination of eigenvalues and eigenvectors

We shall now demonstrate that any square matrix of order n has at least 1 and at most n distinct (real or complex) eigenvalues. To this purpose, let us rewrite the system of Eq. (3.53) as

$$(\tilde{A} - \lambda\tilde{I})X = 0. \tag{3.54}$$

This matrix equation really consists of n homogeneous linear equations in the n unknown elements x_i of X:

$$\left.\begin{array}{l}(a_{11} - \lambda)x_1 + a_{12}x_2 + \cdots + a_{1n}x_n = 0 \\ a_{21}x_1 + (a_{22} - \lambda)x_2 + \cdots + a_{2n}x_n = 0 \\ \qquad \cdots \\ a_{n1}x_1 + a_{n2}x_2 + \cdots + (a_{nn} - \lambda)x_n = 0\end{array}\right\} \qquad (3.55)$$

In order to have a non-zero solution, we recall that the determinant of the coefficients must be zero; that is,

$$\det(\tilde{A} - \lambda \tilde{I}) = \begin{vmatrix} a_{11} - \lambda & a_{12} & \cdots & a_{1n} \\ a_{21} & a_{22} - \lambda & \cdots & a_{2n} \\ \vdots & \vdots & & \vdots \\ a_{n1} & a_{n2} & \cdots & a_{nn} - \lambda \end{vmatrix} = 0. \qquad (3.56)$$

The expansion of the determinant gives an nth order polynomial equation in λ, and we write this as

$$c_0 \lambda^n + c_1 \lambda^{n-1} + c_2 \lambda^{n-2} + \cdots + c_{n-1}\lambda + c_n = 0, \qquad (3.57)$$

where the coefficients c_i are functions of the elements a_{jk} of $\tilde{A}$. Eq. (3.56) or (3.57) is called the characteristic equation corresponding to the matrix $\tilde{A}$. We have thus obtained a very important result: the eigenvalues of a square matrix $\tilde{A}$ are the roots of the corresponding characteristic equation (3.56) or (3.57).

Some of the coefficients c_i can be readily determined; by an inspection of Eq. (3.56) we find

$$c_0 = (-1)^n, \quad c_1 = (-1)^{n-1}(a_{11} + a_{22} + \cdots + a_{nn}), \quad c_n = \det \tilde{A}. \qquad (3.58)$$

Now let us rewrite the characteristic polynomial in terms of its n roots $\lambda_1, \lambda_2, \ldots, \lambda_n$

$$c_0 \lambda^n + c_1 \lambda^{n-1} + c_2 \lambda^{n-2} + \cdots + c_{n-1}\lambda + c_n = (\lambda_1 - \lambda)(\lambda_2 - \lambda)\cdots(\lambda_n - \lambda),$$

then we see that

$$c_1 = (-1)^{n-1}(\lambda_1 + \lambda_2 + \cdots + \lambda_n), \quad c_n = \lambda_1 \lambda_2 \cdots \lambda_n. \qquad (3.59)$$

Comparing this with Eq. (3.58), we obtain the following two important results on the eigenvalues of a matrix:

(1) The sum of the eigenvalues equals the trace (spur) of the matrix:

$$\lambda_1 + \lambda_2 + \cdots + \lambda_n = a_{11} + a_{22} + \cdots + a_{nn} \equiv \operatorname{Tr} \tilde{A}. \qquad (3.60)$$

(2) The product of the eigenvalues equals the determinant of the matrix:

$$\lambda_1 \lambda_2 \cdots \lambda_n = \det \tilde{A}. \qquad (3.61)$$

Once the eigenvalues have been found, corresponding eigenvectors can be found from the system (3.55). Since the system is homogeneous, if X is an eigenvector of $\tilde{A}$, then kX, where k is any constant (not zero), is also an eigenvector of $\tilde{A}$ corresponding to the same eigenvalue. It is very easy to show this. Since $\tilde{A}X = \lambda X$, multiplying by an arbitrary constant k will give $k\tilde{A}X = k\lambda X$. Now $k\tilde{A} = \tilde{A}k$ (every matrix commutes with a scalar), so we have $\tilde{A}(kX) = \lambda(kX)$, showing that kX is also an eigenvector of $\tilde{A}$ with the same eigenvalue λ. But kX is linearly dependent on X, and if we were to count all such eigenvectors separately, we would have an infinite number of them. Such eigenvectors are therefore not counted separately.

A matrix of order n does not necessarily have n linearly independent eigenvectors; some of them may be repeated. (This will happen when the characteristic polynomial has two or more identical roots.) If an eigenvalue occurs m times, m is called the multiplicity of the eigenvalue. The matrix has at most m linearly independent eigenvectors all corresponding to the same eigenvalue. Such linearly independent eigenvectors having the same eigenvalue are said to be degenerate eigenvectors; in this case, m-fold degenerate. We will deal only with those matrices that have n linearly independent eigenvectors and they are diagonalizable matrices.

Example 3.12

Find (a) the eigenvalues and (b) the eigenvectors of the matrix

$$\tilde{A} = \begin{pmatrix} 5 & 4 \\ 1 & 2 \end{pmatrix}.$$

Solution: (a) The eigenvalues: The characteristic equation is

$$\det(\tilde{A} - \lambda\tilde{I}) = \begin{vmatrix} 5-\lambda & 4 \\ 1 & 2-\lambda \end{vmatrix} = \lambda^2 - 7\lambda + 6 = 0$$

which has two roots

$$\lambda_1 = 6 \quad \text{and} \quad \lambda_2 = 1.$$

(b) The eigenvectors: For $\lambda = \lambda_1$ the system (3.55) assumes the form

$$-x_1 + 4x_2 = 0,$$

$$x_1 - 4x_2 = 0.$$

Thus $x_1 = 4x_2$, and

$$X_1 = \begin{pmatrix} 4 \\ 1 \end{pmatrix}$$

is an eigenvector of $\tilde{A}$ corresponding to $\lambda_1 = 6$. In the same way we find the eigenvector corresponding to $\lambda_2 = 1$:

$$X_2 = \begin{pmatrix} 1 \\ -1 \end{pmatrix}.$$

Example 3.13
If $\tilde{A}$ is a non-singular matrix, show that the eigenvalues of $\tilde{A}^{-1}$ are the reciprocals of those of $\tilde{A}$ and every eigenvector of $\tilde{A}$ is also an eigenvector of $\tilde{A}^{-1}$.

Solution: Let λ be an eigenvalue of $\tilde{A}$ corresponding to the eigenvector X, so that

$$\tilde{A}X = \lambda X.$$

Since $\tilde{A}^{-1}$ exists, multiply the above equation from the left by $\tilde{A}^{-1}$

$$\tilde{A}^{-1}\tilde{A}X = \tilde{A}^{-1}\lambda X \Rightarrow X = \lambda \tilde{A}^{-1}X.$$

Since $\tilde{A}$ is non-singular, λ must be non-zero. Now dividing the above equation by λ, we have

$$\tilde{A}^{-1}X = (1/\lambda)X.$$

Since this is true for every value of $\tilde{A}$, the results follows.

Example 3.14
Show that all the eigenvalues of a unitary matrix have unit magnitude.

Solution: Let $\tilde{U}$ be a unitary matrix and X an eigenvector of $\tilde{U}$ with the eigenvalue λ, so that

$$\tilde{U}X = \lambda X.$$

Taking the hermitian conjugate of both sides, we have

$$X^{\dagger}\tilde{U}^{\dagger} = \lambda * X^{\dagger}.$$

Multiplying the first equation from the left by the second equation, we obtain

$$X^{\dagger}\tilde{U}^{\dagger}\tilde{U}X = \lambda\lambda * X^{\dagger}X.$$

Since $\tilde{U}$ is unitary, $\tilde{U}^{\dagger}\tilde{U} = \tilde{I}$, so that the last equation reduces to

$$X^{\dagger}X(|\lambda|^2 - 1) = 0.$$

Now $X^{\dagger}X$ is the square of the norm of X and hence cannot vanish unless X is a null vector and so we must have $|\lambda|^2 = 1$ or $|\lambda| = 1$, proving the desired result.

Example 3.15

Show that similar matrices have the same characteristic polynomial and hence the same eigenvalues. (Another way of stating this is to say that the eigenvalues of a matrix are invariant under similarity transformations.)

Solution: Let $\tilde{A}$ and $\tilde{B}$ be similar matrices. Thus there exists a third matrix $\tilde{S}$ such that $\tilde{B} = \tilde{S}^{-1}\tilde{A}\tilde{S}$. Substituting this into the characteristic polynomial of matrix $\tilde{B}$ which is $|\tilde{B}\lambda - \tilde{I}|$, we obtain

$$|\tilde{B} - \lambda I| = |\tilde{S}^{-1}\tilde{A}\tilde{S} - \lambda\tilde{I}| = |\tilde{S}^{-1}(\tilde{A} - \lambda\tilde{I})\tilde{S}|.$$

Using the properties of determinants, we have

$$|\tilde{S}^{-1}(\tilde{A} - \lambda\tilde{I})\tilde{S}| = |\tilde{S}^{-1}||\tilde{A} - \lambda\tilde{I}||\tilde{S}|.$$

Then it follows that

$$|\tilde{B} - \lambda\tilde{I}| = |\tilde{S}^{-1}(\tilde{A} - \lambda\tilde{I})\tilde{S}| = |\tilde{S}^{-1}||\tilde{A} - \lambda\tilde{I}||\tilde{S}| = |\tilde{A} - \lambda\tilde{I}|,$$

which shows that the characteristic polynomials of $\tilde{A}$ and $\tilde{B}$ are the same; their eigenvalues will also be identical.

Eigenvalues and eigenvectors of hermitian matrices

In quantum mechanics complex variables are unavoidable because of the form of the Schrödinger equation. And all quantum observables are represented by hermitian operators. So physicists are almost always dealing with adjoint matrices, hermitian matrices, and unitary matrices. Why are physicists interested in hermitian matrices? Because they have the following properties: (1) the eigenvalues of a hermitian matrix are real, and (2) its eigenvectors corresponding to distinct eigenvalues are orthogonal, so they can be used as basis vectors. We now proceed to prove these important properties.

(1) the eigenvalues of a hermitian matrix are real.

Let $\tilde{H}$ be a hermitian matrix and X a non-trivial eigenvector corresponding to the eigenvalue λ, so that

$$\tilde{H}X = \lambda X. \tag{3.62}$$

Taking the hermitian conjugate and note that $\tilde{H}^{\dagger} = \tilde{H}$, we have

$$X^{\dagger}\tilde{H} = \lambda^{*}X^{\dagger}. \tag{3.63}$$

Multiplying (3.62) from the left by $X^{\dagger}$, and (3.63) from the right by $X^{\dagger}$, and then subtracting, we get

$$(\lambda - \lambda^{*})X^{\dagger}X = 0. \tag{3.64}$$

Now, since $X^{\dagger}X$ cannot be zero, it follows that $\lambda = \lambda^{*}$, or that λ is real.

(2) The eigenvectors corresponding to distinct eigenvalues are orthogonal.
Let X_1 and X_2 be eigenvectors of $\tilde{H}$ corresponding to the distinct eigenvalues λ_1 and λ_2, respectively, so that

$$\tilde{H}X_1 = \lambda_1 X_1, \tag{3.65}$$

$$\tilde{H}X_2 = \lambda_2 X_2. \tag{3.66}$$

Taking the hermitian conjugate of (3.66) and noting that $\lambda^* = \lambda$, we have

$$X_2^\dagger \tilde{H} = \lambda_2 X_2^\dagger. \tag{3.67}$$

Multiplying (3.65) from the left by $X_2^\dagger$ and (3.67) from the right by X_1, then subtracting, we obtain

$$(\lambda_1 - \lambda_2)X_2^\dagger + X_1 = 0. \tag{3.68}$$

Since $\lambda_1 = \lambda_2$, it follows that $X_2^\dagger X_1 = 0$ or that X_1 and X_2 are orthogonal.

If X is an eigenvector of $\tilde{H}$, any multiple of X, λX, is also an eigenvector of $\tilde{H}$. Thus we can normalize the eigenvector X with a properly chosen scalar λ. This means that the eigenvectors of $\tilde{H}$ corresponding to distinct eigenvalues are orthonormal. Just as the three orthogonal unit coordinate vectors $\hat{e}_1, \hat{e}_2,$ and $\hat{e}_3$ form the basis of a three-dimensional vector space, the orthonormal eigenvectors of $\tilde{H}$ may serve as a basis for a function space.

Diagonalization of a matrix

Let $\tilde{A} = (a_{ij})$ be a square matrix of order n, which has n linearly independent eigenvectors X_i with the corresponding eigenvalues λ_i: $\tilde{A}X_i = \lambda_i X_i$. If we denote the eigenvectors X_i by column vectors with elements $x_{1i}, x_{2i}, \ldots, x_{ni}$, then the eigenvalue equation can be written in matrix form:

$$\begin{pmatrix} a_{11} & a_{12} & \cdots & a_{1n} \\ a_{21} & a_{22} & \cdots & a_{2n} \\ \vdots & \vdots & & \vdots \\ a_{n1} & a_{n2} & \cdots & a_{nn} \end{pmatrix} \begin{pmatrix} x_{1i} \\ x_{2i} \\ \vdots \\ x_{ni} \end{pmatrix} = \lambda_i \begin{pmatrix} x_{1i} \\ x_{2i} \\ \vdots \\ x_{ni} \end{pmatrix}. \tag{3.69}$$

From the above matrix equation we obtain

$$\sum_{k=1}^{n} a_{jk}x_{ki} = \lambda_i x_{ji}. \tag{3.69b}$$

Now we want to diagonalize $\tilde{A}$. To this purpose, we can follow these steps. We first form a matrix $\tilde{S}$ of order $n \times n$ whose columns are the vector X_i, that is,

$$
\tilde{S} = \begin{pmatrix} x_{11} & \cdots & x_{1i} & \cdots & x_{1n} \\ x_{21} & \cdots & x_{2i} & \cdots & x_{2n} \\ \vdots & & \vdots & & \vdots \\ x_{n1} & \cdots & x_{ni} & \cdots & x_{nn} \end{pmatrix}, \quad (\tilde{S})_{ij} = x_{ij}. \tag{3.70}
$$

Since the vectors X_i are linear independent, $\tilde{S}$ is non-singular and $\tilde{S}^{-1}$ exists. We then form a matrix $\tilde{S}^{-1}\tilde{A}\tilde{S}$; this is a diagonal matrix whose diagonal elements are the eigenvalues of $\tilde{A}$.

To show this, we first define a diagonal matrix $\tilde{B}$ whose diagonal elements are λ_i $(i = 1, 2, \ldots, n)$:

$$
\tilde{B} = \begin{pmatrix} \lambda_1 & & & \\ & \lambda_2 & & \\ & & \ddots & \\ & & & \lambda_n \end{pmatrix}, \tag{3.71}
$$

and we then demonstrate that

$$
\tilde{S}^{-1}\tilde{A}\tilde{S} = \tilde{B}. \tag{3.72a}
$$

Eq. (3.72a) can be rewritten by multiplying it from the left by $\tilde{S}$ as

$$
\tilde{A}\tilde{S} = \tilde{S}\tilde{B}. \tag{3.72b}
$$

Consider the left hand side first. Taking the jith element, we obtain

$$
(\tilde{A}\tilde{S})_{ji} = \sum_{k=1}^{n} (\tilde{A})_{jk}(\tilde{S})_{ki} = \sum_{k=1}^{n} a_{jk}x_{ki}. \tag{3.73a}
$$

Similarly, the jith element of the right hand side is

$$
(\tilde{S}\tilde{B})_{ji} = \sum_{k=1}^{n} (\tilde{S})_{jk}(\tilde{B})_{ki} = \sum_{k=1}^{n} x_{jk}\lambda_i\delta_{ki} = \lambda_i x_{ji}. \tag{3.73b}
$$

Eqs. (3.73a) and (3.73b) clearly show the validity of Eq. (3.72a).

It is important to note that the matrix $\tilde{S}$ that is able to diagonalize matrix $\tilde{A}$ is not unique. This is because we could arrange the eigenvectors $X_1, X_2, \ldots, X_n$ in any order to construct $\tilde{S}$.

We summarize the procedure for diagonalizing a diagonalizable $n \times n$ matrix $\tilde{A}$:

Step 1. Find n linearly independent eigenvectors of $\tilde{A}$, $X_1, X_2, \ldots, X_n$.
Step 2. Form the matrix $\tilde{S}$ having $X_1, X_2, \ldots, X_n$ as its column vectors.
Step 3. Find the inverse of $\tilde{S}$, $\tilde{S}^{-1}$.
Step 4. The matrix $\tilde{S}^{-1}\tilde{A}\tilde{S}$ will then be diagonal with $\lambda_1, \lambda_2, \ldots, \lambda_n$ as its successive diagonal elements, where λ_i is the eigenvalue corresponding to X_i.

Example 3.16
Find a matrix $\tilde{S}$ that diagonalizes

$$\tilde{A} = \begin{pmatrix} 3 & -2 & 0 \\ -2 & 3 & 0 \\ 0 & 0 & 5 \end{pmatrix}.$$

Solution: We have first to find the eigenvalues and the corresponding eigenvectors of matrix $\tilde{A}$. The characteristic equation of $\tilde{A}$ is

$$\begin{vmatrix} 3-\lambda & -2 & 0 \\ -2 & 3-\lambda & 0 \\ 0 & 0 & 5-\lambda \end{vmatrix} = (\lambda-1)(\lambda-5)^2 = 0,$$

so that the eigenvalues of $\tilde{A}$ are $\lambda = 1$ and $\lambda = 5$.

By definition

$$\tilde{X} = \begin{pmatrix} x_1 \\ x_2 \\ x_3 \end{pmatrix}$$

is an eigenvector of $\tilde{A}$ corresponding to λ if and only if $\tilde{X}$ is a non-trivial solution of $(\lambda \tilde{I} - \tilde{A})\tilde{X} = 0$, that is, of

$$\begin{pmatrix} \lambda-3 & 2 & 0 \\ 2 & \lambda-3 & 0 \\ 0 & 0 & \lambda-5 \end{pmatrix} \begin{pmatrix} x_1 \\ x_2 \\ x_3 \end{pmatrix} = \begin{pmatrix} 0 \\ 0 \\ 0 \end{pmatrix}.$$

If $\lambda = 5$ the above equation becomes

$$\begin{pmatrix} 2 & 2 & 0 \\ 2 & 2 & 0 \\ 0 & 0 & 0 \end{pmatrix} \begin{pmatrix} x_1 \\ x_2 \\ x_3 \end{pmatrix} = \begin{pmatrix} 0 \\ 0 \\ 0 \end{pmatrix} \quad \text{or} \quad \begin{pmatrix} 2x_1 + 2x_2 + 0x_3 \\ 2x_1 + 2x_2 + 0x_3 \\ 0x_1 + 0x_2 + 0x_3 \end{pmatrix} = \begin{pmatrix} 0 \\ 0 \\ 0 \end{pmatrix}.$$

Solving this system yields

$$x_1 = -s, \quad x_2 = s, \quad x_3 = t,$$

where s and t are arbitrary values. Thus the eigenvectors of $\tilde{A}$ corresponding to $\lambda = 5$ are the non-zero vectors of the form

$$\tilde{X} = \begin{pmatrix} -s \\ s \\ t \end{pmatrix} = \begin{pmatrix} -s \\ s \\ 0 \end{pmatrix} + \begin{pmatrix} 0 \\ 0 \\ t \end{pmatrix} = s\begin{pmatrix} -1 \\ 1 \\ 0 \end{pmatrix} + t\begin{pmatrix} 0 \\ 0 \\ 1 \end{pmatrix}.$$

Since

$$\begin{pmatrix} -1 \\ 1 \\ 0 \end{pmatrix} \quad \text{and} \quad \begin{pmatrix} 0 \\ 0 \\ 1 \end{pmatrix}$$

are linearly independent, they are the eigenvectors corresponding to $\lambda = 5$.

For $\lambda = 1$, we have

$$\begin{pmatrix} -2 & 2 & 0 \\ 2 & -2 & 0 \\ 0 & 0 & -4 \end{pmatrix} \begin{pmatrix} x_1 \\ x_2 \\ x_3 \end{pmatrix} = \begin{pmatrix} 0 \\ 0 \\ 0 \end{pmatrix} \quad \text{or} \quad \begin{pmatrix} -2x_1 + 2x_2 + 0x_3 \\ 2x_1 - 2x_2 + 0x_3 \\ 0x_1 + 0x_2 - 4x_3 \end{pmatrix} = \begin{pmatrix} 0 \\ 0 \\ 0 \end{pmatrix}.$$

Solving this system yields

$$x_1 = t, \quad x_2 = t, \quad x_3 = 0,$$

where t is arbitrary. Thus the eigenvectors corresponding to $\lambda = 1$ are non-zero vectors of the form

$$\tilde{X} = \begin{pmatrix} t \\ t \\ 0 \end{pmatrix} = t \begin{pmatrix} 1 \\ 1 \\ 0 \end{pmatrix}.$$

It is easy to check that the three eigenvectors

$$\tilde{X}_1 = \begin{pmatrix} -1 \\ 1 \\ 0 \end{pmatrix}, \quad \tilde{X}_2 = \begin{pmatrix} 0 \\ 0 \\ 1 \end{pmatrix}, \quad \tilde{X}_3 = \begin{pmatrix} 1 \\ 1 \\ 0 \end{pmatrix},$$

are linearly independent. We now form the matrix $\tilde{S}$ that has $\tilde{X}_1$, $\tilde{X}_2$, and $\tilde{X}_3$ as its column vectors:

$$\tilde{S} = \begin{pmatrix} -1 & 0 & 1 \\ 1 & 0 & 1 \\ 0 & 1 & 0 \end{pmatrix}.$$

The matrix $\tilde{S}^{-1} \tilde{A} \tilde{S}$ is diagonal:

$$\tilde{S}^{-1} \tilde{A} \tilde{S} = \begin{pmatrix} -1/2 & 1/2 & 0 \\ 0 & 0 & 1 \\ 1/2 & 1/2 & 0 \end{pmatrix} \begin{pmatrix} 3 & -2 & 0 \\ -2 & 3 & 0 \\ 0 & 0 & 5 \end{pmatrix} \begin{pmatrix} -1 & 0 & 1 \\ 1 & 0 & 1 \\ 0 & 1 & 0 \end{pmatrix} = \begin{pmatrix} 5 & 0 & 0 \\ 0 & 5 & 0 \\ 0 & 0 & 1 \end{pmatrix}.$$

There is no preferred order for the columns of $\tilde{S}$. If had we written

$$\tilde{S} = \begin{pmatrix} -1 & 1 & 0 \\ 1 & 1 & 0 \\ 0 & 0 & 1 \end{pmatrix}$$

then we would have obtained (verify)

$$\tilde{S}^{-1}\tilde{A}\tilde{S} = \begin{pmatrix} 5 & 0 & 0 \\ 0 & 1 & 0 \\ 0 & 0 & 1 \end{pmatrix}.$$

Example 3.17
Show that the matrix

$$\tilde{A} = \begin{pmatrix} -3 & 2 \\ -2 & 1 \end{pmatrix}$$

is not diagonalizable.

Solution: The characteristic equation of $\tilde{A}$ is

$$\begin{vmatrix} \lambda + 3 & -2 \\ 2 & \lambda - 1 \end{vmatrix} = (\lambda + 1)^2 = 0.$$

Thus $\lambda = -1$ the only eigenvalue of $\tilde{A}$; the eigenvectors corresponding to $\lambda = -1$ are the solutions of

$$\begin{pmatrix} \lambda + 3 & -2 \\ 2 & \lambda - 1 \end{pmatrix} \begin{pmatrix} x_1 \\ x_2 \end{pmatrix} = \begin{pmatrix} 0 \\ 0 \end{pmatrix} \Rightarrow \begin{pmatrix} 2 & -2 \\ 2 & -2 \end{pmatrix} \begin{pmatrix} x_1 \\ x_2 \end{pmatrix} = \begin{pmatrix} 0 \\ 0 \end{pmatrix}$$

from which we have

$$2x_1 - 2x_2 = 0,$$
$$2x_1 - 2x_2 = 0.$$

The solutions to this system are $x_1 = t$, $x_2 = t$; hence the eigenvectors are of the form

$$\begin{pmatrix} t \\ t \end{pmatrix} = t\begin{pmatrix} 1 \\ 1 \end{pmatrix}.$$

A does not have two linearly independent eigenvectors, and is therefore not diagonalizable.

Eigenvectors of commuting matrices

There is a theorem on eigenvectors of commuting matrices that is of great importance in matrix algebra as well as in quantum mechanics. This theorem states that:

Two commuting matrices possess a common set of eigenvectors.

133

We now proceed to prove it. Let $\tilde{A}$ and $\tilde{B}$ be two square matrices, each of order n, which commute with each other, that is,

$$\tilde{A}\tilde{B} - \tilde{B}\tilde{A} = [\tilde{A}, \tilde{B}] = 0.$$

First, let λ be an eigenvalue of $\tilde{A}$ with multiplicity 1, corresponding to the eigenvector X, so that

$$\tilde{A}X = \lambda X. \tag{3.74}$$

Multiplying both sides from the left by $\tilde{B}$

$$\tilde{B}\tilde{A}X = \lambda \tilde{B}X.$$

Because $\tilde{B}\tilde{A} = \tilde{A}\tilde{B}$, we have

$$\tilde{A}(\tilde{B}X) = (\lambda \tilde{B}X).$$

Now $\tilde{B}$ is an $n \times n$ matrix and X is an $n \times 1$ vector; hence $\tilde{B}X$ is also an $n \times 1$ vector. The above equation shows that $\tilde{B}X$ is also an eigenvector of $\tilde{A}$ with the eigenvalue λ. Now X is a non-degenerate eigenvector of $\tilde{A}$, any other vector which is an eigenvector of $\tilde{A}$ with the same eigenvalue as that of X must be multiple of X. Accordingly

$$\tilde{B}X = \mu X,$$

where μ is a scalar. Thus we have proved that:

> If two matrices commute, every non-degenerate eigenvector of one is also an eigenvector of the other, and vice versa.

Next, let λ be an eigenvalue of $\tilde{A}$ with multiplicity k. So $\tilde{A}$ has k linearly independent eigenvectors, say $X_1, X_2, \ldots, X_k$, each corresponding to λ:

$$\tilde{A}X_i = \lambda X_i, \quad 1 \le i \le k.$$

Multiplying both sides from the left by $\tilde{B}$, we obtain

$$\tilde{A}(\tilde{B}X_i) = \lambda(\tilde{B}X_i),$$

which shows again that $\tilde{B}X$ is also an eigenvector of $\tilde{A}$ with the same eigenvalue λ.

Cayley–Hamilton theorem

The Cayley–Hamilton theorem is useful in evaluating the inverse of a square matrix. We now introduce it here. As given by Eq. (3.57), the characteristic equation associated with a square matrix $\tilde{A}$ of order n may be written as a polynomial

$$f(\lambda) = \sum_{i=0}^{n} c_i \lambda^{n-i} = 0,$$

where λ are the eigenvalues given by the characteristic determinant (3.56). If we replace λ in $f(\lambda)$ by the matrix $\tilde{A}$ so that

$$f(\tilde{A}) = \sum_{i=0}^{n} c_i \tilde{A}^{n-i}.$$

The Cayley–Hamilton theorem says that

$$f(\tilde{A}) = 0 \quad \text{or} \quad \sum_{i=0}^{n} c_i \tilde{A}^{n-i} = 0, \tag{3.75}$$

that is, the matrix $\tilde{A}$ satisfies its characteristic equation.

We now formally multiply Eq. (3.75) by $\tilde{A}^{-1}$ so that we obtain

$$\tilde{A}^{-1} f(\tilde{A}) = c_0 \tilde{A}^{n-1} + c_1 \tilde{A}^{n-2} + \cdots + c_{n-1} \tilde{I} + c_n \tilde{A}^{-1} = 0.$$

Solving for $\tilde{A}^{-1}$ gives

$$\tilde{A}^{-1} = -\frac{1}{c_n} \left[\sum_{i=0}^{n-1} c_i \tilde{A}^{n-1-i} \right]; \tag{3.76}$$

we can use this to find $\tilde{A}^{-1}$ (Problem 3.28).

Moment of inertia matrix

We shall see that physically diagonalization amounts to a simplification of the problem by a better choice of variable or coordinate system. As an illustrative example, we consider the moment of inertia matrix $\tilde{I}$ of a rotating rigid body (see Fig. 3.4). A rigid body can be considered to be a many-particle system, with the

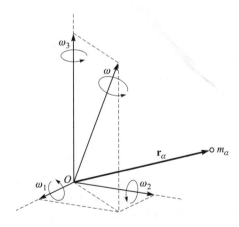

Figure 3.4. A rotating rigid body.

distance between any particle pair constant at all times. Then its angular momentum about the origin O of the coordinate system is

$$\mathbf{L} = \sum_\alpha m_\alpha \mathbf{r}_\alpha \times \mathbf{v}_\alpha = \sum_\alpha m_\alpha \mathbf{r}_\alpha \times (\boldsymbol{\omega} \times \mathbf{r}_\alpha)$$

where the subscript α refers to mass m_a located at $\mathbf{r}_\alpha = (x_{\alpha 1}, x_{\alpha 2}, x_{\alpha 3})$, and $\boldsymbol{\omega}$ the angular velocity of the rigid body.

Expanding the vector triple product by using the vector identity

$$\mathbf{A} \times (\mathbf{B} \times \mathbf{C}) = \mathbf{B}(\mathbf{A} \cdot \mathbf{C}) - \mathbf{C}(\mathbf{A} \cdot \mathbf{B}),$$

we obtain

$$\mathbf{L} = \sum_\alpha m_\alpha \lfloor r_\alpha^2 \boldsymbol{\omega} - \mathbf{r}_\alpha (\mathbf{r}_\alpha \cdot \boldsymbol{\omega}) \rfloor.$$

In terms of the components of the vectors $\mathbf{r}_\alpha$ and $\boldsymbol{\omega}$, the ith component of L_i is

$$L_i = \sum_\alpha m_\alpha \left[\omega_i \sum_{k=1}^{3} x_{\alpha,k}^2 - x_{\alpha,i} \sum_{j=1}^{3} x_{\alpha,j} \omega_j \right]$$

$$= \sum_j \omega_j \sum_\alpha m_\alpha \left[\delta_{ij} \sum_k x_{\alpha,k}^2 - x_{\alpha,i} x_{\alpha,j} \right] = \sum_j I_{ij} \omega_j$$

or

$$\tilde{L} = \tilde{I}\tilde{\omega}.$$

Both $\tilde{L}$ and $\tilde{\omega}$ are three-dimensional column vectors, while $\tilde{I}$ is a 3×3 matrix and is called the moment inertia matrix.

In general, the angular momentum vector $\mathbf{L}$ of a rigid body is not always parallel to its angular velocity $\boldsymbol{\omega}$ and $\tilde{I}$ is not a diagonal matrix. But we can orient the coordinate axes in space so that all the non-diagonal elements I_{ij} $(i \neq j)$ vanish. Such special directions are called the principal axes of inertia. If the angular velocity is along one of these principal axes, the angular momentum and the angular velocity will be parallel.

In many simple cases, especially when symmetry is present, the principal axes of inertia can be found by inspection.

Normal modes of vibrations

Another good illustrative example of the application of matrix methods in classical physics is the longitudinal vibrations of a classical model of a carbon dioxide molecule that has the chemical structure O–C–O. In particular, it provides a good example of the eigenvalues and eigenvectors of an asymmetric real matrix.

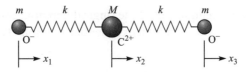

Figure 3.5. A linear symmetrical carbon dioxide molecule.

We can regard a carbon dioxide molecule as equivalent to a set of three particles jointed by elastic springs (Fig. 3.5). Clearly the system will vibrate in some manner in response to an external force. For simplicity we shall consider only longitudinal vibrations, and the interactions of the oxygen molecules with one another will be neglected, so we consider only nearest neighbor interaction. The Lagrangian function L for the system is

$$L = \tfrac{1}{2}m(\dot{x}_1^2 + \dot{x}_3^2) + \tfrac{1}{2}M\dot{x}_2^2 - \tfrac{1}{2}k(x_2 - x_1)^2 - \tfrac{1}{2}k(x_3 - x_2)^2;$$

substituting this into Lagrange's equations

$$\frac{d}{dt}\left(\frac{\partial L}{\partial \dot{x}_i}\right) - \frac{\partial L}{\partial x_i} = 0 \quad (i = 1, 2, 3),$$

we find the equations of motion to be

$$\ddot{x}_1 = -\frac{k}{m}(x_1 - x_2) = -\frac{k}{m}x_1 + \frac{k}{m}x_2,$$

$$\ddot{x}_2 = -\frac{k}{M}(x_2 - x_1) - \frac{k}{M}(x_2 - x_3) = \frac{k}{M}x_1 - \frac{2k}{M}x_2 + \frac{k}{M}x_3,$$

$$\ddot{x}_3 = \frac{k}{m}x_2 - \frac{k}{m}x_3,$$

where the dots denote time derivatives. If we define

$$\tilde{X} = \begin{pmatrix} x_1 \\ x_2 \\ x_3 \end{pmatrix}, \qquad \tilde{A} = \begin{pmatrix} -\dfrac{k}{m} & \dfrac{k}{m} & 0 \\[2mm] -\dfrac{k}{M} & -\dfrac{2k}{M} & \dfrac{k}{M} \\[2mm] 0 & \dfrac{k}{m} & -\dfrac{k}{m} \end{pmatrix}$$

and, furthermore, if we define the derivative of a matrix to be the matrix obtained by differentiating each matrix element, then the above system of differential equations can be written as

$$\ddot{\tilde{X}} = \tilde{A}\tilde{X}.$$

This matrix equation is reminiscent of the single differential equation $\ddot{x} = ax$, with a a constant. The latter always has an exponential solution. This suggests that we try

$$\tilde{X} = \tilde{C}e^{\omega t},$$

where ω is to be determined and

$$\tilde{C} = \begin{pmatrix} C_1 \\ C_2 \\ C_3 \end{pmatrix}$$

is an as yet unknown constant matrix. Substituting this into the above matrix equation, we obtain a matrix-eigenvalue equation

$$\tilde{A}\tilde{C} = \omega^2 \tilde{C}$$

or

$$\begin{pmatrix} -\dfrac{k}{m} & \dfrac{k}{m} & 0 \\[2mm] -\dfrac{k}{M} & -\dfrac{2k}{M} & \dfrac{k}{M} \\[2mm] 0 & \dfrac{k}{m} & -\dfrac{k}{m} \end{pmatrix} \begin{pmatrix} C_1 \\ C_2 \\ C_3 \end{pmatrix} = \omega^2 \begin{pmatrix} C_1 \\ C_2 \\ C_3 \end{pmatrix}. \tag{3.77}$$

Thus the possible values of ω are the square roots of the eigenvalues of the asymmetric matrix $\tilde{A}$ with the corresponding solutions being the eigenvectors of the matrix $\tilde{A}$. The secular equation is

$$\begin{vmatrix} -\dfrac{k}{m} - \omega^2 & \dfrac{k}{m} & 0 \\[2mm] -\dfrac{k}{M} & -\dfrac{2k}{M} - \omega^2 & \dfrac{k}{M} \\[2mm] 0 & \dfrac{k}{m} & -\dfrac{k}{m} - \omega^2 \end{vmatrix} = 0.$$

This leads to

$$\omega^2 \left(-\omega^2 + \frac{k}{m} \right) \left(-\omega^2 + \frac{k}{m} + \frac{2k}{M} \right) = 0.$$

The eigenvalues are

$$\omega^2 = 0, \quad \frac{k}{m}, \quad \text{and} \quad \frac{k}{m} + \frac{2k}{M},$$

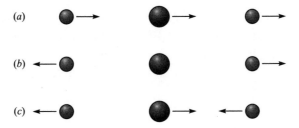

Figure 3.6. Longitudinal vibrations of a carbon dioxide molecule.

all real. The corresponding eigenvectors are determined by substituting the eigenvalues back into Eq. (3.77) one eigenvalue at a time:

(1) Setting $\omega^2 = 0$ in Eq. (3.77) we find that $C_1 = C_2 = C_3$. Thus this mode is not an oscillation at all, but is a pure translation of the system as a whole, no relative motion of the masses (Fig. 3.6(a)).

(2) Setting $\omega^2 = k/m$ in Eq. (3.77), we find $C_2 = 0$ and $C_3 = -C_1$. Thus the center mass M is stationary while the outer masses vibrate in opposite directions with the same amplitude (Fig. 3.6(b)).

(3) Setting $\omega^2 = k/m + 2k/M$ in Eq. (3.77), we find $C_1 = C_3$, and $C_2 = -2C_1(m/M)$. In this mode the two outer masses vibrate in unison and the center mass vibrates oppositely with different amplitude (Fig. 3.6(c)).

Direct product of matrices

Sometimes the direct product of matrices is useful. Given an $m \times m$ matrix $\tilde{A}$ and an $n \times n$ matrix $\tilde{B}$, the direct product of $\tilde{A}$ and $\tilde{B}$ is an $mn \times mn$ matrix, defined by

$$\tilde{C} = \tilde{A} \otimes \tilde{B} = \begin{pmatrix} a_{11}\tilde{B} & a_{12}\tilde{B} & \cdots & a_{1m}\tilde{B} \\ a_{21}\tilde{B} & a_{22}\tilde{B} & \cdots & a_{2m}\tilde{B} \\ \vdots & \vdots & & \vdots \\ a_{m1}\tilde{B} & a_{m2}\tilde{B} & \cdots & a_{mm}\tilde{B} \end{pmatrix}.$$

For example, if

$$\tilde{A} = \begin{pmatrix} a_{11} & a_{12} \\ a_{21} & a_{22} \end{pmatrix}, \quad \tilde{B} = \begin{pmatrix} b_{11} & b_{12} \\ b_{21} & b_{22} \end{pmatrix},$$

then

$$\tilde{A} \otimes \tilde{B} = \begin{pmatrix} a_{11}\tilde{B} & a_{12}\tilde{B} \\ a_{21}\tilde{B} & a_{22}\tilde{B} \end{pmatrix} = \begin{pmatrix} a_{11}b_{11} & a_{11}b_{12} & a_{12}b_{11} & a_{12}b_{12} \\ a_{11}b_{21} & a_{11}b_{22} & a_{12}b_{21} & a_{12}b_{22} \\ a_{21}b_{11} & a_{21}b_{12} & a_{22}b_{11} & a_{22}b_{12} \\ a_{21}b_{21} & a_{21}b_{22} & a_{22}b_{21} & a_{22}b_{22} \end{pmatrix}.$$

Problems

3.1 For the pairs $\tilde{A}$ and $\tilde{B}$ given below, find $\tilde{A} + \tilde{B}$, $\tilde{A}\tilde{B}$, and $\tilde{A}^2$:

$$\tilde{A} = \begin{pmatrix} 1 & 2 \\ 3 & 4 \end{pmatrix}, \qquad \tilde{B} = \begin{pmatrix} 5 & 6 \\ 7 & 8 \end{pmatrix}.$$

3.2 Show that an n-rowed diagonal matrix $\tilde{D}$

$$\tilde{D} = \begin{pmatrix} k & 0 & \cdots & 0 \\ 0 & k & \cdots & 0 \\ \vdots & \vdots & & \vdots \\ & & & k \end{pmatrix}$$

commutes with any n-rowed square matrix $\tilde{A}$: $\tilde{A}\tilde{D} = \tilde{D}\tilde{A} = k\tilde{A}$.

3.3 If $\tilde{A}$, $\tilde{B}$, and $\tilde{C}$ are any matrices such that the addition $\tilde{B} + \tilde{C}$ and the products $\tilde{A}\tilde{B}$ and $\tilde{A}\tilde{C}$ are defined, show that $\tilde{A}(\tilde{B} + \tilde{C}) = \tilde{A}\tilde{B} + \tilde{A}\tilde{C}$. That is, that matrix multiplication is distributive.

3.4 Given

$$\tilde{A} = \begin{pmatrix} 0 & 1 & 0 \\ 1 & 0 & 1 \\ 0 & 1 & 0 \end{pmatrix}, \qquad \tilde{B} = \begin{pmatrix} 1 & 0 & 0 \\ 0 & 1 & 0 \\ 0 & 0 & 1 \end{pmatrix}, \qquad \tilde{C} = \begin{pmatrix} 1 & 0 & 0 \\ 0 & 0 & 0 \\ 0 & 0 & -1 \end{pmatrix},$$

show that $[\tilde{A}, \tilde{B}] = 0$, and $[\tilde{B}, \tilde{C}] = 0$, but that $\tilde{A}$ does not commute with $\tilde{C}$.

3.5 Prove that $(\tilde{A} + \tilde{B})^{\mathrm{T}} = \tilde{A}^{\mathrm{T}} + \tilde{B}^{\mathrm{T}}$.

3.6 Given

$$\tilde{A} = \begin{pmatrix} 2 & -3 \\ 0 & 4 \end{pmatrix}, \qquad \tilde{B} = \begin{pmatrix} -5 & 2 \\ 2 & 1 \end{pmatrix}, \quad \text{and} \quad \tilde{C} = \begin{pmatrix} 0 & 1 & -2 \\ 3 & 0 & 4 \end{pmatrix}:$$

(a) Find $2\tilde{A} - 4\tilde{B}$, $2(\tilde{A} - 2\tilde{B})$
(b) Find $\tilde{A}^{\mathrm{T}}, \tilde{B}^{\mathrm{T}}, (\tilde{B}^{\mathrm{T}})^{\mathrm{T}}$
(c) Find $\tilde{C}^{\mathrm{T}}, (\tilde{C}^{\mathrm{T}})^{\mathrm{T}}$
(d) Is $\tilde{A} + \tilde{C}$ defined?
(e) Is $\tilde{C} + \tilde{C}^{\mathrm{T}}$ defined?
(f) Is $\tilde{A} + \tilde{A}^{\mathrm{T}}$ symmetric?

(g) Is $\tilde{A} - \tilde{A}^T$ antisymmetric?

3.7 Show that the matrix

$$\tilde{A} = \begin{pmatrix} 1 & 4 & 0 \\ 2 & 5 & 0 \\ 3 & 6 & 0 \end{pmatrix}$$

is not invertible.

3.8 Show that if $\tilde{A}$ and $\tilde{B}$ are invertible matrices of the same order, then $\tilde{A}\tilde{B}$ is invertible.

3.9 Given

$$\tilde{A} = \begin{pmatrix} 1 & 2 & 3 \\ 2 & 5 & 3 \\ 1 & 0 & 8 \end{pmatrix},$$

find $\tilde{A}^{-1}$ and check the answer by direct multiplication.

3.10 Prove that if $\tilde{A}$ is a non-singular matrix, then $\det(\tilde{A}^{-1}) = 1/\det(\tilde{A})$.

3.11 If $\tilde{A}$ is an invertible $n \times n$ matrix, show that $\tilde{A}X = 0$ has only the trivial solution.

3.12 Show, by computing a matrix inverse, that the solution to the following system is $x_1 = 4$, $x_2 = 1$:

$$x_1 - x_2 = 3,$$
$$x_1 + x_2 = 5.$$

3.13 Solve the system $\tilde{A}X = \tilde{B}$ if

$$\tilde{A} = \begin{pmatrix} 1 & 0 & 0 \\ 0 & 2 & 0 \\ 0 & 0 & 1 \end{pmatrix}, \qquad \tilde{B} = \begin{pmatrix} 1 \\ 2 \\ 3 \end{pmatrix}.$$

3.14 Given matrix $\tilde{A}$, find A*, A^T, and $A^\dagger$, where

$$\tilde{A} = \begin{pmatrix} 2+3i & 1-i & 5i & -3 \\ 1+i & 6-i & 1+3i & -1-2i \\ 5-6i & 3 & 0 & -4 \end{pmatrix}.$$

3.15 Show that:

(a) The matrix $\tilde{A}\tilde{A}^\dagger$, where $\tilde{A}$ is any matrix, is hermitian.

(b) $(\tilde{A}\tilde{B})^\dagger = \tilde{B}^\dagger \tilde{A}^\dagger$.

(c) If $\tilde{A}, \tilde{B}$ are hermitian, then $\tilde{A}\tilde{B} + \tilde{B}\tilde{A}$ is hermitian.

(d) If $\tilde{A}$ and $\tilde{B}$ are hermitian, then $i(\tilde{A}\tilde{B} - \tilde{B}\tilde{A})$ is hermitian.

3.16 Obtain the most general orthogonal matrix of order 2.
[Hint: use relations (3.34a) and (3.34b).]

3.17. Obtain the most general unitary matrix of order 2.

3.18 If $\tilde{A}\tilde{B} = 0$, show that one of these matrices must have zero determinant.

3.19 Given the Pauli spin matrices (which are very important in quantum mechanics)

$$\sigma_1 = \begin{pmatrix} 0 & 1 \\ 1 & 0 \end{pmatrix}, \quad \sigma_2 = \begin{pmatrix} 0 & -i \\ i & 0 \end{pmatrix}, \quad \sigma_3 = \begin{pmatrix} 1 & 0 \\ 0 & -1 \end{pmatrix},$$

(note that the subscripts x, y, and z are sometimes used instead of 1, 2, and 3). Show that

(a) they are hermitian,

(b) $\sigma_i^2 = \tilde{I}$, $i = 1, 2, 3$

(c) as a result of (a) and (b) they are also unitary, and

(d) $[\sigma_1, \sigma_2] = 2I\sigma_3$ et cycl.

Find the inverses of σ_1, σ_2, σ_3.

3.20 Use a rotation matrix to show that

$$\sin(\theta_1 + \theta_2) = \sin\theta_1 \cos\theta_2 + \sin\theta_2 \cos\theta_1.$$

3.21 Show that: $\mathrm{Tr}\,\tilde{A}\tilde{B} = \mathrm{Tr}\,\tilde{B}\tilde{A}$ and $\mathrm{Tr}\,\tilde{A}\tilde{B}\tilde{C} = \mathrm{Tr}\,\tilde{B}\tilde{C}\tilde{A} = \mathrm{Tr}\,\tilde{C}\tilde{A}\tilde{B}$.

3.22 Show that: (a) the trace and (b) the commutation relation between two matrices are invariant under similarity transformations.

3.23 Determine the eigenvalues and eigenvectors of the matrix

$$\tilde{A} = \begin{pmatrix} a & b \\ -b & a \end{pmatrix}.$$

Given

$$\tilde{A} = \begin{pmatrix} 5 & 7 & -5 \\ 0 & 4 & -1 \\ 2 & 8 & -3 \end{pmatrix},$$

find a matrix $\tilde{S}$ that diagonalizes $\tilde{A}$, and show that $\tilde{S}^{-1}\tilde{A}\tilde{S}$ is diagonal.

3.25 If $\tilde{A}$ and $\tilde{B}$ are square matrices of the same order, then $\det(\tilde{A}\tilde{B}) = \det(\tilde{A})\det(\tilde{B})$. Verify this theorem if

$$\tilde{A} = \begin{pmatrix} 2 & -1 \\ 3 & 2 \end{pmatrix}, \quad \tilde{B} = \begin{pmatrix} 7 & 2 \\ -3 & 4 \end{pmatrix}.$$

3.26 Find a common set of eigenvectors for the two matrices

$$\tilde{A} = \begin{pmatrix} -1 & \sqrt{6} & \sqrt{2} \\ \sqrt{6} & 0 & \sqrt{3} \\ \sqrt{2} & \sqrt{3} & -2 \end{pmatrix}, \quad \tilde{B} = \begin{pmatrix} 10 & \sqrt{6} & -\sqrt{2} \\ \sqrt{6} & 9 & \sqrt{3} \\ -\sqrt{2} & \sqrt{3} & 11 \end{pmatrix}.$$

3.27 Show that two hermitian matrices can be made diagonal if and only if they commute.

3.28 Show the validity of the Cayley–Hamilton theorem by applying it to the matrix

$$\tilde{A} = \begin{pmatrix} 5 & 4 \\ 1 & 2 \end{pmatrix};$$

then use the Cayley–Hamilton theorem to find the inverse of the matrix $\tilde{A}$.

3.29 Given

$$\tilde{A} = \begin{pmatrix} 0 & 1 \\ 1 & 0 \end{pmatrix}, \qquad \tilde{B} = \begin{pmatrix} 0 & -i \\ i & 0 \end{pmatrix},$$

find the direct product of these matrices, and show that it does not commute.

4

Fourier series and integrals

Fourier series are infinite series of sines and cosines which are capable of representing almost any periodic function whether continuous or not. Periodic functions that occur in physics and engineering problems are often very complicated and it is desirable to represent them in terms of simple periodic functions. Therefore the study of Fourier series is a matter of great practical importance for physicists and engineers.

The first part of this chapter deals with Fourier series. Basic concepts, facts, and techniques in connection with Fourier series will be introduced and developed, along with illustrative examples. They are followed by Fourier integrals and Fourier transforms.

Periodic functions

If function $f(x)$ is defined for all x and there is some positive constant P such that

$$f(x + P) = f(x) \tag{4.1}$$

then we say that $f(x)$ is periodic with a period P (Fig. 4.1). From Eq. (4.1) we also

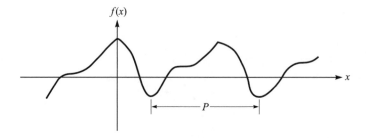

Figure 4.1.   A general periodic function.

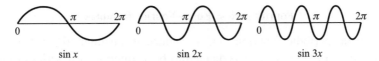

Figure 4.2.   Sine functions.

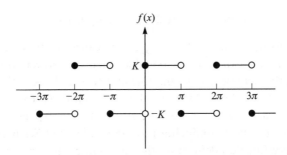

Figure 4.3.   A square wave function.

have, for all x and any integer n,

$$f(x + nP) = f(x).$$

That is, every periodic function has arbitrarily large periods and contains arbitrarily large numbers in its domain. We call P the fundamental (or least) period, or simply the period.

A periodic function need not be defined for all values of its independent variable. For example, $\tan x$ is undefined for the values $x = (\pi/2) + n\pi$. But $\tan x$ is a periodic function in its domain of definition, with π as its fundamental period: $\tan(x + \pi) = \tan x$.

Example 4.1
(a) The period of $\sin x$ is 2π, since $\sin(x + 2\pi)$, $\sin(x + 4\pi)$, $\sin(x + 6\pi), \ldots$ are all equal to $\sin x$, but 2π is the least value of P. And, as shown in Fig. 4.2, the period of $\sin nx$ is $2\pi/n$, where n is a positive integer.

(b) A constant function has any positive number as a period. Since $f(x) = c$ (const.) is defined for all real x, then, for every positive number P, $f(x + P) = c = f(x)$. Hence P is a period of f. Furthermore, f has no fundamental period.

(c)

$$f(x) = \begin{cases} K & \text{for} \quad 2n\pi x \leq (2n+1)\pi \\ -K & \text{for} \quad (2n+1)\pi \leq x < (2n+2)\pi \end{cases} \quad n = 0, \pm 1, \pm 2, \pm 3, \ldots$$

is periodic of period 2π (Fig. 4.3).

145

Fourier series; Euler–Fourier formulas

If the general periodic function $f(x)$ is defined in an interval $-\pi \leq x \leq \pi$, the Fourier series of $f(x)$ in $[-\pi, \pi]$ is defined to be a trigonometric series of the form

$$f(x) = \tfrac{1}{2}a_0 + a_1 \cos x + a_2 \cos 2x + \cdots + a_n \cos nx + \cdots$$

$$+ b_1 \sin x + b_2 \sin 2x + \cdots + b_n \sin nx + \cdots, \qquad (4.2)$$

where the numbers $a_0, a_1, a_2, \ldots, b_1, b_2, b_3, \ldots$ are called the Fourier coefficients of $f(x)$ in $[-\pi, \pi]$. If this expansion is possible, then our power to solve physical problems is greatly increased, since the sine and cosine terms in the series can be handled individually without difficulty. Joseph Fourier (1768–1830), a French mathematician, undertook the systematic study of such expansions. In 1807 he submitted a paper (on heat conduction) to the Academy of Sciences in Paris and claimed that every function defined on the closed interval $[-\pi, \pi]$ could be represented in the form of a series given by Eq. (4.2); he also provided integral formulas for the coefficients a_n and b_n. These integral formulas had been obtained earlier by Clairaut in 1757 and by Euler in 1777. However, Fourier opened a new avenue by claiming that these integral formulas are well defined even for very arbitrary functions and that the resulting coefficients are identical for different functions that are defined within the interval. Fourier's paper was rejected by the Academy on the grounds that it lacked mathematical rigor, because he did not examine the question of the convergence of the series.

The trigonometric series (4.2) is the only series which corresponds to $f(x)$. Questions concerning its convergence and, if it does, the conditions under which it converges to $f(x)$ are many and difficult. These problems were partially answered by Peter Gustave Lejeune Dirichlet (German mathematician, 1805–1859) and will be discussed briefly later.

Now let us assume that the series exists, converges, and may be integrated term by term. Multiplying both sides by $\cos mx$, then integrating the result from $-\pi$ to π, we have

$$\int_{-\pi}^{\pi} f(x) \cos mx \, dx = \frac{a_0}{2} \int_{-\pi}^{\pi} \cos mx \, dx + \sum_{n=1}^{\infty} a_n \int_{-\pi}^{\pi} \cos nx \cos mx \, dx$$

$$+ \sum_{n=1}^{\infty} b_n \int_{-\pi}^{\pi} \sin nx \cos mx \, dx. \qquad (4.3)$$

Now, using the following important properties of sines and cosines:

$$\int_{-\pi}^{\pi} \cos mx \, dx = \int_{-\pi}^{\pi} \sin mx \, dx = 0 \quad \text{if } m = 1, 2, 3, \ldots,$$

$$\int_{-\pi}^{\pi} \cos mx \cos nx \, dx = \int_{-\pi}^{\pi} \sin mx \sin nx \, dx = \begin{cases} 0 & \text{if } n \neq m, \\ \pi & \text{if } n = m, \end{cases}$$

$$\int_{-\pi}^{\pi} \sin mx \cos nx \, dx = 0, \quad \text{for all } m, n > 0,$$

we find that all terms on the right hand side of Eq. (4.3) except one vanish:

$$a_n = \frac{1}{\pi} \int_{-\pi}^{\pi} f(x) \cos nx \, dx, \quad n = \text{integers}; \tag{4.4a}$$

the expression for a_0 can be obtained from the general expression for a_n by setting $n = 0$.

Similarly, if Eq. (4.2) is multiplied through by $\sin mx$ and the result is integrated from $-\pi$ to π, all terms vanish save that involving the square of $\sin nx$, and so we have

$$b_n = \frac{1}{\pi} \int_{-\pi}^{\pi} f(x) \sin nx \, dx. \tag{4.4b}$$

Eqs. (4.4a) and (4.4b) are known as the Euler–Fourier formulas.

From the definition of a definite integral it follows that, if $f(x)$ is single-valued and continuous within the interval $[-\pi, \pi]$ or merely piecewise continuous (continuous except at a finite numbers of finite jumps in the interval), the integrals in Eqs. (4.4) exist and we may compute the Fourier coefficients of $f(x)$ by Eqs. (4.4). If there exists a finite discontinuity in $f(x)$ at the point x_0 (Fig. 4.1), the coefficients a_0, a_n, b_n are determined by integrating first to $x = x_0$ and then from x_0 to π, as

$$a_n = \frac{1}{\pi} \left[\int_{-\pi}^{x_0} f(x) \cos nx \, dx + \int_{x_0}^{\pi} f(x) \cos nx \, dx \right], \tag{4.5a}$$

$$b_n = \frac{1}{\pi} \left[\int_{-\pi}^{x_0} f(x) \sin nx \, dx + \int_{x_0}^{\pi} f(x) \sin nx \, dx \right]. \tag{4.5b}$$

This procedure may be extended to any finite number of discontinuities.

Example 4.2
Find the Fourier series which represents the function

$$f(x) = \begin{cases} -k & -\pi < x < 0 \\ +k & 0 < x < \pi \end{cases} \quad \text{and} \quad f(x + 2\pi) = f(x),$$

in the interval $-\pi \leq x \leq \pi$.

Solution: The Fourier coefficients are readily calculated:

$$a_n = \frac{1}{\pi}\left[\int_{-\pi}^{0}(-k)\cos nx\, dx + \int_{0}^{\pi}k\cos nx\, dx\right]$$

$$= \frac{1}{\pi}\left[-k\,\frac{\sin nx}{n}\bigg|_{-\pi}^{0} + k\,\frac{\sin nx}{n}\bigg|_{0}^{\pi}\right] = 0$$

$$b_n = \frac{1}{\pi}\left[\int_{-\pi}^{0}(-k)\sin nx\, dx + \int_{0}^{\pi}k\sin nx\, dx\right]$$

$$= \frac{1}{\pi}\left[k\,\frac{\cos nx}{n}\bigg|_{-\pi}^{0} - k\,\frac{\cos nx}{n}\bigg|_{0}^{\pi}\right] = \frac{2k}{n\pi}(1 - \cos n\pi)$$

Now $\cos n\pi = -1$ for odd n, and $\cos n\pi = 1$ for even n. Thus

$$b_1 = 4k/\pi, \quad b_2 = 0, \quad b_3 = 4k/3\pi, \quad b_4 = 0, \quad b_5 = 4k/5\pi, \ldots$$

and the corresponding. Fourier series is

$$\frac{4k}{\pi}\left(\sin x + \frac{1}{3}\sin 3x + \frac{1}{5}\sin 5x + \cdots\right).$$

For the special case $k = \pi/2$, the Fourier series becomes

$$2\sin x + \frac{2}{3}\sin 3x + \frac{2}{5}\sin 5x + \cdots.$$

The first two terms are shown in Fig. 4.4, the solid curve is their sum. We will see that as more and more terms in the Fourier series expansion are included, the sum more and more nearly approaches the shape of $f(x)$. This will be further demonstrated by next example.

Example 4.3
Find the Fourier series that represents the function defined by

$$f(t) = \begin{cases} 0, & -\pi < t < 0 \\ \sin t, & 0 < t < \pi \end{cases} \quad \text{in the interval } -\pi < t < \pi.$$

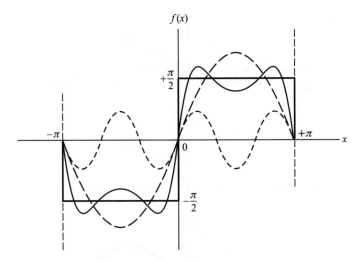

Figure 4.4.   The first two partial sums.

Solution:

$$a_n = \frac{1}{\pi}\left[\int_{-\pi}^{0} 0 \cdot \cos nt \, dt + \int_{0}^{\pi} \sin t \cos nt \, dt\right]$$

$$= -\frac{1}{2\pi}\left[\frac{\cos(1-n)t}{1-n} + \frac{\cos(1+n)t}{1+n}\right]\Bigg|_{0}^{\pi} = \frac{\cos n\pi + 1}{\pi(1-n)^2}, \qquad n \neq 1,$$

$$a_1 = \frac{1}{\pi}\int_{0}^{\pi} \sin t \cos t \, dt = \frac{1}{\pi}\frac{\sin^2 t}{2}\Bigg|_{0}^{\pi} = 0,$$

$$b_n = \frac{1}{\pi}\left[\int_{-\pi}^{0} 0 \cdot \sin nt \, dt + \int_{0}^{\pi} \sin t \sin nt \, dt\right]$$

$$= \frac{1}{2\pi}\left[\frac{\sin(1-n)t}{1-n} - \frac{\sin(1+n)t}{1+n}\right]_{0}^{\pi} = 0$$

$$b_1 = \frac{1}{\pi}\int_{0}^{\pi} \sin^2 t \, dt = \frac{1}{\pi}\left[\frac{t}{2} - \frac{\sin 2t}{4}\right]_{0}^{\pi} = \frac{1}{2}.$$

Accordingly the Fourier expansion of $f(t)$ in $[-\pi, \pi]$ may be written

$$f(t) = \frac{1}{\pi} + \frac{\sin t}{2} - \frac{2}{\pi}\left[\frac{\cos 2t}{3} + \frac{\cos 4t}{15} + \frac{\cos 6t}{35} + \frac{\cos 8t}{63} + \cdots\right].$$

The first three partial sums $S_n (n = 1, 2, 3)$ are shown in Fig. 4.5: $S_1 = 1/\pi$, $S_2 = 1/\pi + \sin t/2$, and $S_3 = 1/\pi + \sin(t)/2 - 2\cos(2t)/3$.

149

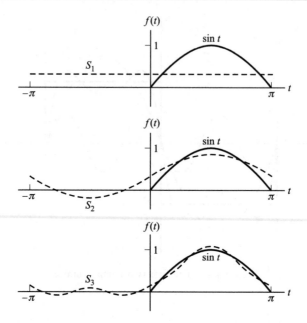

Figure 4.5.   The first three partial sums of the series.

Gibb's phenomena

From Figs. 4.4 and 4.5, two features of the Fourier expansion should be noted:

(*a*) at the points of the discontinuity, the series yields the mean value;

(*b*) in the region immediately adjacent to the points of discontinuity, the expansion overshoots the original function. This effect is known as the *Gibb's phenomena* and occurs in all order of approximation.

Convergence of Fourier series and Dirichlet conditions

The serious question of the convergence of Fourier series still remains: if we determine the Fourier coefficients a_n, b_n of a given function $f(x)$ from Eq. (4.4) and form the Fourier series given on the right hand side of Eq. (4.2), will it converge toward $f(x)$? This question was partially answered by Dirichlet. Here is a restatement of the results of his study, which is often called Dirichlet's theorem:

(1) If $f(x)$ is defined and single-valued except at a finite number of point in $[-\pi, \pi]$,

(2) if $f(x)$ is periodic outside $[-\pi, \pi]$ with period 2π (that is, $f(x + 2\pi) = f(x)$), and

(3) if $f(x)$ and $f'(x)$ are piecewise continuous in $[-\pi, \pi]$,

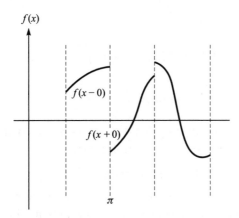

Figure 4.6. A piecewise continuous function.

then the series on the right hand side of Eq. (4.2), with coefficients a_n and b_n given by Eqs. (4.4), converges to

(i) $f(x)$, if x is a point of continuity, or

(ii) $\frac{1}{2}[f(x+0)+f(x-0)]$, if x is a point of discontinuity as shown in Fig. 4.6,

where $f(x+0)$ and $f(x-0)$ are the right and left hand limits of $f(x)$ at x and represent $\lim_{\varepsilon \to 0} f(x+\varepsilon)$ and $\lim_{\varepsilon \to 0} f(x-\varepsilon)$ respectively, where $\varepsilon > 0$.

The proof of Dirichlet's theorem is quite technical and is omitted in this treatment. The reader should remember that the Dirichlet conditions (1), (2), and (3) imposed on $f(x)$ are sufficient but not necessary. That is, if the above conditions are satisfied the convergence is guaranteed; but if they are not satisfied, the series may or may not converge. The Dirichlet conditions are generally satisfied in practice.

Half-range Fourier series

Unnecessary work in determining Fourier coefficients of a function can be avoided if the function is odd or even. A function $f(x)$ is called odd if $f(-x) = -f(x)$ and even if $f(x)\, f(-x) = f(x)$. It is easy to show that in the Fourier series corresponding to an odd function $f_o(x)$, only sine terms can be present in the series expansion in the interval $-\pi < x < \pi$, for

$$a_n = \frac{1}{\pi}\int_{-\pi}^{\pi} f_o(x)\cos nx\,dx = \frac{1}{\pi}\left[\int_{-\pi}^{0} f_o(x)\cos nx\,dx + \int_{0}^{\pi} f_o(x)\cos nx\,dx\right]$$

$$= \frac{1}{\pi}\left[-\int_{0}^{\pi} f_o(x)\cos nx\,dx + \int_{0}^{\pi} f_o(x)\cos nx\,dx\right] = 0 \qquad n = 0,1,2,\ldots, \quad (4.6a)$$

151

but

$$b_n = \frac{1}{\pi}\left[\int_{-\pi}^{0} f_0(x)\sin nx\, dx + \int_{0}^{\pi} f_0(x)\sin nx\, dx\right]$$

$$= \frac{2}{\pi}\int_{0}^{\pi} f_0(x)\sin nx\, dx \qquad n = 1, 2, 3, \dots . \tag{4.6b}$$

Here we have made use of the fact that $\cos(-nx) = \cos nx$ and $\sin(-nx) = -\sin nx$. Accordingly, the Fourier series becomes

$$f_0(x) = b_1 \sin x + b_2 \sin 2x + \cdots .$$

Similarly, in the Fourier series corresponding to an even function $f_e(x)$, only cosine terms (and possibly a constant) can be present. Because in this case, $f_e(x)\sin nx$ is an odd function and accordingly $b_n = 0$ and the a_n are given by

$$a_n = \frac{2}{\pi}\int_{0}^{\pi} f_e(x)\cos nx\, dx \qquad n = 0, 1, 2, \dots . \tag{4.7}$$

Note that the Fourier coefficients a_n and b_n, Eqs. (4.6) and (4.7) are computed in the interval $(0, \pi)$ which is *half* of the interval $(-\pi, \pi)$. Thus, the Fourier sine or cosine series in this case is often called a half-range Fourier series.

Any arbitrary function (neither even nor odd) can be expressed as a combination of $f_e(x)$ and $f_0(x)$ as

$$f(x) = \tfrac{1}{2}[f(x) + f(-x)] + \tfrac{1}{2}[f(x) - f(-x)] = f_e(x) + f_0(x).$$

When a half-range series corresponding to a given function is desired, the function is generally defined in the interval $(0, \pi)$ and then the function is specified as odd or even, so that it is clearly defined in the other half of the interval $(-\pi, 0)$.

Change of interval

A Fourier expansion is not restricted to such intervals as $-\pi < x < \pi$ and $0 < x < \pi$. In many problems the period of the function to be expanded may be some other interval, say $2L$. How then can the Fourier series developed above be applied to the representation of periodic functions of arbitrary period? The problem is not a difficult one, for basically all that is involved is to change the variable. Let

$$z = \frac{\pi}{L}x \tag{4.8a}$$

then

$$f(z) = f(\pi x/L) = F(x). \tag{4.8b}$$

Thus, if $f(z)$ is expanded in the interval $-\pi < z < \pi$, the coefficients being determined by expressions of the form of Eqs. (4.4a) and (4.4b), the coefficients for the expansion of $F(x)$ in the interval $-L < x < L$ may be obtained merely by substituting Eqs. (4.8) into these expressions. We have then

$$a_n = \frac{1}{L} \int_{-L}^{L} F(x) \cos \frac{n\pi}{L} x \, dx \qquad n = 0, 1, 2, 3, \ldots, \tag{4.9a}$$

$$b_n = \frac{1}{L} \int_{-L}^{L} F(x) \sin \frac{n\pi}{L} x \, dx, \qquad n = 1, 2, 3, \ldots. \tag{4.9b}$$

The possibility of having expanding functions in which the period is other than 2π increases the usefulness of Fourier expansion. As an example, consider the value of L, it is obvious that the larger the value of L, the larger the basic period of the function being expanded. As $L \to \infty$, the function would not be periodic at all. We will see later that in such cases the Fourier series becomes a Fourier integral.

Parseval's identity

Parseval's identity states that:

$$\frac{1}{2L} \int_{-L}^{L} [f(x)]^2 dx = \left(\frac{a_0}{2}\right)^2 + \frac{1}{2} \sum_{n=1}^{\infty} (a_n^2 + b_n^2), \tag{4.10}$$

if a_n and b_n are coefficients of the Fourier series of $f(x)$ and if $f(x)$ satisfies the Dirichlet conditions.

It is easy to prove this identity. Assuming that the Fourier series corresponding to $f(x)$ converges to $f(x)$

$$f(x) = \frac{a_0}{2} + \sum_{n=1}^{\infty} \left(a_n \cos \frac{n\pi x}{L} + b_n \sin \frac{n\pi x}{L}\right).$$

Multiplying by $f(x)$ and integrating term by term from $-L$ to L, we obtain

$$\int_{-L}^{L} [f(x)]^2 dx = \frac{a_0}{2} \int_{-L}^{L} f(x) dx$$

$$+ \sum_{n=1}^{\infty} \left\{ a_n \int_{-L}^{L} f(x) \cos \frac{n\pi x}{L} dx + b_n \int_{-L}^{L} f(x) \sin \frac{n\pi x}{L} dx \right\}$$

$$= \frac{a_0^2}{2} L + L \sum_{n=1}^{\infty} (a_n^2 + b_n^2), \tag{4.11}$$

where we have used the results

$$\int_{-L}^{L} f(x) \cos \frac{n\pi x}{L} dx = La_n, \quad \int_{-L}^{L} f(x) \sin \frac{n\pi x}{L} dx = Lb_n, \quad \int_{-L}^{L} f(x) dx = La_0.$$

The required result follows on dividing both sides of Eq. (4.11) by L.

Parseval's identity shows a relation between the average of the square of $f(x)$ and the coefficients in the Fourier series for $f(x)$:

the average of $\{f(x)\}^2$ is $\int_{-L}^{L} [f(x)]^2 dx / 2L$;
the average of $(a_0/2)$ is $(a_0/2)^2$;
the average of $(a_n \cos nx)$ is $a_n^2/2$;
the average of $(b_n \sin nx)$ is $b_n^2/2$.

Example 4.4

Expand $f(x) = x, 0 < x < 2$, in a half-range cosine series, then write Parseval's identity corresponding to this Fourier cosine series.

Solution: We first extend the definition of $f(x)$ to that of the even function of period 4 shown in Fig. 4.7. Then $2L = 4, L = 2$. Thus $b_n = 0$ and

$$a_n = \frac{2}{L} \int_0^L f(x) \cos \frac{n\pi x}{L} dx = \frac{2}{2} \int_0^2 f(x) \cos \frac{n\pi x}{2} dx$$

$$= \left[x \cdot \left(\frac{2}{n\pi} \sin \frac{n\pi x}{2} \right) - 1 \cdot \left(\frac{-4}{n^2\pi^2} \cos \frac{n\pi x}{2} \right) \right]_0^2$$

$$= \frac{-4}{n^2\pi^2} (\cos n\pi - 1) \qquad \text{if } n \neq 0.$$

If $n = 0$,

$$a_0 = \int_0^L x \, dx = 2.$$

Then

$$f(x) = 1 + \sum_{n=1}^{\infty} \frac{4}{n^2\pi^2} (\cos n\pi - 1) \cos \frac{n\pi x}{2}.$$

We now write Parseval's identity. We first compute the average of $[f(x)]^2$:

the average of $[f(x)]^2 = \frac{1}{2} \int_{-2}^{2} \{f(x)\}^2 dx = \frac{1}{2} \int_{-2}^{2} x^2 dx = \frac{8}{3}$,

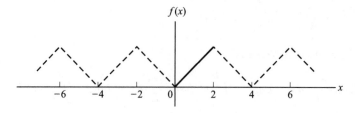

$f(x)$

$-6 \quad -4 \quad -2 \quad 0 \quad 2 \quad 4 \quad 6 \qquad x$

Figure 4.7.

154

then the average

$$\frac{a_0^2}{2} + \sum_{n=1}^{\infty} (a_n^2 + b_n^2) = \frac{(2)^2}{2} + \sum_{n=1}^{\infty} \frac{16}{n^4 \pi^4} (\cos n\pi - 1)^2.$$

Parseval's identity now becomes

$$\frac{8}{3} = 2 + \frac{64}{\pi^4} \left(\frac{1}{1^4} + \frac{1}{3^4} + \frac{1}{5^4} + \cdots \right),$$

or

$$\frac{1}{1^4} + \frac{1}{3^4} + \frac{1}{5^4} + \cdots = \frac{\pi^4}{96}$$

which shows that we can use Parseval's identity to find the sum of an infinite series. With the help of the above result, we can find the sum S of the following series:

$$\frac{1}{1^4} + \frac{1}{2^4} + \frac{1}{3^4} + \frac{1}{4^4} + \cdots + \frac{1}{n^4} + \cdots.$$

$$S = \frac{1}{1^4} + \frac{1}{2^4} + \frac{1}{3^4} + \frac{1}{4^4} + \cdots = \left(\frac{1}{1^4} + \frac{1}{3^4} + \frac{1}{5^4} + \cdots \right) + \left(\frac{1}{2^4} + \frac{1}{4^4} + \frac{1}{6^4} + \cdots \right)$$

$$= \left(\frac{1}{1^4} + \frac{1}{3^4} + \frac{1}{5^4} + \cdots \right) + \frac{1}{2^4} \left(\frac{1}{1^4} + \frac{1}{2^4} + \frac{1}{3^4} + \frac{1}{4^4} + \cdots \right)$$

$$= \frac{\pi^4}{96} + \frac{S}{16}$$

from which we find $S = \pi^4/90$.

Alternative forms of Fourier series

Up to this point the Fourier series of a function has been written as an infinite series of sines and cosines, Eq. (4.2):

$$f(x) = \frac{a_0}{2} + \sum_{n=1}^{\infty} \left(a_n \cos \frac{n\pi x}{L} + b_n \sin \frac{n\pi x}{L} \right).$$

This can be converted into other forms. In this section, we just discuss two alternative forms. Let us first write, with $\pi/L = \alpha$

$$a_n \cos n\alpha x + b_n \sin n\alpha x = \sqrt{a_n^2 + b_n^2} \left(\frac{a_n}{\sqrt{a_n^2 + b_n^2}} \cos n\alpha x + \frac{b_n}{\sqrt{a_n^2 + b_n^2}} \sin n\alpha x \right).$$

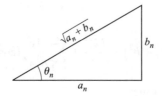

Figure 4.8.

Now let (see Fig. 4.8)

$$\cos \theta_n = \frac{a_n}{\sqrt{a_n^2 + b_n^2}}, \quad \sin \theta_n = \frac{b_n}{\sqrt{a_n^2 + b_n^2}}, \quad \text{so} \quad \theta_n = \tan^{-1}\left(\frac{b_n}{a_n}\right),$$

$$C_n = \sqrt{a_n^2 + b_n^2}, \quad C_0 = \tfrac{1}{2}a_0,$$

then we have the trigonometric identity

$$a_n \cos n\alpha x + b_n \sin n\alpha x = C_n \cos(n\alpha x - \theta_n),$$

and accordingly the Fourier series becomes

$$f(x) = C_0 + \sum_{n=1}^{\infty} C_n \cos(n\alpha x - \theta_n). \tag{4.12}$$

In this new form, the Fourier series represents a periodic function as a sum of sinusoidal components having different frequencies. The sinusoidal component of frequency $n\alpha$ is called the nth harmonic of the periodic function. The first harmonic is commonly called the fundamental component. The angles θ_n and the coefficients C_n are known as the phase angle and amplitude.

Using Euler's identities $e^{\pm i\theta} = \cos\theta \pm i\sin\theta$ where $i^2 = -1$, the Fourier series for $f(x)$ can be converted into complex form

$$f(x) = \sum_{n=-\infty}^{\infty} c_n e^{in\pi x/L}, \tag{4.13a}$$

where

$$c_{\pm n} = a_n \mp i b_n$$

$$= \frac{1}{2L} \int_{-L}^{L} f(x) e^{-in\pi x/L} \, dx, \quad \text{for } n > 0. \tag{4.13b}$$

Eq. (4.13a) is obtained on the understanding that the Dirichlet conditions are satisfied and that $f(x)$ is continuous at x. If $f(x)$ is discontinuous at x, the left hand side of Eq. (4.13a) should be replaced by $[f(x+0) + f(x-0)]/2$.

The exponential form (4.13a) can be considered as a basic form in its own right: it is not obtained by transformation from the trigonometric form, rather it is constructed directly from the given function. Furthermore, in the complex representation defined by Eqs. (4.13a) and (4.13b), a certain symmetry between the expressions for a function and for its Fourier coefficients is evident. In fact the expressions (4.13a) and (4.13b) are of essentially the same structure, as the following correlation reveals:

$$x \sim L, \; f(x) \sim c_n \equiv c(n), \quad e^{in\pi x/L} \sim e^{-in\pi x/L}, \quad \sum_{n=-\infty}^{\infty} (\;) \sim \frac{1}{2L} \int_{-L}^{L} (\;) dx.$$

This duality is worthy of note, and as our development proceeds to the Fourier integral, it will become more striking and fundamental.

Integration and differentiation of a Fourier series

The Fourier series of a function $f(x)$ may always be integrated term-by-term to give a new series which converges to the integral of $f(x)$. If $f(x)$ is a continuous function of x for all x, and is periodic (of period 2π) outside the interval $-\pi < x < \pi$, then term-by-term differentiation of the Fourier series of $f(x)$ leads to the Fourier series of $f'(x)$, provided $f'(x)$ satisfies Dirichlet's conditions.

Vibrating strings

The equation of motion of transverse vibration

There are numerous applications of Fourier series to solutions of boundary value problems. Here we consider one of them, namely vibrating strings. Let a string of length L be held fixed between two points $(0, 0)$ and $(L, 0)$ on the x-axis, and then given a transverse displacement parallel to the y-axis. Its subsequent motion, with no external forces acting on it, is to be considered; this is described by finding the displacement y as a function of x and t (if we consider only vibration in one plane, and take the xy plane as the plane of vibration). We will assume that ρ, the mass per unit length is uniform over the entire length of the string, and that the string is perfectly flexible, so that it can transmit tension but not bending or shearing forces.

As the string is drawn aside from its position of rest along the x-axis, the resulting increase in length causes an increase in tension, denoted by P. This tension at any point along the string is always in the direction of the tangent to the string at that point. As shown in Fig. 4.9, a force $P(x)A$ acts at the left hand side of an element ds, and a force $P(x + dx)A$ acts at the right hand side, where A is the cross-sectional area of the string. If α is the inclination to the horizontal, then

$$F_x \cong AP\cos(\alpha + d\alpha) - AP\cos\alpha, \quad F_y \cong AP\sin(\alpha + d\alpha) - AP\sin\alpha.$$

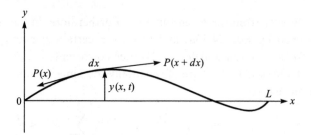

Figure 4.9. A vibrating string.

We limit the displacement to small values, so that we may set

$$\cos \alpha = 1 - \alpha^2/2, \quad \sin \alpha \cong \alpha \cong \tan \alpha = dy/dx,$$

then

$$F_y = AP\left[\left(\frac{dy}{dx}\right)_{x+dx} - \left(\frac{dy}{dx}\right)_x\right] = AP\frac{d^2y}{dx^2}\,dx.$$

Using Newton's second law, the equation of motion of transverse vibration of the element becomes

$$\rho A dx \frac{\partial^2 y}{\partial t^2} = AP\frac{\partial^2 y}{\partial x^2}\,dx, \quad \text{or} \quad \frac{\partial^2 y}{\partial x^2} = \frac{1}{v^2}\frac{\partial^2 y}{\partial t^2}, \quad v = \sqrt{P/\rho}.$$

Thus the transverse displacement of the string satisfies the partial differential wave equation

$$\frac{\partial^2 y}{\partial x^2} = \frac{1}{v^2}\frac{\partial^2 y}{\partial t^2}, \quad 0 < x < L, \quad t > 0 \tag{4.14}$$

with the following boundary conditions: $y(0, t) = y(L, t) = 0$, $\partial y/\partial t = 0$, $y(x, 0) = f(x)$; where $f(x)$ describes the initial shape (position) of the string, and v is the velocity of propagation of the wave along the string.

Solution of the wave equation

To solve this boundary value problem, let us try the method of separation variables:

$$y(x, t) = X(x)T(t). \tag{4.15}$$

Substituting this into Eq. (4.14) yields

$$(1/X)(d^2 X/dx^2) = (1/v^2 T)(d^2 T/dt^2).$$

Since the left hand side is a function of x only and the right hand side is a function of time only, they must be equal to a common separation constant, which we will call $-\lambda^2$. Then we have

$$d^2X/dx^2 = -\lambda^2 X, \quad X(0) = X(L) = 0 \qquad (4.16\text{a})$$

and

$$d^2T/dt^2 = -\lambda^2 v^2 T \quad dT/dt = 0 \quad \text{at } t = 0. \qquad (4.16\text{b})$$

Both of these equations are typical eigenvalue problems: we have a differential equation containing a parameter λ, and we seek solutions satisfying certain boundary conditions. If there are special values of λ for which non-trivial solutions exist, we call these eigenvalues, and the corresponding solutions eigensolutions or eigenfunctions.

The general solution of Eq. (4.16a) can be written as

$$X(x) = A_1 \sin(\lambda x) + B_1 \cos(\lambda x).$$

Applying the boundary conditions

$$X(0) = 0 \Rightarrow B_1 = 0,$$

and

$$X(L) = 0 \Rightarrow A_1 \sin(\lambda L) = 0$$

$A_1 = 0$ is the trivial solution $X = 0$ (so $y = 0$); hence we must have $\sin(\lambda L) = 0$, that is,

$$\lambda L = n\pi, \quad n = 1, 2, \ldots,$$

and we obtain a series of eigenvalues

$$\lambda_n = n\pi/L, \quad n = 1, 2, \ldots$$

and the corresponding eigenfunctions

$$X_n(x) = \sin(n\pi/L)x, \quad n = 1, 2, \ldots.$$

To solve Eq. (4.16b) for $T(t)$ we must use one of the values λ_n found above. The general solution is of the form

$$T(t) = A_2 \cos(\lambda_n vt) + B_2 \sin(\lambda_n vt).$$

The boundary condition leads to $B_2 = 0$.

The general solution of Eq. (4.14) is hence a linear superposition of the solutions of the form

$$y(x, t) = \sum_{n=1}^{\infty} A_n \sin(n\pi x/L) \cos(n\pi vt/L); \qquad (4.17)$$

the A_n are as yet undetermined constants. To find A_n, we use the boundary condition $y(x,t) = f(x)$ at $t = 0$, so that Eq. (4.17) reduces to

$$f(x) = \sum_{n=1}^{\infty} A_n \sin(n\pi x/L).$$

Do you recognize the infinite series on the right hand side? It is a Fourier sine series. To find A_n, multiply both sides by $\sin(m\pi x/L)$ and then integrate with respect to x from 0 to L and we obtain

$$A_m = \frac{2}{L} \int_0^L f(x) \sin(m\pi x/L) dx, \qquad m = 1, 2, \ldots$$

where we have used the relation

$$\int_0^L \sin(m\pi x/L) \sin(n\pi x/L) dx = \frac{L}{2} \delta_{mn}.$$

Eq. (4.17) now gives

$$y(x,t) = \sum_{n=1}^{\infty} \left[\frac{2}{L} \int_0^L f(x) \sin\frac{n\pi x}{L} dx \right] \sin\frac{n\pi x}{L} \cos\frac{n\pi v t}{L}. \qquad (4.18)$$

The terms in this series represent the natural modes of vibration. The frequency of the nth normal mode f_n is obtained from the term involving $\cos(n\pi v t/L)$ and is given by

$$2\pi f_n = n\pi v/L \quad \text{or} \quad f_n = nv/2L.$$

All frequencies are integer multiples of the lowest frequency f_1. We call f_1 the fundamental frequency or first harmonic, and f_2 and f_3 the second and third harmonics (or first and second overtones) and so on.

RLC circuit

Another good example of application of Fourier series is an *RLC* circuit driven by a variable voltage $E(t)$ which is periodic but not necessarily sinusoidal (see Fig. 4.10). We want to find the current $I(t)$ flowing in the circuit at time t.

According to Kirchhoff's second law for circuits, the impressed voltage $E(t)$ equals the sum of the voltage drops across the circuit components. That is,

$$L\frac{dI}{dt} + RI + \frac{Q}{C} = E(t),$$

where Q is the total charge in the capacitor C. But $I = dQ/dt$, thus differentiating the above differential equation once we obtain

$$L\frac{d^2 I}{dt^2} + R\frac{dI}{dt} + \frac{1}{C}I = \frac{dE}{dt}.$$

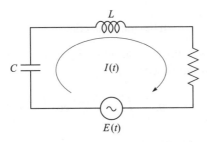

Figure 4.10. The *RLC* circuit.

Under steady-state conditions the current $I(t)$ is also periodic, with the same period P as for $E(t)$. Let us assume that both $E(t)$ and $I(t)$ possess Fourier expansions and let us write them in their complex forms:

$$E(t) = \sum_{n=-\infty}^{\infty} E_n e^{in\omega t}, \qquad I(t) = \sum_{n=-\infty}^{\infty} c_n e^{in\omega t} \qquad (\omega = 2\pi/P).$$

Furthermore, we assume that the series can be differentiated term by term. Thus

$$\frac{dE}{dt} = \sum_{n=-\infty}^{\infty} in\omega E_n e^{in\omega t}, \quad \frac{dI}{dt} = \sum_{n=-\infty}^{\infty} in\omega c_n e^{in\omega t}, \quad \frac{d^2 I}{dt^2} = \sum_{n=-\infty}^{\infty} (-n^2\omega^2) c_n e^{in\omega t}.$$

Substituting these into the last (second-order) differential equation and equating the coefficients with the same exponential $e^{in\alpha t}$, we obtain

$$\left(-n^2\omega^2 L + in\omega R + 1/C\right)c_n = in\omega E_n.$$

Solving for c_n

$$c_n = \frac{in\omega/L}{[(1/CL)^2 - n^2\omega^2] + i(R/L)n\omega} E_n.$$

Note that $1/LC$ is the natural frequency of the circuit and R/L is the attenuation factor of the circuit. The Fourier coefficients for $E(t)$ are given by

$$E_n = \frac{1}{P} \int_{-P/2}^{P/2} E(t) e^{-in\omega t} dt.$$

The current $I(t)$ in the circuit is given by

$$I(t) = \sum_{n=-\infty}^{\infty} c_n e^{in\omega t}.$$

Orthogonal functions

Many of the properties of Fourier series considered above depend on orthogonal properties of sine and cosine functions

$$\int_0^L \sin\frac{m\pi x}{L}\sin\frac{n\pi x}{L}\,dx = 0, \quad \int_0^L \cos\frac{m\pi x}{L}\cos\frac{n\pi x}{L}\,dx = 0 \quad (m \neq n).$$

In this section we seek to generalize this orthogonal property. To do so we first recall some elementary properties of real vectors in three-dimensional space.

Two vectors $\mathbf{A}$ and $\mathbf{B}$ are called orthogonal if $\mathbf{A} \cdot \mathbf{B} = 0$. Although not geometrically or physically obvious, we generalize these ideas to think of a function, say $A(x)$, as being an infinite-dimensional vector (a vector with an infinity of components), the value of each component being specified by substituting a particular value of x taken from some interval (a, b), and two functions, $A(x)$ and $B(x)$ are orthogonal in (a, b) if

$$\int_a^b A(x)B(x)dx = 0. \tag{4.19}$$

The left-side of Eq. (4.19) is called the scalar product of $A(x)$ and $B(x)$ and denoted by, in the Dirac bracket notation, $\langle A(x)|B(x)\rangle$. The first factor in the bracket notation is referred to as the bra and the second factor as the ket, so together they comprise the bracket.

A vector $\mathbf{A}$ is called a unit vector or normalized vector if its magnitude is unity: $\mathbf{A} \cdot \mathbf{A} = A^2 = 1$. Extending this concept, we say that the function $A(x)$ is normal or normalized in (a, b) if

$$\langle A(x)|A(x)\rangle = \int_a^b A(x)A(x)dx = 1. \tag{4.20}$$

If we have a set of functions $\varphi_i(x), i = 1, 2, 3, \ldots,$ having the properties

$$\langle \varphi_m(x)|\varphi_n(x)\rangle = \int_a^b \varphi_m(x)\varphi_n(x)dx = \delta_{mn}, \tag{4.20a}$$

where δ_{nm} is the Kronecker delta symbol, we then call such a set of functions an orthonormal set in (a, b). For example, the set of functions $\varphi_m(x) = (2/\pi)^{1/2}\sin(mx), m = 1, 2, 3, \ldots$ is an orthonormal set in the interval $0 \leq x \leq \pi$.

Just as in three-dimensional vector space, any vector $\mathbf{A}$ can be expanded in the form $\mathbf{A} = A_1\hat{e}_1 + A_2\hat{e}_2 + A_3\hat{e}_3$, we can consider a set of orthonormal functions φ_i as base vectors and expand a function $f(x)$ in terms of them, that is,

$$f(x) = \sum_{n=1}^{\infty} c_n\varphi_n(x) \quad a \leq x \leq b; \tag{4.21}$$

the series on the right hand side is called an orthonormal series; such series are generalizations of Fourier series. Assuming that the series on the right converges to $f(x)$, we can then multiply both sides by $\varphi_m(x)$ and integrate both sides from a to b to obtain

$$c_m = \langle f(x)|\varphi_m(x)\rangle = \int_a^b f(x)\varphi_m(x)dx; \qquad (4.21a)$$

c_m can be called the generalized Fourier coefficients.

Multiple Fourier series

A Fourier expansion of a function of two or three variables is often very useful in many applications. Let us consider the case of a function of two variables, say $f(x, y)$. For example, we can expand $f(x, y)$ into a double Fourier sine series

$$f(x, y) = \sum_{m=1}^{\infty} \sum_{n=1}^{\infty} B_{mn} \sin\frac{m\pi x}{L_1} \sin\frac{n\pi y}{L_2}, \qquad (4.22)$$

where

$$B_{mn} = \frac{4}{L_1 L_2} \int_0^{L_1} \int_0^{L_2} f(x, y) \sin\frac{m\pi x}{L_1} \sin\frac{n\pi y}{L_2} dxdy. \qquad (4.22a)$$

Similar expansions can be made for cosine series and for series having both sines and cosines.

To obtain the coefficients B_{mn}, let us rewrite $f(x, y)$ as

$$f(x, y) = \sum_{m=1}^{\infty} C_m \sin\frac{m\pi x}{L_1}, \qquad (4.23)$$

where

$$C_m = \sum_{n=1}^{\infty} B_{mn} \sin\frac{n\pi y}{L_2}. \qquad (4.23a)$$

Now we can consider Eq. (4.23) as a Fourier series in which y is kept constant so that the Fourier coefficients C_m are given by

$$C_m = \frac{2}{L_1} \int_0^{L_1} f(x, y) \sin\frac{m\pi x}{L_1} dx. \qquad (4.24)$$

On noting that C_m is a function of y, we see that Eq. (4.23a) can be considered as a Fourier series for which the coefficients B_{mn} are given by

$$B_{mn} = \frac{2}{L_2} \int_0^{L_2} C_m \sin\frac{n\pi y}{L_2} dy.$$

Substituting Eq. (4.24) for C_m into the above equation, we see that B_{mn} is given by Eq. (4.22a).

Similar results can be obtained for cosine series or for series containing both sines and cosines. Furthermore, these ideas can be generalized to triple Fourier series, etc. They are very useful in solving, for example, wave propagation and heat conduction problems in two or three dimensions. Because they lie outside of the scope of this book, we have to omit these interesting applications.

Fourier integrals and Fourier transforms

The properties of Fourier series that we have thus far developed are adequate for handling the expansion of any periodic function that satisfies the Dirichlet conditions. But many problems in physics and engineering do not involve periodic functions, and it is therefore desirable to generalize the Fourier series method to include non-periodic functions. A non-periodic function can be considered as a limit of a given periodic function whose period becomes infinite, as shown in Examples 4.5 and 4.6.

Example 4.5
Consider the periodic functions $f_L(x)$

$$f_L(x) = \begin{cases} 0 \text{ when} & -L/2 < x < -1 \\ 1 \text{ when} & -1 < x < 1, \\ 0 \text{ when} & 1 < x < L/2 \end{cases}$$

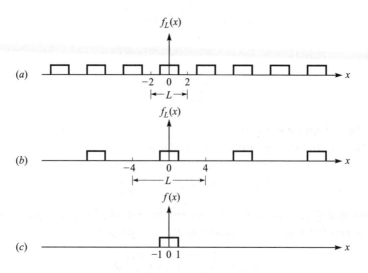

Figure 4.11. Square wave function: (a) $L = 4$; (b) $L = 8$; (c) $L \to \infty$.

which has period $L > 2$. Fig. 4.11(a) shows the function when $L = 4$. If L is increased to 8, the function looks like the one shown in Fig. 4.11(b). As $L \to \infty$ we obtain a non-periodic function $f(x)$, as shown in Fig. 4.11(c):

$$f(x) = \begin{cases} 1 & -1 < x < 1 \\ 0 & \text{otherwise} \end{cases}.$$

Example 4.6
Consider the periodic function $g_L(x)$ (Fig. 4.12(a)):

$$g_L(x) = e^{-|x|} \quad \text{when} \quad -L/2 < x < L/2.$$

As $L \to \infty$ we obtain a non-periodic function $g(x)$: $g(x) = \lim_{L \to \infty} g_L(x)$ (Fig. 4.12(b)).

By investigating the limit that is approached by a Fourier series as the period of the given function becomes infinite, a suitable representation for non-periodic functions can perhaps be obtained. To this end, let us write the Fourier series representing a periodic function $f(x)$ in complex form:

$$f(x) = \sum_{n=-\infty}^{\infty} c_n e^{i\omega x}, \tag{4.25}$$

$$c_n = \frac{1}{2L} \int_{-L}^{L} f(x) e^{-i\omega x} dx \tag{4.26}$$

where ω denotes $n\pi/L$

$$\omega = \frac{n\pi}{L}, \quad n \text{ positive or negative.} \tag{4.27}$$

The transition $L \to \infty$ is a little tricky since c_n apparently approaches zero, but these coefficients should not approach zero. We can ask for help from Eq. (4.27), from which we have

$$\Delta\omega = (\pi/L)\Delta n,$$

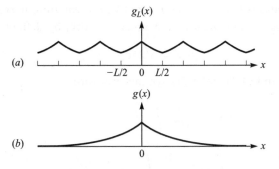

Figure 4.12.   Sawtooth wave functions: (a) $-L/2 < x < L/2$; (b) $L \to \infty$.

and the 'adjacent' values of ω are obtained by setting $\Delta n = 1$, which corresponds to

$$(L/\pi)\Delta\omega = 1.$$

Then we can multiply each term of the Fourier series by $(L/\pi)\Delta\omega$ and obtain

$$f(x) = \sum_{n=-\infty}^{\infty} \left(\frac{L}{\pi}c_n\right)e^{i\omega x}\Delta\omega,$$

where

$$\frac{L}{\pi}c_n = \frac{1}{2\pi}\int_{-L}^{L} f(x)e^{-i\omega x}dx.$$

The troublesome factor $1/L$ has disappeared. Switching completely to the ω notation and writing $(L/\pi)c_n = c_L(\omega)$, we obtain

$$c_L(\omega) = \frac{1}{2\pi}\int_{-L}^{L} f(x)e^{-i\omega x}dx$$

and

$$f(x) = \sum_{L\omega/\pi=-\infty}^{\infty} c_L(\omega)e^{i\omega x}\Delta\omega.$$

In the limit as $L \to \infty$, the ωs are distributed continuously instead of discretely, $\Delta\omega \to d\omega$ and this sum is exactly the definition of an integral. Thus the last equations become

$$c(\omega) = \lim_{L\to\infty} c_L(\omega) = \frac{1}{2\pi}\int_{-\infty}^{\infty} f(x)e^{-i\omega x}dx \qquad (4.28)$$

and

$$f(x) = \int_{-\infty}^{\infty} c(\omega)e^{i\omega x}d\omega. \qquad (4.29)$$

This set of formulas is known as the Fourier transformation, in somewhat different form. It is easy to put them in a symmetrical form by defining

$$g(\omega) = \sqrt{2\pi}c(-\omega),$$

then Eqs. (4.28) and (4.29) take the symmetrical form

$$g(\omega) = \frac{1}{\sqrt{2\pi}}\int_{-\infty}^{\infty} f(x')e^{-i\omega x'}dx', \qquad (4.30)$$

$$f(x) = \frac{1}{\sqrt{2\pi}}\int_{-\infty}^{\infty} g(\omega)e^{i\omega x}d\omega. \qquad (4.31)$$

The function $g(\omega)$ is called the Fourier transform of $f(x)$ and is written $g(\omega) = F\{f(x)\}$. Eq. (4.31) is the inverse Fourier transform of $g(\omega)$ and is written $f(x) = F^{-1}\{g(\omega)\}$; sometimes it is also called the Fourier integral representation of $f(x)$. The exponential function $e^{-i\omega x}$ is sometimes called the kernel of transformation.

It is clear that $g(\omega)$ is defined only if $f(x)$ satisfies certain restrictions. For instance, $f(x)$ should be integrable in some finite region. In practice, this means that $f(x)$ has, at worst, jump discontinuities or mild infinite discontinuities. Also, the integral should converge at infinity. This would require that $f(x) \to 0$ as $x \to \pm\infty$.

A very common sufficient condition is the requirement that $f(x)$ is absolutely integrable. That is, the integral

$$\int_{-\infty}^{\infty} |f(x)|dx$$

exists. Since $|f(x)e^{-i\omega x}| = |f(x)|$, it follows that the integral for $g(\omega)$ is absolutely convergent; therefore it is convergent.

It is obvious that $g(\omega)$ is, in general, a complex function of the real variable ω. So if $f(x)$ is real, then

$$g(-\omega) = g^*(\omega).$$

There are two immediate corollaries to this property:

(1) $f(x)$ is even, $g(\omega)$ is real;
(2) if $f(x)$ is odd, $g(\omega)$ is purely imaginary.

Other, less symmetrical forms of the Fourier integral can be obtained by working directly with the sine and cosine series, instead of with the exponential functions.

Example 4.7
Consider the Gaussian probability function $f(x) = Ne^{-\alpha x^2}$, where N and α are constant. Find its Fourier transform $g(\omega)$, then graph $f(x)$ and $g(\omega)$.

Solution: Its Fourier transform is given by

$$g(\omega) = \frac{1}{\sqrt{2\pi}} \int_{-\infty}^{\infty} f(x)e^{-i\omega x}\, dx = \frac{N}{\sqrt{2\pi}} \int_{-\infty}^{\infty} e^{-\alpha x^2} e^{-i\omega x} dx.$$

This integral can be simplified by a change of variable. First, we note that

$$-\alpha x^2 - i\omega x = -(x\sqrt{\alpha} + i\omega/2\sqrt{\alpha})^2 - \omega^2/4\alpha,$$

and then make the change of variable $x\sqrt{\alpha} + i\omega/2\sqrt{\alpha} = u$ to obtain

$$g(\omega) = \frac{N}{\sqrt{2\pi\alpha}} e^{-\omega^2/4\alpha} \int_{-\infty}^{\infty} e^{-u^2}\, du = \frac{N}{\sqrt{2\alpha}} e^{-\omega^2/4\alpha}.$$

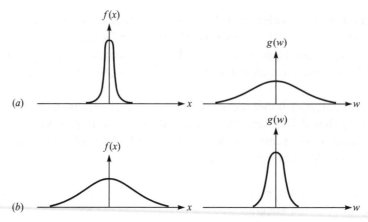

Figure 4.13. Gaussian probability function: (a) large α; (b) small α.

It is easy to see that $g(\omega)$ is also a Gaussian probability function with a peak at the origin, monotonically decreasing as $\omega \to \pm\infty$. Furthermore, for large α, $f(x)$ is sharply peaked but $g(\omega)$ is flattened, and vice versa as shown in Fig. 4.13. It is interesting to note that this is a general feature of Fourier transforms. We shall see later that in quantum mechanical applications it is related to the Heisenberg uncertainty principle.

The original function $f(x)$ can be retrieved from Eq. (4.31) which takes the form

$$\frac{1}{\sqrt{2\pi}} \int_{-\infty}^{\infty} g(\omega) e^{i\omega x} d\omega = \frac{1}{\sqrt{2\pi}} \frac{N}{\sqrt{2\alpha}} \int_{-\infty}^{\infty} e^{-\omega^2/4\alpha} e^{i\omega x} d\omega$$

$$= \frac{1}{\sqrt{2\pi}} \frac{N}{\sqrt{2\alpha}} \int_{-\infty}^{\infty} e^{-\alpha'\omega^2} e^{-i\omega x'} d\omega$$

in which we have set $\alpha' = 1/4\alpha$, and $x' = -x$. The last integral can be evaluated by the same technique, and we finally find

$$\frac{1}{\sqrt{2\pi}} \int_{-\infty}^{\infty} g(\omega) e^{i\omega x} d\omega = \frac{1}{\sqrt{2\pi}} \frac{N}{\sqrt{2\alpha}} \int_{-\infty}^{\infty} e^{-\alpha'\omega^2} e^{-i\omega x'} d\omega$$

$$= \frac{N}{\sqrt{2\alpha}} \sqrt{2\alpha} e^{-\alpha x^2}$$

$$= N e^{-\alpha x^2} = f(x).$$

Example 4.8
Given the box function which can represent a single pulse

$$f(x) = \begin{cases} 1 & |x|a \\ 0 & |x| > a \end{cases}$$

find the Fourier transform of $f(x)$, $g(\omega)$; then graph $f(x)$ and $g(\omega)$ for $a = 3$.

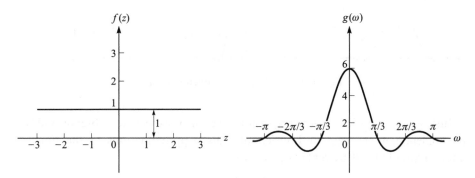

Figure 4.14.   The box function.

Solution: The Fourier transform of $f(x)$ is, as shown in Fig. 4.14,

$$g(\omega) = \frac{1}{\sqrt{2\pi}} \int_{-\infty}^{\infty} f(x') e^{-i\omega x'} dx' = \frac{1}{\sqrt{2\pi}} \int_{-a}^{a} (1) e^{-i\omega x'} dx' = \frac{1}{\sqrt{2\pi}} \frac{e^{-i\omega x'}}{-i\omega} \bigg|_{-a}^{a}$$

$$= \sqrt{\frac{2}{\pi}} \frac{\sin \omega a}{\omega}, \quad \omega \neq 0.$$

For $\omega = 0$, we obtain $g(\omega) = \sqrt{2/\pi}\, a$.

The Fourier integral representation of $f(x)$ is

$$f(x) = \frac{1}{\sqrt{2\pi}} \int_{-\infty}^{\infty} g(\omega) e^{i\omega x} d\omega = \frac{1}{2\pi} \int_{-\infty}^{\infty} \frac{2 \sin \omega a}{\omega} e^{i\omega x} d\omega.$$

Now

$$\int_{-\infty}^{\infty} \frac{\sin \omega a}{\omega} e^{i\omega x} d\omega = \int_{-\infty}^{\infty} \frac{\sin \omega a \cos \omega x}{\omega} d\omega + i \int_{-\infty}^{\infty} \frac{\sin \omega a \sin \omega x}{\omega} d\omega.$$

The integrand in the second integral is odd and so the integral is zero. Thus we have

$$f(x) = \frac{1}{\sqrt{2\pi}} \int_{-\infty}^{\infty} g(\omega) e^{i\omega x} d\omega = \frac{1}{\pi} \int_{-\infty}^{\infty} \frac{\sin \omega a \cos \omega x}{\omega} d\omega = \frac{2}{\pi} \int_{0}^{\infty} \frac{\sin \omega a \cos \omega x}{\omega} d\omega;$$

the last step follows since the integrand is an even function of ω.

It is very difficult to evaluate the last integral. But a known property of $f(x)$ will help us. We know that $f(x)$ is equal to 1 for $|x| \leq a$, and equal to 0 for $|x| > a$. Thus we can write

$$\frac{2}{\pi} \int_{0}^{\infty} \frac{\sin \omega a \cos \omega x}{\omega} d\omega = \begin{cases} 1 & |x|a \\ 0 & |x| > a \end{cases}$$

169

Figure 4.15. The Gibb's phenomenon.

Just as in Fourier series expansion, we also expect to observe Gibb's phenomenon in the case of Fourier integrals. Approximations to the Fourier integral are obtained by replacing ∞ by α:

$$\int_0^\alpha \frac{\sin \omega \cos \omega x}{\omega} \, d\omega,$$

where we have set $a = 1$. Fig. 4.15 shows oscillations near the points of discontinuity of $f(x)$. We might expect these oscillations to disappear as $\alpha \to \infty$, but they are just shifted closer to the points $x = \pm 1$.

Example 4.9
Consider now a harmonic wave of frequency ω_0, $e^{i\omega_0 t}$, which is chopped to a lifetime of $2T$ seconds (Fig. 4.16(a)):

$$f(t) = \begin{cases} e^{i\omega_0 t} & -T \leq t \leq T \\ 0 & |t| > 0 \end{cases}.$$

The chopping process will introduce many new frequencies in varying amounts, given by the Fourier transform. Then we have, according to Eq. (4.30),

$$g(\omega) = (2\pi)^{-1/2} \int_{-T}^{T} e^{i\omega_0 t} e^{-i\omega t} dt = (2\pi)^{-1/2} \int_{-T}^{T} e^{i(\omega_0 - \omega)t} dt$$

$$= (2\pi)^{-1/2} \frac{e^{i(\omega_0 - \omega)t}}{i(\omega_0 - \omega)} \Big|_{-T}^{T} = (2/\pi)^{1/2} T \frac{\sin(\omega_0 - \omega)T}{(\omega_0 - \omega)T}.$$

This function is plotted schematically in Fig. 4.16(b). (Note that $\lim_{x \to 0} (\sin x/x) = 1$.) The most striking aspect of this graph is that, although the principal contribution comes from the frequencies in the neighborhood of ω_0, an infinite number of frequencies are presented. Nature provides an example of this kind of chopping in the emission of photons during electronic and nuclear transitions in atoms. The light emitted from an atom consists of regular vibrations that last for a finite time of the order of 10^{-9} s or longer. When light is examined by a spectroscope (which measures the wavelengths and, hence, the frequencies) we find that there is an irreducible minimum frequency spread for each spectrum line. This is known as the natural line width of the radiation.

The relative percentage of frequencies, other than the basic one, present depends on the shape of the pulse, and the spread of frequencies depends on

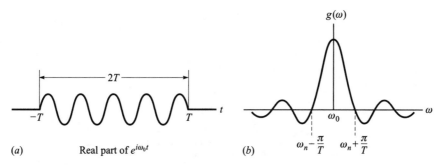

Figure 4.16. (a) A chopped harmonic wave $e^{i\omega_0 t}$ that lasts a finite time $2T$. (b) Fourier transform of $e^{(i\omega_0 t)}$, $|t| < T$, and 0 otherwise.

the time T of the duration of the pulse. As T becomes larger the central peak becomes higher and the width $\Delta\omega(= 2\pi/T)$ becomes smaller. Considering only the spread of frequencies in the central peak we have

$$\Delta\omega = 2\pi/T, \quad \text{or} \quad T\Delta = 1.$$

Multiplying by the Planck constant h and replacing T by Δt, we have the relation

$$\Delta t \Delta E = h. \tag{4.32}$$

A wave train that lasts a finite time also has a finite extension in space. Thus the radiation emitted by an atom in 10^{-9} s has an extension equal to $3 \times 10^8 \times 10^{-9} = 3 \times 10^{-1}$ m. A Fourier analysis of this pulse in the space domain will yield a graph identical to Fig. 4.11(b), with the wave numbers clustered around $k_0(= 2\pi/\lambda_0 = \omega_0/v)$. If the wave train is of length $2a$, the spread in wave number will be given by $a\Delta k = 2\pi$, as shown below. This time we are chopping an infinite plane wave front with a shutter such that the length of the packet is $2a$, where $2a = 2vT$, and $2T$ is the time interval that the shutter is open. Thus

$$\psi(x) = \begin{cases} e^{ik_0 x}, & -a \leq x \leq a \\ 0, & |x| > a \end{cases}.$$

Then

$$\phi(k) = (2\pi)^{-1/2} \int_{-\infty}^{\infty} \psi(x) e^{-ikx} dx = (2\pi)^{-1/2} \int_{-a}^{a} \psi(x) e^{-ikx} dx$$

$$= (2/\pi)^{1/2} a \frac{\sin(k_0 - k)a}{(k_0 - k)a}.$$

This function is plotted in Fig. 4.17: it is identical to Fig. 4.16(b), but here it is the wave vector (or the momentum) that takes on a spread of values around k_0. The breadth of the central peak is $\Delta k = 2\pi/a$, or $a\Delta k = 2\pi$.

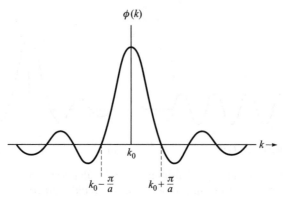

Figure 4.17. Fourier transform of e^{ikx}, $|x| \le a$.

Fourier sine and cosine transforms

If $f(x)$ is an odd function, the Fourier transforms reduce to

$$g(\omega) = \sqrt{\frac{2}{\pi}} \int_0^\infty f(x') \sin \omega x' dx', \qquad f(x) = \sqrt{\frac{2}{\pi}} \int_0^\infty g(\omega) \sin \omega x d\omega. \qquad (4.33a)$$

Similarly, if $f(x)$ is an even function, then we have Fourier cosine transformations:

$$g(\omega) = \sqrt{\frac{2}{\pi}} \int_0^\infty f(x') \cos \omega x' dx', \qquad f(x) = \sqrt{\frac{2}{\pi}} \int_0^\infty g(\omega) \cos \omega x d\omega. \qquad (4.33b)$$

To demonstrate these results, we first expand the exponential function on the right hand side of Eq. (4.30)

$$g(\omega) = \frac{1}{\sqrt{2\pi}} \int_{-\infty}^\infty f(x') e^{-i\omega x'} dx'$$

$$= \frac{1}{\sqrt{2\pi}} \int_{-\infty}^\infty f(x') \cos \omega x' dx' - \frac{i}{\sqrt{2\pi}} \int_{-\infty}^\infty f(x') \sin \omega x' dx'.$$

If $f(x)$ is even, then $f(x) \cos \omega x$ is even and $f(x) \sin \omega x$ is odd. Thus the second integral on the right hand side of the last equation is zero and we have

$$g(\omega) = \frac{1}{\sqrt{2\pi}} \int_{-\infty}^\infty f(x') \cos \omega x' dx' = \sqrt{\frac{2}{\pi}} \int_0^\infty f(x') \cos \omega x' dx';$$

$g(\omega)$ is an even function, since $g(-\omega) = g(\omega)$. Next from Eq. (4.31) we have

$$f(x) = \frac{1}{\sqrt{2\pi}} \int_{-\infty}^\infty g(\omega) e^{i\omega x} d\omega$$

$$= \frac{1}{\sqrt{2\pi}} \int_{-\infty}^\infty g(\omega) \cos \omega x d\omega + \frac{i}{\sqrt{2\pi}} \int_{-\infty}^\infty g(\omega) \sin \omega x d\omega.$$

Since $g(\omega)$ is even, so $g(\omega) \sin \omega x$ is odd and the second integral on the right hand side of the last equation is zero, and we have

$$f(x) = \frac{1}{\sqrt{2\pi}} \int_{-\infty}^{\infty} g(\omega) \cos \omega x d\omega = \sqrt{\frac{2}{\pi}} \int_{0}^{\infty} g(\omega) \cos \omega x d\omega.$$

Similarly, we can prove Fourier sine transforms by replacing the cosine by the sine.

Heisenberg's uncertainty principle

We have demonstrated in above examples that if $f(x)$ is sharply peaked, then $g(\omega)$ is flattened, and vice versa. This is a general feature in the theory of Fourier transforms and has important consequences for all instances of wave propagation. In electronics we understand now why we use a wide-band amplification in order to reproduce a sharp pulse without distortion.

In quantum mechanical applications this general feature of the theory of Fourier transforms is related to the Heisenberg uncertainty principle. We saw in Example 4.9 that the spread of the Fourier transform in k space (Δk) times its spread in coordinate space (a) is equal to 2π ($a\Delta k \cong 2\pi$). This result is of special importance because of the connection between values of k and momentum p: $p = \hbar k$ (where $\hbar$ is the Planck constant h divided by 2π). A particle localized in space must be represented by a superposition of waves with different momenta. As a result, the position and momentum of a particle cannot be measured *simultaneously* with infinite precision; the product of 'uncertainty in the position determination' and 'uncertainty in the momentum determination' is governed by the relation $\Delta x \Delta p \cong h(a\hbar \Delta k \cong 2\pi\hbar = h,$ or $\Delta x \Delta p \cong h, \Delta x = a)$. This statement is called Heisenberg's uncertainty principle. If position is known better, knowledge of the momentum must be unavoidably reduced proportionally, and vice versa. A complete knowledge of one, say k (and so p), is possible only when there is complete ignorance of the other. We can see this in physical terms. A wave with a unique value of k is infinitely long. A particle represented by an infinitely long wave (a free particle) cannot have a definite position, since the particle can be anywhere along its length. Hence the position uncertainty is infinite in order that the uncertainty in k is zero.

Equation (4.32) represents Heisenberg's uncertainty principle in a different form. It states that we cannot know with infinite precision the exact energy of a quantum system at every moment in time. In order to measure the energy of a quantum system with good accuracy, one must carry out such a measurement for a sufficiently long time. In other words, if the dynamical state exists only for a time of order Δt, then the energy of the state cannot be defined to a precision better than $h/\Delta t$.

We should not look upon the uncertainty principle as being merely an unfortunate limitation on our ability to know nature with infinite precision. We can use it to our advantage. For example, when combining the time–energy uncertainty relation with Einstein's mass–energy relation ($E = mc^2$) we obtain the relation $\Delta m \Delta t \cong h/c^2$. This result is very useful in our quest to understand the universe, in particular, the origin of matter.

Wave packets and group velocity

Energy (that is, a signal or information) is transmitted by groups of waves, not a single wave. Phase velocity may be greater than the speed of light c, 'group velocity' is always less than c. The wave groups with which energy is transmitted from place to place are called wave packets. Let us first consider a simple case where we have two waves φ_1 and φ_2: each has the same amplitude but differs slightly in frequency and wavelength,

$$\varphi_1(x, t) = A\cos(\omega t - kx),$$

$$\varphi_2(x, t) = A\cos[(\omega + \Delta\omega)t - (k + \Delta k)x],$$

where $\Delta\omega \ll \omega$ and $\Delta k \ll k$. Each represents a pure sinusoidal wave extending to infinite along the x-axis. Together they give a resultant wave

$$\varphi = \varphi_1 + \varphi_2$$
$$= A\{\cos(\omega t - kx) + \cos[(\omega + \Delta\omega)t - (k + \Delta k)x]\}.$$

Using the trigonometrical identity

$$\cos A + \cos B = 2\cos\frac{A + B}{2}\cos\frac{A - B}{2},$$

we can rewrite φ as

$$\varphi = 2\cos\frac{2\omega t - 2kx + \Delta\omega t - \Delta kx}{2}\cos\frac{-\Delta\omega t + \Delta kx}{2}$$
$$= 2\cos\tfrac{1}{2}(\Delta\omega t - \Delta kx)\cos(\omega t - kx).$$

This represents an oscillation of the original frequency ω, but with a modulated amplitude as shown in Fig. 4.18. A given segment of the wave system, such as AB, can be regarded as a 'wave packet' and moves with a velocity v_g (not yet determined). This segment contains a large number of oscillations of the primary wave that moves with the velocity v. And the velocity v_g with which the modulated amplitude propagates is called the group velocity and can be determined by the requirement that the phase of the modulated amplitude be constant. Thus

$$v_g = dx/dt = \Delta\omega/\Delta k \rightarrow d\omega/dk.$$

174

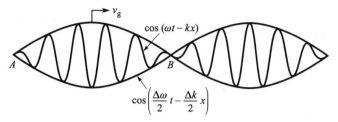

Figure 4.18.   Superposition of two waves.

The modulation of the wave is repeated indefinitely in the case of superposition of two almost equal waves. We now use the Fourier technique to demonstrate that any isolated packet of oscillatory disturbance of frequency ω can be described in terms of a combination of infinite trains of frequencies distributed around ω. Let us first superpose a system of n waves

$$\psi(x,t) = \sum_{j=1}^{n} A_j e^{i(k_j x - \omega_j t)},$$

where A_j denotes the amplitudes of the individual waves. As n approaches infinity, the frequencies become continuously distributed. Thus we can replace the summation with an integration, and obtain

$$\psi(x,t) = \int_{-\infty}^{\infty} A(k) e^{i(kx - \omega t)} dk; \tag{4.34}$$

the amplitude $A(k)$ is often called the distribution function of the wave. For $\psi(x,t)$ to represent a wave packet traveling with a characteristic group velocity, it is necessary that the range of propagation vectors included in the superposition be fairly small. Thus, we assume that the amplitude $A(k) \neq 0$ only for a small range of values about a particular k_0 of k:

$$A(k) \neq 0, \quad k_0 - \varepsilon < k < k_0 + \varepsilon, \quad \varepsilon \ll k_0.$$

The behavior in time of the wave packet is determined by the way in which the angular frequency ω depends upon the wave number k: $\omega = \omega(k)$, known as the law of dispersion. If ω varies slowly with k, then $\omega(k)$ can be expanded in a power series about k_0:

$$\omega(k) = \omega(k_0) + \frac{d\omega}{dk}\bigg|_0 (k - k_0) + \cdots = \omega_0 + \omega'(k - k_0) + O\left[(k - k_0)^2\right],$$

where

$$\omega_0 = \omega(k_0), \quad \text{and} \quad \omega' = \frac{d\omega}{dk}\bigg|_0$$

and the subscript zero means 'evaluated' at $k = k_0$. Now the argument of the exponential in Eq. (4.34) can be rewritten as

$$\omega t - kx = (\omega_0 t - k_0 x) + \omega'(k - k_0)t - (k - k_0)x$$
$$= (\omega_0 t - k_0 x) + (k - k_0)(\omega' t - x)$$

and Eq. (4.34) becomes

$$\psi(x, t) = \exp[i(k_0 x - \omega_0 t)] \int_{k_0 - \varepsilon}^{k_0 + \varepsilon} A(k) \exp[i(k - k_0)(x - \omega' t)]dk. \qquad (4.35)$$

If we take $k - k_0$ as the new integration variable y and assume $A(k)$ to be a slowly varying function of k in the integration interval 2ε, then Eq. (4.35) becomes

$$\psi(x, t) \cong \exp[i(k_0 x - \omega_0 t)] \int_{k_0 - \varepsilon}^{k_0 + \varepsilon} A(k_0 + y) \exp[i(x - \omega' t)y]dy.$$

Integration, transformation, and the approximation $A(k_0 + y) \cong A(k_0)$ lead to the result

$$\psi(x, t) = B(x, t) \exp[i(k_0 x - \omega_0 t)] \qquad (4.36)$$

with

$$B(x, t) = 2A(k_0) \frac{\sin[\Delta k(x - \omega' t)]}{x - \omega' t}. \qquad (4.37)$$

As the argument of the sine contains the small quantity Δk, $B(x, t)$ varies slowly depending on time t and coordinate x. Therefore, we can regard $B(x, t)$ as the small amplitude of an approximately monochromatic wave and $k_0 x - \omega_0 t$ as its phase. If we multiply the numerator and denominator on the right hand side of Eq. (4.37) by Δk and let

$$z = \Delta k(x - \omega' t)$$

then $B(x, t)$ becomes

$$B(x, t) = 2A(k_0)\Delta k \frac{\sin z}{z}$$

and we see that the variation in amplitude is determined by the factor $\sin (z)/z$. This has the properties

$$\lim_{z \to 0} \frac{\sin z}{z} = 1 \quad \text{for} \quad z = 0$$

and

$$\frac{\sin z}{z} = 0 \quad \text{for} \quad z = \pm\pi, \pm 2\pi, \ldots.$$

If we further increase the absolute value of z, the function $\sin(z)/z$ runs alternately through maxima and minima, the function values of which are small compared with the principal maximum at $z = 0$, and quickly converges to zero. Therefore, we can conclude that superposition generates a wave packet whose amplitude is non-zero only in a finite region, and is described by $\sin(z)/z$ (see Fig. 4.19).

The modulating factor $\sin(z)/z$ of the amplitude assumes the maximum value 1 as $z \to 0$. Recall that $z = \Delta k(x - \omega' t)$, thus for $z = 0$, we have

$$x - \omega' t = 0,$$

which means that the maximum of the amplitude is a plane propagating with velocity

$$\frac{dx}{dt} = \omega' = \left.\frac{d\omega}{dk}\right|_0,$$

that is, ω' is the group velocity, the velocity of the whole wave packet.

The concept of a wave packet also plays an important role in quantum mechanics. The idea of associating a wave-like property with the electron and other material particles was first proposed by Louis Victor de Broglie (1892–1987) in 1925. His work was motivated by the mystery of the Bohr orbits. After Rutherford's successful α-particle scattering experiments, a planetary-type nuclear atom, with electrons orbiting around the nucleus, was in favor with most physicists. But, according to classical electromagnetic theory, a charge undergoing continuous centripetal acceleration emits electromagnetic radiation

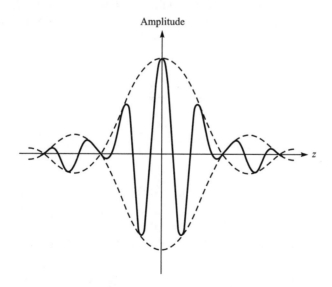

Figure 4.19.   A wave packet.

continuously and so the electron would lose energy continuously and it would spiral into the nucleus after just a fraction of a second. This does not occur. Furthermore, atoms do not radiate unless excited, and when radiation does occur its spectrum consists of discrete frequencies rather than the continuum of frequencies predicted by the classical electromagnetic theory. In 1913 Niels Bohr (1885–1962) proposed a theory which successfully explained the radiation spectra of the hydrogen atom. According to Bohr's postulates, an atom can exist in certain allowed stationary states without radiation. Only when an electron makes a transition between two allowed stationary states, does it emit or absorb radiation. The possible stationary states are those in which the angular momentum of the electron about the nucleus is quantized, that is, $mvr = n\hbar$, where v is the speed of the electron in the nth orbit and r is its radius. Bohr didn't clearly describe this quantum condition. De Broglie attempted to explain it by fitting a standing wave around the circumference of each orbit. Thus de Broglie proposed that $n\lambda = 2\pi r$, where λ is the wavelength associated with the nth orbit. Combining this with Bohr's quantum condition we immediately obtain

$$\lambda = \frac{h}{mv} = \frac{h}{p}.$$

De Broglie proposed that any material particle of total energy E and momentum p is accompanied by a wave whose wavelength is given by $\lambda = h/p$ and whose frequency is given by the Planck formula $\nu = E/h$. Today we call these waves de Broglie waves or matter waves. The physical nature of these matter waves was not clearly described by de Broglie, we shall not ask what these matter waves are – this is addressed in most textbooks on quantum mechanics. Let us ask just one question: what is the (phase) velocity of such a matter wave? If we denote this velocity by u, then

$$u = \lambda\nu = \frac{E}{p} = \frac{1}{p}\sqrt{p^2c^2 + m_0^2c^4}$$

$$= c\sqrt{1 + (m_0c/p)^2} = \frac{c^2}{v} \quad \left(p = \frac{m_0v}{\sqrt{1 - v^2/c^2}}\right),$$

which shows that for a particle with $m_0 > 0$ the wave velocity u is always greater than c, the speed of light in a vacuum. Instead of individual waves, de Broglie suggested that we can think of particles inside a wave packet, synthesized from a number of individual waves of different frequencies, with the entire packet traveling with the particle velocity v.

De Broglie's matter wave idea is one of the cornerstones of quantum mechanics.

178

Heat conduction

We now consider an application of Fourier integrals in classical physics. A semi-infinite thin bar ($x \geq 0$), whose surface is insulated, has an initial temperature equal to $f(x)$. The temperature of the end $x = 0$ is suddenly dropped to and maintained at zero. The problem is to find the temperature $T(x, t)$ at any point x at time t. First we have to set up the boundary value problem for heat conduction, and then seek the general solution that will give the temperature $T(x, t)$ at any point x at time t.

Head conduction equation

To establish the equation for heat conduction in a conducting medium we need first to find the heat flux (the amount of heat per unit area per unit time) across a surface. Suppose we have a flat sheet of thickness Δn, which has temperature T on one side and $T + \Delta T$ on the other side (Fig. 4.20). The heat flux which flows from the side of high temperature to the side of low temperature is directly proportional to the difference in temperature ΔT and inversely proportional to the thickness Δn. That is, the heat flux from I to II is equal to

$$-K \frac{\Delta T}{\Delta n},$$

where K, the constant of proportionality, is called the thermal conductivity of the conducting medium. The minus sign is due to the fact that if $\Delta T > 0$ the heat actually flows from II to I. In the limit of $\Delta n \to 0$, the heat flux across from II to I can be written

$$-K \frac{\partial T}{\partial n} = -K\nabla T.$$

The quantity $\partial T / \partial n$ is called the gradient of T which in vector form is ∇T.

We are now ready to derive the equation for heat conduction. Let V be an arbitrary volume lying within the solid and bounded by surface S. The total

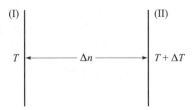

Figure 4.20.   Heat flux through a thin sheet.

179

amount of heat entering S per unit time is

$$\iint_S (K\nabla T) \cdot \hat{n}\, dS,$$

where $\hat{n}$ is an outward unit vector normal to element surface area dS. Using the divergence theorem, this can be written as

$$\iint_S (K\nabla T) \cdot \hat{n}\, dS = \iiint_V \nabla \cdot (K\nabla T)\, dV. \tag{4.38}$$

Now the heat contained in V is given by

$$\iiint_V c\rho T\, dV,$$

where c and ρ are respectively the specific heat capacity and density of the solid. Then the time rate of increase of heat is

$$\frac{\partial}{\partial t} \iiint_V c\rho T\, dV = \iiint_V c\rho \frac{\partial T}{\partial t}\, dV. \tag{4.39}$$

Equating the right hand sides of Eqs. (4.38) and (4.39) yields

$$\iiint_V \left[c\rho \frac{\partial T}{\partial t} - \nabla \cdot (K\nabla T) \right] dV = 0.$$

Since V is arbitrary, the integrand (assumed continuous) must be identically zero:

$$c\rho \frac{\partial T}{\partial t} = \nabla \cdot (K\nabla T)$$

or if K, c, ρ are constants

$$\frac{\partial T}{\partial t} = k\nabla \cdot \nabla T = k\nabla^2 T, \tag{4.40}$$

where $k = K/c\rho$. This is the required equation for heat conduction and was first developed by Fourier in 1822. For the semiinfinite thin bar, the boundary conditions are

$$T(x,0) = f(x), \; T(0,t) = 0, \qquad |T(x,t)| < M, \tag{4.41}$$

where the last condition means that the temperature must be bounded for physical reasons.

A solution of Eq. (4.40) can be obtained by separation of variables, that is by letting

$$T = X(x)H(t).$$

Then

$$XH' = kX''H \quad \text{or} \quad X''/X = H'/kH.$$

Each side must be a constant which we call $-\lambda^2$. (If we use $+\lambda^2$, the resulting solution does not satisfy the boundedness condition for real values of λ.) Then

$$X'' + \lambda^2 X = 0, \quad H' + \lambda^2 kH = 0$$

with the solutions

$$X(x) = A_1 \cos \lambda x + B_1 \sin \lambda x, \quad H(t) = C_1 e^{-k\lambda^2 t}.$$

A solution to Eq. (4.40) is thus given by

$$T(x,t) = C_1 e^{-k\lambda^2 t}(A_1 \cos \lambda x + B_1 \sin \lambda x)$$

$$= e^{-k\lambda^2 t}(A \cos \lambda x + B \sin \lambda x).$$

From the second of the boundary conditions (4.41) we find $A = 0$ and so $T(x,t)$ reduces to

$$T(x,t) = Be^{-k\lambda^2 t} \sin \lambda x.$$

Since there is no restriction on the value of λ, we can replace B by a function $B(\lambda)$ and integrate over λ from 0 to ∞ and still have a solution:

$$T(x,t) = \int_0^\infty B(\lambda) e^{-k\lambda^2 t} \sin \lambda x \, d\lambda. \tag{4.42}$$

Using the first of boundary conditions (4.41) we find

$$f(x) = \int_0^\infty B(\lambda) \sin \lambda x \, d\lambda.$$

Then by the Fourier sine transform we find

$$B(\lambda) = \frac{2}{\pi} \int_0^\infty f(x) \sin \lambda x \, dx = \frac{2}{\pi} \int_0^\infty f(u) \sin \lambda u \, du$$

and the temperature distribution along the semiinfinite thin bar is

$$T(x,t) = \frac{2}{\pi} \int_0^\infty \int_0^\infty f(u) e^{-k\lambda^2 t} \sin \lambda u \sin \lambda x \, d\lambda \, du. \tag{4.43}$$

Using the relation

$$\sin \lambda u \sin \lambda x = \tfrac{1}{2}[\cos \lambda(u - x) - \cos \lambda(u + x)],$$

Eq. (4.43) can be rewritten

$$T(x,t) = \frac{1}{\pi} \int_0^\infty \int_0^\infty f(u) e^{-k\lambda^2 t}[\cos \lambda(u - x) - \cos \lambda(u + x)] \, d\lambda \, du$$

$$= \frac{1}{\pi} \int_0^\infty f(u) \left[\int_0^\infty e^{-k\lambda^2 t} \cos \lambda(u - x) \, d\lambda - \int_0^\infty e^{-k\lambda^2 t} \cos \lambda(u + x) \, d\lambda \right] du.$$

Using the integral

$$\int_0^\infty e^{-\alpha\lambda^2}\cos\beta\lambda d\lambda = \frac{1}{2}\sqrt{\frac{\pi}{\alpha}}e^{-\beta^2/4\alpha},$$

we find

$$T(x,t) = \frac{1}{2\sqrt{\pi kt}}\left[\int_0^\infty f(u)e^{-(u-x)^2/4kt}du - \int_0^\infty f(u)e^{-(u+x)^2/4kt}du\right].$$

Letting $(u-x)/2\sqrt{kt} = w$ in the first integral and $(u+x)/2\sqrt{kt} = w$ in the second integral, we obtain

$$T(x,t) = \frac{1}{\sqrt{\pi}}\left[\int_{-x/2\sqrt{kt}}^\infty e^{-w^2}f(2w\sqrt{kt}+x)dw - \int_{x/2\sqrt{kt}}^\infty e^{-w^2}f(2w\sqrt{kt}-x)dw\right].$$

Fourier transforms for functions of several variables

We can extend the development of Fourier transforms to a function of several variables, such as $f(x,y,z)$. If we first decompose the function into a Fourier integral with respect to x, we obtain

$$f(x,y,z) = \frac{1}{\sqrt{2\pi}}\int_{-\infty}^\infty \gamma(\omega_x,y,z)e^{i\omega_x x}d\omega_x,$$

where γ is the Fourier transform. Similarly, we can decompose the function with respect to y and z to obtain

$$f(x,y,z) = \frac{1}{(2\pi)^{2/3}}\int_{-\infty}^\infty g(\omega_x,\omega_y,\omega_z)e^{i(\omega_x x+\omega_y y+\omega_z z)}d\omega_x d\omega_y d\omega_z,$$

with

$$g(\omega_x,\omega_y,\omega_z) = \frac{1}{(2\pi)^{2/3}}\int_{-\infty}^\infty f(x,y,z)e^{-i(\omega_x x+\omega_y y+\omega_z z)}dxdydz.$$

We can regard $\omega_x,\omega_y,\omega_z$ as the components of a vector ω whose magnitude is

$$\omega = \sqrt{\omega_x^2 + \omega_y^2 + \omega_z^2},$$

then we express the above results in terms of the vector ω:

$$f(\mathbf{r}) = \frac{1}{(2\pi)^{2/3}}\int_{-\infty}^\infty g(\omega)e^{i\omega\cdot\mathbf{r}}d\omega, \tag{4.44}$$

$$g(\omega) = \frac{1}{(2\pi)^{2/3}}\int_{-\infty}^\infty f(\mathbf{r})e^{-(i\omega\cdot\mathbf{r})}d\mathbf{r}. \tag{4.45}$$

The Fourier integral and the delta function

The delta function is a very useful tool in physics, but it is not a function in the usual mathematical sense. The need for this strange 'function' arises naturally from the Fourier integrals. Let us go back to Eqs. (4.30) and (4.31) and substitute $g(\omega)$ into $f(x)$; we then have

$$f(x) = \frac{1}{2\pi} \int_{-\infty}^{\infty} d\omega \int_{-\infty}^{\infty} dx' f(x') e^{i\omega(x-x')}.$$

Interchanging the order of integration gives

$$f(x) = \int_{-\infty}^{\infty} dx' f(x') \frac{1}{2\pi} \int_{-\infty}^{\infty} d\omega e^{i\omega(x-x')}. \tag{4.46}$$

If the above equation holds for any function $f(x)$, then this tells us something remarkable about the integral

$$\frac{1}{2\pi} \int_{-\infty}^{\infty} d\omega e^{i\omega(x-x')}$$

considered as a function of x'. It vanishes everywhere except at $x' = x$, and its integral with respect to x' over any interval including x is unity. That is, we may think of this function as having an infinitely high, infinitely narrow peak at $x = x'$. Such a strange function is called Dirac's delta function (first introduced by Paul A. M. Dirac):

$$\delta(x - x') = \frac{1}{2\pi} \int_{-\infty}^{\infty} d\omega e^{i\omega(x-x')}. \tag{4.47}$$

Equation (4.46) then becomes

$$f(x) = \int_{-\infty}^{\infty} f(x') \delta(x - x') dx'. \tag{4.48}$$

Equation (4.47) is an integral representation of the delta function. We summarize its properties below:

$$\delta(x - x') = 0, \quad \text{if } x' \neq x; \tag{4.49a}$$

$$\int_a^b \delta(x - x') dx' = \begin{cases} 0, & \text{if } x > b \text{ or } x < a \\ 1, & \text{if } a < x < b \end{cases}; \tag{4.49b}$$

$$f(x) = \int_{-\infty}^{\infty} f(x') \delta(x - x') dx'. \tag{4.49c}$$

It is often convenient to place the origin at the singular point, in which case the delta function may be written as

$$\delta(x) = \frac{1}{2\pi}\int_{-\infty}^{\infty} dw\, e^{iwx}. \tag{4.50}$$

To examine the behavior of the function for both small and large x, we use an alternative representation of this function obtained by integrating as follows:

$$\delta(x) = \frac{1}{2\pi}\lim_{a\to\infty}\int_{-a}^{a} e^{iwx}\, dw = \lim_{a\to\infty}\frac{1}{2\pi}\left[\frac{e^{iax} - e^{-iax}}{ix}\right] = \lim_{a\to\infty}\frac{\sin ax}{\pi x}, \tag{4.51}$$

where a is positive and real. We see immediately that $\delta(-x) = \delta(x)$. To examine its behavior for small x, we consider the limit as x goes to zero:

$$\lim_{x\to 0}\frac{\sin ax}{\pi x} = \frac{a}{\pi}\lim_{x\to 0}\frac{\sin ax}{ax} = \frac{a}{\pi}.$$

Thus, $\delta(0) = \lim_{a\to\infty}(a/\pi) \to \infty$, or the amplitude becomes infinite at the singularity. For large $|x|$, we see that $\sin(ax)/x$ oscillates with period $2\pi/a$, and its amplitude falls off as $1/|x|$. But in the limit as a goes to infinity, the period becomes infinitesimally narrow so that the function approaches zero everywhere except for the infinite spike of infinitesimal width at the singularity. What is the integral of Eq. (4.51) over all space?

$$\int_{-\infty}^{\infty}\lim_{a\to\infty}\frac{\sin ax}{\pi x}\, dx = \lim_{a\to\infty}\frac{2}{\pi}\int_{0}^{\infty}\frac{\sin ax}{\pi x}\, dx = \frac{2}{\pi}\frac{\pi}{2} = 1.$$

Thus, the delta function may be thought of as a spike function which has unit area but a non-zero amplitude at the point of singularity, where the amplitude becomes infinite. No ordinary mathematical function with these properties exists. How do we end up with such an improper function? It occurs because the change of order of integration in Eq. (4.46) is not permissible. In spite of this, the Dirac delta function is a most convenient function to use symbolically. For in applications the delta function always occurs under an integral sign. Carrying out this integration, using the formal properties of the delta function, is really equivalent to inverting the order of integration once more, thus getting back to a mathematically correct expression. Thus, using Eq. (4.49) we have

$$\int_{-\infty}^{\infty} f(x)\delta(x - x')\, dx = f(x'),$$

but, on substituting Eq. (4.47) for the delta function, the integral on the left hand side becomes

$$\int_{-\infty}^{\infty} f(x)\left\{\frac{1}{\sqrt{2\pi}}\int_{-\infty}^{\infty} dw\, e^{iw(x-x')}\right\} dx$$

or, using the property $\delta(-x) = \delta(x)$,

$$\int_{-\infty}^{\infty} f(x)\left\{\frac{1}{\sqrt{2\pi}}\int_{-\infty}^{\infty} dw e^{-iw(x-x')}\right\}dx$$

and *changing the order of integration*, we have

$$\int_{-\infty}^{\infty} f(x)\left\{\frac{1}{\sqrt{2\pi}}\int_{-\infty}^{\infty} dw e^{-iwx}\right\}e^{iwx'}dx.$$

Comparing this expression with Eqs. (4.30) and (4.31), we see at once that this double integral is equal to $f(x')$, the correct mathematical expression.

It is important to keep in mind that the delta function cannot be the end result of a calculation and has meaning only so long as a subsequent integration over its argument is carried out.

We can easily verify the following most frequently required properties of the delta function:

If $a < b$

$$\int_{a}^{b} f(x)\delta(x - x')dx = \begin{cases} f(x'), & \text{if } a < x' < b \\ 0, & \text{if } x' < a \text{ or } x' < b \end{cases}, \tag{4.52a}$$

$$\delta(-x) = \delta(x), \tag{4.52b}$$

$$\delta'(x) = -\delta'(-x), \quad \delta'(x) = d\delta(x)/dx, \tag{4.52c}$$

$$x\delta(x) = 0, \tag{4.52d}$$

$$\delta(ax) = a^{-1}\delta(x), \quad a > 0, \tag{4.52e}$$

$$\delta(x^2 - a^2) = (2a)^{-1}[\delta(x - a) + \delta(x + a)], \quad a > 0, \tag{4.52f}$$

$$\int \delta(a - x)\delta(x - b)dx = \delta(a - b), \tag{4.52g}$$

$$f(x)\delta(x - a) = f(a)\delta(x - a). \tag{4.52h}$$

Each of the first six of these listed properties can be established by multiplying both sides by a continuous, differentiable function $f(x)$ and then integrating over x. For example, multiplying $x\delta'(x)$ by $f(x)$ and integrating over x gives

$$\int f(x)x\delta'(x)dx = -\int \delta(x)\frac{d}{dx}[xf(x)]dx$$

$$= -\int \delta(x)[f(x) + xf'(x)]dx = -\int f(x)\delta(x)dx.$$

Thus $x\delta(x)$ has the same effect when it is a factor in an integrand as has $-\delta(x)$.

Parseval's identity for Fourier integrals

We arrived earlier at Parseval's identity for Fourier series. An analogy exists for Fourier integrals. If $g(\alpha)$ and $G(\alpha)$ are Fourier transforms of $f(x)$ and $F(x)$ respectively, we can show that

$$\int_{-\infty}^{\infty} f(x)F^*(x)dx = \frac{1}{2\pi}\int_{-\infty}^{\infty} g(\alpha)G^*(\alpha)d\alpha, \tag{4.54}$$

where $F^*(x)$ is the complex conjugate of $F(x)$. In particular, if $F(x) = f(x)$ and hence $G(\alpha) = g(\alpha)$, then we have

$$\int_{-\infty}^{\infty} |f(x)|^2 dx = \int_{-\infty}^{\infty} |g(\alpha)|d\alpha. \tag{4.54}$$

Equation (4.53), or the more general Eq. (4.54), is known as the Parseval's identity for Fourier integrals. Its proof is straightforward:

$$\int_{-\infty}^{\infty} f(x)F^*(x)dx = \int_{-\infty}^{\infty}\left[\frac{1}{\sqrt{2\pi}}\int_{-\infty}^{\infty} g(\alpha)e^{-i\alpha x}d\alpha\right]$$

$$\times\left[\frac{1}{\sqrt{2\pi}}\int_{-\infty}^{\infty} G^*(\alpha')e^{i\alpha' x}d\alpha'\right]dx$$

$$= \int_{-\infty}^{\infty} d\alpha\int_{-\infty}^{\infty} d\alpha' g(\alpha)G^*(\alpha')\left[\frac{1}{2\pi}\int_{-\infty}^{\infty} e^{ix(\alpha-\alpha')}dx\right]$$

$$= \int_{-\infty}^{\infty} d\alpha g(\alpha)\int_{-\infty}^{\infty} d\alpha' G^*(\alpha')\delta(\alpha'-\alpha) = \int_{-\infty}^{\infty} g(\alpha)G^*(\alpha)d\alpha.$$

Parseval's identity is very useful in understanding the physical interpretation of the transform function $g(\alpha)$ when the physical significance of $f(x)$ is known. The following example will show this.

Example 4.10
Consider the following function, as shown in Fig. 4.21, which might represent the current in an antenna, or the electric field in a radiated wave, or displacement of a damped harmonic oscillator:

$$f(t) = \begin{cases} 0 & t < 0 \\ e^{-t/T}\sin\omega_0 t & t > 0 \end{cases}.$$

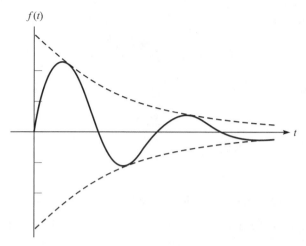

Figure 4.21. A damped sine wave.

Its Fourier transform $g(\omega)$ is

$$g(\omega) = \frac{1}{\sqrt{2\pi}} \int_{-\infty}^{\infty} f(t)e^{-i\omega t}\,dt$$

$$= \frac{1}{\sqrt{2\pi}} \int_{-\infty}^{\infty} e^{-t/T}e^{-i\omega t}\sin\omega_0 t\,dt$$

$$= \frac{1}{2\sqrt{2\pi}}\left(\frac{1}{\omega+\omega_0 - i/T} - \frac{1}{\omega - \omega_0 - i/T}\right).$$

If $f(t)$ is a radiated electric field, the radiated power is proportional to $|f(t)|^2$ and the total energy radiated is proportional to $\int_0^{\infty} |f(t)|^2\,dt$. This is equal to $\int_0^{\infty} |g(\omega)|^2\,d\omega$ by Parseval's identity. Then $|g(\omega)|^2$ must be the energy radiated per unit frequency interval.

Parseval's identity can be used to evaluate some definite integrals. As an example, let us revisit Example 4.8, where the given function is

$$f(x) = \begin{cases} 1 & |x| < a \\ 0 & |x| > a \end{cases}$$

and its Fourier transform is

$$g(\omega) = \sqrt{\frac{2}{\pi}}\frac{\sin\omega a}{\omega}.$$

By Parseval's identity, we have

$$\int_{-\infty}^{\infty} \{f(x)\}^2\,dx = \int_{-\infty}^{\infty} \{g(\omega)\}^2\,d\omega.$$

187

This is equivalent to

$$\int_{-a}^{a} (1)^2 dx = \int_{-\infty}^{\infty} \frac{2}{\pi} \frac{\sin^2 \omega a}{\omega^2} d\omega,$$

from which we find

$$\int_{0}^{\infty} \frac{2}{\pi} \frac{\sin^2 \omega a}{\omega^2} d\omega = \frac{\pi a}{2}.$$

The convolution theorem for Fourier transforms

The convolution of the functions $f(x)$ and $H(x)$, denoted by $f * H$, is defined by

$$f * H = \int_{-\infty}^{\infty} f(u)H(x-u)du. \qquad (4.55)$$

If $g(\omega)$ and $G(\omega)$ are Fourier transforms of $f(x)$ and $H(x)$ respectively, we can show that

$$\frac{1}{2\pi} \int_{-\infty}^{\infty} g(\omega)G(\omega)e^{i\omega x} d\omega = \int_{-\infty}^{\infty} f(u)H(x-u)du. \qquad (4.56)$$

This is known as the convolution theorem for Fourier transforms. It means that the Fourier transform of the product $g(\omega)G(\omega)$, the left hand side of Eq. (55), is the convolution of the original function.

The proof is not difficult. We have, by definition of the Fourier transform,

$$g(\omega) = \frac{1}{\sqrt{2\pi}} \int_{-\infty}^{\infty} f(x)e^{-i\omega x} dx, \qquad G(\omega) = \frac{1}{\sqrt{2\pi}} \int_{-\infty}^{\infty} H(x')e^{-i\omega x'} dx'.$$

Then

$$g(\omega)G(\omega) = \frac{1}{2\pi} \int_{-\infty}^{\infty} \int_{-\infty}^{\infty} f(x)H(x')e^{-i\omega(x+x')} dx dx'. \qquad (4.57)$$

Let $x + x' = u$ in the double integral of Eq. (4.57) and we wish to transform from (x, x') to (x, u). We thus have

$$dx dx' = \frac{\partial(x, x')}{\partial(x, u)} du dx,$$

where the Jacobian of the transformation is

$$\frac{\partial(x, x')}{\partial(x, u)} = \begin{vmatrix} \dfrac{\partial x}{\partial x} & \dfrac{\partial x}{\partial u} \\ \dfrac{\partial x'}{\partial x} & \dfrac{\partial x'}{\partial u} \end{vmatrix} = \begin{vmatrix} 1 & 0 \\ 0 & 1 \end{vmatrix} = 1.$$

Thus Eq. (4.57) becomes

$$g(\omega)G(\omega) = \frac{1}{2\pi} \int_{-\infty}^{\infty} \int_{-\infty}^{\infty} f(x)H(u-x)e^{-i\omega u} dx du$$

$$= \frac{1}{2\pi} \int_{-\infty}^{\infty} e^{-i\omega u} \left[\int_{-\infty}^{\infty} f(x)H(u-x) du \right] dx$$

$$= F\left\{ \int_{-\infty}^{\infty} f(x)H(u-x) du \right\} = F\{f * H\}. \qquad (4.58)$$

From this we have equivalently

$$f * H = F^{-1}\{g(\omega)G(\omega)\} = (1/2\pi) \int_{-\infty}^{\infty} e^{i\omega x} g(\omega)G(\omega),$$

which is Eq. (4.56).

Equation (4.58) can be rewritten as

$$F\{f\}F\{H\} = F\{f * H\} \qquad (g = F\{f\}, G = F\{H\}),$$

which states that *the Fourier transform of the convolution of f(x) and H(x) is equal to the product of the Fourier transforms of f(x) and H(x)*. This statement is often taken as the convolution theorem.

The convolution obeys the commutative, associative and distributive laws of algebra that is, if we have functions f_1, f_2, f_3 then

$$\left. \begin{array}{ll} f_1 * f_2 = f_2 * f_1 & \text{commutative;} \\ f_1 * (f_2 * f_3) = (f_1 * f_2) * f_3 & \text{associative;} \\ f_1 * (f_2 + f_3) = f_1 * f_2 + f_1 * f_3 & \text{distributive.} \end{array} \right\} \qquad (4.59)$$

It is not difficult to prove these relations. For example, to prove the commutative law, we first have

$$f_1 * f_2 \equiv \int_{-\infty}^{\infty} f_1(u)f_2(x-u) du.$$

Now let $x - u = v$, then

$$f_1 * f_2 \equiv \int_{-\infty}^{\infty} f_1(u)f_2(x-u) du$$

$$= \int_{-\infty}^{\infty} f_1(x-v)f_2(v) dv = f_2 * f_1.$$

Example 4.11

Solve the integral equation $y(x) = f(x) + \int_{-\infty}^{\infty} y(u)r(x-u) du$, where $f(x)$ and $r(x)$ are given, and the Fourier transforms of $y(x)$, $f(x)$ and $r(x)$ exist.

Solution: Let us denote the Fourier transforms of $y(x)$, $f(x)$ and $r(x)$ by $Y(\omega), F(\omega)$, and $R(\omega)$ respectively. Taking the Fourier transform of both sides of the given integral equation, we have by the convolution theorem

$$Y(\omega) = F(\omega) + Y(\omega)R(\omega) \quad \text{or} \quad Y(\omega) = \frac{F(\omega)}{1 - R(\omega)}.$$

Calculations of Fourier transforms

Fourier transforms can often be used to transform a differential equation which is difficult to solve into a simpler equation that can be solved relatively easy. In order to use the transform methods to solve first- and second-order differential equations, the transforms of first- and second-order derivatives are needed. By taking the Fourier transform with respect to the variable x, we can show that

$$\left.\begin{array}{ll}
(a) & F\left(\dfrac{\partial u}{\partial x}\right) = i\alpha F(u), \\[3mm]
(b) & F\left(\dfrac{\partial^2 u}{\partial x^2}\right) = -\alpha^2 F(u), \\[3mm]
(c) & F\left(\dfrac{\partial u}{\partial t}\right) = \dfrac{\partial}{\partial t} F(u).
\end{array}\right\} \tag{4.60}$$

Proof: (a) By definition we have

$$F\left(\frac{\partial u}{\partial x}\right) = \int_{-\infty}^{\infty} \frac{\partial u}{\partial x} e^{-i\alpha x} dx,$$

where the factor $1/\sqrt{2\pi}$ has been dropped. Using integration by parts, we obtain

$$F\left(\frac{\partial u}{\partial x}\right) = \int_{-\infty}^{\infty} \frac{\partial u}{\partial x} e^{-i\alpha x} dx$$

$$= ue^{-i\alpha x}\Big|_{-\infty}^{\infty} + i\alpha \int_{-\infty}^{\infty} ue^{-i\alpha x} dx$$

$$= i\alpha F(u).$$

(b) Let $u = \partial v/\partial x$ in (a), then

$$F\left(\frac{\partial^2 v}{\partial x^2}\right) = i\alpha F\left(\frac{\partial v}{\partial x}\right) = (i\alpha)^2 F(v).$$

Now if we formally replace v by u we have

$$F\left(\frac{\partial^2 u}{\partial x^2}\right) = -\alpha^2 F(u),$$

190

provided that u and $\partial u/\partial x \to 0$ as $x \to \pm\infty$. In general, we can show that

$$F\left(\frac{\partial^n u}{\partial x^n}\right) = (i\alpha)^n F(u)$$

if $u, \partial u/\partial x, \ldots, \partial^{n-1} u/\partial x^{n-1} \to \pm\infty$ as $x \to \pm\infty$.

(c) By definition

$$F\left(\frac{\partial u}{\partial t}\right) = \int_{-\infty}^{\infty} \frac{\partial u}{\partial t} e^{-i\alpha x} dx = \frac{\partial}{\partial t} \int_{-\infty}^{\infty} u e^{-i\alpha x} dx = \frac{\partial}{\partial t} F(u).$$

Example 4.12

Solve the inhomogeneous differential equation

$$\left(\frac{d^2}{dx^2} + p\frac{d}{dx} + q\right) f(x) = R(x), \quad -\infty \le x \le \infty,$$

where p and q are constants.

Solution: We transform both sides

$$F\left\{\frac{d^2 f}{dx^2} + p\frac{df}{dx} + qf\right\} = [(i\alpha)^2 + p(i\alpha) + q]F\{f(x)\}$$

$$= F\{R(x)\}.$$

If we denote the Fourier transforms of $f(x)$ and $R(x)$ by $g(\alpha)$ and $G(\alpha)$, respectively,

$$F\{f(x)\} = g(\alpha), \quad F\{R(x)\} = G(\alpha),$$

we have

$$(-\alpha^2 + ip\alpha + q)g(\alpha) = G(\alpha), \quad \text{or} \quad g(\alpha) = G(\alpha)/(-\alpha^2 + ip\alpha + q)$$

and hence

$$f(x) = \frac{1}{\sqrt{2\pi}} \int_{-\infty}^{\infty} e^{i\alpha x} g(\alpha) d\alpha$$

$$= \frac{1}{\sqrt{2\pi}} \int_{-\infty}^{\infty} e^{i\alpha x} \frac{G(\alpha)}{-\alpha^2 + ip\alpha + q} d\alpha.$$

We will not gain anything if we do not know how to evaluate this complex integral. This is not a difficult problem in the theory of functions of complex variables (see Chapter 7).

The delta function and the Green's function method

The Green's function method is a very useful technique in the solution of partial differential equations. It is usually used when boundary conditions, rather than initial conditions, are specified. To appreciate its usefulness, let us consider the inhomogeneous differential equation

$$L(x)f(x) - \lambda f(x) = R(x) \tag{4.61}$$

over a domain D, with L an arbitrary differential operator, and λ a given constant. Suppose we can expand $f(x)$ and $R(x)$ in eigenfunctions u_n of the operator $L(Lu_n = \lambda_n u_n)$:

$$f(x) = \sum_n c_n u_n(x), \quad R(x) = \sum_n d_n u_n(x).$$

Substituting these into Eq. (4.61) we obtain

$$\sum_n c_n(\lambda_n - \lambda)u_n(x) = \sum_n d_n u_n(x).$$

Since the eigenfunctions $u_n(x)$ are linearly independent, we must have

$$c_n(\lambda_n - \lambda) = d_n \quad \text{or} \quad c_n = d_n/(\lambda_n - \lambda).$$

Moreover,

$$d_n = \int_D u_n^* R(x)dx.$$

Now we may write c_n as

$$c_n = \frac{1}{\lambda_n - \lambda} \int_D u_n^* R(x)dx,$$

therefore

$$f(x) = \sum_n \frac{u_n}{\lambda_n - \lambda} \int_D u_n^*(x')R(x')dx'.$$

This expression may be written in the form

$$f(x) = \int_D G(x, x')R(x')dx', \tag{4.62}$$

where $G(x, x')$ is given by

$$G(x, x') = \sum_n \frac{u_n(x)u_n^*(x')}{\lambda_n - \lambda} \tag{4.63}$$

and is called the Green's function. Some authors prefer to write $G(x, x'; \lambda)$ to emphasize the dependence of G on λ as well as on x and x'.

What is the differential equation obeyed by $G(x, x')$? Suppose $f(x')$ in Eq. (4.62) is taken to be $\delta(x' - x_0)$, then we obtain

$$f(x) = \int_D G(x, x')\delta(x' - x_0)dx = G(x, x_0).$$

Therefore $G(x, x')$ is the solution of

$$LG(x, x') - \lambda G(x, x') = \delta(x - x'), \tag{4.64}$$

subject to the appropriate boundary conditions. Eq. (4.64) shows clearly that *the Green's function is the solution of the problem for a unit point 'source'* $R(x) = \delta(x - x')$.

Example 4.13
Find the solution to the differential equation

$$\frac{d^2u}{dx^2} - k^2u = f(x) \tag{4.65}$$

on the interval $0 \leq x \leq l$, with $u(0) = u(l) = 0$, for a general function $f(x)$.

Solution: We first solve the differential equation which $G(x, x')$ obeys:

$$\frac{d^2G(x, x')}{dx^2} - k^2G(x, x') = \delta(x - x'). \tag{4.66}$$

For x equal to anything but x' (that is, for $x < x'$ or $x > x'$), $\delta(x - x') = 0$ and we have

$$\frac{d^2G_<(x, x')}{dx^2} - k^2G_<(x, x') = 0 \qquad (x < x'),$$

$$\frac{d^2G_>(x, x')}{dx^2} - k^2G_>(x, x') = 0 \qquad (x > x').$$

Therefore, for $x < x'$

$$G_< = Ae^{kx} + Be^{-kx}.$$

By the boundary condition $u(0) = 0$ we find $A + B = 0$, and $G_<$ reduces to

$$G_< = A(e^{kx} - e^{-kx}); \tag{4.67a}$$

similarly, for $x > x'$

$$G_> = Ce^{kx} + De^{-kx}.$$

By the boundary condition $u(l) = 0$ we find $Ce^{kl} + De^{-kl} = 0$, and $G_>$ can be rewritten as

$$G_> = C'[e^{k(x-l)} - e^{-k(x-l)}], \qquad (4.67b)$$

where $C' = Ce^{kl}$.

How do we determine the constants A and C'? First, continuity of G at $x = x'$ gives

$$A(e^{kx} - e^{-kx}) = C'(e^{k(x-l)} - e^{-k(x-l)}). \qquad (4.68)$$

A second constraint is obtained by integrating Eq. (4.61) from $x' - \varepsilon$ to $x' + \varepsilon$, where ε is infinitesimal:

$$\int_{x'-\varepsilon}^{x'+\varepsilon} \left[\frac{d^2 G}{dx^2} - k^2 G\right] dx = \int_{x'-\varepsilon}^{x'+\varepsilon} \delta(x - x') dx = 1. \qquad (4.69)$$

But

$$\int_{x'-\varepsilon}^{x'+\varepsilon} k^2 G dx = k^2(G_> - G_<) = 0,$$

where the last step is required by the continuity of G. Accordingly, Eq. (4.64) reduces to

$$\int_{x'-\varepsilon}^{x'+\varepsilon} \frac{d^2 G}{dx^2} dx = \frac{dG_>}{dx} - \frac{dG_<}{dx} = 1. \qquad (4.70)$$

Now

$$\frac{dG_<}{dx}\bigg|_{x=x'} = Ak(e^{kx'} + e^{-kx'})$$

and

$$\frac{dG_>}{dx}\bigg|_{x=x'} = C'k[e^{k(x'-l)} + e^{-k(x'-l)}].$$

Substituting these into Eq. (4.70) yields

$$C'k(e^{k(x'-l)} + e^{-k(x'-l)}) - Ak(e^{kx'} + e^{-kx'}) = 1. \qquad (4.71)$$

We can solve Eqs. (4.68) and (4.71) for the constants A and C'. After some algebraic manipulation, the solution is

$$A = \frac{1}{2k} \frac{\sinh k(x' - l)}{\sinh kl}, \qquad C' = \frac{1}{2k} \frac{\sinh kx'}{\sinh kl}$$

and the Green's function is

$$G(x, x') = \frac{1}{k} \frac{\sinh kx \sinh k(x' - l)}{\sinh kl}, \tag{4.72}$$

which can be combined with $f(x)$ to obtain $u(x)$:

$$u(x) = \int_0^l G(x, x') f(x') dx'.$$

Problems

4.1 (a) Find the period of the function $f(x) = \cos(x/3) + \cos(x/4)$.

    (b) Show that, if the function $f(t) = \cos \omega_1 t + \cos \omega_2 t$ is periodic with a period T, then the ratio ω_1/ω_2 must be a rational number.

4.2 Show that if $f(x + P) = f(x)$, then

$$\int_{a-P/2}^{a+P/2} f(x) dx = \int_{-P/2}^{P/2} f(x) dx, \qquad \int_P^{P+x} f(x) dx = \int_0^x f(x) dx.$$

4.3 (a) Using the result of Example 4.2, prove that

$$1 - \frac{1}{3} + \frac{1}{5} - \frac{1}{7} + - \cdots = \frac{\pi}{4}.$$

    (b) Using the result of Example 4.3, prove that

$$\frac{1}{1 \times 3} - \frac{1}{3 \times 5} + \frac{1}{5 \times 7} - + \cdots = \frac{\pi - 2}{4}.$$

4.4 Find the Fourier series which represents the function $f(x) = |x|$ in the interval $-\pi \le x \le \pi$.

4.5 Find the Fourier series which represents the function $f(x) = x$ in the interval $-\pi \le x \le \pi$.

4.6 Find the Fourier series which represents the function $f(x) = x^2$ in the interval $-\pi \le x \le \pi$.

4.7 Represent $f(x) = x, 0 < x < 2$, as: (a) in a half-range sine series, (b) a half-range cosine series.

4.8 Represent $f(x) = \sin x$, $0 < x < \pi$, as a Fourier cosine series.

4.9 (a) Show that the function $f(x)$ of period 2 which is equal to x on $(-1, 1)$ can be represented by the following Fourier series

$$-\frac{i}{\pi} \left(e^{i\pi x} - e^{-i\pi x} - \frac{1}{2} e^{2i\pi x} + \frac{1}{2} e^{-2i\pi x} + \frac{1}{3} e^{3i\pi x} - \frac{1}{3} e^{-3i\pi x} + \cdots \right).$$

    (b) Write Parseval's identity corresponding to the Fourier series of (a).

    (c) Determine from (b) the sum S of the series $1 + \frac{1}{4} + \frac{1}{9} + \cdots = \sum_{n=1}^{\infty} 1/n^2$.

4.10 Find the exponential form of the Fourier series of the function whose definition in one period is $f(x) = e^{-x}, -1 < x < 1$.

4.11 (a) Show that the set of functions

$$1, \quad \sin\frac{\pi x}{L}, \quad \cos\frac{\pi x}{L}, \quad \sin\frac{2\pi x}{L}, \quad \cos\frac{2\pi x}{L}, \quad \sin\frac{3\pi x}{L}, \quad \cos\frac{3\pi x}{L}, \ldots$$

form an orthogonal set in the interval $(-L, L)$.

(b) Determine the corresponding normalizing constants for the set in (a) so that the set is orthonormal in $(-L, L)$.

4.12 Express $f(x, y) = xy$ as a Fourier series for $0 \le x \le 1, 0 \le y \le 2$.

4.13 Steady-state heat conduction in a rectangular plate: Consider steady-state heat conduction in a flat plate having temperature values prescribed on the sides (Fig. 4.22). The boundary value problem modeling this is:

$$\frac{\partial^2 u}{\partial x^2} + \frac{\partial^2 u}{\partial y^2} = 0, \qquad 0 < x < \alpha, \quad 0 < y < \beta;$$

$$u(x, 0) = u(x, \beta) = 0, \qquad 0 < x < \alpha;$$

$$u(0, y) = 0, u(\alpha, y) = T, \qquad 0 < y < \beta.$$

Determine the temperature at any point of the plate.

4.14 Derive and solve the following eigenvalue problem which occurs in the theory of a vibrating square membrane whose sides, of length L, are kept fixed:

$$\frac{\partial^2 w}{\partial x^2} + \frac{\partial^2 w}{\partial y^2} + \lambda w = 0,$$

$$w(0, y) = w(L, y) = 0 \quad (0 \le y \le L),$$

$$w(x, 0) = w(x, L) = 0 \quad (0 \le y \le L).$$

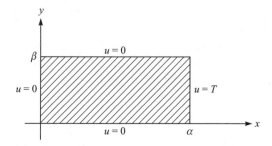

Figure 4.22. Flat plate with prescribed temperature.

4.15 Show that the Fourier integral can be written in the form

$$f(x) = \frac{1}{\pi} \int_0^\infty d\omega \int_{-\infty}^\infty f(x') \cos \omega (x - x') dx'.$$

4.16 Starting with the form obtained in Problem 4.15, show that the Fourier integral can be written in the form

$$f(x) = \int_0^\infty \{A(\omega) \cos \omega x + B(\omega) \sin \omega x\} d\omega,$$

where

$$A(\omega) = \frac{1}{\pi} \int_{-\infty}^\infty f(x) \cos \omega x \, dx, \qquad B(\omega) = \frac{1}{\pi} \int_{-\infty}^\infty f(x) \sin \omega x \, dx.$$

4.17 (a) Find the Fourier transform of

$$f(x) = \begin{cases} 1 - x^2 & |x| < 1 \\ 0 & |x| > 1 \end{cases}.$$

(b) Evaluate

$$\int_0^\infty \frac{x \cos x - \sin x}{x^3} \cos \frac{x}{2} dx.$$

4.18 (a) Find the Fourier cosine transform of $f(x) = e^{-mx}, m > 0$.

(b) Use the result in (a) to show that

$$\int_0^\infty \frac{\cos px}{x^2 + \alpha^2} dx = \frac{\pi}{2\alpha} e^{-p\alpha} \quad (p > 0, \alpha > 0).$$

4.19 Solve the integral equation

$$\int_0^\infty f(x) \sin \alpha x \, dx = \begin{cases} 1 - \alpha & 0 \le \alpha \le 1 \\ 0 & \alpha > 1 \end{cases}.$$

4.20 Find a bounded solution to Laplace's equation $\nabla^2 u(x, y) = 0$ for the half-plane $y > 0$ if u takes on the value of $f(x)$ on the x-axis:

$$\frac{\partial^2 u}{\partial x^2} + \frac{\partial^2 u}{\partial y^2} = 0, \qquad u(x, 0) = f(x), \qquad |u(x, y)| < M.$$

4.21 Show that the following two functions are valid representations of the delta function, where ε is positive and real:

(a) $\delta(x) = \dfrac{1}{\sqrt{\pi}} \lim_{\varepsilon \to 0} \dfrac{1}{\sqrt{\varepsilon}} e^{-x^2/\varepsilon}$

(b) $\delta(x) = \dfrac{1}{\pi} \lim_{\varepsilon \to 0} \dfrac{\varepsilon}{x^2 + \varepsilon^2}.$

4.22 Verify the following properties of the delta function:
(a) $\delta(x) = \delta(-x)$,
(b) $x\delta(x) = 0$,
(c) $\delta'(-x) = -\delta'(x)$,
(d) $x\delta'(x) = -\delta(x)$,
(e) $c\delta(cx) = \delta(x)$, $\quad c > 0$.

4.23 Solve the integral equation for $y(x)$

$$\int_{-\infty}^{\infty} \frac{y(u)du}{(x-u)^2 + a^2} = \frac{1}{x^2 + b^2} \quad 0 < a < b.$$

4.24 Use Fourier transforms to solve the boundary value problem

$$\frac{\partial u}{\partial t} = k\frac{\partial^2 u}{\partial x^2}, \qquad u(x,0) = f(x), \qquad |u(x,t)| < M,$$

where $-\infty < x < \infty, t > 0$.

4.25 Obtain a solution to the equation of a driven harmonic oscillator

$$\ddot{x}(t) + 2\beta\dot{x}(t) + \omega_0^2 x(t0 = R(t),$$

where β and ω_0 are positive and real constants.

5

Linear vector spaces

Linear vector space is to quantum mechanics what calculus is to classical mechanics. In this chapter the essential ideas of linear vector spaces will be discussed. The reader is already familiar with vector calculus in three-dimensional Euclidean space E_3 (Chapter 1). We therefore present our discussion as a generalization of elementary vector calculus. The presentation will be, however, slightly abstract and more formal than the discussion of vectors in Chapter 1. Any reader who is not already familiar with this sort of discussion should be patient with the first few sections. You will then be amply repaid by finding the rest of this chapter relatively easy reading.

Euclidean n-space E_n

In the study of vector analysis in E_3, an ordered triple of numbers (a_1, a_2, a_3) has two different geometric interpretations. It represents a point in space, with a_1, a_2, a_3 being its coordinates; it also represents a vector, with a_1, a_2, and a_3 being its components along the three coordinate axes (Fig. 5.1). This idea of using triples of numbers to locate points in three-dimensional space was first introduced in the mid-seventeenth century. By the latter part of the nineteenth century physicists and mathematicians began to use the quadruples of numbers (a_1, a_2, a_3, a_4) as points in four-dimensional space, quintuples $(a_1, a_2, a_3, a_4, a_5)$ as points in five-dimensional space etc. We now extend this to n-dimensional space E_n, where n is a positive integer. Although our geometric visualization doesn't extend beyond three-dimensional space, we can extend many familiar ideas beyond three-dimensional space by working with analytic or numerical properties of points and vectors rather than their geometric properties.

For two- or three-dimensional space, we use the terms 'ordered pair' and 'ordered triple.' When $n > 3$, we use the term 'ordered-n-tuplet' for a sequence

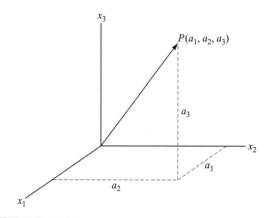

Figure 5.1.   A space point P whose position vector is **A**.

of n numbers, real or complex, $(a_1, a_2, a_3, \ldots, a_n)$; they will be viewed either as a generalized point or a generalized vector in a n-dimensional space E_n.

Two vectors $\mathbf{u} = (u_1, u_2, \ldots, u_n)$ and $\mathbf{v} = (v_1, v_2, \ldots, v_n)$ in E_n are called equal if

$$u_i = v_i, \quad i = 1, 2, \ldots, n \tag{5.1}$$

The sum $\mathbf{u} + \mathbf{v}$ is defined by

$$\mathbf{u} + \mathbf{v} = (u_1 + v_1, u_2 + v_2, \ldots, u_n + v_n) \tag{5.2}$$

and if k is any scalar, the scalar multiple $k\mathbf{u}$ is defined by

$$k\mathbf{u} = (ku_1, ku_2, \ldots, ku_n). \tag{5.3}$$

If $u = (u_1, u_2, \ldots, u_n)$ is any vector in E_n, its negative is given by

$$-\mathbf{u} = (-u_1, -u_2, \ldots, -u_n) \tag{5.4}$$

and the subtraction of vectors in E_n can be considered as addition: $\mathbf{v} - \mathbf{u} = \mathbf{v} + (-\mathbf{u})$. The null (zero) vector in E_n is defined to be the vector $0 = (0, 0, \ldots, 0)$.

The addition and scalar multiplication of vectors in E_n have the following arithmetic properties:

$$\mathbf{u} + \mathbf{v} = \mathbf{v} + \mathbf{u}, \tag{5.5a}$$

$$\mathbf{u} + (\mathbf{v} + \mathbf{w}) = (\mathbf{u} + \mathbf{v}) + \mathbf{w}, \tag{5.5b}$$

$$\mathbf{u} + 0 = 0 + \mathbf{u} = \mathbf{u}, \tag{5.5c}$$

$$a(b\mathbf{u}) = (ab)\mathbf{u}, \tag{5.5d}$$

$$a(\mathbf{u} + \mathbf{v}) = a\mathbf{u} + a\mathbf{v}, \tag{5.5e}$$

$$(a + b)\mathbf{u} = a\mathbf{u} + b\mathbf{u}, \tag{5.5f}$$

where $\mathbf{u}, \mathbf{v}, \mathbf{w}$ are vectors in E_n and a and b are scalars.

We usually define the inner product of two vectors in E_3 in terms of lengths of the vectors and the angle between the vectors: $\mathbf{A} \cdot \mathbf{B} = AB \cos \theta, \theta = \angle(\mathbf{A}, \mathbf{B})$. We do not define the inner product in E_n in the same manner. However, the inner product in E_3 has a second equivalent expression in terms of components: $\mathbf{A} \cdot \mathbf{B} = A_1 B_1 + A_2 B_2 + A_3 B_3$. We choose to define a similar formula for the general case. We made this choice because of the further generalization that will be outlined in the next section. Thus, for any two vectors $\mathbf{u} = (u_1, u_2, \ldots, u_n)$ and $\mathbf{v} = (v_1, v_2, \ldots, v_n)$ in E_n, the inner (or dot) product $\mathbf{u} \cdot \mathbf{v}$ is defined by

$$\mathbf{u} \cdot \mathbf{v} = u_1^* v_1 + u_2^* v_2 + \cdots + u_n^* v_n \tag{5.6}$$

where the asterisk denotes complex conjugation. $\mathbf{u}$ is often called the prefactor and $\mathbf{v}$ the post-factor. The inner product is linear with respect to the post-factor, and anti-linear with respect to the prefactor:

$$\mathbf{u} \cdot (a\mathbf{v} + b\mathbf{w}) = a\mathbf{u} \cdot \mathbf{v} + b\mathbf{u} \cdot \mathbf{w}, \quad (a\mathbf{u} + b\mathbf{v}) \cdot \mathbf{w} = a^*(\mathbf{u} \cdot \mathbf{v}) + b^*(\mathbf{u} \cdot \mathbf{w}).$$

We expect the inner product for the general case also to have the following three main features:

$$\mathbf{u} \cdot \mathbf{v} = (\mathbf{v} \cdot \mathbf{u})^* \tag{5.7a}$$

$$\mathbf{u} \cdot (a\mathbf{v} + b\mathbf{w}) = a\mathbf{u} \cdot \mathbf{v} + b\mathbf{u} \cdot \mathbf{w} \tag{5.7b}$$

$$\mathbf{u} \cdot \mathbf{u} \geq 0 \ (= 0, \text{ if and only if } \mathbf{u} = 0). \tag{5.7c}$$

Many of the familiar ideas from E_2 and E_3 have been carried over, so it is common to refer to E_n with the operations of addition, scalar multiplication, and with the inner product that we have defined here as Euclidean n-space.

General linear vector spaces

We now generalize the concept of vector space still further: a set of 'objects' (or elements) obeying a set of axioms, which will be chosen by abstracting the most important properties of vectors in E_n, forms a linear vector space V_n with the objects called vectors. Before introducing the requisite axioms, we first adapt a notation for our general vectors: general vectors are designated by the symbol $| \ \rangle$, which we call, following Dirac, ket vectors; the conjugates of ket vectors are denoted by the symbol $\langle \ |$, the bra vectors. However, for simplicity, we shall refer in the future to the ket vectors $| \ \rangle$ simply as vectors, and to the $\langle \ |$s as conjugate vectors. We now proceed to define two basic operations on these vectors: addition and multiplication by scalars.

> By addition we mean a rule for forming the sum, denoted $|\psi_1\rangle + |\psi_2\rangle$, for any pair of vectors $|\psi_1\rangle$ and $|\psi_2\rangle$.
> By scalar multiplication we mean a rule for associating with each scalar k and each vector $|\psi\rangle$ a new vector $k|\psi\rangle$.

We now proceed to generalize the concept of a vector space. An arbitrary set of n objects $|1\rangle, |2\rangle, |3\rangle, \ldots, |\phi\rangle, \ldots, |\varphi\rangle$ form a linear vector V_n if these objects, called vectors, meet the following axioms or properties:

A.1 If $|\phi\rangle$ and $|\varphi\rangle$ are objects in V_n and k is a scalar, then $|\phi\rangle + |\varphi\rangle$ and $k|\phi\rangle$ are in V_n, a feature called closure.

A.2 $|\phi\rangle + |\varphi\rangle = |\varphi\rangle + |\phi\rangle$; that is, addition is commutative.

A.3 $(|\phi\rangle + |\varphi\rangle) + |\psi\rangle = |\phi\rangle + (|\varphi\rangle + |\psi\rangle)$; that is, addition is associative.

A.4 $k(|\phi\rangle + |\varphi\rangle) = k|\phi\rangle + k|\varphi\rangle$; that is, scalar multiplication is distributive in the vectors.

A.5 $(k + \alpha)|\phi\rangle = k|\phi\rangle + \alpha|\phi\rangle$; that is, scalar multiplication is distributive in the scalars.

A.6 $k(\alpha|\phi\rangle) = k\alpha|\phi\rangle$; that is, scalar multiplication is associative.

A.7 There exists a null vector $|0\rangle$ in V_n such that $|\phi\rangle + |0\rangle = |\phi\rangle$ for all $|\phi\rangle$ in V_n.

A.8 For every vector $|\phi\rangle$ in V_n, there exists an inverse under addition, $|-\phi\rangle$ such that $|\phi\rangle + |-\phi\rangle = |0\rangle$.

The set of numbers $a, b, \ldots$ used in scalar multiplication of vectors is called the field over which the vector field is defined. If the field consists of real numbers, we have a real vector field; if they are complex, we have a complex field. *Note that the vectors themselves are neither real nor complex, the nature of the vectors is not specified. Vectors can be any kinds of objects; all that is required is that the vector space axioms be satisfied.* Thus we purposely do not use the symbol **V** to denote the vectors as the first step to turn the reader away from the limited concept of the vector as a directed line segment. Instead, we use Dirac's ket and bra symbols, $| \ \rangle$ and $\langle \ |$, to denote generic vectors.

The familiar three-dimensional space of position vectors E_3 is an example of a vector space over the field of real numbers. Let us now examine two simple examples.

Example 5.1

Let V be any plane through the origin in E_3. We wish to show that the points in the plane V form a vector space under the addition and scalar multiplication operations for vector in E_3.

Solution: Since E_3 itself is a vector space under the addition and scalar multiplication operations, thus Axioms A.2, A.3, A.4, A.5, and A.6 hold for all points in E_3 and consequently for all points in the plane V. We therefore need only show that Axioms A.1, A.7, and A.8 are satisfied.

Now the plane V, passing through the origin, has an equation of the form

$$ax_1 + bx_2 + cx_3 = 0.$$

Hence, if $\mathbf{u} = (u_1, u_2, u_3)$ and $\mathbf{v} = (v_1, v_2, v_3)$ are points in V, then we have

$$au_1 + bu_2 + cu_3 = 0 \quad \text{and} \quad av_1 + bv_2 + cv_3 = 0.$$

Addition gives

$$a(u_1 + v_1) + b(u_2 + v_2) + c(u_3 + v_3) = 0,$$

which shows that the point $\mathbf{u} + \mathbf{v}$ also lies in the plane V. This proves that Axiom A.1 is satisfied. Multiplying $au_1 + bu_2 + cu_3 = 0$ through by -1 gives

$$a(-u_1) + b(-u_2) + c(-u_3) = 0,$$

that is, the point $-\mathbf{u} = (-u_1, -u_2, -u_3)$ lies in V. This establishes Axiom A.8. The verification of Axiom A.7 is left as an exercise.

Example 5.2
Let V be the set of all $m \times n$ matrices with real elements. We know how to add matrices and multiply matrices by scalars. The corresponding rules obey closure, associativity and distributive requirements. The null matrix has all zeros in it, and the inverse under matrix addition is the matrix with all elements negated. Thus the set of all $m \times n$ matrices, together with the operations of matrix addition and scalar multiplication, is a vector space. We shall denote this vector space by the symbol M_{mn}.

Subspaces

Consider a vector space V. If W is a subset of V and forms a vector space under the addition and scalar multiplication, then W is called a subspace of V. For example, lines and planes passing through the origin form vector spaces and they are subspaces of E_3.

Example 5.3
We can show that the set of all 2×2 matrices having zero on the main diagonal is a subspace of the vector space M_{22} of all 2×2 matrices.

Solution: To prove this, let

$$\tilde{X} = \begin{pmatrix} 0 & x_{12} \\ x_{21} & 0 \end{pmatrix} \qquad \tilde{Y} = \begin{pmatrix} 0 & y_{12} \\ y_{21} & 0 \end{pmatrix}$$

be two matrices in W and k any scalar. Then

$$k\tilde{X} = \begin{pmatrix} 0 & x_{12} \\ kx_{21} & 0 \end{pmatrix} \quad \text{and} \quad \tilde{X} + \tilde{Y} = \begin{pmatrix} 0 & x_{12} + y_{12} \\ x_{21} + y_{21} & 0 \end{pmatrix}$$

and thus they lie in W. We leave the verification of other axioms as exercises.

Linear combination

A vector $|W\rangle$ is a linear combination of the vectors $|v_1\rangle, |v_2\rangle, \ldots, |v_r\rangle$ if it can be expressed in the form

$$|W\rangle = k_1|v_1\rangle + k_2|v_2\rangle + \cdots + k_r|v_r\rangle,$$

where $k_1, k_2, \ldots, k_r$ are scalars. For example, it is easy to show that the vector $|W\rangle = (9, 2, 7)$ in E_3 is a linear combination of $|v_1\rangle = (1, 2, -1)$ and $|v_2\rangle = (6, 4, 2)$. To see this, let us write

$$(9, 2, 7) = k_1(1, 2, -1) + k_2(6, 4, 2)$$

or

$$(9, 2, 7) = (k_1 + 6k_2, 2k_1 + 4k_2, -k_1 + 2k_2).$$

Equating corresponding components gives

$$k_1 + 6k_2 = 9, \quad 2k_1 + 4k_2 = 2, \quad -k_1 + 2k_2 = 7.$$

Solving this system yields $k_1 = -3$ and $k_2 = 2$ so that

$$|W\rangle = -3|v_1\rangle + 2|v_2\rangle.$$

Linear independence, bases, and dimensionality

Consider a set of vectors $|1\rangle, |2\rangle, \ldots, |r\rangle, \ldots |n\rangle$ in a linear vector space V. If every vector in V is expressible as a linear combination of $|1\rangle, |2\rangle, \ldots, |r\rangle, \ldots, |n\rangle$, then we say that these vectors *span* the vector space V, and they are called the *base vectors* or *basis* of the vector space V. For example, the three unit vectors $e_1 = (1, 0, 0), e_2 = (0, 1, 0)$, and $e_3 = (0, 0, 1)$ span E_3 because every vector in E_3 is expressible as a linear combination of e_1, e_2, and e_3. But the following three vectors in E_3 do not span E_3: $|1\rangle = (1, 1, 2), |2\rangle = (1, 0, 1)$, and $|3\rangle = (2, 1, 3)$.

Base vectors are very useful in a variety of problems since it is often possible to study a vector space by first studying the vectors in a base set, then extending the results to the rest of the vector space. Therefore it is desirable to keep the spanning set as small as possible. Finding the spanning sets for a vector space depends upon the notion of linear independence.

We say that a finite set of n vectors $|1\rangle, |2\rangle, \ldots, |r\rangle, \ldots, |n\rangle$, none of which is a null vector, is linearly independent if no set of non-zero numbers a_k exists such that

$$\sum_{k=1}^{n} a_k|k\rangle = |0\rangle. \tag{5.8}$$

In other words, the set of vectors is linearly independent if it is impossible to construct the null vector from a linear combination of the vectors except when all the coefficients vanish. For example, non-zero vectors $|1\rangle$ and $|2\rangle$ of E_2 that lie along the same coordinate axis, say x_1, are not linearly independent, since we can write one as a multiple of the other: $|1\rangle = a|2\rangle$, where a is a scalar which may be positive or negative. That is, $|1\rangle$ and $|2\rangle$ depend on each other and so they are not linearly independent. Now let us move the term $a|2\rangle$ to the left hand side and the result is the null vector: $|1\rangle - a|2\rangle = |0\rangle$. Thus, for these two vectors $|1\rangle$ and $|2\rangle$ in E_2, we can find two non-zero numbers $(1, -a)$ such that Eq. (5.8) is satisfied, and so they are not linearly independent.

On the other hand, the n vectors $|1\rangle, |2\rangle, \ldots, |r\rangle, \ldots, |n\rangle$ are linearly dependent if it is possible to find scalars $a_1, a_2, \ldots, a_n$, at least two of which are non-zero, such that Eq. (5.8) is satisfied. Let us say $a_9 \neq 0$. Then we could express $|9\rangle$ in terms of the other vectors

$$|9\rangle = \sum_{i=1, \neq 9}^{n} \frac{-a_i}{a_9} |i\rangle.$$

That is, the n vectors in the set are linearly dependent if any one of them can be expressed as a linear combination of the remaining $n - 1$ vectors.

Example 5.4
The set of three vectors $|1\rangle = (2, -1, 0, 3)$, $|2\rangle = (1, 2, 5, -1)$, $|3\rangle = (7, -1, 5, 8)$ is linearly dependent, since $3|1\rangle + |2\rangle - |3\rangle = |0\rangle$.

Example 5.5
The set of three unit vectors $|e_1\rangle = (1, 0, 0)$, $|e_2\rangle = (0, 1, 0)$, and $|e_3\rangle = (0, 0, 1)$ in E_3 is linearly independent. To see this, let us start with Eq. (5.8) which now takes the form

$$a_1|e_1\rangle + a_2|e_2\rangle + a_3|e_3\rangle = |0\rangle$$

or

$$a_1(1, 0, 0) + a_2(0, 1, 0) + a_3(0, 0, 1) = (0, 0, 0)$$

from which we obtain

$$(a_1, a_2, a_3) = (0, 0, 0);$$

the set of three unit vectors $|e_1\rangle, |e_2\rangle$, and $|e_3\rangle$ is therefore linearly independent.

Example 5.6
The set S of the following four matrices

$$|1\rangle = \begin{pmatrix} 1 & 0 \\ 0 & 0 \end{pmatrix}, \quad |2\rangle = \begin{pmatrix} 0 & 1 \\ 0 & 0 \end{pmatrix}, \quad |3\rangle = \begin{pmatrix} 0 & 0 \\ 1 & 0 \end{pmatrix}, \quad |4\rangle = \begin{pmatrix} 0 & 0 \\ 0 & 1 \end{pmatrix},$$

is a basis for the vector space M_{22} of 2×2 matrices. To see that S spans M_{22}, note that a typical 2×2 vector (matrix) can be written as

$$\begin{pmatrix} a & b \\ c & d \end{pmatrix} = a\begin{pmatrix} 1 & 0 \\ 0 & 0 \end{pmatrix} + b\begin{pmatrix} 0 & 1 \\ 0 & 0 \end{pmatrix} + c\begin{pmatrix} 0 & 0 \\ 1 & 0 \end{pmatrix} + d\begin{pmatrix} 0 & 0 \\ 0 & 1 \end{pmatrix}$$

$$= a|1\rangle + b|2\rangle + c|3\rangle + d|4\rangle.$$

To see that S is linearly independent, assume that

$$a|1\rangle + b|2\rangle + c|3\rangle + d|4\rangle = |0\rangle,$$

that is,

$$a\begin{pmatrix} 1 & 0 \\ 0 & 0 \end{pmatrix} + b\begin{pmatrix} 0 & 1 \\ 0 & 0 \end{pmatrix} + c\begin{pmatrix} 0 & 0 \\ 1 & 0 \end{pmatrix} + d\begin{pmatrix} 0 & 0 \\ 0 & 1 \end{pmatrix} = \begin{pmatrix} 0 & 0 \\ 0 & 0 \end{pmatrix},$$

from which we find $a = b = c = d = 0$ so that S is linearly independent.

We now come to the dimensionality of a vector space. We think of space around us as three-dimensional. How do we extend the notion of dimension to a linear vector space? Recall that the three-dimensional Euclidean space E_3 is spanned by the three base vectors: $e_1 = (1, 0, 0)$, $e_2 = (0, 1, 0)$, $e_3 = (0, 0, 1)$. Similarly, the dimension n of a vector space V is defined to be the number n of linearly independent base vectors that span the vector space V. The vector space will be denoted by $V_n(R)$ if the field is real and by $V_n(C)$ if the field is complex. For example, as shown in Example 5.6, 2×2 matrices form a four-dimensional vector space whose base vectors are

$$|1\rangle = \begin{pmatrix} 1 & 0 \\ 0 & 0 \end{pmatrix}, \quad |2\rangle = \begin{pmatrix} 0 & 1 \\ 0 & 0 \end{pmatrix}, \quad |3\rangle = \begin{pmatrix} 0 & 0 \\ 1 & 0 \end{pmatrix}, \quad |4\rangle = \begin{pmatrix} 0 & 0 \\ 0 & 1 \end{pmatrix};$$

since any arbitrary 2×2 matrix can be written in terms of these:

$$\begin{pmatrix} a & b \\ c & d \end{pmatrix} = a|1\rangle + b|2\rangle + c|3\rangle + d|4\rangle.$$

If the scalars a, b, c, d are real, we have a real four-dimensional space, if they are complex we have a complex four-dimensional space.

Inner product spaces (unitary spaces)

In this section the structure of the vector space will be greatly enriched by the addition of a numerical function, the inner product (or scalar product). Linear vector spaces in which an inner product is defined are called inner-product spaces (or unitary spaces). The study of inner-product spaces enables us to make a real juncture with physics.

In our earlier discussion, the inner product of two vectors in E_n was defined by Eq. (5.6), a generalization of the inner product of two vectors in E_3. In a general linear vector space, an inner product is defined axiomatically analogously with the inner product on E_n. Thus given two vectors $|U\rangle$ and $|W\rangle$

$$|U\rangle = \sum_{i=1}^{n} u_i |i\rangle, \quad |W\rangle = \sum_{i=1}^{n} w_i |i\rangle, \tag{5.9}$$

where $|U\rangle$ and $|W\rangle$ are expressed in terms of the n base vectors $|i\rangle$, the inner product, denoted by the symbol $\langle U|W\rangle$, is defined to be

$$\langle U|W\rangle = \sum_{i=1}^{n} \sum_{j=1}^{n} u_i^* w_j \langle i|j\rangle. \tag{5.10}$$

$\langle U|$ is often called the pre-factor and $|W\rangle$ the post-factor. The inner product obeys the following rules (or axioms):

B.1 $\langle U|W\rangle = \langle W|U\rangle^*$ (skew-symmetry);
B.2 $\langle U|U\rangle \geq 0$, $= 0$ if and only if $|U\rangle = |0\rangle$ (positive semidefiniteness);
B.3 $\langle U|X\rangle + |W\rangle) = \langle U|X\rangle + \langle U|W\rangle$ (additivity);
B.4 $\langle aU|W\rangle = a^*\langle U|W\rangle$, $\langle U|bW\rangle = b\langle U|W\rangle$ (homogeneity);

where a and b are scalars and the asterisk (*) denotes complex conjugation. Note that Axiom B.1 is different from the one for the inner product on E_3: the inner product on a general linear vector space depends on the order of the two factors for a complex vector space. In a real vector space E_3, the complex conjugation in Axioms B.1 and B.4 adds nothing and may be ignored. In either case, real or complex, Axiom B.1 implies that $\langle U|U\rangle$ is real, so the inequality in Axiom B.2 makes sense.

The inner product is linear with respect to the post-factor:

$$\langle U|aW + bX\rangle = a\langle U|W\rangle + b\langle U|X\rangle,$$

and anti-linear with respect to the prefactor,

$$\langle aU + bX|W\rangle = a^*\langle U|W\rangle + b^*\langle X|W\rangle.$$

Two vectors are said to be orthogonal if their inner product vanishes. And we will refer to the quantity $\langle U|U\rangle^{1/2} = \| U \|$ as the norm or length of the vector. A normalized vector, having unit norm, is a unit vector. Any given non-zero vector may be normalized by dividing it by its length. An orthonormal basis is a set of basis vectors that are all of unit norm and pair-wise orthogonal. It is very handy to have an orthonormal set of vectors as a basis for a vector space, so for $\langle i|j\rangle$ in Eq. (5.10) we shall assume

$$\langle i|j\rangle = \delta_{ij} = \begin{cases} 1 & \text{for } i = j \\ 0 & \text{for } i \neq j \end{cases},$$

then Eq. (5.10) reduces to

$$\langle U|W\rangle = \sum_i \sum_j u_i^* w_j \delta_{ij} = \sum_i u_i^* \left(\sum_j w_j \delta_{ij} \right) = \sum_i u_i^* w_i. \qquad (5.11)$$

Note that Axiom B.2 implies that if a vector $|U\rangle$ is orthogonal to every vector of the vector space, then $|U\rangle = 0$: since $\langle U| \ \rangle = 0$ for all $| \ \rangle$ belongs to the vector space, so we have in particular $\langle U|U\rangle = 0$.

We will show shortly that we may construct an orthonormal basis from an arbitrary basis using a technique known as the Gram–Schmidt orthogonalization process.

Example 5.7

Let $|U\rangle = (3 - 4i)|1\rangle + (5 - 6i)|2\rangle$ and $|W\rangle = (1 - i)|1\rangle + (2 - 3i)|2\rangle$ be two vectors expanded in terms of an orthonormal basis $|1\rangle$ and $|2\rangle$. Then we have, using Eq. (5.10):

$$\langle U|U\rangle = (3 + 4i)(3 - 4i) + (5 + 6i)(5 - 6i) = 86,$$

$$\langle W|W\rangle = (1 + i)(1 - i) + (2 + 3i)(2 - 3i) = 15,$$

$$\langle U|W\rangle = (3 + 4i)(1 - i) + (5 + 6i)(2 - 3i) = 35 - 2i = \langle W|U\rangle^*.$$

Example 5.8

If $\tilde{A}$ and $\tilde{B}$ are two matrices, where

$$\tilde{A} = \begin{pmatrix} a_{11} & a_{12} \\ a_{21} & a_{22} \end{pmatrix}, \qquad \tilde{B} = \begin{pmatrix} b_{11} & b_{12} \\ b_{21} & b_{22} \end{pmatrix},$$

then the following formula defines an inner product on M_{22}:

$$\langle \tilde{A}|\tilde{B}\rangle = a_{11}b_{11} + a_{12}b_{12} + a_{21}b_{21} + a_{22}b_{22}.$$

To see this, let us first expand $\tilde{A}$ and $\tilde{B}$ in terms of the following base vectors

$$|1\rangle = \begin{pmatrix} 1 & 0 \\ 0 & 0 \end{pmatrix}, \quad |2\rangle = \begin{pmatrix} 0 & 1 \\ 0 & 0 \end{pmatrix}, \quad |3\rangle = \begin{pmatrix} 0 & 0 \\ 1 & 0 \end{pmatrix}, \quad |4\rangle = \begin{pmatrix} 0 & 0 \\ 0 & 1 \end{pmatrix};$$

$$\tilde{A} = a_{11}|1\rangle + a_{12}|2\rangle + a_{21}|3\rangle + a_{22}|4\rangle, \quad \tilde{B} = b_{11}|1\rangle + b_{12}|2\rangle + b_{21}|3\rangle + b_{22}|4\rangle.$$

The result follows easily from the defining formula (5.10).

Example 5.9

Consider the vector $|U\rangle$, in a certain orthonormal basis, with components

$$|U\rangle = \begin{pmatrix} 1 + i \\ \sqrt{3} + i \end{pmatrix}, \qquad i = \sqrt{-1}.$$

We now expand it in a new orthonormal basis $|e_1\rangle, |e_2\rangle$ with components

$$|e_1\rangle = \frac{1}{\sqrt{2}}\begin{pmatrix} 1 \\ 1 \end{pmatrix}, \quad |e_2\rangle = \frac{1}{\sqrt{2}}\begin{pmatrix} 1 \\ -1 \end{pmatrix}.$$

To do this, let us write

$$|U\rangle = u_1|e_1\rangle + u_2|e_2\rangle$$

and determine u_1 and u_2. To determine u_1, we take the inner product of both sides with $\langle e_1|$:

$$u_1 = \langle e_1|U\rangle = \frac{1}{\sqrt{2}}(1 \quad 1)\begin{pmatrix} 1+i \\ \sqrt{3}+i \end{pmatrix} = \frac{1}{\sqrt{2}}(1 + \sqrt{3} + 2i);$$

likewise,

$$u_2 = \frac{1}{\sqrt{2}}(1 - \sqrt{3}).$$

As a check on the calculation, let us compute the norm squared of the vector and see if it equals $|1 + i|^2 + |\sqrt{3} + i|^2 = 6$. We find

$$|u_1|^2 + |u_2|^2 = \frac{1}{2}(1 + 3 + 2\sqrt{3} + 4 + 1 + 3 - 2\sqrt{3}) = 6.$$

The Gram–Schmidt orthogonalization process

We now take up the Gram–Schmidt orthogonalization method for converting a linearly independent basis into an orthonormal one. The basic idea can be clearly illustrated in the following steps. Let $|1\rangle, |2\rangle, \ldots, |i\rangle, \ldots$ be a linearly independent basis. To get an orthonormal basis out of these, we do the following:

Step 1.   Rescale the first vector by its own length, so it becomes a unit vector. This will be the first basis vector.

$$|e_1\rangle = \frac{|1\rangle}{||1\rangle|},$$

where $||1\rangle| = \sqrt{\langle 1 | 1\rangle}$. Clearly

$$\langle e_1 | e_1\rangle = \frac{\langle 1 | 1\rangle}{||1\rangle|} = 1.$$

Step 2.   To construct the second member of the orthonormal basis, we subtract from the second vector $|2\rangle$ its projection along the first, leaving behind only the part perpendicular to the first.

$$|II\rangle = |2\rangle - |e_1\rangle\langle e_1|2\rangle.$$

Clearly

$$\langle e_1|II\rangle = \langle e_1|2\rangle - \langle e_1|e_1\rangle\langle e_1|2\rangle = 0, \quad \text{i.e.,} \quad (II|\perp|e_1\rangle.$$

Dividing $|II\rangle$ by its norm (length), we now have the second basis vector and it is orthogonal to the first base vector $|e_1\rangle$ and of unit length.

Step 3.   To construct the third member of the orthonormal basis, consider

$$|III\rangle = |3\rangle - |e_1\rangle\langle e_1|III\rangle - |e_2\rangle_2|III\rangle$$

which is orthogonal to both $|e_1\rangle$ and $|e_2\rangle$. Dividing by its norm we get $|e_3\rangle$.

Continuing in this way, we will obtain an orthonormal basis $|e_1\rangle, |e_2\rangle, \ldots, |e_n\rangle$.

The Cauchy–Schwarz inequality

If $\mathbf{A}$ and $\mathbf{B}$ are non-zero vectors in E_3, then the dot product gives $\mathbf{A} \cdot \mathbf{B} = AB\cos\theta$, where θ is the angle between the vectors. If we square both sides and use the fact that $\cos^2\theta \leq 1$, we obtain the inequality

$$(\mathbf{A} \cdot \mathbf{B})^2 \leq A^2B^2 \quad \text{or} \quad |\mathbf{A} \cdot \mathbf{B}| \leq AB.$$

This is known as the Cauchy–Schwarz inequality. There is an inequality corresponding to the Cauchy–Schwarz inequality in any inner-product space that obeys Axioms B.1–B.4, which can be stated as

$$|\langle U|W\rangle| \leq |U||W|, \quad |U| = \sqrt{\langle U|U\rangle} \quad \text{etc.,} \tag{5.13}$$

where $|U\rangle$ and $|W\rangle$ are two non-zero vectors in an inner-product space.

This can be proved as follows. We first note that, for any scalar α, the following inequality holds

$$0 \leq |\langle U + \alpha W|U + \alpha W\rangle|^2 = \langle U + \alpha W|U + \alpha W\rangle$$

$$= \langle U|U\rangle + \langle \alpha W|U\rangle + \langle U|\alpha W\rangle + \langle \alpha W|\alpha W\rangle$$

$$= |U|^2 + \alpha^*\langle V|U\rangle + \alpha\langle U|W\rangle + |\alpha|^2|W|^2.$$

Now let $\alpha = \lambda\langle U|W\rangle^*/|\langle U|W\rangle|$, with λ real. This is possible if $|W\rangle \neq 0$, but if $\langle U|W\rangle = 0$, then Cauchy–Schwarz inequality is trivial. Making this substitution in the above, we have

$$0 \leq |U|^2 + 2\lambda|\langle U|W\rangle| + \lambda^2|W|^2.$$

This is a quadratic expression in the real variable λ with real coefficients. Therefore, the discriminant must be less than or equal to zero:

$$4|\langle U|W\rangle|^2 - 4|U|^2|W|^2 \leq 0$$

or

$$|\langle U|W\rangle| \leq |U||W|,$$

which is the Cauchy–Schwarz inequality.

From the Cauchy–Schwarz inequality follows another important inequality, known as the triangle inequality,

$$|U + W| \leq |U| + |W|. \tag{5.14}$$

The proof of this is very straightforward. For any pair of vectors, we have

$$|U + W|^2 = \langle U + W|U + W\rangle = |U|^2 + |W|^2 + \langle U|W\rangle + \langle W|U\rangle$$

$$\leq |U|^2 + |W|^2 + 2|\langle U|W\rangle|$$

$$\leq |U|^2 + |W|^2 + 2|U||W|(|U|^2 + |W|^2)$$

from which it follows that

$$|U + W| \leq |U| + |W|.$$

If V denotes the vector space of real continuous functions on the interval $a \leq x \leq b$, and f and g are any real continuous functions, then the following is an inner product on V:

$$\langle f|g\rangle = \int_a^b f(x)g(x)dx.$$

The Cauchy–Schwarz inequality now gives

$$\left(\int_a^b f(x)g(x)dx\right)^2 \leq \int_a^b f^2(x)dx \int_a^b g^2(x)dx$$

or in Dirac notation

$$|\langle f|g\rangle|^2 \leq |f|^2|g|^2.$$

Dual vectors and dual spaces

We begin with a technical point regarding the inner product $\langle u|v\rangle$. If we set

$$|v\rangle = \alpha|w\rangle + \beta|z\rangle,$$

then

$$\langle u|v\rangle = \alpha\langle u|w\rangle + \beta\langle u|z\rangle$$

is a linear function of α and β. However, if we set

$$|u\rangle = \alpha|w\rangle + \beta|z\rangle,$$

211

then

$$\langle u|v\rangle = \langle v|u\rangle^* = \alpha^*\langle v|w\rangle^* + \beta^*\langle v|z\rangle^* = \alpha^*\langle w|v\rangle + \beta^*\langle z|v\rangle$$

is no longer a linear function of α and β. To remove this asymmetry, we can introduce, besides the ket vectors $|\,\rangle$, bra vectors $\langle\,|$ which form a different vector space. We will assume that there is a one-to-one correspondence between ket vectors $|\,\rangle$, and bra vectors $\langle\,|$. Thus there are two vector spaces, the space of kets and a dual space of bras. A pair of vectors in which each is in correspondence with the other will be called a pair of dual vectors. Thus, for example, $\langle v|$ is the dual vector of $|v\rangle$. Note they always carry the same identification label.

We now define the multiplication of ket vectors by bra vectors by requiring

$$\langle u|\cdot|v\rangle \equiv \langle u|v\rangle.$$

Setting

$$\langle u| = \langle w|\alpha^* + \langle z|\beta^*,$$

we have

$$\langle u|v\rangle = \alpha^*\langle w|v\rangle + \beta^*\langle z|v\rangle,$$

the same result we obtained above, and we see that $\langle w|\alpha^* + \langle z|\beta^*$ is the dual vector of $\alpha|w\rangle + \beta|z\rangle$.

From the above discussion, it is obvious that inner products are really defined only between bras and kets and hence from elements of two distinct but related vector spaces. There is a basis of vectors $|i\rangle$ for expanding kets and a similar basis $\langle i|$ for expanding bras. The basis ket $|t\rangle$ is represented in the basis we are using by a column vector with all zeros except for a 1 in the ith row, while the basis $\langle i|$ is a row vector with all zeros except for a 1 in the ith column.

Linear operators

A useful concept in the study of linear vector spaces is that of a linear transformation, from which the concept of a linear operator emerges naturally. It is instructive first to review the concept of transformation or mapping. Given vector spaces V and W and function T, if T associates each vector in V with a unique vector in W, we say T maps V into W, and write $T\colon V \to W$. If T associates the vector $|w\rangle$ in W with the vector $|v\rangle$ in V, we say that $|w\rangle$ is the *image* of $|v\rangle$ under T and write $|w\rangle = T|v\rangle$. Further, T is a linear transformation if:

(a) $T(|u\rangle + |v\rangle) = T|u\rangle + T|v\rangle$ for all vectors $|u\rangle$ and $|v\rangle$ in V.
(b) $T(k|v\rangle) = kT|v\rangle$ for all vectors $|v\rangle$ in V and all scalars k.

We can illustrate this with a very simple example. If $|v\rangle = (x, y)$ is a vector in E_2, then $T(|v\rangle) = (x, x + y, x - y)$ defines a function (a transformation) that maps

E_2 into E_3. In particular, if $|v\rangle = (1,1)$, then the image of $|v\rangle$ under $\underset{\sim}{T}$ is $\underset{\sim}{T}$ $(|v\rangle) = (1,2,0)$. It is easy to see that the transformation is linear. If $|u\rangle = (x_1, y_1)$ and $|v\rangle = (x_2, y_2)$, then

$$|u\rangle + |v\rangle = (x_1 + x_2, y_1 + y_2),$$

so that

$$\underset{\sim}{T}(|u\rangle + |v\rangle) = (x_1 + x_2, (x_1 + x_2) + (y_1 + y_2), (x_1 + x_2) - (y_1 + y_2))$$

$$= (x_1, x_1 + y_1, x_1 - y_1) + (x_2, x_2 + y_2, x_2 - y_2)$$

$$= \underset{\sim}{T}(|u\rangle) + \underset{\sim}{T}(|v\rangle)$$

and if k is a scalar, then

$$\underset{\sim}{T}(k|u\rangle) = (kx_1, kx_1 + ky_1, kx_1 - ky_1) = k(x_1, x_1 + y_1, x_1 - y_1) = k\underset{\sim}{T}(|u\rangle).$$

Thus $\underset{\sim}{T}$ is a linear transformation.

If $\underset{\sim}{T}$ maps the vector space onto itself ($\underset{\sim}{T}: V \to V$), then it is called a linear operator on V. In E_3 a rotation of the entire space about a fixed axis is an example of an operation that maps the space onto itself. We saw in Chapter 3 that rotation can be represented by a matrix with elements λ_{ij} ($i, j = 1, 2, 3$); if x_1, x_2, x_3 are the components of an arbitrary vector in E_3 before the transformation and x_1', x_2', x_3' the components of the transformed vector, then

$$\left.\begin{array}{l} x_1' = \lambda_{11}x_1 + \lambda_{12}x_2 + \lambda_{13}x_3, \\ x_2' = \lambda_{21}x_1 + \lambda_{22}x_2 + \lambda_{23}x_3, \\ x_3' = \lambda_{31}x_1 + \lambda_{32}x_2 + \lambda_{33}x_3. \end{array}\right\} \qquad (5.15)$$

In matrix form we have

$$\tilde{x}' = \tilde{\lambda}(\theta)\tilde{x}, \qquad (5.16)$$

where θ is the angle of rotation, and

$$\tilde{x}' = \begin{pmatrix} x_1' \\ x_2' \\ x_3' \end{pmatrix}, \quad \tilde{x} = \begin{pmatrix} x_1 \\ x_2 \\ x_3 \end{pmatrix}, \quad \text{and} \quad \tilde{\lambda}(\theta) = \begin{pmatrix} \lambda_{11} & \lambda_{12} & \lambda_{13} \\ \lambda_{21} & \lambda_{22} & \lambda_{23} \\ \lambda_{31} & \lambda_{32} & \lambda_{33} \end{pmatrix}.$$

In particular, if the rotation is carried out about x_3-axis, $\tilde{\lambda}(\theta)$ has the following form:

$$\tilde{\lambda}(\theta) = \begin{pmatrix} \cos\theta & -\sin\theta & 0 \\ \sin\theta & \cos\theta & 0 \\ 0 & 0 & 1 \end{pmatrix}.$$

Eq. (5.16) determines the vector $\mathbf{x}'$ if the vector $\mathbf{x}$ is given, and $\tilde{\lambda}(\theta)$ is the operator (matrix representation of the rotation operator) which turns $\mathbf{x}$ into $\mathbf{x}'$.

Loosely speaking, an operator is any mathematical entity which operates on any vector in V and turns it into another vector in V. Abstractly, an operator $\underline{L}$ is a mapping that assigns to a vector $|v\rangle$ in a linear vector space V another vector $|u\rangle$ in V: $|u\rangle = \underline{L}|v\rangle$. The set of vectors $|v\rangle$ for which the mapping is defined, that is, the set of vectors $|v\rangle$ for which $\underline{L}|v\rangle$ has meaning, is called the domain of $\underline{L}$. The set of vectors $|u\rangle$ in the domain expressible as $|u\rangle = \underline{L}|v\rangle$ is called the range of the operator. An operator $\underline{L}$ is linear if the mapping is such that for any vectors $|u\rangle, |w\rangle$ in the domain of $\underline{L}$ and for arbitrary scalars α, β, the vector $\alpha|u\rangle + \beta|w\rangle$ is in the domain of $\underline{L}$ and

$$\underline{L}(\alpha|u\rangle + \beta|w\rangle) = \alpha\underline{L}|u\rangle + \beta\underline{L}|w\rangle.$$

A linear operator is bounded if its domain is the entire space V and if there exists a single constant C such that

$$|\underline{L}|v\rangle| < C||v\rangle|$$

for all $|v\rangle$ in V. We shall consider linear bounded operators only.

Matrix representation of operators

Linear bounded operators may be represented by matrix. The matrix will have a finite or an infinite number of rows according to whether the dimension of V is finite or infinite. To show this, let $|1\rangle, |2\rangle, \ldots$ be an orthonormal basis in V; then every vector $|\varphi\rangle$ in V may be written in the form

$$|\varphi\rangle = \alpha_1|1\rangle + \alpha_2|2\rangle + \cdots.$$

Since $\underline{L}|\rangle$ is also in V, we may write

$$\underline{L}|\varphi\rangle = \beta_1|1\rangle + \beta_2|2\rangle + \cdots.$$

But

$$\underline{L}|\varphi\rangle = \alpha_1\underline{L}|1\rangle + \alpha_2\underline{L}|2\rangle + \cdots,$$

so

$$\beta_1|1\rangle + \beta_2|2\rangle + \cdots = \alpha_1\underline{L}|1\rangle + \alpha_2\underline{L}|2\rangle + \cdots.$$

Taking the inner product of both sides with $\langle 1|$ we obtain

$$\beta_1 = \langle 1|\underline{L}|1\rangle\alpha_1 + \langle 1|\underline{L}|2\rangle\alpha_2 = \gamma_{11}\alpha_1 + \gamma_{12}\alpha_2 + \cdots;$$

214

Similarly

$$\beta_2 = \langle 2|\underset{\sim}{L}|1\rangle\alpha_1 + \langle 2|\underset{\sim}{L}|2\rangle\alpha_2 = \gamma_{21}\alpha_1 + \gamma_{22}\alpha_2 + \cdots,$$

$$\beta_3 = \langle 3|\underset{\sim}{L}|1\rangle\alpha_1 + \langle 3|\underset{\sim}{L}|2\rangle\alpha_2 = \gamma_{31}\alpha_1 + \gamma_{32}\alpha_2 + \cdots.$$

In general, we have

$$\beta_i = \sum_j \gamma_{ij}\alpha_j,$$

where

$$\gamma_{ij} = \langle i|\underset{\sim}{L}|j\rangle. \tag{5.17}$$

Consequently, in terms of the vectors $|1\rangle, |2\rangle, \ldots$ as a basis, operator $\underset{\sim}{L}$ is represented by the matrix whose elements are γ_{ij}.

A matrix representing $\underset{\sim}{L}$ can be found by using any basis, not necessarily an orthonormal one. Of course, a change in the basis changes the matrix representing $\underset{\sim}{L}$.

The algebra of linear operators

Let $\underset{\sim}{A}$ and $\underset{\sim}{B}$ be two operators defined in a linear vector space V of vectors $|\rangle$. The equation $\underset{\sim}{A} = \underset{\sim}{B}$ will be understood in the sense that

$$\underset{\sim}{A}|\rangle = \underset{\sim}{B}|\rangle \quad \text{for all } |\rangle \in V.$$

We define the addition and multiplication of linear operators as

$$\underset{\sim}{C} = \underset{\sim}{A} + \underset{\sim}{B} \quad \text{and} \quad \underset{\sim}{D} = \underset{\sim}{A}\underset{\sim}{B}$$

if for any $|\rangle$

$$\underset{\sim}{C}|\rangle = (\underset{\sim}{A} + \underset{\sim}{B})|\rangle = \underset{\sim}{A}|\rangle + \underset{\sim}{B}|\rangle,$$

$$\underset{\sim}{D}|\rangle = (\underset{\sim}{A}\underset{\sim}{B})|\rangle = \underset{\sim}{A}(\underset{\sim}{B}|\rangle).$$

Note that $\underset{\sim}{A} + \underset{\sim}{B}$ and $\underset{\sim}{A}\underset{\sim}{B}$ are themselves linear operators.

Example 5.10

(a) $(\underset{\sim}{A}\underset{\sim}{B})(\alpha|u\rangle + \beta|v\rangle) = \underset{\sim}{A}[\alpha(\underset{\sim}{B}|u\rangle) + \beta(\underset{\sim}{B}|v\rangle)] = \alpha(\underset{\sim}{A}\underset{\sim}{B})|u\rangle + \beta(\underset{\sim}{A}\underset{\sim}{B})|v\rangle,$

(b) $\underset{\sim}{C}(\underset{\sim}{A} + \underset{\sim}{B})|v\rangle = \underset{\sim}{C}(\underset{\sim}{A}|v\rangle + \underset{\sim}{B}|v\rangle) = \underset{\sim}{C}\underset{\sim}{A}|\rangle + \underset{\sim}{C}\underset{\sim}{B}|\rangle,$

which shows that

$$\underset{\sim}{C}(\underset{\sim}{A} + \underset{\sim}{B}) = \underset{\sim}{C}\underset{\sim}{A} + \underset{\sim}{C}\underset{\sim}{B}.$$

In general $A B \neq B A$. The difference $A B - B A$ is called the commutator of A and B and is denoted by the symbol $[A, B]$:

$$[A, B] \equiv A B - B A. \tag{5.18}$$

An operator whose commutator vanishes is called a commuting operator.

The operator equation

$$B = \alpha A = A \alpha$$

is equivalent to the vector equation

$$B| \rangle = \alpha A | \rangle \quad \text{for any } | \rangle.$$

And the vector equation

$$A| \rangle = \alpha | \rangle$$

is equivalent to the operator equation

$$A = \alpha E$$

where E is the identity (or unit) operator:

$$E| \rangle = | \rangle \quad \text{for any } | \rangle.$$

It is obvious that the equation $A = \alpha$ is meaningless.

Example 5.11

To illustrate the non-commuting nature of operators, let $A = x$, $B = d/dx$. Then

$$A Bf(x) = x \frac{d}{dx} f(x),$$

and

$$B Af(x) = \frac{d}{dx} xf(x) = \left(\frac{dx}{dx} \right) f + x \frac{df}{dx} = f + x \frac{df}{dx} = (E + A B)f.$$

Thus,

$$(A B - B A)f(x) = -Ef(x)$$

or

$$\left[x, \frac{d}{dx} \right] = x \frac{d}{dx} - \frac{d}{dx} x = -E.$$

Having defined the product of two operators, we can also define an operator raised to a certain power. For example

$$A^m| \rangle = \underbrace{A \quad A \quad \cdots \quad A}_{m \, \text{factor}} | \rangle.$$

By combining the operations of addition and multiplication, functions of operators can be formed. We can also define functions of operators by their power series expansions. For example, $e^{\underset{\sim}{A}}$ formally means

$$e^{\underset{\sim}{A}} \equiv 1 + \underset{\sim}{A} + \frac{1}{2!}\underset{\sim}{A}^2 + \frac{1}{3!}\underset{\sim}{A}^3 + \cdots.$$

A function of a linear operator is a linear operator.

Given an operator $\underset{\sim}{A}$ that acts on vector $|\rangle$, we can define the action of the same operator on vector $\langle|$. We shall use the convention of operating on $\langle|$ from the right. Then the action of $\underset{\sim}{A}$ on a vector $\langle|$ is defined by requiring that for any $|u\rangle$ and $\langle v|$, we have

$$\{\langle u|\underset{\sim}{A}\}|v\rangle \equiv \langle u|\{\underset{\sim}{A}|v\rangle\} = \langle u|\underset{\sim}{A}|v\rangle.$$

We may write $\alpha|v\rangle = |\alpha v\rangle$ and the corresponding bra as $\langle \alpha v|$. However, it is important to note that $\langle \alpha v| = A^*\langle v|$.

Eigenvalues and eigenvectors of an operator

The result of operating on a vector with an operator $\underset{\sim}{A}$ is, in general, a different vector. But there may be some vector $|v\rangle$ with the property that operating with $\underset{\sim}{A}$ on it yields the same vector $|v\rangle$ multiplied by a scalar, say α:

$$\underset{\sim}{A}|v\rangle = \alpha|v\rangle.$$

This is called the eigenvalue equation for the operator $\underset{\sim}{A}$, and the vector $|v\rangle$ is called an eigenvector of $\underset{\sim}{A}$ belonging to the eigenvalue α. A linear operator has, in general, several eigenvalues and eigenvectors, which can be distinguished by a subscript

$$\underset{\sim}{A}|v_k\rangle = \alpha_k|v_k\rangle.$$

The set $\{\alpha_k\}$ of all the eigenvalues taken together constitutes the spectrum of the operator. The eigenvalues may be discrete, continuous, or partly discrete and partly continuous. In general, an eigenvector belongs to only one eigenvalue. If several linearly independent eigenvectors belong to the same eigenvalue, the eigenvalue is said to be degenerate, and the degree of degeneracy is given by the number of linearly independent eigenvectors.

Some special operators

Certain operators with rather special properties play very important roles in physics. We now consider some of them below.

The inverse of an operator

The operator X satisfying $XA = E$ is called the left inverse of A and we denote it by A_L^{-1}. Thus, $A_L^{-1} A \equiv E$. Similarly, the right inverse of A is defined by the equation

$$A A_R^{-1} \equiv E .$$

In general, A_L^{-1} or A_R^{-1}, or both, may not be unique and even may not exist at all. However, if both A_L^{-1} and A_R^{-1} exist, then they are unique and equal to each other:

$$A_L^{-1} = A_R^{-1} \equiv A^{-1},$$

and

$$AA^{-1} = A^{-1}A = E . \tag{5.19}$$

A^{-1} is called the operator inverse to A. Obviously, an operator is the inverse of another if the corresponding matrices are.

An operator for which an inverse exists is said to be non-singular, whereas one for which no inverse exists is singular. A necessary and sufficient condition for an operator A to be non-singular is that corresponding to each vector $|u\rangle$, there should be a unique vector $|v\rangle$ such that $|u\rangle = A |v\rangle$.

The inverse of a linear operator is a linear operator. The proof is simple: let

$$|u_1\rangle = A |v_1\rangle, \quad |u_2\rangle = A |v_2\rangle.$$

Then

$$|v_1\rangle = A^{-1}|u_1\rangle, \quad |v_2\rangle = A^{-1}|u_2\rangle$$

so that

$$c_1|v_1\rangle = c_1 A^{-1}|u_1\rangle, \quad c_2|v_2\rangle = c_2 A^{-1}|u_2\rangle.$$

Thus,

$$A^{-1}[c_1|u_1\rangle + c_2|u_2\rangle] = A^{-1}[c_1 A |v_1\rangle + c_2 A |v_2\rangle]$$

$$= A^{-1}A[c_1|v_1\rangle + c_2|v_2\rangle]$$

$$= c_1|v_1\rangle + c_2|v_2\rangle$$

or

$$A^{-1}[c_1|u_1\rangle + c_2|u_2\rangle] = c_1 A^{-1}|u_1\rangle + c_2 A^{-1}|u_2\rangle.$$

218

The inverse of a product of operators is the product of the inverse in the reverse order

$$(\underset{\sim}{A}\underset{\sim}{B})^{-1} = \underset{\sim}{B}^{-1}\underset{\sim}{A}^{-1}. \qquad (5.20)$$

The proof is straightforward: we have

$$\underset{\sim}{A}\underset{\sim}{B}(\underset{\sim}{A}\underset{\sim}{B})^{-1} = \underset{\sim}{E}.$$

Multiplying successively from the left by $\underset{\sim}{A}^{-1}$ and $\underset{\sim}{B}^{-1}$, we obtain

$$(\underset{\sim}{A}\underset{\sim}{B})^{-1} = \underset{\sim}{B}^{-1}\underset{\sim}{A}^{-1},$$

which is identical to Eq. (5.20).

The adjoint operators

Assuming that V is an inner-product space, then the operator $\underset{\sim}{X}$ satisfying the relation

$$\langle u|\underset{\sim}{X}|v\rangle = \langle v|\underset{\sim}{A}|u\rangle^* \quad \text{for any } |u\rangle, \quad |v\rangle \in V$$

is called the adjoint operator of $\underset{\sim}{A}$ and is denoted by $\underset{\sim}{A}^+$. Thus

$$\langle u|\underset{\sim}{A}^+|v\rangle \equiv \langle v|\underset{\sim}{A}|u\rangle^* \quad \text{for any } |u\rangle, \quad |v\rangle \in V. \qquad (5.21)$$

We first note that $\langle \,|\underset{\sim}{A}^+$ is a dual vector of $\underset{\sim}{A}|\,\rangle$. Next, it is obvious that

$$(\underset{\sim}{A}^+)^+ = \underset{\sim}{A}. \qquad (5.22).$$

To see this, let $\underset{\sim}{A}^+ = \underset{\sim}{B}$, then $(\underset{\sim}{A}^+)^+$ becomes $\underset{\sim}{B}^+$, and from Eq. (5.21) we find

$$\langle v|\underset{\sim}{B}^+|u\rangle = \langle u|\underset{\sim}{B}|v\rangle^*, \quad \text{for any } |u\rangle, |v\rangle \in V.$$

But

$$\langle u|\underset{\sim}{B}|v\rangle^* = \langle u|\underset{\sim}{A}^+|v\rangle^* = \langle v|\underset{\sim}{A}|u\rangle.$$

Thus

$$\langle v|\underset{\sim}{B}^+|u\rangle = \langle u|\underset{\sim}{B}|v\rangle^* = \langle v|\underset{\sim}{A}|u\rangle$$

from which we find

$$(\underset{\sim}{A}^+)^+ = \underset{\sim}{A}.$$

It is also easy to show that

$$(\underset{\sim}{A}B)^+ = B^+\underset{\sim}{A}^+. \tag{5.23}$$

For any $|u\rangle$, $|v\rangle$, $\langle v|\underset{\sim}{B}^+$ and $\underset{\sim}{B}|v\rangle$ is a pair of dual vectors; $\langle u|\underset{\sim}{A}^+$ and $\underset{\sim}{A}|u\rangle$ is also a pair of dual vectors. Thus we have

$$\langle v|\underset{\sim}{B}^+\underset{\sim}{A}^+|u\rangle = \{\langle v|\underset{\sim}{B}^+\}\{\underset{\sim}{A}^+|u\rangle\} = [\{\langle u|\underset{\sim}{A}\}\{\underset{\sim}{B}|v\rangle\}]^*$$

$$= \langle u|\underset{\sim}{A}\,\underset{\sim}{B}|v\rangle^* = \langle v|(\underset{\sim}{A}\,\underset{\sim}{B})^+|u\rangle$$

and therefore

$$(\underset{\sim}{A}\,\underset{\sim}{B})^+ = B^+\underset{\sim}{A}^+.$$

Hermitian operators

An operator $\underset{\sim}{H}$ that is equal to its adjoint, that is, that obeys the relation

$$\underset{\sim}{H} = \underset{\sim}{H}^+ \tag{5.24}$$

is called Hermitian or self-adjoint. And $\underset{\sim}{H}$ is anti-Hermitian if

$$\underset{\sim}{H} = -\underset{\sim}{H}^+.$$

Hermitian operators have the following important properties:

(1) The eigenvalues are real: Let $\underset{\sim}{H}$ be the Hermitian operator and let $|v\rangle$ be an eigenvector belonging to the eigenvalue α:

$$\underset{\sim}{H}|v\rangle = \alpha|v\rangle.$$

By definition, we have

$$\langle v|\underset{\sim}{A}|v\rangle = \langle v|\underset{\sim}{A}|v\rangle^*,$$

that is,

$$(\alpha^* - \alpha)\langle v|v\rangle = 0.$$

Since $\langle v|v\rangle \neq 0$, we have

$$\alpha^* = \alpha.$$

(2) Eigenvectors belonging to different eigenvalues are orthogonal: Let $|u\rangle$ and $|v\rangle$ be eigenvectors of $\underset{\sim}{H}$ belonging to the eigenvalues α and β respectively:

$$\underset{\sim}{H}|u\rangle = \alpha|u\rangle, \quad \underset{\sim}{H}|v\rangle = \beta|v\rangle.$$

Then

$$\langle u|H|v\rangle = \langle v|H|u\rangle^*.$$

That is,

$$(\alpha - \beta)\langle v|u\rangle = 0 \quad (\text{since } \alpha^* = \alpha).$$

But $\alpha \neq \beta$, so that

$$\langle v|u\rangle = 0.$$

(3) The set of all eigenvectors of a Hermitian operator forms a complete set: The eigenvectors are orthogonal, and since we can normalize them, this means that the eigenvectors form an orthonormal set and serve as a basis for the vector space.

Unitary operators

A linear operator U is unitary if it preserves the Hermitian character of an operator under a similarity transformation:

$$(U A U^{-1})^+ = U A U^{-1},$$

where

$$A^+ = A.$$

But, according to Eq. (5.23)

$$(U A U^{-1})^+ = (U^{-1})^+ A U^+,$$

thus, we have

$$(U^{-1})^+ A U^+ = U A U^{-1}.$$

Multiplying from the left by U^+ and from the right by U, we obtain

$$U^+(U^{-1})^+ A U^+ U = U^+ U A;$$

this reduces to

$$A(U^+ U) = (U^+ U)A,$$

since

$$U^+(U^{-1})^+ = (U^{-1} U)^+ = E.$$

Thus

$$U^+ U = E$$

or

$$U^+ = U^{-1}. \tag{5.25}$$

We often use Eq. (5.25) for the definition of the unitary operator.

Unitary operators have the remarkable property that transformation by a unitary operator preserves the inner product of the vectors. This is easy to see: under the operation U, a vector $|v\rangle$ is transformed into the vector $|v'\rangle = U|v\rangle$. Thus, if two vectors $|v\rangle$ and $|u\rangle$ are transformed by the same unitary operator U, then

$$\langle u'|v'\rangle = \langle Uu|Uv\rangle = \langle u|U^+Uv\rangle = \langle u|v\rangle,$$

that is, the inner product is preserved. In particular, it leaves the norm of a vector unchanged. Thus, a unitary transformation in a linear vector space is analogous to a rotation in the physical space (which also preserves the lengths of vectors and the inner products).

Corresponding to every unitary operator U, we can define a Hermitian operator H and vice versa by

$$U = e^{i\varepsilon H}, \tag{5.26}$$

where ε is a parameter. Obviously

$$U^+ = e^{(i\varepsilon H)^+} / e^{-i\varepsilon H} = U^{-1}.$$

A unitary operator possesses the following properties:

(1) The eigenvalues are unimodular; that is, if $U|v\rangle = \alpha|v\rangle$, then $|\alpha| = 1$.
(2) Eigenvectors belonging to different eigenvalues are orthogonal.
(3) The product of unitary operators is unitary.

The projection operators

A symbol of the type of $|u\rangle\langle v|$ is quite useful: it has all the properties of a linear operator, multiplied from the right by a ket $|\ \rangle$, it gives $|u\rangle$ whose magnitude is $\langle v|\ \rangle$; and multiplied from the left by a bra $\langle\ |$ it gives $\langle v|$ whose magnitude is $\langle|u\rangle$. The linearity of $|u\rangle\langle v|$ results from the linear properties of the inner product. We also have

$$\{|u\rangle\langle v|\}^+ = |v\rangle\langle u|.$$

The operator $P_j = |j\rangle\langle j|$ is a very particular example of projection operator. To see its effect on an arbitrary vector $|u\rangle$, let us expand $|u\rangle$:

$$|u\rangle = \sum_{j=1}^{n} u_j |j\rangle, \quad u_j = \langle j|u\rangle. \tag{5.27}$$

We may write the above as

$$|u\rangle = \left(\sum_{j=1}^{n} |j\rangle\langle j| \right) |u\rangle,$$

which is true for all $|u\rangle$. Thus the object in the brackets must be identified with the identity operator:

$$I = \sum_{j=1}^{n} |j\rangle\langle j| = \sum_{j=1}^{n} P_j. \tag{5.28}$$

Now we will see that the effect of this particular projection operator on $|u\rangle$ is to produce a new vector whose direction is along the basis vector $|j\rangle$ and whose magnitude is $\langle j|u\rangle$:

$$P_j|u\rangle = |j\rangle\langle j|u\rangle = |j\rangle u_j.$$

We see that whatever $|u\rangle$ is, $P_j|u\rangle$ is a multiple of $|j\rangle$ with a coefficient u_j which is the component of $|u\rangle$ along $|j\rangle$. Eq. (5.28) says that the sum of the projections of a vector along all the n directions equals the vector itself.

When $P_j = |j\rangle\langle j|$ acts on $|j\rangle$, it reproduces that vector. On the other hand, since the other basis vectors are orthogonal to $|j\rangle$, a projection operation on any one of them gives zero (null vector). The basis vectors are therefore eigenvectors of P_k with the property

$$P_k|j\rangle = \delta_{kj}|j\rangle, \quad (j,k = 1, \ldots, n).$$

In this orthonormal basis the projection operators have the matrix form

$$P_1 = \begin{pmatrix} 1 & 0 & 0 & \cdots \\ 0 & 0 & 0 & \cdots \\ 0 & 0 & 0 & \cdots \\ \vdots & \vdots & \vdots & \ddots \end{pmatrix}, \quad P_2 = \begin{pmatrix} 0 & 0 & 0 & \cdots \\ 0 & 1 & 0 & \cdots \\ 0 & 0 & 0 & \cdots \\ \vdots & \vdots & \vdots & \ddots \end{pmatrix}, \quad P_N = \begin{pmatrix} 0 & 0 & 0 & \cdots \\ 0 & 0 & 0 & \cdots \\ 0 & 0 & 0 & \cdots \\ \vdots & \vdots & \vdots & \ddots \\ & & & & 1 \end{pmatrix}.$$

Projection operators can also act on bras in the same way:

$$\langle u|P_j = \langle u|j\rangle\langle j| = u_j^* \langle j|.$$

Change of basis

The choice of base vectors (basis) is largely arbitrary and different representations are physically equally acceptable. How do we change from one orthonormal set of base vectors $|\varphi_1\rangle, |\varphi_2\rangle, \ldots, |\varepsilon_n\rangle$ to another such set $|\xi_1\rangle, |\xi_2\rangle, \ldots, |\xi_n\rangle$? In other words, how do we generate the orthonomal set $|\xi_1\rangle, |\xi_2\rangle, \ldots, |\xi_n\rangle$ from the old set $|\varphi_1\rangle, |\varphi_2\rangle, \ldots, |\varphi_n\rangle$? This task can be accomplished by a unitary transformation:

$$|\xi_i\rangle = \underset{\sim}{U}|\varphi_i\rangle \quad (i = 1, 2, \ldots, n). \tag{5.29}$$

Then given a vector $|X\rangle = \sum_{i=1}^{n} a_i|\varphi_i\rangle$, it will be transformed into $|X'\rangle$:

$$|X'\rangle = \underset{\sim}{U}|X\rangle = \underset{\sim}{U}\sum_{i=1}^{n} a_i|\varphi_i\rangle = \sum_{i=1}^{n} \underset{\sim}{U} a_i|\varphi_i\rangle = \sum_{i=1}^{n} a_i|\xi_i\rangle.$$

We can see that the operator $\underset{\sim}{U}$ possesses an inverse $\underset{\sim}{U}^{-1}$ which is defined by the equation

$$|\varphi_i\rangle = \underset{\sim}{U}^{-1}|\xi_i\rangle \quad (i = 1, 2, \ldots, n).$$

The operator $\underset{\sim}{U}$ is unitary; for, if $|X\rangle = \sum_{i=1}^{n} a_i|\varphi_i\rangle$ and $|Y\rangle = \sum_{i=1}^{n} b_i|\varphi_i\rangle$, then

$$\langle X|Y\rangle = \sum_{i,j=1}^{n} a_i^* b_j \langle \varphi_i|\varphi_j\rangle = \sum_{i=1}^{n} a_i^* b_i, \qquad \langle UX|UY\rangle = \sum_{i,j=1}^{n} a_i^* b_j \langle \xi_i|\xi_j\rangle = \sum_{i=1}^{n} a_i^* b_i.$$

Hence

$$U^{-1} = U^+.$$

The inner product of two vectors is independent of the choice of basis which spans the vector space, since unitary transformations leave all inner products invariant. In quantum mechanics inner products give physically observable quantities, such as expectation values, probabilities, etc.

It is also clear that the matrix representation of an operator is different in a different basis. To find the effect of a change of basis on the matrix representation of an operator, let us consider the transformation of the vector $|X\rangle$ into $|Y\rangle$ by the operator $\underset{\sim}{A}$:

$$|Y\rangle = \underset{\sim}{A}|(X\rangle. \tag{5.30}$$

Referred to the basis $|\varphi_1\rangle$, $|\varphi_2\rangle$, $\ldots$, $|\varphi\rangle$, $|X\rangle$ and $|Y\rangle$ are given by $|X\rangle = \sum_{i=1}^{n} a_i|\varphi_i\rangle$ and $|Y\rangle = \sum_{i=1}^{n} b_i|\varphi_i\rangle$, and the equation $|Y\rangle = \underset{\sim}{A}|X\rangle$ becomes

$$\sum_{i=1}^{n} b_i|\varphi_i\rangle = \underset{\sim}{A}\sum_{j=1}^{n} a_j|\varphi_j\rangle.$$

Multiplying both sides from the left by the bra vector $\langle \varphi_i |$ we find

$$b_i = \sum_{j=1}^{n} a_j \langle \varphi_i | \underset{\sim}{A} | \varphi_j \rangle = \sum_{j=1}^{n} a_j A_{ij}. \tag{5.31}$$

Referred to the basis $|\xi_1\rangle, |\xi_2\rangle, \ldots, |\xi_n\rangle$ the same vectors $|X\rangle$ and $|Y\rangle$ are $|X\rangle = \sum_{i=1}^{n} a_i' |\xi_i\rangle$, and $|Y\rangle = \sum_{i=1}^{n} b_i' |\xi_i\rangle$, and Eqs. (5.31) are replaced by

$$b_i' = \sum_{j=1}^{n} a_j' \langle \xi_i | \underset{\sim}{A} | \xi_j \rangle = \sum_{j=1}^{n} a_j' A_{ij}',$$

where $A_{ij}' = \langle \xi_i | \underset{\sim}{A} | \xi_j \rangle$, which is related to A_{ij} by the following relation:

$$A_{ij}' = \langle \xi_i | \underset{\sim}{A} | \xi_j \rangle = \langle U\varphi_i | \underset{\sim}{A} | U\varphi_j \rangle = \langle \varphi_i | U^* \underset{\sim}{A} U | \varphi_j \rangle = (U^* \underset{\sim}{A} U)_{ij}$$

or using the rule for matrix multiplication

$$A_{ij}' = \langle \xi_i | \underset{\sim}{A} | \xi_j \rangle = (U^* \underset{\sim}{A} U)_{ij} = \sum_{r=1}^{n} \sum_{s=1}^{n} U_{ir}^* A_{rs} U_{sj}. \tag{5.32}$$

From Eqs. (5.32) we can find the matrix representation of an operator with respect to a new basis.

If the operator $\underset{\sim}{A}$ transforms vector $|X\rangle$ into vector $|Y\rangle$ which is vector $|X\rangle$ itself multiplied by a scalar λ: $|Y\rangle = \lambda |X\rangle$, then Eq. (5.30) becomes an eigenvalue equation:

$$\underset{\sim}{A} | X \rangle = \lambda | X \rangle.$$

Commuting operators

In general, operators do not commute. But commuting operators do exist and they are of importance in quantum mechanics. As Hermitian operators play a dominant role in quantum mechanics, and the eigenvalues and the eigenvectors of a Hermitian operator are real and form a complete set, respectively, we shall concentrate on Hermitian operators. It is straightforward to prove that

> Two commuting Hermitian operators possess a complete ortho-
> normal set of common eigenvectors, and vice versa.

If $\underset{\sim}{A}$ and $\underset{\sim}{A} |v\rangle = \alpha|v\rangle$ are two commuting Hermitian operators, and if

$$\underset{\sim}{A} | v \rangle = \alpha | v \rangle, \tag{5.33}$$

then we have to show that

$$\underset{\sim}{B} | v \rangle = \beta | v \rangle. \tag{5.34}$$

Multiplying Eq. (5.33) from the left by $\underset{\sim}{B}$, we obtain

$$\underset{\sim}{B}(\underset{\sim}{A}|v\rangle) = \alpha(\underset{\sim}{B}|v\rangle),$$

which using the fact $\underset{\sim}{A}\underset{\sim}{B} = \underset{\sim}{B}\underset{\sim}{A}$, can be rewritten as

$$\underset{\sim}{A}(\underset{\sim}{B}|v\rangle) = \alpha(\underset{\sim}{B}|v\rangle).$$

Thus, $\underset{\sim}{B}|v\rangle$ is an eigenvector of $\underset{\sim}{A}$ belonging to eigenvalue α. If α is non-degenerate, then $\underset{\sim}{B}|v\rangle$ should be linearly dependent on $|v\rangle$, so that

$$a(\underset{\sim}{B}|v\rangle) + b|v\rangle = 0, \quad \text{with} \quad a \neq 0 \quad \text{and} \quad b \neq 0.$$

It follows that

$$\underset{\sim}{B}|v\rangle = -(b/a)|v\rangle = \beta|v\rangle.$$

If A is degenerate, then the matter becomes a little complicated. We now state the results without proof. There are three possibilities:

(1) The degenerate eigenvectors (that is, the linearly independent eigenvectors belonging to a degenerate eigenvalue) of $\underset{\sim}{A}$ are degenerate eigenvectors of $\underset{\sim}{B}$ also.
(2) The degenerate eigenvectors of $\underset{\sim}{A}$ belong to different eigenvalues of $\underset{\sim}{B}$. In this case, we say that the degeneracy is removed by the Hermitian operator $\underset{\sim}{B}$.
(3) Every degenerate eigenvector of $\underset{\sim}{A}$ is not an eigenvector of $\underset{\sim}{B}$. But there are linear combinations of the degenerate eigenvectors, as many in number as the degrees of degeneracy, which are degenerate eigenvectors of $\underset{\sim}{A}$ but are non-degenerate eigenvectors of $\underset{\sim}{B}$. Of course, the degeneracy is removed by $\underset{\sim}{B}$.

Function spaces

We have seen that functions can be elements of a vector space. We now return to this theme for a more detailed analysis. Consider the set of all functions that are continuous on some interval. Two such functions can be added together to construct a third function $h(x)$:

$$h(x) = f(x) + g(x), \quad a \leq x \leq b,$$

where the plus symbol has the usual operational meaning of 'add the value of f at the point x to the value of g at the same point.'

A function $f(x)$ can also be multiplied by a number k to give the function $p(x)$:

$$p(x) = k \cdot f(x), \quad a \leq x \leq b.$$

The centred dot, the multiplication symbol, is again understood in the conventional meaning of 'multiply by k the value of $f(x)$ at the point x.'

It is evident that the following conditions are satisfied:

(a) By adding two continuous functions, we obtain a continuous function.

(b) The multiplication by a scalar of a continuous function yields again a continuous function.

(c) The function that is identically zero for $a \leq x \leq b$ is continuous, and its addition to any other function does not alter this function.

(d) For any function $f(x)$ there exists a function $(-1)f(x)$, which satisfies

$$f(x) + [(-1)f(x)] = 0.$$

Comparing these statements with the axioms for linear vector spaces (Axioms A.1–A.8), we see clearly that the set of all continuous functions defined on some interval forms a linear vector space; this is called a function space. We shall consider the entire set of values of a function $f(x)$ as representing a vector $|f\rangle$ of this abstract vector space F (F stands for function space). In other words, we shall treat the number $f(x)$ at the point x as the component with 'index x' of an abstract vector $|f\rangle$. This is quite similar to what we did in the case of finite-dimensional spaces when we associated a component a_i of a vector with each value of the index i. The only difference is that this index assumed a discrete set of values 1, 2, etc., up to N (for N-dimensional space), whereas the argument x of a function $f(x)$ is a continuous variable. In other words, the function $f(x)$ has an infinite number of components, namely the values it takes in the continuum of points labeled by the real variable x. However, two questions may be raised.

The first question concerns the orthonormal basis. The components of a vector are defined with respect to some basis and we do not know which basis has been (or could be) chosen in the function space. Unfortunately, we have to postpone the answer to this question. Let us merely note that, once a basis has been chosen, we work only with the components of a vector. Therefore, provided we do not change to other basis vectors, we need not be concerned about the particular basis that has been chosen.

The second question is how to define an inner product in an infinite-dimensional vector space. Suppose the function $f(x)$ describes the displacement of a string clamped at $x = 0$ and $x = L$. We divide the interval of length L into N equal parts and measure the displacements $f(x_i) \equiv f_i$ at N point $x_i, i = 1, 2, \ldots, N$. At fixed N, the functions are elements of a finite N-dimensional vector space. An inner product is defined by the expression

$$\langle f|g\rangle = \sum_{i=1}^{N} f_i g_i.$$

For a vibrating string, the space is real and there is no need to conjugate anything.

To improve the description, we can increase the number N. However, as $N \to \infty$ by increasing the number of points without limit, the inner product diverges as we subdivide further and further. The way out of this is to modify the definition by a positive prefactor $\Delta = L/N$ which does not violate any of the axioms for the inner product. But now

$$\langle f|g \rangle = \lim_{\Delta \to 0} \sum_{i=1}^{N} f_i g_i \Delta \to \int_0^L f(x)g(x)dx,$$

by the usual definition of an integral. Thus the inner product of two functions is the integral of their product. Two functions are orthogonal if this inner product vanishes, and a function is normalized if the integral of its square equals unity. Thus we can speak of an orthonormal set of functions in a function space just as in finite dimensions. The following is an example of such a set of functions defined in the interval $0 \le x \le L$ and vanishing at the end points:

$$|e_m\rangle \to m(x) = \sqrt{\frac{2}{L}} \sin \frac{m\pi x}{L}, \qquad m = 1, 2, \ldots, \infty,$$

$$\langle e_m|e_n \rangle = \frac{2}{L} \int_0^L \sin \frac{m\pi x}{L} \sin \frac{n\pi x}{L} dx = \delta_{mn}.$$

For the details, see 'Vibrating strings' of Chapter 4.

In quantum mechanics we often deal with complex functions and our definition of the inner product must then modified. We define the inner product of $f(x)$ and $g(x)$ as

$$\langle f|g \rangle = \int_0^L f^*(x)g(x)dx,$$

where f^* is the complex conjugate of f. An orthonormal set for this case is

$$m(x) = \frac{1}{\sqrt{2\pi}} e^{imx}, \qquad m = 0, \pm 1, \pm 2, \ldots,$$

which spans the space of all functions of period 2π with finite norm. A linear vector space with a complex-type inner product is called a *Hilbert space*.

Where and how did we get the orthonormal functions? In general, by solving the eigenvalue equation of some Hermitian operator. We give a simple example here. Consider the derivative operator $D = d(\ )/dx$:

$$Df(x) = df(x)/dx, \quad D|f\rangle = d|f\rangle/dx.$$

However, D is not Hermitian, because it does not meet the condition:

$$\int_0^L f^*(x) \frac{dg(x)}{dx} dx = \left(\int_0^L g^*(x) \frac{df(x)}{dx} dx \right)^*.$$

Here is why:

$$\left(\int_0^L g^*(x) \frac{df(x)}{dx} dx \right)^* = \int_0^L g(x) \frac{df^*(x)}{dx} dx$$

$$= g f^* \Big|_0^L - \int_0^L f^*(x) \frac{dg(x)}{dx} dx.$$

It is easy to see that hermiticity of D is lost on two counts. First we have the term coming from the end points. Second the integral has the wrong sign. We can fix both of these by doing the following:

(a) Use operator $-iD$. The extra i will change sign under conjugation and kill the minus sign in front of the integral.
(b) Restrict the functions to those that are periodic: $f(0) = f(L)$.

Thus, $-iD$ is a Hermitian operator on period functions. Now we have

$$-i \frac{df(x)}{dx} = \lambda f(x),$$

where λ is the eigenvalue. Simple integration gives

$$f(x) = A e^{i\lambda x}.$$

Now the periodicity requirement gives

$$e^{i\lambda L} = e^{i\lambda 0} = 1$$

from which it follows that

$$\lambda = 2\pi m / L, \qquad m = 0, \pm 1, \pm 2,$$

and the normalization condition gives

$$A = \frac{1}{\sqrt{L}}.$$

Hence the set of orthonormal eigenvectors is given by

$$f_m(x) = \frac{1}{\sqrt{L}} e^{2\pi i m x / L}.$$

In quantum mechanics the eigenvalue equation is the Schrödinger equation and the Hermitian operator is the Hamiltonian operator. Quantum mechanically, a system with n degrees of freedom which is classically specified by n generalized coordinates $q_1, \ldots, q_2, q_n$ is specified at a fixed instant of time by a wave function $\psi(q_1, q_2, \ldots, q_n)$ whose norm is unity, that is,

$$\langle \psi | \psi \rangle = \int |\psi(q_1, q_2, \ldots, q_n)|^2 dq_1, dq_2, \ldots, dq_n = 1,$$

the integration being over the accessible values of the coordinates $q_1, q_2, \ldots, q_n$. The set of all such wave functions with unit norm spans a Hilbert space H. Every possible state of the system is represented by a function in this Hilbert space, and conversely, every vector in this Hilbert space represents a possible state of the system. In addition to depending on the coordinates $q_1, q_2, \ldots, q_n$, the wave function depends also on the time t, but the dependence on the qs and on t are essentially different. The Hilbert space H is formed with respect to the spatial coordinates $q_1, q_2, \ldots, q_n$ only, for example, the inner product is formed with respect to the qs only, and one wave function $\psi(q_1, q_2, \ldots, q_n)$ states its complete spatial dependence. On the other hand the states of the system at different instants of time $t_1, t_2, \ldots$ are given by the different wave functions $\psi_1(q_1, q_2, \ldots, q_n), \psi_2(q_1, q_2, \ldots, q_n) \ldots$ of the Hilbert space.

Problems

5.1 Prove the three main properties of the dot product given by Eq. (5.7).

5.2 Show that the points on a line V passing through the origin in E_3 form a linear vector space under the addition and scalar multiplication operations for vectors in E_3.

Hint: The points of V satisfy parametric equations of the form

$$x_1 = at, \quad x_2 = bt, \quad x_3 = ct, \quad -\infty < t < \infty.$$

5.3 Do all Hermitian 2×2 matrices form a vector space under addition? Is there any requirement on the scalars that multiply them?

5.4 Let V be the set of all points (x_1, x_2) in E_2 that lie in the first quadrant; that is, such that $x_1 \geq 0$ and $x_2 \geq 0$. Show that the set V fails to be a vector space under the operations of addition and scalar multiplication.

Hint: Consider $\mathbf{u} = (1, 1)$ which lies in V. Now form the scalar multiplication $(-1)\mathbf{u} = (-1, -1)$; where is this point located?

5.5 Show that the set W of all 2×2 matrices having zeros on the principal diagonal is a subspace of the vector space M_{22} of all 2×2 matrices.

5.6 Show that $|W\rangle = (4, -1, 8)$ is not a linear combination of $|U\rangle = (1, 2, -1)$ and $|V\rangle = (6, 4, 2)$.

5.7 Show that the following three vectors in E_3 cannot serve as base vectors of E_3:

$$|1\rangle = (1, 1, 2), |2\rangle = (1, 0, 1), \quad \text{and} \quad |3\rangle = (2, 1, 3).$$

5.8 Determine which of the following lie in the space spanned by $|f\rangle = \cos^2 x$ and $|g\rangle = \sin^2 x$: (a) $\cos 2x$; (b) $3 + x^2$; (c) 1; (d) $\sin x$.

5.9 Determine whether the three vectors

$$|1\rangle = (1, -2, 3), \quad |2\rangle = (5, 6, -1), \quad |3\rangle = (3, 2, 1)$$

are linearly dependent or independent.

230

5.10 Given the following three vectors from the vector space of real 2×2 matrices:

$$|1\rangle = \begin{pmatrix} 0 & 1 \\ 0 & 0 \end{pmatrix}, \quad |2\rangle = \begin{pmatrix} 1 & 1 \\ 0 & 1 \end{pmatrix}, \quad |3\rangle = \begin{pmatrix} -2 & -1 \\ 0 & -2 \end{pmatrix},$$

determine whether they are linearly dependent or independent.

5.11 If $S = \{|1\rangle, |2\rangle, \ldots, |n\rangle\}$ is a basis for a vector space V, show that every set with more than n vectors is linearly dependent.

5.12 Show that any two bases for a finite-dimensional vector space have the same number of vectors.

5.13 Consider the vector space E_3 with the Euclidean inner product. Apply the Gram–Schmidt process to transform the basis

$$|1\rangle = (1,1,1), \quad |2\rangle = (0,1,1), \quad |3\rangle = (0,0,1)$$

into an orthonormal basis.

5.14 Consider the two linearly independent vectors of Example 5.10:

$$|U\rangle = (3 - 4i)|1\rangle + (5 - 6i)|2\rangle,$$

$$|W\rangle = (1 - i)|1\rangle + (2 - 3i)|2\rangle,$$

where $|1\rangle$ and $|2\rangle$ are an orthonormal basis. Apply the Gram–Schmidt process to transform the two vectors into an orthonormal basis.

5.15 Show that the eigenvalue of the square of an operator is the square of the eigenvalue of the operator.

5.16 Show that if, for a given $\underset{\sim}{A}$, both operators $\underset{\sim}{A}_L^{-1}$ and $\underset{\sim}{A}_R^{-1}$ exist, then

$$\underset{\sim}{A}_L^{-1} = \underset{\sim}{A}_R^{-1} \equiv \underset{\sim}{A}^{-1}.$$

5.17 Show that if a unitary operator $\underset{\sim}{U}$ can be written in the form $\underset{\sim}{U} = 1 + ie\,\underset{\sim}{F}$, where e is a real infinitesimally small number, then the operator $\underset{\sim}{F}$ is Hermitian.

5.18 Show that the differential operator

$$\underset{\sim}{p} = \frac{\hbar}{i} \frac{d}{dx}$$

is linear and Hermitian in the space of all differentiable wave functions $\phi(x)$ that, say, vanish at both ends of an interval (a, b).

5.19 The translation operator $T(a)$ is defined to be such that $T(a)\phi(x) = \phi(x + a)$. Show that:

(a) $T(a)$ may be expressed in terms of the operator

$$\underset{\sim}{p} = \frac{\hbar}{i} \frac{d}{dx};$$

(b) $T(a)$ is unitary.

5.21 Verify that:

(a) $\dfrac{2}{L} \displaystyle\int_0^L \sin\dfrac{m\pi x}{L} \sin\dfrac{n\pi x}{L} \, dx = \delta_{mn}$.

(b) $\dfrac{1}{\sqrt{2\pi}} \displaystyle\int_0^{2\pi} e^{i(m-n)} \, dx = \delta_{mn}$.

6

Functions of a complex variable

The theory of functions of a complex variable is a basic part of mathematical analysis. It provides some of the very useful mathematical tools for physicists and engineers. In this chapter a brief introduction to complex variables is presented which is intended to acquaint the reader with at least the rudiments of this important subject.

Complex numbers

The number system as we know it today is a result of gradual development. The natural numbers (positive integers 1, 2, ...) were first used in counting. Negative integers and zero (that is, 0, $-1, -2, \ldots$) then arose to permit solutions of equations such as $x + 3 = 2$. In order to solve equations such as $bx = a$ for all integers a and b where $b \neq 0$, rational numbers (or fractions) were introduced. Irrational numbers are numbers which cannot be expressed as a/b, with a and b integers and $b \neq 0$, such as $\sqrt{2} = 1.41423, \pi = 3.14159$

Rational and irrational numbers are all real numbers. However, the real number system is still incomplete. For example, there is no real number x which satisfies the algebraic equation $x^2 + 1 = 0$: $x = \sqrt{-1}$. The problem is that we do not know what to make of $\sqrt{-1}$ because there is no real number whose square is -1. Euler introduced the symbol $i = \sqrt{-1}$ in 1777 years later Gauss used the notation $a + ib$ to denote a complex number, where a and b are real numbers. Today, $i = \sqrt{-1}$ is called the unit imaginary number.

In terms of i, the answer to equation $x^2 + 1 = 0$ is $x = i$. It is postulated that i will behave like a real number in all manipulations involving addition and multiplication.

We now introduce a general complex number, in Cartesian form

$$z = x + iy \tag{6.1}$$

233

and refer to x and y as its real and imaginary parts and denote them by the symbols Re z and Im z, respectively. Thus if $z = -3 + 2i$, then Re $z = -3$ and Im $z = +2$.

A number with just $y \neq 0$ is called a pure imaginary number.

The complex conjugate, or briefly conjugate, of the complex number $z = x + iy$ is

$$z^* = x - iy \qquad (6.2)$$

and is called 'z-star'. Sometimes we write it $\bar{z}$ and call it 'z-bar'. Complex conjugation can be viewed as the process of replacing i by $-i$ within the complex number.

Basic operations with complex numbers

Two complex numbers $z_1 = x_1 + iy_1$ and $z_2 = x_2 + iy_2$ are equal if and only if $x_1 = x_2$ and $y_1 = y_2$.

In performing operations with complex numbers we can proceed as in the algebra of real numbers, replacing i^2 by -1 when it occurs. Given two complex numbers z_1 and z_2 where $z_1 = a + ib, z_2 = c + id$, the basic rules obeyed by complex numbers are the following:

(1) Addition:

$$z_1 + z_2 = (a + ib) + (c + id) = (a + c) + i(b + d).$$

(2) Subtraction:

$$z_1 - z_2 = (a + ib) - (c + id) = (a - c) + i(b - d).$$

(3) Multiplication:

$$z_1 z_2 = (a + ib)(c + id) = (ac - bd) + i(ad - bc).$$

(4) Division:

$$\frac{z_1}{z_2} = \frac{a + ib}{c + id} = \frac{(a + ib)(c - id)}{(c + id)(c - id)} = \frac{ac + bd}{c^2 + d^2} + i\frac{bc - ad}{c^2 + d^2}.$$

Polar form of complex numbers

All real numbers can be visualized as points on a straight line (the x-axis). A complex number, containing two real numbers, can be represented by a point in a two-dimensional xy plane, known as the z plane or the complex plane (also known as the Gauss plane or Argand diagram). The complex variable $z = x + iy$ and its complex conjugation z^* are labeled in Fig. 6.1.

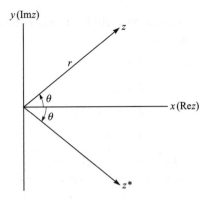

Figure 6.1.   The complex plane.

The complex variable can also be represented by the plane polar coordinates (r, θ):

$$z = r(\cos \theta + i \sin \theta).$$

With the help of Euler's formula

$$e^{i\theta} = \cos \theta + i \sin \theta,$$

we can rewrite the last equation in polar form:

$$z = r(\cos \theta + i \sin \theta) = re^{i\theta}, \quad r = \sqrt{x^2 + y^2} = \sqrt{zz^*}. \qquad (6.3)$$

r is called the modulus or absolute value of z, denoted by $|z|$ or mod z; and θ is called the phase or argument of z and it is denoted by arg z. For any complex number $z \neq 0$ there corresponds only one value of θ in $0 \leq \theta \leq 2\pi$. The absolute value of z has the following properties. If $z_1, z_2, \ldots, z_m$ are complex numbers, then we have:

(1) $|z_1 z_2 \cdots z_m| = |z_1||z_2| \cdots |z_m|$.

(2) $\left| \dfrac{z_1}{z_2} \right| = \dfrac{|z_1|}{|z_2|}, \quad z_2 \neq 0.$

(3) $|z_1 + z_2 + \cdots + z_m| \leq |z_1| + |z_2| + \cdots + |z_m|$.

(4) $|z_1 \pm z_2| \geq |z_1| - |z_2|$.

Complex numbers $z = re^{i\theta}$ with $r = 1$ have $|z| = 1$ and are called unimodular.

235

We may imagine them as lying on a circle of unit radius in the complex plane. Special points on this circle are

$$\theta = 0 \quad (1)$$
$$\theta = \pi/2 \quad (i)$$
$$\theta = \pi \quad (-1)$$
$$\theta = -\pi/2 \quad (-i).$$

The reader should know these points at all times.

Sometimes it is easier to use the polar form in manipulations. For example, to multiply two complex numbers, we multiply their moduli and add their phases; to divide, we divide by the modulus and subtract the phase of the denominator:

$$zz_1 = (re^{i\theta})(r_1 e^{i\theta_1}) = rr_1 e^{i(\theta + \theta_1)}, \quad \frac{z}{z_1} = \frac{re^{i\theta}}{r_1 e^{i\theta_1}} = \frac{r}{r_1} e^{i(\theta - \theta_1)}.$$

On the other hand to add two complex numbers we have to go back to the Cartesian forms, add the components and revert to the polar form.

If we view a complex number z as a vector, then the multiplication of z by $e^{i\alpha}$ (where α is real) can be interpreted as a rotation of z counterclockwise through angle α; and we can consider $e^{i\alpha}$ as an operator which acts on z to produce this rotation. Similarly, the multiplication of two complex numbers represents a rotation and a change of length: $z_1 = r_1 e^{i\theta_1}, z_2 = r_2 e^{i\theta_2}, z_1 z_2 = r_1 r_2 e^{i(\theta_1 + \theta_2)}$; the new complex number has length $r_1 r_2$ and phase $\theta_1 + \theta_2$.

Example 6.1
Find $(1 + i)^8$.

Solution: We first write z in polar form: $z = 1 + i = r(\cos\theta + i\sin\theta)$, from which we find $r = \sqrt{2}, \theta = \pi/4$. Then

$$z = \sqrt{2}(\cos\pi/4 + i\sin\pi/4) = \sqrt{2}e^{i\pi/4}.$$

Thus

$$(1 + i)^8 = (\sqrt{2}e^{i\pi/4})^8 = 16e^{2\pi i} = 16.$$

Example 6.2
Show that

$$\left(\frac{1 + \sqrt{3}i}{1 - \sqrt{3}i}\right)^{10} = -\frac{1}{2} + i\frac{\sqrt{3}}{2}.$$

$$\left(\frac{1+i\sqrt{3}}{1-i\sqrt{3}}\right)^{10} = \left(\frac{2e^{\pi i/3}}{2e^{-\pi i/3}}\right)^{10} = \left(e^{2\pi i/3}\right)^{10} = e^{20\pi i/3}$$

$$= e^{6\pi i}e^{2\pi i/3} = 1[\cos(2\pi/3) + i\sin(2\pi/3)] = -\frac{1}{2} + i\frac{\sqrt{3}}{2}.$$

De Moivre's theorem and roots of complex numbers

If $z_1 = r_1 e^{i\theta_1}$ and $z_2 = r_2 e^{i\theta_2}$, then

$$z_1 z_2 = r_1 r_2 e^{i(\theta_1 + \theta_2)} = r_1 r_2[\cos(\theta_1 + \theta_2) + i\sin(\theta_1 + \theta_2)].$$

A generalization of this leads to

$$z_1 z_2 \cdots z_n = r_1 r_2 \cdots r_n e^{i(\theta_1 + \theta_2 + \cdots + \theta_n)}$$

$$= r_1 r_2 \cdots r_n[\cos(\theta_1 + \theta_2 + \cdots + \theta_n) + i\sin(\theta_1 + \theta_2 + \cdots + \theta_n)];$$

if $z_1 = z_2 = \cdots = z_n = z$ this becomes

$$z^n = (re^{i\theta})^n = r^n[\cos(n\theta) + i\sin(n\theta)],$$

from which it follows that

$$(\cos\theta + i\sin\theta)^n = \cos(n\theta) + i\sin(n\theta), \tag{6.4}$$

a result known as De Moivre's theorem. Thus we now have a general rule for calculating the nth power of a complex number z. We first write z in polar form $z = r(\cos\theta + i\sin\theta)$, then

$$z^n = r^n(\cos\theta + i\sin\theta)^n = r^n[\cos n\theta + i\sin n\theta]. \tag{6.5}$$

The general rule for calculating the nth root of a complex number can now be derived without difficulty. A number w is called an nth root of a complex number z if $w^n = z$, and we write $w = z^{1/n}$. If $z = r(\cos\theta + i\sin\theta)$, then the complex number

$$w_0 = \sqrt[n]{r}\left(\cos\frac{\theta}{n} + i\sin\frac{\theta}{n}\right)$$

is definitely the nth root of z because $w_0^n = z$. But the numbers

$$w_k = \sqrt[n]{r}\left(\cos\frac{\theta + 2\pi k}{n} + i\sin\frac{\theta + 2\pi k}{n}\right), \qquad k = 1, 2, \ldots, (n-1),$$

are also nth roots of z because $w_k^n = z$. Thus the general rule for calculating the nth root of a complex number is

$$w = \sqrt[n]{r}\left(\cos\frac{\theta + 2\pi k}{n} + i\sin\frac{\theta + 2\pi k}{n}\right), \qquad k = 0, 1, 2, \ldots, (n-1). \tag{6.6}$$

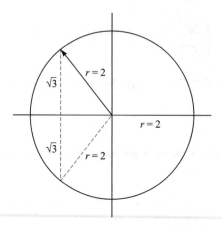

Figure 6.2.   The cube roots of 8.

It is customary to call the number corresponding to $k = 0$ (that is, w_0) the princi-pal root of z.

The nth roots of a complex number z are always located at the vertices of a regular polygon of n sides inscribed in a circle of radius $\sqrt[n]{r}$ about the origin.

Example 6.3
Find the cube roots of 8.

Solution: In this case $z = 8 + i0 = r(\cos\theta + i\sin\theta), r = 2$ and the principal argu-ment $\theta = 0$. Formula (6.6) then yields

$$\sqrt[3]{8} = 2\left(\cos\frac{2k\pi}{3} + i\sin\frac{2k\pi}{3}\right), \qquad k = 0, 1, 2.$$

These roots are plotted in Fig. 6.2:

$$
\begin{aligned}
2 & \qquad (k = 0, \theta = 0°), \\
-1 + i\sqrt{3} & \qquad (k = 1, \theta = 120°), \\
-1 - i\sqrt{3} & \qquad (k = 2, \theta = 240°).
\end{aligned}
$$

Functions of a complex variable

Complex numbers $z = x + iy$ become variables if x or y (or both) vary. Then functions of a complex variable may be formed. If to each value which a complex variable z can assume there corresponds one or more values of a complex variable w, we say that w is a function of z and write $w = f(z)$ or $w = g(z)$, etc. The variable z is sometimes called an independent variable, and then w is a dependent variable. If only one value of w corresponds to each value of z, we say that w is a single-valued function of z or that $f(z)$ is single-valued; and if more than one value of w corresponds to each value of z, w is then a multiple-valued function of z. For example, $w = z^2$ is a single-valued function of z, but $w = \sqrt{z}$ is a double-valued function of z. In this chapter, whenever we speak of a function we shall mean a single-valued function, unless otherwise stated.

Mapping

Note that w is also a complex variable and so can be written in the form

$$w = u + iv = f(x + iy), \tag{6.7}$$

where u and v are real. By equating real and imaginary parts this is seen to be equivalent to

$$u = u(x, y), \quad v = v(x, y). \tag{6.8}$$

If $w = f(z)$ is a single-valued function of z, then to each point of the complex z plane, there corresponds a point in the complex w plane. If $f(z)$ is multiple-valued, a point in the z plane is mapped in general into more than one point. The following two examples show the idea of mapping clearly.

Example 6.4
Map $w = z^2 = r^2 e^{2i\theta}$.

Solution: This is single-valued function. The mapping is unique, but not one-to-one. It is a two-to-one mapping, since z and $-z$ give the same square. For example as shown in Fig. 6.3, $z = -2 + i$ and $z = 2 - i$ are mapped to the same point $w = 3 - 4i$; and $z = 1 - 3i$ and $-1 + 3i$ are mapped into the same point $w = -8 - 6i$.

The line joining the points $P(-2, 1)$ and $Q(1, -3)$ in the z-plane is mapped by $w = z^2$ into a curve joining the image points $P'(3, -4)$ and $Q'(-8, -6)$. It is not

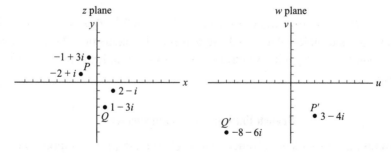

Figure 6.3.   The mapping function $w = z^2$.

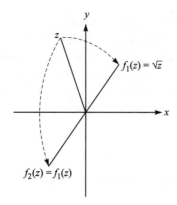

Figure 6.4.   The mapping function $w = \sqrt{z}$.

very difficult to determine the equation of this curve. We first need the equation of the line joining P and Q in the z plane. The parametric equations of the line joining P and Q are given by

$$\frac{x - (-2)}{1 - (-2)} = \frac{y - 1}{-3 - 1} = t \quad \text{or} \quad x = 3t - 2, y = 1 - 4t.$$

The equation of the line PQ is then given by $z = 3t - 2 + i(1 - 4t)$. The curve in the w plane into which the line PQ is mapped has the equation

$$w = z^2 = [3t - 2 + i(1 - 4t)]^2 = 3 - 4t - 7t^2 + i(-4 + 22t - 24t^2),$$

from which we obtain

$$u = 3 - 4t - 7t^2, \qquad v = -4 + 22t - 24t^2.$$

By assigning various values to the parameter t, this curve may be graphed.

Sometimes it is convenient to superimpose the z and w planes. Then the images of various points are located on the same plane and the function $w = f(z)$ may be said to transform the complex plane to itself (or a part of itself).

Example 6.5
Map $w = f(z) = \sqrt{z}, z = re^{i\theta}$.

Solution: There are two square roots: $f_1(re^{i\theta}) = \sqrt{r}e^{i\theta/2}$, $f_2 = -f_1 = \sqrt{r}e^{i(\theta+2\pi)/2}$. The function is double-valued, and the mapping is one-to-two. This is shown in Fig. 6.4, where for simplicity we have used the same complex plane for both z and $w = f(z)$.

Branch lines and Riemann surfaces

We now take a close look at the function $w = \sqrt{z}$ of Example 6.5. Suppose we allow z to make a complete counterclockwise motion around the origin starting from point

240

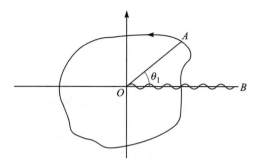

Figure 6.5.   Branch cut for the function $w = \sqrt{z}$.

A, as shown in Fig. 6.5. At A, $\theta = \theta_1$ and $w = \sqrt{r}e^{i\theta/2}$. After a complete circuit back to A, $\theta = \theta_1 + 2\pi$ and $w = \sqrt{r}e^{i(\theta+2\pi)/2} = -\sqrt{r}e^{i\theta/2}$. However, by making a second complete circuit back to A, $\theta = \theta_1 + 4\pi$, and so $w = \sqrt{r}e^{i(\theta+4\pi)/2} = \sqrt{r}e^{i\theta/2}$; that is, we obtain the same value of w with which we started.

We can describe the above by stating that if $0 \le \theta < 2\pi$ we are on one branch of the multiple-valued function $\sqrt{z}$, while if $2\pi \le \theta < 4\pi$ we are on the other branch of the function. It is clear that each branch of the function is single-valued. In order to keep the function single-valued, we set up an artificial barrier such as OB (the wavy line) which we agree not to cross. This artificial barrier is called a branch line or branch cut, point O is called a branch point. Any other line from O can be used for a branch line.

Riemann (George Friedrich Bernhard Riemann, 1826–1866) suggested another way to achieve the purpose of the branch line described above. Imagine the z plane consists of two sheets superimposed on each other. We now cut the two sheets along OB and join the lower edge of the bottom sheet to the upper edge of the top sheet. Then on starting in the bottom sheet and making one complete circuit about O we arrive in the top sheet. We must now imagine the other cut edges to be joined together (independent of the first join and actually disregarding its existence) so that by continuing the circuit we go from the top sheet back to the bottom sheet.

The collection of two sheets is called a Riemann surface corresponding to the function $\sqrt{z}$. Each sheet corresponds to a branch of the function and on each sheet the function is singled-valued. The concept of Riemann surfaces has the advantage that the various values of multiple-valued functions are obtained in a continuous fashion.

The differential calculus of functions of a complex variable

Limits and continuity

The definitions of limits and continuity for functions of a complex variable are similar to those for a real variable. We say that $f(z)$ has limit w_0 as z approaches

241

z_0, which is written as

$$\lim_{z \to z_0} f(z) = w_0, \qquad (6.9)$$

if

(a) $f(z)$ is defined and single-valued in a neighborhood of $z = z_0$, with the possible exception of the point z_0 itself; and

(b) given any positive number ε (however small), there exists a positive number δ such that $|f(z) - w_0| < \varepsilon$ whenever $0 < |z - z_0| < \delta$.

The limit must be independent of the manner in which z approaches z_0.

Example 6.6
(a) If $f(z) = z^2$, prove that $\lim_{z \to z_0} f(z) = z_0^2$
(b) Find $\lim_{z \to z_0} f(z)$ if

$$f(z) = \begin{cases} z^2 & z \neq z_0 \\ 0 & z = z_0 \end{cases}.$$

Solution: (a) We must show that given any $\varepsilon > 0$ we can find δ (depending in general on ε) such that $|z^2 - z_0^2| < \varepsilon$ whenever $0 < |z - z_0| < \delta$.

Now if $\delta \leq 1$, then $0 < |z - z_0| < \delta$ implies that

$$|z - z_0||z + z_0| < \delta|z + z_0| = \delta|z - z_0 + 2z_0|,$$

$$|z^2 - z_0^2| < \delta(|z - z_0| + 2|z_0|) < \delta(1 + 2|z_0|).$$

Taking δ as 1 or $\varepsilon/(1 + 2|z_0|)$, whichever is smaller, we then have $|z^2 - z_0^2| < \varepsilon$ whenever $0 < |z - z_0| < \delta$, and the required result is proved.

(b) There is no difference between this problem and that in part (a), since in both cases we exclude $z = z_0$ from consideration. Hence $\lim_{z \to z_0} f(z) = z_0^2$. Note that the limit of $f(z)$ as $z \to z_0$ has nothing to do with the value of $f(z)$ at z_0.

A function $f(z)$ is said to be continuous at z_0 if, given any $\varepsilon > 0$, there exists a $\delta > 0$ such that $|f(z) - f(z_0)| < \varepsilon$ whenever $0 < |z - z_0| < \delta$. This implies three conditions that must be met in order that $f(z)$ be continuous at $z = z_0$:

(1) $\lim_{z \to z_0} f(z) = w_0$ must exist;
(2) $f(z_0)$ must exist, that is, $f(z)$ is defined at z_0;
(3) $w_0 = f(z_0)$.

For example, complex polynomials, $\alpha_0 + \alpha_1 z^1 + \alpha_2 z^2 + \alpha_n z^n$ (where α_i may be complex), are continuous everywhere. Quotients of polynomials are continuous whenever the denominator does not vanish. The following example provides further illustration.

242

A function $f(z)$ is said to be continuous in a region R of the z plane if it is continuous at all points of R.

Points in the z plane where $f(z)$ fails to be continuous are called discontinuities of $f(z)$, and $f(z)$ is said to be discontinuous at these points. If $\lim_{z \to z_0} f(z)$ exists but is not equal to $f(z_0)$, we call the point z_0 a removable discontinuity, since by redefining $f(z_0)$ to be the same as $\lim_{z \to z_0} f(z)$ the function becomes continuous.

To examine the continuity of $f(z)$ at $z = \infty$, we let $z = 1/w$ and examine the continuity of $f(1/w)$ at $w = 0$.

Derivatives and analytic functions

Given a continuous, single-valued function of a complex variable $f(z)$ in some region R of the z plane, the derivative $f'(z) (\equiv df/dz)$ at some fixed point z_0 in R is defined as

$$f'(z_0) = \lim_{\Delta z \to 0} \frac{f(z_0 + \Delta z) - f(z_0)}{\Delta z}, \tag{6.10}$$

provided the limit exists independently of the manner in which $\Delta z \to 0$. Here $\Delta z = z - z_0$, and z is any point of some neighborhood of z_0. If $f'(z)$ exists at z_0 and every point z in some neighborhood of z_0, then $f(z)$ is said to be analytic at z_0. And $f(z)$ is analytic in a region R of the complex z plane if it is analytic at every point in R.

In order to be analytic, $f(z)$ must be single-valued and continuous. It is straightforward to see this. In view of Eq. (6.10), whenever $f'(z_0)$ exists, then

$$\lim_{\Delta z \to 0} [f(z_0 + \Delta z) - f(z_0)] = \lim_{\Delta z \to 0} \frac{f(z_0 + \Delta z) - f(z_0)}{\Delta z} \lim_{\Delta z \to 0} \Delta z = 0$$

that is,

$$\lim_{\Delta z \to 0} f(z) = f(z_0).$$

Thus f is necessarily continuous at any point z_0 where its derivative exists. But the converse is not necessarily true, as the following example shows.

Example 6.7
The function $f(z) = z^*$ is continuous at z_0, but dz^*/dz does not exist anywhere. By definition,

$$\frac{dz^*}{dz} = \lim_{\Delta z \to 0} \frac{(z + \Delta z)^* - z^*}{\Delta z} = \lim_{\Delta x, \Delta y \to 0} \frac{(x + iy + \Delta x + i\Delta y)^* - (x + iy)^*}{\Delta x + i\Delta y}$$

$$= \lim_{\Delta x, \Delta y \to 0} \frac{x - iy + \Delta x - i\Delta y - (x - iy)}{\Delta x + i\Delta y} = \lim_{\Delta x, \Delta y \to 0} \frac{\Delta x - i\Delta y}{\Delta x + i\Delta y}.$$

If $\Delta y = 0$, the required limit is $\lim_{\Delta x \to 0} \Delta x / \Delta x = 1$. On the other hand, if $\Delta x = 0$, the required limit is -1. Then since the limit depends on the manner in which $\Delta z \to 0$, the derivative does not exist and so $f(z) = z^*$ is non-analytic everywhere.

Example 6.8
Given $f(z) = 2z^2 - 1$, find $f'(z)$ at $z_0 = 1 - i$.

Solution:

$$f'(z_0) = f'(1 - i) = \lim_{z \to 1-i} \frac{(2z^2 - 1) - [2(1 - i)^2 - 1]}{z - (1 - i)}$$

$$= \lim_{z \to 1-i} \frac{2[z - (1 - i)][z + (1 - i)]}{z - (1 - i)}$$

$$= \lim_{z \to 1-i} 2[z + (1 - i)] = 4(1 - i).$$

The rules for differentiating sums, products, and quotients are, in general, the same for complex functions as for real-valued functions. That is, if $f'(z_0)$ and $g'(z_0)$ exist, then:

(1) $(f + g)'(z_0) = f'(z_0) + g'(z_0)$;
(2) $(fg)'(z_0) = f'(z_0)g(z_0) + f(z_0)g'(z_0)$;

(3) $\left(\dfrac{f}{g}\right)'(z_0) = \dfrac{g(z_0)f'(z_0) - f(z_0)g'(z_0)}{g(z_0)^2}$,   if $g'(z_0) \neq 0$.

The Cauchy–Riemann conditions

We call $f(z)$ analytic at z_0, if $f'(z)$ exists for all z in some δ neighborhood of z_0; and $f(z)$ is analytic in a region R if it is analytic at every point of R. Cauchy and Riemann provided us with a simple but extremely important test for the analyticity of $f(z)$. To deduce the Cauchy–Riemann conditions for the analyticity of $f(z)$, let us return to Eq. (6.10):

$$f'(z_0) = \lim_{\Delta z \to 0} \frac{f(z_0 + \Delta z) - f(z_0)}{\Delta z}.$$

If we write $f(z) = u(x, y) + iv(x, y)$, this becomes

$$f'(z) = \lim_{\Delta x, \Delta y \to 0} \frac{u(x + \Delta x, y + \Delta y) - u(x, y) + i(\text{same for } v)}{\Delta x + i\Delta y}.$$

There are of course an infinite number of ways to approach a point z on a two-dimensional surface. Let us consider two possible approaches – along x and along

y. Suppose we first take the x route, so y is fixed as we change x, that is, $\Delta y = 0$ and $\Delta x \to 0$, and we have

$$f'(z) = \lim_{\Delta x \to 0}\left[\frac{u(x+\Delta x, y) - u(x, y)}{\Delta x} + i\frac{v(x+\Delta x, y) - v(x, y)}{\Delta x}\right] = \frac{\partial u}{\partial x} + i\frac{\partial v}{\partial x}.$$

We next take the y route, and we have

$$f'(z) = \lim_{\Delta y \to 0}\left[\frac{u(x, y+\Delta y) - u(x, y)}{i\Delta y} + i\frac{v(x, y+\Delta y) - v(x, y)}{i\Delta y}\right] = -i\frac{\partial u}{\partial y} + \frac{\partial v}{\partial y}.$$

Now $f(z)$ cannot possibly be analytic unless the two derivatives are identical. Thus a necessary condition for $f(z)$ to be analytic is

$$\frac{\partial u}{\partial x} + i\frac{\partial v}{\partial x} = -i\frac{\partial u}{\partial y} + \frac{\partial v}{\partial y},$$

from which we obtain

$$\frac{\partial u}{\partial x} = \frac{\partial v}{\partial y} \quad \text{and} \quad \frac{\partial u}{\partial y} = -\frac{\partial v}{\partial x}. \tag{6.11}$$

These are the Cauchy–Riemann conditions, named after the French mathematician A. L. Cauchy (1789–1857) who discovered them, and the German mathematician Riemann who made them fundamental in his development of the theory of analytic functions. Thus if the function $f(z) = u(x, y) + iv(x, y)$ is analytic in a region R, then $u(x, y)$ and $v(x, y)$ satisfy the Cauchy–Riemann conditions at all points of R.

Example 6.9
If $f(z) = z^2 = x^2 - y^2 + 2ixy$, then $f'(z)$ exists for all z: $f'(z) = 2z$, and

$$\frac{\partial u}{\partial x} = 2x = \frac{\partial v}{\partial y}, \quad \text{and} \quad \frac{\partial u}{\partial y} = -2y = -\frac{\partial v}{\partial x}.$$

Thus, the Cauchy–Riemann equations (6.11) hold in this example at all points z.
We can also find examples in which $u(x, y)$ and $v(x, y)$ satisfy the Cauchy–Riemann conditions (6.11) at $z = z_0$, but $f'(z_0)$ doesn't exist. One such example is the following:

$$f(z) = u(x, y) + iv(x, y) = \begin{cases} z^5/|z|^4 & \text{if } z \neq 0 \\ 0 & \text{if } z = 0 \end{cases}.$$

The reader can show that $u(x, y)$ and $v(x, y)$ satisfy the Cauchy–Riemann conditions (6.11) at $z = 0$, but that $f'(0)$ does not exist. Thus $f(z)$ is not analytic at $z = 0$. The proof is straightforward, but very tedious.

However, the Cauchy–Riemann conditions do imply analyticity provided an additional hypothesis is added:

> Given $f(z) = u(x,y) + iv(x,y)$, if $u(x,y)$ and $v(x,y)$ are continuous with continuous first partial derivatives and satisfy the Cauchy–Riemann conditions (11) at all points in a region R, then $f(z)$ is analytic in R.

To prove this, we need the following result from the calculus of real-valued functions of two variables: If $h(x,y), \partial h/\partial x$, and $\partial h/\partial y$ are continuous in some region R about (x_0, y_0), then there exists a function $H(\Delta x, \Delta y)$ such that $H(\Delta x, \Delta y) \to 0$ as $(\Delta x, \Delta y) \to (0,0)$ and

$$h(x_0 + \Delta x, y_0 + \Delta y) - h(x_0, y_0) = \frac{\partial h(x_0, y_0)}{\partial x} \Delta x + \frac{\partial h(x_0, y_0)}{\partial y} \Delta y$$
$$+ H(\Delta x, \Delta y)\sqrt{(\Delta x)^2 + (\Delta y)^2}.$$

Let us return to

$$\lim_{\Delta z \to 0} \frac{f(z_0 + \Delta z) - f(z_0)}{\Delta z},$$

where z_0 is any point in region R and $\Delta z = \Delta x + i\Delta y$. Now we can write

$$f(z_0 + \Delta z) - f(z_0) = [u(x_0 + \Delta x, y_0 + \Delta y) - u(x_0, y_0)]$$
$$+ i[v(x_0 + \Delta x, y_0 + \Delta y) - v(x_0, y_0)]$$
$$= \frac{\partial u(x_0 y_0)}{\partial x} \Delta x + \frac{\partial u(x_0 y_0)}{\partial y} \Delta y + H(\Delta x, \Delta y)\sqrt{(\Delta x)^2 + (\Delta y)^2}$$
$$+ i\left[\frac{\partial v(x_0 y_0)}{\partial x} \Delta x + \frac{\partial v(x_0 y_0)}{\partial y} \Delta y\right.$$
$$\left. + G(\Delta x, \Delta y)\sqrt{(\Delta x)^2 + (\Delta y)^2}\right],$$

where $H(\Delta x, \Delta y) \to 0$ and $G(\Delta x, \Delta y) \to 0$ as $(\Delta x, \Delta y) \to (0,0)$.

Using the Cauchy–Riemann conditions and some algebraic manipulation we obtain

$$f(z_0 + \Delta z) - f(z_0) = \left[\frac{\partial u(x_0, y_0)}{\partial x} + i\frac{\partial v(x_0, y_0)}{\partial x}\right](\Delta x + i\Delta y)$$
$$+ [H(\Delta x, \Delta y) + iG(\Delta x, \Delta y)]\sqrt{(\Delta x)^2 + (\Delta y)^2}$$

and

$$\frac{f(z_0 + \Delta z) - f(z_0)}{\Delta z} = \frac{\partial u(x_0, y_0)}{\partial x} + i\frac{\partial v(x_0, y_0)}{\partial x}$$

$$+ [H(\Delta x \cdot \Delta y) + iG(\Delta x \cdot \Delta y)]\frac{\sqrt{(\Delta x)^2 + (\Delta y)^2}}{\Delta x + i\Delta y}.$$

But

$$\left|\frac{\sqrt{(\Delta x)^2 + (\Delta y)^2}}{\Delta x + i\Delta y}\right| = 1.$$

Thus, as $\Delta z \to 0$, we have $(\Delta x, \Delta y) \to (0,0)$ and

$$\lim_{\Delta z \to 0} \frac{f(z_0 + \Delta z) - f(z_0)}{\Delta z} = \frac{\partial u(x_0, y_0)}{\partial x} + i\frac{\partial v(x_0, y_0)}{\partial x},$$

which shows that the limit and so $f'(z_0)$ exist. Since $f(z)$ is differentiable at all points in region R, $f(z)$ is analytic at z_0 which is any point in R.

The Cauchy–Riemann equations turn out to be both necessary and sufficient conditions that $f(z) = u(x, y) + iv(x, y)$ be analytic. Analytic functions are also called regular or holomorphic functions. If $f(z)$ is analytic everywhere in the finite z complex plane, it is called an entire function. A function $f(z)$ is said to be singular at $z = z_0$, if it is not differentiable there; the point z_0 is called a singular point of $f(z)$.

Harmonic functions

If $f(z) = u(x, y) + iv(x, y)$ is analytic in some region of the z plane, then at every point of the region the Cauchy–Riemann conditions are satisfied:

$$\frac{\partial u}{\partial x} = \frac{\partial v}{\partial y}, \quad \text{and} \quad \frac{\partial u}{\partial y} = -\frac{\partial v}{\partial x},$$

and therefore

$$\frac{\partial^2 u}{\partial x^2} = \frac{\partial^2 v}{\partial x \partial y}, \quad \text{and} \quad \frac{\partial^2 u}{\partial y^2} = -\frac{\partial^2 v}{\partial y \partial x},$$

provided these second derivatives exist. In fact, one can show that if $f(z)$ is analytic in some region R, all its derivatives exist and are continuous in R. Equating the two cross terms, we obtain

$$\frac{\partial^2 u}{\partial x^2} + \frac{\partial^2 u}{\partial y^2} = 0 \qquad (6.12a)$$

throughout the region R.

247

Similarly, by differentiating the first of the Cauch–Riemann equations with respect to y, the second with respect to x, and subtracting we obtain

$$\frac{\partial^2 v}{\partial x^2} + \frac{\partial^2 v}{\partial y^2} = 0. \tag{6.12b}$$

Eqs. (6.12a) and (6.12b) are Laplace's partial differential equations in two independent variables x and y. Any function that has continuous partial derivatives of second order and that satisfies Laplace's equation is called a harmonic function.

We have shown that if $f(z) = u(x, y) + iv(x, y)$ is analytic, then both u and v are harmonic functions. They are called conjugate harmonic functions. This is a different use of the word conjugate from that employed in determining z^*.

Given one of two conjugate harmonic functions, the Cauchy–Riemann equations (6.11) can be used to find the other.

Singular points

A point at which $f(z)$ fails to be analytic is called a singular point or a singularity of $f(z)$; the Cauchy–Riemann conditions break down at a singularity. Various types of singular points exist.

(1) Isolated singular points: The point $z = z_0$ is called an isolated singular point of $f(z)$ if we can find $\delta > 0$ such that the circle $|z - z_0| = \delta$ encloses no singular point other than z_0. If no such δ can be found, we call z_0 a non-isolated singularity.

(2) Poles: If we can find a positive integer n such that $\lim_{z \to z_0} (z - z_0)^n f(z) = A \neq 0$, then $z = z_0$ is called a pole of order n. If $n = 1$, z_0 is called a simple pole. As an example, $f(z) = 1/(z - 2)$ has a simple pole at $z = 2$. But $f(z) = 1/(z - 2)^3$ has a pole of order 3 at $z = 2$.

(3) Branch point: A function has a branch point at z_0 if, upon encircling z_0 and returning to the starting point, the function does not return to the starting value. Thus the function is multiple-valued. An example is $f(z) = \sqrt{z}$, which has a branch point at $z = 0$.

(4) Removable singularities: The singular point z_0 is called a removable singularity of $f(z)$ if $\lim_{z \to z_0} f(z)$ exists. For example, the singular point at $z = 0$ of $f(z) = \sin(z)/z$ is a removable singularity, since $\lim_{z \to 0} \sin(z)/z = 1$.

(5) Essential singularities: A function has an essential singularity at a point z_0 if it has poles of arbitrarily high order which cannot be eliminated by multiplication by $(z - z_0)^n$, which for any finite choice of n. An example is the function $f(z) = e^{1/(z-2)}$, which has an essential singularity at $z = 2$.

(6) Singularities at infinity: The singularity of $f(z)$ at $z = \infty$ is the same type as that of $f(1/w)$ at $w = 0$. For example, $f(z) = z^2$ has a pole of order 2 at $z = \infty$, since $f(1/w) = w^{-2}$ has a pole of order 2 at $w = 0$.

Elementary functions of z

The exponential function e^z (or $\exp(z)$)

The exponential function is of fundamental importance, not only for its own sake, but also as a basis for defining all the other elementary functions. In its definition we seek to preserve as many of the characteristic properties of the real exponential function e^x as possible. Specifically, we desire that:

(a) e^z is single-valued and analytic.
(b) $de^z/dz = e^z$.
(c) e^z reduces to e^x when Im $z = 0$.

Recall that if we approach the point z along the x-axis (that is, $\Delta y = 0$, $\Delta x \to 0$), the derivative of an analytic function $f'(z)$ can be written in the form

$$f'(z) = \frac{df}{dz} = \frac{\partial u}{\partial x} + i\frac{\partial v}{\partial x}.$$

If we let

$$e^z = u + iv,$$

then to satisfy (b) we must have

$$\frac{\partial u}{\partial x} + i\frac{\partial v}{\partial x} = u + iv.$$

Equating real and imaginary parts gives

$$\frac{\partial u}{\partial x} = u, \tag{6.13}$$

$$\frac{\partial v}{\partial x} = v. \tag{6.14}$$

Eq. (6.13) will be satisfied if we write

$$u = e^x \phi(y), \tag{6.15}$$

where $\phi(y)$ is any function of y. Moreover, since e^z is to be analytic, u and v must satisfy the Cauchy–Riemann equations (6.11). Then using the second of Eqs. (6.11), Eq. (6.14) becomes

$$-\frac{\partial u}{\partial y} = v.$$

Differentiating this with respect to y, we obtain

$$\frac{\partial^2 u}{\partial y^2} = -\frac{\partial v}{\partial y}$$

$$= -\frac{\partial u}{\partial x} \quad \text{(with the aid of the first of Eqs. (6.11)).}$$

Finally, using Eq. (6.13), this becomes

$$\frac{\partial^2 u}{\partial y^2} = -u,$$

which, on substituting Eq. (6.15), becomes

$$e^x \phi''(y) = -e^x \phi(y) \quad \text{or} \quad \phi''(y) = -\phi(y).$$

This is a simple linear differential equation whose solution is of the form

$$\phi(y) = A \cos y + B \sin y.$$

Then

$$u = e^x \phi(y)$$

$$= e^x (A \cos y + B \sin y)$$

and

$$v = -\frac{\partial u}{\partial y} = -e^x(-A \sin y + B \cos y).$$

Therefore

$$e^z = u + iv = e^x[(A \cos y + B \sin y) + i(A \sin y - B \cos y)].$$

If this is to reduce to e^x when $y = 0$, according to (c), we must have

$$e^x = e^x(A - iB)$$

from which we find

$$A = 1 \quad \text{and} \quad B = 0.$$

Finally we find

$$e^z = e^{x+iy} = e^x(\cos y + i \sin y). \tag{6.16}$$

This expression meets our requirements (a), (b), and (c); hence we adopt it as the definition of e^z. It is analytic at each point in the entire z plane, so it is an entire function. Moreover, it satisfies the relation

$$e^{z_1} e^{z_2} = e^{z_1 + z_2}. \tag{6.17}$$

It is important to note that the right hand side of Eq. (6.16) is in standard polar form with the modulus of e^z given by e^x and an argument by y:

$$\text{mod } e^z \equiv |e^z| = e^x \quad \text{and} \quad \arg e^z = y.$$

From Eq. (6.16) we obtain the Euler formula: $e^{iy} = \cos y + i \sin y$. Now let $y = 2\pi$, and since $\cos 2\pi = 1$ and $\sin 2\pi = 0$, the Euler formula gives

$$e^{2\pi i} = 1.$$

Similarly,

$$e^{\pm \pi i} = -1, \quad e^{\pm \pi i/2} = \pm i.$$

Combining this with Eq. (6.17), we find

$$e^{z+2\pi i} = e^z e^{2\pi i} = e^z,$$

which shows that e^z is periodic with the imaginary period $2\pi i$. Thus

$$e^{z \pm 2\pi n I} = e^z \quad (n = 0, 1, 2, \ldots). \tag{6.18}$$

Because of the periodicity all the values that $w = f(z) = e^z$ can assume are already assumed in the strip $-\pi < y \leq \pi$. This infinite strip is called the fundamental region of e^z.

Trigonometric and hyperbolic functions

From the Euler formula we obtain

$$\cos x = \frac{1}{2}(e^{ix} + e^{-ix}), \quad \sin x = \frac{1}{2i}(e^{ix} - e^{-ix}) \quad (x \text{ real}).$$

This suggests the following definitions for complex z:

$$\cos z = \frac{1}{2}(e^{iz} + e^{-iz}), \quad \sin z = \frac{1}{2i}(e^{iz} - e^{-iz}). \tag{6.19}$$

The other trigonometric functions are defined in the usual way:

$$\tan z = \frac{\sin z}{\cos z}, \quad \cot z = \frac{\cos z}{\sin z}, \quad \sec z = \frac{1}{\cos z}, \quad \operatorname{cosec} z = \frac{1}{\sin z},$$

whenever the denominators are not zero.

From these definitions it is easy to establish the validity of such familiar formulas as:

$$\sin(-z) = -\sin z, \cos(-z) = \cos z, \quad \text{and} \quad \cos^2 z + \sin^2 z = 1,$$

$$\cos(z_1 \pm z_2) = \cos z_1 \cos z_2 \mp \sin z_1 \sin z_2, \quad \sin(z_1 \pm z_2) = \sin z_1 \cos z_2 \pm \cos z_1 \sin z_2$$

$$\frac{d \cos z}{dz} = -\sin z, \quad \frac{d \sin z}{dz} = \cos z.$$

Since e^z is analytic for all z, the same is true for the function $\sin z$ and $\cos z$. The functions $\tan z$ and $\sec z$ are analytic except at the points where $\cos z$ is zero, and $\cot z$ and $\operatorname{cosec} z$ are analytic except at the points where $\sin z$ is zero. The functions $\cos z$ and $\sec z$ are even, and the other functions are odd. Since the exponential function is periodic, the trigonometric functions are also periodic, and we have

$$\cos(z \pm 2n\pi) = \cos z, \quad \sin(z \pm 2n\pi) = \sin z,$$

$$\tan(z \pm 2n\pi) = \tan z, \quad \cot(z \pm 2n\pi) = \cot z,$$

where $n = 0, 1, \ldots$.

Another important property also carries over: $\sin z$ and $\cos z$ have the same zeros as the corresponding real-valued functions:

$$\sin z = 0 \quad \text{if and only if} \quad z = n\pi \quad (n \text{ integer});$$

$$\cos z = 0 \quad \text{if and only if} \quad z = (2n+1)\pi/2 \quad (n \text{ integer}).$$

We can also write these functions in the form $u(x, y) + iv(x, y)$. As an example, we give the details for $\cos z$. From Eq. (6.19) we have

$$\cos z = \frac{1}{2}(e^{iz} + e^{-iz}) = \frac{1}{2}(e^{i(x+iy)} + e^{-i(x+iy)}) = \frac{1}{2}(e^{-y}e^{ix} + e^{y}e^{-ix})$$

$$= \frac{1}{2}[e^{-y}(\cos x + i \sin x) + e^{y}(\cos x - i \sin x)]$$

$$= \cos x \frac{e^{y} + e^{-y}}{2} - i \sin x \frac{e^{y} - e^{-y}}{2}$$

or, using the definitions of the hyperbolic functions of real variables

$$\cos z = \cos(x + iy) = \cos x \cosh y - i \sin x \sinh y;$$

similarly,

$$\sin z = \sin(x + iy) = \sin x \cosh y + i \cos x \sinh y.$$

In particular, taking $x = 0$ in these last two formulas, we find

$$\cos(iy) = \cosh y, \quad \sin(iy) = i \sinh y.$$

There is a big difference between the complex and real sine and cosine functions. The real functions are bounded between -1 and $+1$, but the complex functions can take on arbitrarily large values. For example, if y is real, then $\cos iy = \frac{1}{2}(e^{-y} + e^{y}) \to \infty$ as $y \to \infty$ or $y \to -\infty$.

The logarithmic function $w = \ln z$

The real natural logarithm $y = \ln x$ is defined as the inverse of the exponential function $e^{y} = x$. For the complex logarithm, we take the same approach and define $w = \ln z$ which is taken to mean that

$$e^{w} = z \tag{6.20}$$

for each $z \neq 0$.

Setting $w = u + iv$ and $z = re^{i\theta} = |z|e^{i\theta}$ we have

$$e^w = e^{u+iv} = e^u e^{iv} = re^{i\theta}.$$

It follows that

$$e^u = r = |z| \quad \text{or} \quad u = \ln r = \ln|z|$$

and

$$v = \theta = \arg z.$$

Therefore

$$w = \ln z = \ln r + i\theta = \ln|z| + i \arg z.$$

Since the argument of z is determined only in multiples of 2π, the complex natural logarithm is infinitely many-valued. If we let θ_1 be the principal argument of z, that is, the particular argument of z which lies in the interval $0 \le \theta < 2\pi$, then we can rewrite the last equation in the form

$$\ln z = \ln|z| + i(\theta + 2n\pi) \qquad n = 0, \pm 1, \pm 2, \ldots . \qquad (6.21)$$

For any particular value of n, a unique branch of the function is determined, and the logarithm becomes effectively single-valued. If $n = 0$, the resulting branch of the logarithmic function is called the principal value. Any particular branch of the logarithmic function is analytic, for we have by differentiating the definitive relation $z = e^w$,

$$dz/dw = e^w = z \quad \text{or} \quad dw/dz = d(\ln z)/dz = 1/z.$$

For a particular value of n the derivative of $\ln z$ thus exists for all $z \ne 0$.

For the real logarithm, $y = \ln x$ makes sense when $x > 0$. Now we can take a natural logarithm of a negative number, as shown in the following example.

Example 6.10
$\ln -4 = \ln|-4| + i \arg(-4) = \ln 4 + i(\pi + 2n\pi)$; its principal value is $\ln 4 + i(\pi)$, a complex number. This explains why the logarithm of a negative number makes no sense in real variable.

Hyperbolic functions

We conclude this section on "elementary functions" by mentioning briefly the hyperbolic functions; they are defined at points where the denominator does not vanish:

$$\sinh z = \frac{1}{2}(e^z - e^{-z}), \quad \cosh z = \frac{1}{2}(e^z + e^{-z}),$$

$$\tanh z = \sinh z/\cosh z, \quad \coth z = \cosh z/\sinh z,$$

$$\operatorname{sech} z = 1/\cosh z, \quad \operatorname{cosech} z = 1/\sinh z.$$

Since e^z and e^{-z} are entire functions, $\sinh z$ and $\cosh z$ are also entire functions. The singularities of $\tanh z$ and $\operatorname{sech} z$ occur at the zeros of $\cosh z$, and the singularities of $\coth z$ and $\operatorname{cosech} z$ occur at the zeros of $\sinh z$.

As with the trigonometric functions, basic identities and derivative formulas carry over in the same form to the complex hyperbolic functions (just replace x by z). Hence we shall not list them here.

Complex integration

Complex integration is very important. For example, in applications we often encounter real integrals which cannot be evaluated by the usual methods, but we can get help and relief from complex integration. In theory, the method of complex integration yields proofs of some basic properties of analytic functions, which would be very difficult to prove without using complex integration.

The most fundamental result in complex integration is Cauchy's integral theorem, from which the important Cauchy integral formula follows. These will be the subject of this section.

Line integrals in the complex plane

As in real integrals, the indefinite integral $\int f(z)dz$ stands for any function whose derivative is $f(z)$. The definite integral of real calculus is now replaced by integrals of a complex function along a curve. Why? To see this, we can express z in terms of a real parameter t: $z(t) = x(t) + iy(t)$, where, say, $a \leq t \leq b$. Now as t varies from a to b, the point (x, y) describes a curve in the plane. We say this curve is smooth if there exists a tangent vector at all points on the curve; this means that dx/dt and dy/dt are continuous and do not vanish simultaneously for $a < t < b$.

Let C be such a smooth curve in the complex z plane (Fig. 6.6), and we shall assume that C has a finite length (mathematicians call C a rectifiable curve). Let $f(z)$ be continuous at all points of C. Subdivide C into n parts by means of points $z_1, z_2, \ldots, z_{n-1}$, chosen arbitrarily, and let $a = z_0, b = z_n$. On each arc joining z_{k-1} to z_k ($k = 1, 2, \ldots, n$) choose a point w_k (possibly $w_k = z_{k-1}$ or $w_k = z_k$) and form the sum

$$S_n = \sum_{k=1}^{n} f(w_k)\Delta z_k \qquad \Delta z_k = z_k - z_{k-1}.$$

Now let the number of subdivisions n increase in such a way that the largest of the chord lengths $|\Delta z_k|$ approaches zero. Then the sum S_n approaches a limit. If this limit exists and has the same value no matter how the z_js and w_js are chosen, then

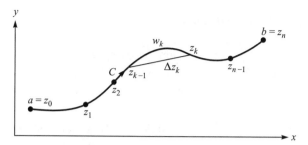

Figure 6.6.   Complex line integral.

this limit is called the integral of $f(z)$ along C and is denoted by

$$\int_C f(z)dz \quad \text{or} \quad \int_a^b f(z)dz. \tag{6.22}$$

This is often called a contour integral (with contour C) or a line integral of $f(z)$. Some authors reserve the name contour integral for the special case in which C is a closed curve (so end a and end b coincide), and denote it by the symbol $\oint f(z)dz$.

We now state, without proof, a basic theorem regarding the existence of the contour integral: *If C is piecewise smooth and $f(z)$ is continuous on C, then $\int_C f(z)dz$ exists.*

If $f(z) = u(x, y) + iv(x, y)$, the complex line integral can be expressed in terms of real line integrals as

$$\int_C f(z)dz = \int_C (u + iv)(dx + idy) = \int_C (udx - vdy) + i \int_C (vdx + udy), \tag{6.23}$$

where curve C may be open or closed but the direction of integration must be specified in either case. Reversing the direction of integration results in the change of sign of the integral. Complex integrals are, therefore, reducible to curvilinear real integrals and possess the following properties:

(1) $\int_C [f(z) + g(z)]dz = \int_C f(z)dz + \int_C g(z)dz$;

(2) $\int_C kf(z)dz = k \int_C f(z)dz$, k = any constant (real or complex);

(3) $\int_a^b f(z)dz = -\int_b^a f(z)dz$;

(4) $\int_a^b f(z)dz = \int_a^m f(z)d + \int_m^b f(z)dz$;

(5) $|\int_C f(z)dz| \leq ML$, where $M = \max|f(z)|$ on C, and L is the length of C.

Property (5) is very useful, because in working with complex line integrals it is often necessary to establish bounds on their absolute values. We now give a brief

255

proof. Let us go back to the definition:

$$\int_C f(z)dz = \lim_{n\to\infty} \sum_{k=1}^{n} f(w_k)\Delta z_k.$$

Now

$$\left|\sum_{k=1}^{n} f(w_k)\Delta z_k\right| \le \sum_{k=1}^{n} |f(w_k)||\Delta z_k| \le M\sum_{k=1}^{n} |\Delta z_k| \le ML,$$

where we have used the fact that $|f(z)| \le M$ for all points z on C and that $\sum|\Delta z_k|$ represents the sum of all the chord lengths joining z_{k-1} and z_k, and that this sum is not greater than the length L of C. Now taking the limit of both sides, and property (5) follows. It is possible to show, more generally, that

$$\left|\int_C f(z)dz\right| \le \int_C |f(z)||dz|. \tag{6.24}$$

Example 6.11
Evaluate the integral $\int_C (z^*)^2 dz$, where C is a straight line joining the points $z = 0$ and $z = 1 + 2i$.

Solution: Since

$$(z^*)^2 = (x - iy)^2 = x^2 - y^2 - 2xyi,$$

we have

$$\int_C (z^*)^2 dz = \int_C [(x^2 - y^2)dx + 2xydy] + i\int_C [-2xydx + (x^2 - y^2)dy].$$

But the Cartesian equation of C is $y = 2x$, and the above integral therefore becomes

$$\int_C (z^*)^2 dz = \int_0^1 5x^2 dx + i\int_0^1 (-10x^2)dx = 5/3 - i10/3.$$

Example 6.12
Evaluate the integral

$$\int_C \frac{dz}{(z - z_0)^{n+1}},$$

where C is a circle of radius r and center at z_0, and n is an integer.

Solution: For convenience, let $z - z_0 = re^{i\theta}$, where θ ranges from 0 to 2π as z ranges around the circle (Fig. 6.7). Then $dz = rie^{i\theta}d\theta$, and the integral becomes

$$\int_0^{2\pi} \frac{rie^{i\theta}d\theta}{r^{n+1}e^{i(n+1)\theta}} = \frac{i}{r^n}\int_0^{2\pi} e^{-in\theta}d\theta.$$

If $n = 0$, this reduces to

$$i\int_0^{2\pi} d\theta = 2\pi i$$

and if $n \neq 0$, we have

$$\frac{i}{r^n}\int_0^{2\pi} (\cos n\theta - i\sin n\theta)d\theta = 0.$$

This is an important and useful result to which we will refer later.

Cauchy's integral theorem

Cauchy's integral theorem has various theoretical and practical consequences. It states that if $f(z)$ is analytic in a simply-connected region (domain) and on its boundary C, then

$$\oint_C f(z)dz = 0. \tag{6.25}$$

What do we mean by a simply-connected region? A region R (mathematicians prefer the term 'domain') is called simply-connected if any simple closed curve which lies in R can be shrunk to a point without leaving R. That is, a simply-connected region has no hole in it (Fig. 6.7(a)); this is not true for a multiply-connected region. The multiply-connected regions of Fig. 6.7(b) and (c) have respectively one and three holes in them.

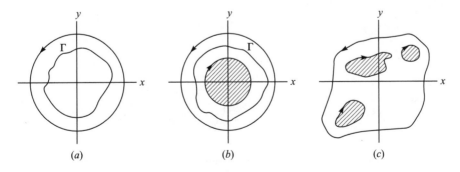

(a)   (b)   (c)

Figure 6.7.   Simply-connected and doubly-connected regions.

Although a rigorous proof of Cauchy's integral theorem is quite demanding and beyond the scope of this book, we shall sketch the main ideas. Note that the integral can be expressed in terms of two-dimensional vector fields $\mathbf{A}$ and $\mathbf{B}$:

$$\oint_C f(z)dz = \oint_C (udx - vdy) + i\int_C (vdx + udy)$$

$$= \oint_C \mathbf{A}(\mathbf{r}) \cdot d\mathbf{r} + i\oint_C \mathbf{B}(\mathbf{r}) \cdot d\mathbf{r},$$

where

$$\mathbf{A}(\mathbf{r}) = u\hat{e}_1 - v\hat{e}_2, \quad \mathbf{B}(\mathbf{r}) = v\hat{e}_1 + u\hat{e}_2.$$

Applying Stokes' theorem, we obtain

$$\oint_C f(z)dz = \iint_R d\mathbf{a} \cdot (\nabla \times \mathbf{A} + i\nabla \times \mathbf{B})$$

$$= \iint_R dxdy\left[-\left(\frac{\partial v}{\partial x} + \frac{\partial u}{\partial y}\right) + i\left(\frac{\partial u}{\partial x} - \frac{\partial v}{\partial y}\right)\right],$$

where R is the region enclosed by C. Since $f(x)$ satisfies the Cauchy–Riemann conditions, both the real and the imaginary parts of the integral are zero, thus proving Cauchy's integral theorem.

Cauchy's theorem is also valid for multiply-connected regions. For simplicity we consider a doubly-connected region (Fig. 6.8). $f(z)$ is analytic in and on the boundary of the region R between two simple closed curves C_1 and C_2. Construct a cross-cut AF. Then the region bounded by $ABDEAFGHFA$ is simply-connected so by Cauchy's theorem

$$\oint_C f(z)dz = \oint_{ABDEAFGHFA} f(z)dz = 0$$

or

$$\int_{ABDEA} f(z)dz + \int_{AF} f(z)dz + \int_{FGHF} f(z)dz + \int_{FA} f(z)dz = 0.$$

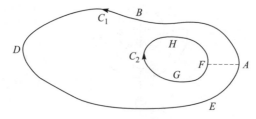

Figure 6.8. Proof of Cauchy's theorem for a doubly-connected region.

But $\int_{AF} f(z)dz = -\int_{FA} f(z)dz$, therefore this becomes

$$\int_{ABDEA} f(z)dzy + \int_{FGHF} f(z)dzy = 0$$

or

$$\oint_C f(z)dz = \oint_{C_1} f(z)dz + \oint_{C_2} f(z)dz = 0, \qquad (6.26)$$

where both C_1 and C_2 are traversed in the positive direction (in the sense that an observer walking on the boundary always has the region R on his left). Note that curves C_1 and C_2 are in opposite directions.

If we reverse the direction of C_2 (now C_2 is also counterclockwise, that is, both C_1 and C_2 are in the same direction.), we have

$$\oint_{C_1} f(z)dz - \oint_{C_2} f(z)dz = 0 \quad \text{or} \quad \oint_{C_2} f(z)dz = \oint_{C_1} f(z)dz.$$

Because of Cauchy's theorem, an integration contour can be moved across any region of the complex plane over which the integrand is analytic without changing the value of the integral. It cannot be moved across a hole (the shaded area) or a singularity (the dot), but it can be made to collapse around one, as shown in Fig. 6.9. As a result, an integration contour C enclosing n holes or singularities can be replaced by n separated closed contours C_i, each enclosing a hole or a singularity:

$$\oint_C f(z)dz = \sum_{k=1}^{n} \oint_{C_i} f(z)dz$$

which is a generalization of Eq. (6.26) to multiply-connected regions.

There is a converse of the Cauchy's theorem, known as Morera's theorem. We now state it without proof:

Morera's theorem:
If $f(z)$ is continuous in a simply-connected region R and the Cauchy's theorem is valid around every simple closed curve C in R, then $f(z)$ is analytic in R.

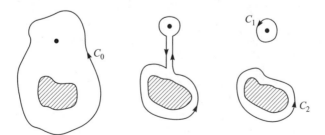

Figure 6.9.   Collapsing a contour around a hole and a singularity.

259

Example 6.13
Evaluate $\oint_C dz/(z-a)$ where C is any simple closed curve and $z=a$ is (*a*) outside C, (*b*) inside C.

Solution: (*a*) If a is outside C, then $f(z)=1/(z-a)$ is analytic everywhere inside and on C. Hence by Cauchy's theorem

$$\oint_C dz/(z-a)=0.$$

(*b*) If a is inside C and Γ is a circle of radius $\in$ with center at $z=a$ so that Γ is inside C (Fig. 6.10). Then by Eq. (6.26) we have

$$\oint_C dz/(z-a)=\oint_\Gamma dz/(z-a).$$

Now on Γ, $|z-a|=\varepsilon$, or $z-a=\varepsilon e^{i\theta}$, then $dz=i\varepsilon e^{i\theta}\,d\theta$, and

$$\oint_\Gamma \frac{dz}{z-a}=\int_0^{2\pi}\frac{i\varepsilon e^{i\theta}d\theta}{\varepsilon e^{i\theta}}=i\int_0^{2\pi}d\theta=2\pi i.$$

Cauchy's integral formulas

One of the most important consequences of Cauchy's integral theorem is what is known as Cauchy's integral formula. It may be stated as follows.

If $f(z)$ is analytic in a simply-connected region R, and z_0 is any point in the interior of R which is enclosed by a simple closed curve C, then

$$f(z_0)=\frac{1}{2\pi i}\oint_C \frac{f(z)}{z-z_0}dz, \tag{6.27}$$

the integration around C being taken in the positive sense (counterclockwise).

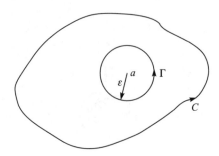

Figure 6.10.

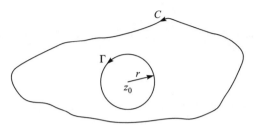

Figure 6.11. Cauchy's integral formula.

To prove this, let Γ be a small circle with center at z_0 and radius r (Fig. 6.11), then by Eq. (6.26) we have

$$\oint_C \frac{f(z)}{z - z_0}\, dz = \oint_\Gamma \frac{f(z)}{z - z_0}\, dz.$$

Now $|z - z_0| = r$ or $z - z_0 = re^{i\theta}, 0 \le \theta < 2\pi$. Then $dz = ire^{i\theta}\, d\theta$ and the integral on the right becomes

$$\oint_\Gamma \frac{f(z)}{z - z_0}\, dz = \int_0^{2\pi} \frac{f(z_0 + re^{i\theta})ire^{i\theta}}{re^{i\theta}}\, d\theta = i\int_0^{2\pi} f(z_0 + re^{i\theta})\, d\theta.$$

Taking the limit of both sides and making use of the continuity of $f(z)$, we have

$$\oint_C \frac{f(z)}{z - z_0}\, dz = \lim_{r \to 0} \int_0^{2\pi} f(z_0 + re^{i\theta})\, d\theta$$

$$= i\int_0^{2\pi} \lim_{r \to 0} f(z_0 + re^{i\theta})\, d\theta = i\int_0^{2\pi} f(z_0)\, d\theta = 2\pi i f(z_0),$$

from which we obtain

$$f(z_0) = \frac{1}{2\pi i}\oint_C \frac{f(z)}{z - z_0}\, dz \quad \text{q.e.d.}$$

Cauchy's integral formula is also true for multiply-connected regions, but we shall leave its proof as an exercise.

It is useful to write Cauchy's integral formula (6.27) in the form

$$f(z) = \frac{1}{2\pi i}\oint_C \frac{f(z')\, dz'}{z' - z}$$

to emphasize the fact that z can be any point inside the close curve C.

Cauchy's integral formula is very useful in evaluating integrals, as shown in the following example.

Example 6.14

Evaluate the integral $\oint_C e^z dz/(z^2 + 1)$, if C is a circle of unit radius with center at (a) $z = i$ and (b) $z = -i$.

Solution: (a) We first rewrite the integral in the form

$$\oint_C \left(\frac{e^z}{z+i}\right) \frac{dz}{z-i},$$

then we see that $f(z) = e^z/(z+i)$ and $z_0 = i$. Moreover, the function $f(z)$ is analytic everywhere within and on the given circle of unit radius around $z = i$. By Cauchy's integral formula we have

$$\oint_C \left(\frac{e^z}{z+i}\right) \frac{dz}{z-i} = 2\pi i f(i) = 2\pi i \frac{e^i}{2i} = \pi(\cos 1 + i \sin 1).$$

(b) We find $z_0 = -i$ and $f(z) = e^z/(z-i)$. Cauchy's integral formula gives

$$\oint_C \left(\frac{e^z}{z-i}\right) \frac{dz}{z+i} = -\pi(\cos 1 - i \sin 1).$$

Cauchy's integral formula for higher derivatives

Using Cauchy's integral formula, we can show that an analytic function $f(z)$ has derivatives of all orders given by the following formula:

$$f^{(n)}(z_0) = \frac{n!}{2\pi i} \oint_C \frac{f(z) dz}{(z - z_0)^{n+1}}, \tag{6.28}$$

where C is any simple closed curve around z_0 and $f(z)$ is analytic on and inside C. Note that this formula implies that each derivative of $f(z)$ is itself analytic, since it possesses a derivative.

We now prove the formula (6.28) by induction on n. That is, we first prove the formula for $n = 1$:

$$f'(z_0) = \frac{1}{2\pi i} \oint_C \frac{f(z) dz}{(z - z_0)^2}.$$

As shown in Fig. 6.12, both z_0 and $z_0 + h$ lie in R, and

$$f'(z_0) = \lim_{h \to 0} \frac{f(z_0 + h) - f(z_0)}{h}.$$

Using Cauchy's integral formula we obtain

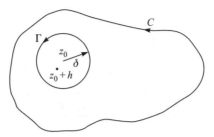

Figure 6.12.

$$f'(z_0) = \lim_{h \to 0} \frac{f(z_0 + h) - f(z_0)}{h}$$

$$= \lim_{h \to 0} \frac{1}{2\pi i h} \oint_C \left\{ \frac{1}{z - (z_0 + h)} - \frac{1}{z - z_0} \right\} f(z) dz.$$

Now

$$\frac{1}{h} \left[\frac{1}{z - (z_0 + h)} - \frac{1}{z - z_0} \right] = \frac{1}{(z - z_0)^2} + \frac{h}{(z - z_0 - h)(z - z_0)^2}.$$

Thus,

$$f'(z_0) = \frac{1}{2\pi i} \oint_C \frac{f(z)}{(z - z_0)^2} \, dz + \frac{1}{2\pi i} \lim_{h \to 0} h \oint_C \frac{f(z)}{(z - z_0 - h)(z - z_0)^2} dz.$$

The proof follows if the limit on the right hand side approaches zero as $h \to 0$. To show this, let us draw a small circle Γ of radius δ centered at z_0 (Fig. 6.12), then

$$\frac{1}{2\pi i} \lim_{h \to 0} h \oint_C \frac{f(z)}{(z - z_0 - h)(z - z_0)^2} \, dz = \frac{1}{2\pi i} \lim_{h \to 0} h \oint_\Gamma \frac{f(z)}{(z - z_0 - h)(z - z_0)^2} \, dz.$$

Now choose h so small (in absolute value) that $z_0 + h$ lies in Γ and $|h| < \delta/2$, and the equation for Γ is $|z - z_0| = \delta$. Thus, we have $|z - z_0 - h| \geq |z - z_0| - |h| > \delta - \delta/2 = \delta/2$. Next, as $f(z)$ is analytic in R, we can find a positive number M such that $|f(z)| \leq M$. And the length of Γ is $2\pi\delta$. Thus,

$$\left| \frac{h}{2\pi i} \oint_\Gamma \frac{f(z) dz}{(z - z_0 - h)(z - z_0)^2} \right| \leq \frac{|h|}{2\pi} \frac{M(2\pi\delta)}{(\delta/2)(\delta^2)} = \frac{2|h|M}{\delta^2} \to 0 \quad \text{as} \quad h \to 0,$$

proving the formula for $f'(z_0)$.

For $n = 2$, we begin with

$$\frac{f'(z_0 + h) - f'(z_0)}{h} = \frac{1}{2\pi i h} \oint_C \left\{ \frac{1}{(z - z_0 h)^2} - \frac{1}{(z - z_0)^2} \right\} f(z) dz$$

$$= \frac{2!}{2\pi i} \oint_C \frac{f(z)}{(z - z_0)^3} dz + \frac{h}{2\pi i} \oint_C \frac{3(z - z_0) - 2h}{(z - z_0 - h)^2 (z - z_0)^3} f(z) dz.$$

The result follows on taking the limit as $h \to 0$ if the last term approaches zero. The proof is similar to that for the case $n = 1$, for using the fact that the integral around C equals the integral around Γ, we have

$$\left| \frac{h}{2\pi i} \oint_\Gamma \frac{3(z - z_0) - 2h}{(z - z_0 - h)^2 (z - z_0)^3} f(z) dz \right| \le \frac{|h|}{2\pi} \frac{M(2\pi\delta)}{(\delta/2)^2 \delta^3} = \frac{4|h|M}{\delta^4},$$

assuming M exists such that $|[3(z - z_0) - 2h] f(z)| < M$.

In a similar manner we can establish the results for $n = 3, 4, \ldots$. We leave it to the reader to complete the proof by establishing the formula for $f^{(n+1)}(z_0)$, assuming that $f^{(n)}(z_0)$ is true.

Sometimes Cauchy's integral formula for higher derivatives can be used to evaluate integrals, as illustrated by the following example.

Example 6.15
Evaluate

$$\oint_C \frac{e^{2z}}{(z + 1)^4} dz,$$

where C is any simple closed path not passing through -1. Consider two cases:

(a) C does not enclose -1. Then $e^{2z}/(z + 1)^4$ is analytic on and inside C, and the integral is zero by Cauchy's integral theorem.

(b) C encloses -1. Now Cauchy's integral formula for higher derivatives applies.

Solution: Let $f(z) = e^{2z}$, then

$$f^{(3)}(-1) = \frac{3!}{2\pi i} \oint_C \frac{e^{2z}}{(z + 1)^4} dz.$$

Now $f^{(3)}(-1) = 8e^{-2}$, hence

$$\oint_C \frac{e^{2z}}{(z + 1)^4} dz = \frac{2\pi i}{3!} f^{(3)}(-1) = \frac{8\pi}{3} e^{-2} i.$$

Series representations of analytic functions

We now turn to a very important notion: series representations of analytic functions. As a prelude we must discuss the notion of convergence of complex series. Most of the definitions and theorems relating to infinite series of real terms can be applied with little or no change to series whose terms are complex.

Complex sequences

A complex sequence is an ordered list which assigns to each positive integer n a complex number z_n:

$$z_1, z_2, \ldots, z_n, \ldots.$$

The numbers z_n are called the terms of the sequence. For example, both $i, i^2, \ldots, i^n, \ldots$ or $1 + i, (1 + i)/2, (1 + i)/4, (1 + i)/8, \ldots$ are complex sequences. The nth term of the second sequence is $(1 + i)/2^{n-1}$. A sequence $z_1, z_2, \ldots, z_n, \ldots$ is said to be convergent with the limit l (or simply to converge to the number l) if, given $\varepsilon > 0$, we can find a positive integer N such that $|z_n - l| < \varepsilon$ for each $n \geq N$ (Fig. 6.13). Then we write

$$\lim_{n \to \infty} z_n = l.$$

In words, or geometrically, this means that each term z_n with $n > N$ (that is, $z_N, z_{N+1}, z_{N+2}, \ldots$) lies in the open circular region of radius ε with center at l. In general, N depends on the choice of ε. Here is an illustrative example.

Example 6.17
Using the definition, show that $\lim_{n \to \infty}(1 + z/n) = 1$ for all z.

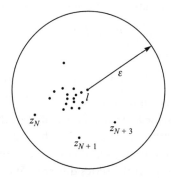

Figure 6.13. Convergent complex sequence.

265

Solution: Given any number $\varepsilon > 0$, we must find N such that

$$\left|1 + \frac{z}{n} - 1\right| < \varepsilon, \quad \text{for all} \quad n > N$$

from which we find

$$|z/n| < \varepsilon$$

or

$$|z|/n < \varepsilon \quad \text{if} \quad n > |z|/\varepsilon \equiv N.$$

Setting $z_n = x_n + iy_n$, we may consider a complex sequence $z_1, z_2, \ldots, z_n$ in terms of real sequences, the sequence of the real parts and the sequence of the imaginary parts: $x_1, x_2, \ldots, x_n$, and $y_1, y_2, \ldots, y_n$. If the sequence of the real parts converges to the number A, and the sequence of the imaginary parts converges to the number B, then the complex sequence $z_1, z_2, \ldots, z_n$ converges to the limit $A + iB$, as illustrated by the following example.

Example 6.18
Consider the complex sequence whose nth term is

$$z_n = \frac{n^2 - 2n + 3}{3n^2 - 4} + i\frac{2n - 1}{2n + 1}.$$

Setting $z_n = x_n + iy_n$, we find

$$x_n = \frac{n^2 - 2n + 3}{3n^2 - 4} = \frac{1 - (2/n) + (3/n^2)}{3 - 4/n^2} \quad \text{and} \quad y_n = \frac{2n - 1}{2n + 1} = \frac{2 - 1/n}{2 + 1/n}.$$

As $n \to \infty, x_n \to 1/3$ and $y_n \to 1$, thus, $z_n \to 1/3 + i$.

Complex series

We are interested in complex series whose terms are complex functions

$$f_1(z) + f_2(z) + f_3(z) + \cdots + f_n(z) + \cdots . \tag{6.29}$$

The sum of the first n terms is

$$S_n(z) = f_1(z) + f_2(z) + f_3(z) + \cdots + f_n(z),$$

which is called the nth partial sum of the series (6.29). The sum of the remaining terms after the nth term is called the remainder of the series.

We can now associate with the series (6.29) the sequence of its partial sums $S_1, S_2, \ldots$. If this sequence of partial sums is convergent, then the series converges; and if the sequence diverges, then the series diverges. We can put this in a formal way. The series (6.29) is said to converge to the sum $S(z)$ in a region R if for any

266

$\varepsilon > 0$ there exists an integer N depending in general on ε and on the particular value of z under consideration such that

$$|S_n(z) - S(z)| < \varepsilon \quad \text{for all } n > N$$

and we write

$$\lim_{n \to \infty} S_n(z) = S(z).$$

The difference $S_n(z) - S(z)$ is just the remainder after n terms, $R_n(z)$; thus the definition of convergence requires that $|R_n(z)| \to 0$ as $n \to \infty$.

If the absolute values of the terms in (6.29) form a convergent series

$$|f_1(z)| + |f_2(z)| + |f_3(z)| + \cdots + |f_n(z)| + \cdots$$

then series (6.29) is said to be absolutely convergent. If series (6.29) converges but is not absolutely convergent, it is said to be conditionally convergent. The terms of an absolutely convergent series can be rearranged in any manner whatsoever without affecting the sum of the series whereas rearranging the terms of a conditionally convergent series may alter the sum of the series or even cause the series to diverge.

As with complex sequences, questions about complex series can also be reduced to questions about real series, the series of the real part and the series of the imaginary part. From the definition of convergence it is not difficult to prove the following theorem:

A necessary and sufficient condition that the series of complex terms

$$f_1(z) + f_2(z) + f_3(z) + \cdots + f_n(z) + \cdots$$

should convergence is that the series of the real parts and the series of the imaginary parts of these terms should each converge. Moreover, if

$$\sum_{n=1}^{\infty} \operatorname{Re} f_n \quad \text{and} \quad \sum_{n=1}^{\infty} \operatorname{Im} f_n$$

converge to the respective functions $R(z)$ and $I(z)$, then the given series converges to $R(z) + I(z)$, and the series $f_1(z) + f_2(z) + f_3(z) + \cdots + f_n(z) + \cdots$ converges to $R(z) + iI(z)$.

Of all the tests for the convergence of infinite series, the most useful is probably the familiar *ratio test*, which applies to real series as well as complex series.

Ratio test

Given the series $f_1(z) + f_2(z) + f_3(z) + \cdots + f_n(z) + \cdots$, the series converges absolutely if

$$0 < |r(z)| = \lim_{n \to \infty} \left| \frac{f_{n+1}(z)}{f_n(z)} \right| < 1 \tag{6.30}$$

and diverges if $|r(z)| > 1$. When $|r(z)| = 1$, the ratio test provides no information about the convergence or divergence of the series.

Example 6.19

Consider the complex series

$$\sum_n S_n = \sum_{n=0}^{\infty} (2^{-n} + ie^{-n}) = \sum_{n=0}^{\infty} 2^{-n} + i \sum_{n=0}^{\infty} e^{-n}.$$

The ratio tests on the real and imaginary parts show that both converge:

$$\lim_{n \to \infty} \left| \frac{2^{-(n+1)}}{2^{-n}} \right| = \frac{1}{2}, \text{ which is positive and less than 1;}$$

$$\lim_{n \to \infty} \left| \frac{e^{-(n+1)}}{e^{-n}} \right| = \frac{1}{e}, \text{ which is also positive and less than 1.}$$

One can prove that the full series converges to

$$\sum_{n=1}^{\infty} S_n = \frac{1}{1 - 1/2} + i \frac{1}{1 - e^{-1}}.$$

Uniform convergence and the Weierstrass M-test

To establish conditions, under which series can legitimately be integrated or differentiated term by term, the concept of uniform convergence is required:

> *A series of functions is said to converge uniformly to the function $S(z)$ in a region R, either open or closed, if corresponding to an arbitrary $\varepsilon < 0$ there exists an integral N, depending on ε but not on z, such that for every value of z in R*

$$|S(z) - S_n(z)| < \varepsilon \quad \text{for all } n > N.$$

One of the tests for uniform convergence is the Weierstrass *M*-test (a sufficient test).

If a sequence of positive constants $\{M_n\}$ exists such that $|f_n(z)| \leq M_n$ for all positive integers n and for all values of z in a given region R, and if the series

$$M_1 + M_2 + \cdots + M_n + \cdots$$

is convergent, then the series

$$f_1(z) + f_2(z) + f_3(z) + \cdots + f_n(z) + \cdots$$

converges uniformly in R.

As an illustrative example, we use it to test for uniform convergence of the series

$$\sum_{n=1}^{\infty} u_n = \sum_{n=1}^{\infty} \frac{z^n}{n\sqrt{n+1}}$$

in the region $|z| \leq 1$. Now

$$|u_n| = \frac{|z|^n}{n\sqrt{n+1}} \leq \frac{1}{n^{3/2}}$$

if $|z| \leq 1$. Calling $M_n = 1/n^{3/2}$, we see that $\sum M_n$ converges, as it is a p series with $p = 3/2$. Hence by Wierstrass M-test the given series converges uniformly (and absolutely) in the indicated region $|z| \leq 1$.

Power series and Taylor series

Power series are one of the most important tools of complex analysis, as power series with non-zero radii of convergence represent analytic functions. As an example, the power series

$$S = \sum_{n=0}^{\infty} a_n z^n \tag{6.31}$$

clearly defines an analytic function as long as the series converge. We will only be interested in absolute convergence. Thus we have

$$\lim_{n \to \infty} \left| \frac{a_{n+1} z^{n+1}}{a_n z^n} \right| < 1 \quad \text{or} \quad |z| < R = \lim_{n \to \infty} \frac{|a_n|}{|a_{n+1}|},$$

where R is the radius of convergence since the series converges for all z lying strictly inside a circle of radius R centered at the origin. Similarly, the series

$$S = \sum_{n=0}^{\infty} a_n (z - z_0)^n$$

converges within a circle of radius R centered at z_0.

Notice that the Eq. (6.31) is just a Taylor series at the origin of a function with $f^n(0) = a_n n!$. Every choice we make for the infinite variables a_n defines a new function with its own set of derivatives at the origin. Of course we can go beyond the origin, and expand a function in a Taylor series centered at $z = z_0$. Thus in the complex analysis there is a Taylor expansion for every analytic function. This is the question addressed by *Taylor's theorem* (named after the English mathematician Brook Taylor, 1685–1731):

If $f(z)$ is analytic throughout a region R bounded by a simple closed curve C, and if z and a are both interior to C, then $f(z)$ can be expanded in a Taylor series centered at $z = a$ for $|z - a| < R$:

$$f(z) = f(a) + f'(a)(z - a) + f''(a)\frac{(z - a)^2}{2!} + \cdots$$

$$+ f^n(a)\frac{(z - a)^{n-1}}{n!} + R_n, \tag{6.32}$$

where the remainder R_n is given by

$$R_n(z) = (z - a)^n \frac{1}{2\pi i} \oint_C \frac{f(w)dw}{(w - a)^n(w - z)}.$$

Proof: To prove this, we first rewrite Cauchy's integral formula as

$$f(z) = \frac{1}{2\pi i} \oint_C \frac{f(w)dw}{w - z} = \frac{1}{2\pi i} \oint_C \frac{f(w)}{w - a}\left[\frac{1}{1 - (z - a)/(w - a)}\right]dw. \tag{6.33}$$

For later use we note that since w is on C while z is inside C,

$$\left|\frac{z - a}{w - a}\right| < 1.$$

From the geometric progression

$$1 + q + q^2 + \cdots + q^n = \frac{1 - q^{n+1}}{1 - q} = \frac{1}{1 - q} - \frac{q^{n+1}}{1 - q}$$

we obtain the relation

$$\frac{1}{1 - q} = 1 + q + \cdots + q^n + \frac{q^{n+1}}{1 - q}.$$

By setting $q = (z - a)/(w - a)$ we find

$$\frac{1}{1 - [(z-a)/(w-a)]} = 1 + \frac{z-a}{w-a} + \left(\frac{z-a}{w-a}\right)^2 + \cdots + \left(\frac{z-a}{w-a}\right)^n$$
$$+ \frac{[(z-a)/(w-a)]^{n+1}}{(w-z)/(w-a)}.$$

We insert this into Eq. (6.33). Since z and a are constant, we may take the powers of $(z - a)$ out from under the integral sign, and then Eq. (6.33) takes the form

$$f(z) = \frac{1}{2\pi i} \oint_C \frac{f(w)dw}{w-a} + \frac{z-a}{2\pi i} \oint_C \frac{f(w)dw}{(w-a)^2} + \cdots + \frac{(z-a)^n}{2\pi i} \oint_C \frac{f(w)dw}{(w-a)^{n+1}} + R_n(z).$$

Using Eq. (6.28), we may write this expansion in the form

$$f(z) = f(a) + \frac{z-a}{1!}f'(a) + \frac{(z-a)^2}{2!}f''(a) + \cdots + \frac{(z-a)^n}{n!}f^n(a) + R_n(z),$$

where

$$R_n(z) = (z-a)^n \frac{1}{2\pi i} \oint_C \frac{f(w)dw}{(w-a)^n(w-z)}.$$

Clearly, the expansion will converge and represent $f(z)$ if and only if $\lim_{n\to\infty} R_n(z) = 0$. This is easy to prove. Note that w is on C while z is inside C, so we have $|w - z| > 0$. Now $f(z)$ is analytic inside C and on C, so it follows that the absolute value of $f(w)/(w - z)$ is bounded, say,

$$\left|\frac{f(w)}{w-z}\right| < M$$

for all w on C. Let r be the radius of C, then $|w - a| = r$ for all w on C, and C has the length $2\pi r$. Hence we obtain

$$|R_n| = \frac{|z-a|^n}{2\pi} \left|\oint_C \frac{f(w)dw}{(w-a)^n(w-z)}\right| < \frac{|z-a|^n}{2\pi} M \frac{1}{r^n} 2\pi r$$
$$= Mr\left|\frac{z-a}{r}\right|^n \to 0 \quad \text{as} \quad n \to \infty.$$

Thus

$$f(z) = f(a) + \frac{z-a}{1!}f'(a) + \frac{(z-a)^2}{2!}f''(a) + \cdots + \frac{(z-a)^n}{n!}f^n(a)$$

is a valid representation of $f(z)$ at all points in the interior of any circle with its center at a and within which $f(z)$ is analytic. This is called the Taylor series of $f(z)$ with center at a. And the particular case where $a = 0$ is called the Maclaurin series of $f(z)$ [Colin Maclaurin 1698–1746, Scots mathematician].

The Taylor series of $f(z)$ converges to $f(z)$ only within a circular region around the point $z = a$, the circle of convergence; and it diverges everywhere outside this circle.

Taylor series of elementary functions

Taylor series of analytic functions are quite similar to the familiar Taylor series of real functions. Replacing the real variable in the latter series by a complex variable we may 'continue' real functions analytically to the complex domain. The following is a list of Taylor series of elementary functions: in the case of multiple-valued functions, the principal branch is used.

$$e^z = \sum_{n=0}^{\infty} \frac{z^n}{n!} = 1 + z + \frac{z^2}{2!} + \cdots, \qquad |z| < \infty,$$

$$\sin z = \sum_{n=0}^{\infty} (-1)^n \frac{z^{2n+1}}{(2n+1)!} = z - \frac{z^3}{3!} + \frac{z^5}{5!} - + \cdots, \qquad |z| < \infty,$$

$$\cos z = \sum_{n=0}^{\infty} (-1)^n \frac{z^{2n}}{(2n)!} = 1 - \frac{z^2}{2!} + \frac{z^4}{4!} - + \cdots, \qquad |z| < \infty,$$

$$\sinh z = \sum_{n=0}^{\infty} \frac{z^{2n+1}}{(2n+1)!} = z + \frac{z^3}{3!} + \frac{z^5}{5!} + \cdots, \qquad |z| < \infty,$$

$$\cosh z = \sum_{n=0}^{\infty} \frac{z^{2n}}{(2n)!} = 1 + \frac{z^2}{2!} + \frac{z^4}{4!} + \cdots, \qquad |z| < \infty,$$

$$\ln(1+z) = \sum_{n=0}^{\infty} \frac{(-1)^{n+1} z^n}{n} = z - \frac{z^2}{2} + \frac{z^3}{3} - + \cdots, \qquad |z| < 1.$$

Example 6.20
Expand $(1-z)^{-1}$ about a.

Solution:

$$\frac{1}{1-z} = \frac{1}{(1-a)-(z-a)} = \frac{1}{1-a}\frac{1}{1-(z-a)/(1-a)} = \frac{1}{1-a}\sum_{n=0}^{\infty}\left(\frac{z-a}{1-a}\right)^n.$$

We have established two surprising properties of complex analytic functions:

(1) *They have derivatives of all order.*
(2) *They can always be represented by Taylor series.*

This is not true in general for real functions; there are real functions which have derivatives of all orders but cannot be represented by a power series.

Example 6.21

Expand $\ln(a+z)$ about a.

Solution: Suppose we know the Maclaurin series, then

$$\ln(1+z)=\ln(1+a+z-a)=\ln(1+a)\left(1+\frac{z-a}{1+a}\right)=\ln(1+a)+\ln\left(1+\frac{z-a}{1+a}\right)$$

$$=\ln(1+a)+\left(\frac{z-a}{1+a}\right)-\frac{1}{2}\left(\frac{z-a}{1+a}\right)^2+\frac{1}{3}\left(\frac{z-a}{1+a}\right)^3-+\cdots.$$

Example 6.22

Let $f(z)=\ln(1+z)$, and consider that branch which has the value zero when $z=0$.

    (*a*) Expand $f(z)$ in a Taylor series about $z=0$, and determine the region of convergence.

    (*b*) Expand $\ln[(1+z)/(1-z)]$ in a Taylor series about $z=0$.

Solution: (*a*)

$$f(z)=\ln(1+z) \qquad\qquad f(0)=0$$

$$f'(z)=(1+z)^{-1} \qquad\qquad f'(0)=1$$

$$f''(z)=-(1+z)^{-2} \qquad\qquad f''(0)=-1$$

$$f'''(z)=2(1+z)^{-3} \qquad\qquad f'''(0)=2!$$

$$\vdots \qquad\qquad\qquad\qquad \vdots$$

$$f^{(n+1)}(z)=(-1)^n n!(1+n)^{(n+1)} \qquad f^{(n+1)}(0)=(-1)^n n!.$$

Then

$$f(z)=\ln(1+z)=f(0)+f'(0)z+\frac{f''(0)}{2!}z^2+\frac{f'''(0)}{3!}z^3+\cdots$$

$$=z-\frac{z^2}{2}+\frac{z^3}{3}-\frac{z^4}{4}+-\cdots.$$

The nth term is $u_n=(-1)^{n-1}z^n/n$. The ratio test gives

$$\lim_{n\to\infty}\left|\frac{u_{n+1}}{u_n}\right|=\lim_{n\to\infty}\left|\frac{nz}{n+1}\right|=|z|$$

and the series converges for $|z|<1$.

273

(b) $\ln[(1+z)/(1-z)] = \ln(1+z) - \ln(1-z)$. Next, replacing z by $-z$ in Taylor's expansion for $\ln(1+z)$, we have

$$\ln(1-z) = -z - \frac{z^2}{2} - \frac{z^3}{3} - \frac{z^4}{4} - \cdots.$$

Then by subtraction, we obtain

$$\ln\frac{1+z}{1-z} = 2\left(z + \frac{z^3}{3} + \frac{z^5}{5} + \cdots\right) = \sum_{n=0}^{\infty} \frac{2z^{2n+1}}{2n+1}.$$

Laurent series

In many applications it is necessary to expand a function $f(z)$ around points where or in the neighborhood of which the function is not analytic. The Taylor series is not applicable in such cases. A new type of series known as the Laurent series is required. The following is a representation which is valid in an annular ring bounded by two concentric circles of C_1 and C_2 such that $f(z)$ is single-valued and analytic in the annulus and at each point of C_1 and C_2, see Fig. 6.14. The function $f(z)$ may have singular points outside C_1 and inside C_2. Hermann Laurent (1841–1908, French mathematician) proved that, at any point in the annular ring bounded by the circles, $f(z)$ can be represented by the series

$$f(z) = \sum_{n=-\infty}^{\infty} a_n(z-a)^n \tag{6.34}$$

where

$$a_n = \frac{1}{2\pi i} \oint_C \frac{f(w)\,dw}{(w-a)^{n+1}}, \quad n = 0, \pm 1, \pm 2, \ldots, \tag{6.35}$$

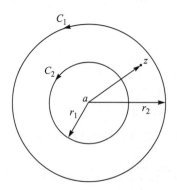

Figure 6.14.   Laurent theorem.

each integral being taken in the counterclockwise sense around curve C lying in the annular ring and encircling its inner boundary (that is, C is any concentric circle between C_1 and C_2).

To prove this, let z be an arbitrary point of the annular ring. Then by Cauchy's integral formula we have

$$f(z) = \frac{1}{2\pi i} \oint_{C_1} \frac{f(w)dw}{w - z} + \frac{1}{2\pi i} \oint_{C_2} \frac{f(w)dw}{w - z},$$

where C_2 is traversed in the counterclockwise direction and C_2 is traversed in the clockwise direction, in order that the entire integration is in the positive direction. Reversing the sign of the integral around C_2 and also changing the direction of integration from clockwise to counterclockwise, we obtain

$$f(z) = \frac{1}{2\pi i} \oint_{C_1} \frac{f(w)dw}{w - z} - \frac{1}{2\pi i} \oint_{C_2} \frac{f(w)dw}{w - z}.$$

Now

$$1/(w - z) = [1/(w - a)]\{1/[1 - (z - a)/(w - a)]\},$$

$$-1/(w - z) = 1/(z - w) = [1/(z - a)]\{1/[1 - (w - a)/(z - a)]\}.$$

Substituting these into $f(z)$ we obtain:

$$f(z) = \frac{1}{2\pi i} \oint_{C_1} \frac{f(w)dw}{w - z} - \frac{1}{2\pi i} \oint_{C_2} \frac{f(w)dw}{w - z}$$

$$= \frac{1}{2\pi i} \oint_{C_1} \frac{f(w)}{w - a} \left[\frac{1}{1 - (z - a)/(w - a)} \right] dw$$

$$+ \frac{1}{2\pi i} \oint_{C_2} \frac{f(w)}{z - a} \left[\frac{1}{1 - (w - a)/(z - a)} \right] dw.$$

Now in each of these integrals we apply the identity

$$\frac{1}{1 - q} = 1 + q + q^2 + \cdots + q^{n-1} + \frac{q^n}{1 - q}$$

275

to the last factor. Then

$$f(z) = \frac{1}{2\pi i} \oint_{C_1} \frac{f(w)}{w-a}\left[1 + \frac{z-a}{w-a} + \cdots + \left(\frac{z-a}{w-a}\right)^{n-1} + \frac{(z-a)^n/(w-a)^n}{1-(z-a)/(w-a)}\right]dw$$

$$+ \frac{1}{2\pi i} \oint_{C_2} \frac{f(w)}{z-a}\left[1 + \frac{w-a}{z-a} + \cdots + \left(\frac{w-a}{z-a}\right)^{n-1} + \frac{(w-a)^n/(z-a)^n}{1-(w-a)/(z-a)}\right]dw$$

$$= \frac{1}{2\pi i} \oint_{C_1} \frac{f(w)dw}{w-a} + \frac{z-a}{2\pi i} \oint_{C_2} \frac{f(w)dw}{(w-a)^2} + \cdots + \frac{(z-a)^{n-1}}{2\pi i} \oint_{C_2} \frac{f(w)dw}{(w-a)^n} + R_{n1}$$

$$+ \frac{1}{2\pi i(z-a)} \oint_{C_2} f(w)dw + \frac{1}{2\pi i(z-a)^2} \oint_{C_1} (w-a)f(w)dw + \cdots$$

$$+ \frac{1}{2\pi i(z-a)^n} \oint_{C_1} (w-a)^{n-1}f(w)dw + R_{n2},$$

where

$$R_{n1} = \frac{(z-a)^n}{2\pi i} \oint_{C_1} \frac{f(w)dw}{(w-a)^n(w-z)},$$

$$R_{n2} = \frac{1}{2\pi i(z-a)^n} \oint_{C_2} \frac{(w-a)^n f(w)dw}{z-w}.$$

The theorem will be established if we can show that $\lim_{n\to\infty} R_{n2} = 0$ and $\lim_{n\to\infty} R_{n1} = 0$. The proof of $\lim_{n\to\infty} R_{n1} = 0$ has already been given in the derivation of the Taylor series. To prove the second limit, we note that for values of w on C_2

$$|w-a| = r_1, |z-a| = \rho \quad \text{say,} \quad |z-w| = |(z-a)-(w-a)| \geq \rho - r_1,$$

and

$$|(f(w)| \leq M,$$

where M is the maximum of $|f(w)|$ on C_2. Thus

$$|R_{n2}| = \left|\frac{1}{2\pi i(z-a)^n} \oint_{C_2} \frac{(w-a)^n f(w)dw}{z-w}\right| \leq \frac{1}{|2\pi||z-a|^n} \oint_{C_2} \frac{|w-a|^n|f(w)||dw|}{|z-w|}$$

or

$$|R_{n2}| \leq \frac{r_1^n M}{2\pi \rho^n(\rho - r_1)} \oint_{C_2} |dw| = \frac{M}{2\pi}\left(\frac{r_1}{\rho}\right)^n \frac{2\pi r_1}{\rho - r_1}.$$

Since $r_1/\rho < 1$, the last expression approaches zero as $n \to \infty$. Hence $\lim_{n \to \infty} R_{n2} = 0$ and we have

$$f(z) = \frac{1}{2\pi i} \oint_{C_1} \frac{f(w)dw}{w - a} + \left[\frac{1}{2\pi i} \oint_{C_1} \frac{f(w)dw}{(w - a)^2}\right](z - a)$$

$$+ \left[\frac{1}{2\pi i} \oint_{C_1} \frac{f(w)dw}{(w - a)^3}\right](z - a)^2 + \cdots$$

$$+ \left[\frac{1}{2\pi i} \oint_{C_2} f(w)dw\right]\frac{1}{z - a} + \left[\frac{1}{2\pi i} \oint_{C_2} (w - a)f(w)dw\right]\frac{1}{(z - a)^2} + \cdots.$$

Since $f(z)$ is analytic throughout the region between C_1 and C_2, the paths of integration C_1 and C_2 can be replaced by any other curve C within this region and enclosing C_2. And the resulting integrals are precisely the coefficients a_n given by Eq. (6.35). This proves the Laurent theorem.

It should be noted that the coefficients of the positive powers $(z - a)$ in the Laurent expansion, while identical in form with the integrals of Eq. (6.28), cannot be replaced by the derivative expressions

$$\frac{f^n(a)}{n!}$$

as they were in the derivation of Taylor series, since $f(z)$ is not analytic throughout the entire interior of C_2 (or C), and hence Cauchy's generalized integral formula cannot be applied.

In many instances the Laurent expansion of a function is not found through the use of the formula (6.34), but rather by algebraic manipulations suggested by the nature of the function. In particular, in dealing with quotients of polynomials it is often advantageous to express them in terms of partial fractions and then expand the various denominators in series of the appropriate form through the use of the binomial expansion, which we assume the reader is familiar with:

$$(s + t)^n = s^n + ns^{n-1}t + \frac{n(n - 1)}{2!}s^{n-2}t^2 + \frac{n(n - 1)(n - 2)}{3!}s^{n-3}t^3 + \cdots.$$

This expansion is valid for all values of n if $|s| > |t|$. If $|s| \leq |t|$ the expansion is valid only if n is a non-negative integer.

That such procedures are correct follows from the fact that *the Laurent expansion of a function over a given annular ring is unique.* That is, if an expansion of the Laurent type is found by any process, it must be the Laurent expansion.

Example 6.23
Find the Laurent expansion of the function $f(z) = (7z - 2)/[(z + 1)z(z - 2)]$ in the annulus $1 < |z + 1| < 3$.

Solution: We first apply the method of partial fractions to $f(z)$ and obtain

$$f(z) = \frac{-3}{z+1} + \frac{1}{z} + \frac{2}{z-2}.$$

Now the center of the given annulus is $z = -1$, so the series we are seeking must be one involving powers of $z + 1$. This means that we have to modify the second and third terms in the partial fraction representation of $f(z)$:

$$f(z) = \frac{-3}{z+1} + \frac{1}{(z+1)-1} + \frac{2}{(z+1)-3},$$

but the series for $[(z+1) - 3]^{-1}$ converges only where $|z+1| > 3$, whereas we require an expansion valid for $|z+1| < 3$. Hence we rewrite the third term in the other order:

$$f(z) = \frac{-3}{z+1} + \frac{1}{(z+1)-1} + \frac{2}{-3+(z+1)}$$

$$= -3(z+1)^{-1} + [(z+1) - 1]^{-1} + 2[-3 + (z+1)]^{-1}$$

$$= \cdots + (z+1)^{-2} - 2(z+1)^{-1} - \frac{2}{3} - \frac{2}{9}(z+1)$$

$$- \frac{2}{27}(z+1)^2 - \cdots, \qquad 1 < |z+1| < 3.$$

Example 6.24

Given the following two functions:

$$(a)\ e^{3z}(z+1)^{-3}, \qquad (b)\ (z+2)\sin\frac{1}{z+2},$$

find Laurent series about the singularity for each of the functions, name the singularity, and give the region of convergence.

Solution: (a) $z = -1$ is a triple pole (pole of order 3). Let $z + 1 = u$, then $z = u - 1$ and

$$\frac{e^{3z}}{(z+1)^3} = \frac{e^{3(u-1)}}{u^3} = e^{-3}\frac{e^{3u}}{u^3} = \frac{e^{-3}}{u^3}\left(1 + 3u + \frac{(3u)^2}{2!} + \frac{(3u)^3}{3!} + \frac{(3u)^4}{4!} + \cdots\right)$$

$$= e^{-3}\left(\frac{1}{(z+1)^3} + \frac{3}{(z+1)^2} + \frac{9}{2(z+1)} + \frac{9}{2} + \frac{27(z+1)}{8} + \cdots\right).$$

The series converges for all values of $z \neq -1$.

(b) $z = -2$ is an essential singularity. Let $z + 2 = u$, then $z = u - 2$, and

$$(z+2)\sin\frac{1}{z+2} = u\sin\frac{1}{u} = u\left(\frac{1}{u} - \frac{1}{3!u^3} + \frac{1}{5!u^5} + \cdots\right)$$

$$= 1 - \frac{1}{6(z+2)^2} + \frac{1}{120(z+2)^4} - + \cdots.$$

The series converges for all values of $z \neq -2$.

Integration by the method of residues

We now turn to integration by the method of residues which is useful in evaluating both real and complex integrals. We first discuss briefly the theory of residues, then apply it to evaluate certain types of real definite integrals occurring in physics and engineering.

Residues

If $f(z)$ is single-valued and analytic in a neighborhood of a point $z = a$, then, by Cauchy's integral theorem,

$$\oint_C f(z)dz = 0$$

for any contour in that neighborhood. But if $f(z)$ has a pole or an isolated essential singularity at $z = a$ and lies in the interior of C, then the above integral will, in general, be different from zero. In this case we may represent $f(z)$ by a Laurent series:

$$f(z) = \sum_{n=-\infty}^{\infty} a_n(z-a)^n = a_0 + a_1(z-a) + a_2(z-a)^2 + \cdots + \frac{a_{-1}}{z-a} + \frac{a_{-2}}{(z-a)^2} + \cdots,$$

where

$$a_n = \frac{1}{2\pi i}\oint_C \frac{f(z)}{(z-a)^{n+1}}dz, \qquad n = 0, \pm 1, \pm 2, \ldots.$$

The sum of all the terms containing negative powers, namely $a_{-1}/(z-a) + a_{-2}/(z-a)^2 + \cdots$, is called the principal part of $f(z)$ at $z = a$. In the special case $n = -1$, we have

$$a_{-1} = \frac{1}{2\pi i}\oint_C f(z)dz$$

or

$$\oint_C f(z)dz = 2\pi i a_{-1}, \tag{6.36}$$

279

the integration being taken in the counterclockwise sense around a simple closed curve C that lies in the region $0 < |z - a| < D$ and contains the point $z = a$, where D is the distance from a to the nearest singular point of $f(z)$. The coefficient a_{-1} is called the residue of $f(z)$ at $z = a$, and we shall use the notation

$$a_{-1} = \operatorname*{Res}_{z=a} f(z). \tag{6.37}$$

We have seen that Laurent expansions can be obtained by various methods, without using the integral formulas for the coefficients. Hence, we may determine the residue by one of those methods and then use the formula (6.36) to evaluate contour integrals. To illustrate this, let us consider the following simple example.

Example 6.25
Integrate the function $f(z) = z^{-4} \sin z$ around the unit circle C in the counterclockwise sense.

Solution: Using

$$\sin z = \sum_{n=0}^{\infty} (-1)^n \frac{z^{2n+1}}{(2n + 1)!} = z - \frac{z^3}{3!} + \frac{z^5}{5!} - + \cdots,$$

we obtain the Laurent series

$$f(z) = \frac{\sin z}{z^4} = \frac{1}{z^3} - \frac{1}{3!z} + \frac{z}{5!} - \frac{z^3}{7!} + - \cdots.$$

We see that $f(z)$ has a pole of third order at $z = 0$, the corresponding residue is $a_{-1} = -1/3!$, and from Eq. (6.36) it follows that

$$\oint \frac{\sin z}{z^4} dz = 2\pi i a_{-1} = -i\frac{\pi}{3}.$$

There is a simple standard method for determining the residue in the case of a pole. If $f(z)$ has a simple pole at a point $z = a$, the corresponding Laurent series is of the form

$$f(z) = \sum_{n=-1}^{\infty} a_n(z - a)^n = a_0 + a_1(z - a) + a_2(z - a)^2 + \cdots + \frac{a_{-1}}{z - a},$$

where $a_{-1} \neq 0$. Multiplying both sides by $z - a$, we have

$$(z - a)f(z) = (z - a)[a_0 + a_1(z - a) + \cdots] + a_{-1}$$

and from this we have

$$\operatorname*{Res}_{z=a} f(z) = a_{-1} = \lim_{z \to a} (z - a)f(z). \tag{6.38}$$

Another useful formula is obtained as follows. If $f(z)$ can be put in the form

$$f(z) = \frac{p(z)}{q(z)},$$

where $p(z)$ and $q(z)$ are analytic at $z = a$, $p(z) \neq 0$, and $q(z) = 0$ at $z = a$ (that is, $q(z)$ has a simple zero at $z = a$). Consequently, $q(z)$ can be expanded in a Taylor series of the form

$$q(z) = (z - a)q'(a) + \frac{(z - a)^2}{2!}q''(a) + \cdots.$$

Hence

$$\operatorname*{Res}_{z=a} f(z) = \lim_{z \to a}(z - a)\frac{p(z)}{q(z)} = \lim_{z \to a} \frac{(z - a)p(z)}{(z - a)[q'(a) + (z - a)q''(a)/2 + \cdots]} = \frac{p(a)}{q'(a)}.$$

$$(6.39)$$

Example 6.26
The function $f(z) = (4 - 3z)/(z^2 - z)$ is analytic except at $z = 0$ and $z = 1$ where it has simple poles. Find the residues at these poles.

Solution: We have $p(z) = 4 - 3z$, $q(z) = z^2 - z$. Then from Eq. (6.39) we obtain

$$\operatorname*{Res}_{z=0} f(z) = \left[\frac{4 - 3z}{2z - 1}\right]_{z=0} = -4, \qquad \operatorname*{Res}_{z=1} f(z) = \left[\frac{4 - 3z}{2z - 1}\right]_{z=1} = 1.$$

We now consider poles of higher orders. If $f(z)$ has a pole of order $m > 1$ at a point $z = a$, the corresponding Laurent series is of the form

$$f(z) = a_0 + a_1(z - a) + a_2(z - a)^2 + \cdots + \frac{a_{-1}}{z - a} + \frac{a_{-2}}{(z - a)^2} + \cdots + \frac{a_{-m}}{(z - a)^m},$$

where $a_{-m} \neq 0$ and the series converges in some neighborhood of $z = a$, except at the point itself. By multiplying both sides by $(z - a)^m$ we obtain

$$(z - a)^m f(z) = a_{-m} + a_{-m+1}(z - a) + a_{-m+2}(z - a)^2 + \cdots + a_{-m+(m-1)}(z - a)^{(m-1)}$$

$$+ (z - a)^m[a_0 + a_1(z - a) + \cdots].$$

This represents the Taylor series about $z = a$ of the analytic function on the left hand side. Differentiating both sides $(m - 1)$ times with respect to z, we have

$$\frac{d^{m-1}}{dz^{m-1}}[(z - a)^m f(z)] = (m - 1)!a_{-1} + m(m - 1)\cdots 2a_0(z - a) + \cdots.$$

Thus on letting $z \to a$

$$\lim_{z \to a} \frac{d^{m-1}}{dz^{m-1}}[(z-a)^m f(z)] = (m-1)!a_{-1},$$

that is,

$$\operatorname*{Res}_{z=a} f(z) = \frac{1}{(m-1)!} \lim_{z \to a} \left\{ \frac{d^{m-1}}{dz^{m-1}}[(z-a)^m f(z)] \right\}. \tag{6.40}$$

Of course, in the case of a rational function $f(z)$ the residues can also be determined from the representation of $f(z)$ in terms of partial fractions.

The residue theorem

So far we have employed the residue method to evaluate contour integrals whose integrands have only a single singularity inside the contour of integration. Now consider a simple closed curve C containing in its interior a number of isolated singularities of a function $f(z)$. If around each singular point we draw a circle so small that it encloses no other singular points (Fig. 6.15), these small circles, together with the curve C, form the boundary of a multiply-connected region in which $f(z)$ is everywhere analytic and to which Cauchy's theorem can therefore be applied. This gives

$$\frac{1}{2\pi i} \left[\oint_C f(z)dz + \oint_{C_1} f(z)dz + \cdots + \oint_{C_m} f(z)dz \right] = 0.$$

If we reverse the direction of integration around each of the circles and change the sign of each integral to compensate, this can be written

$$\frac{1}{2\pi i} \oint_C f(z)dz = \frac{1}{2\pi i} \oint_{C_1} f(z)dz + \frac{1}{2\pi i} \oint_{C_2} f(z)dz + \cdots + \frac{1}{2\pi i} \oint_{C_m} f(z)dz,$$

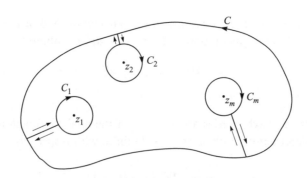

Figure 6.15.   Residue theorem.

282

where all the integrals are now to be taken in the counterclockwise sense. But the integrals on the right are, by definition, just the residues of $f(z)$ at the various isolated singularities within C. Hence we have established an important theorem, the residue theorem:

> If $f(z)$ is analytic inside a simple closed curve C and on C, except at a finite number of singular points $a_1, a_2, \ldots, a_m$ in the interior of C, then
>
> $$\oint_C f(z)\,dz = 2\pi i \sum_{j=1}^{m} \operatorname*{Res}_{z=a_j} f(z) = 2\pi i\,(r_1 + r_2 + \cdots + r_m), \quad (6.41)$$
>
> where r_j is the residue of $f(z)$ at the singular point a_j.

Example 6.27
The function $f(z) = (4 - 3z)/(z^2 - z)$ has simple poles at $z = 0$ and $z = 1$; the residues are -4 and 1, respectively (cf. Example 6.26). Therefore

$$\oint_C \frac{4 - 3z}{z^2 - z}\,dz = 2\pi i(-4 + 1) = -6\pi i$$

for every simple closed curve C which encloses the points 0 and 1, and

$$\oint_C \frac{4 - 3z}{z^2 - z}\,dz = 2\pi i(-4) = -8\pi i$$

for any simple closed curve C for which $z = 0$ lies inside C and $z = 1$ lies outside, the integrations being taken in the counterclockwise sense.

Evaluation of real definite integrals

The residue theorem yields a simple and elegant method for evaluating certain classes of complicated real definite integrals. One serious restriction is that the contour must be closed. But many integrals of practical interest involve integration over open curves. Their paths of integration must be closed before the residue theorem can be applied. So our ability to evaluate such an integral depends crucially on how the contour is closed, since it requires knowledge of the additional contributions from the added parts of the closed contour. A number of techniques are known for closing open contours. The following types are most common in practice.

Improper integrals of the rational function $\displaystyle\int_{-\infty}^{\infty} f(x)\,dx$

The improper integral has the meaning

$$\int_{-\infty}^{\infty} f(x)\,dx = \lim_{a \to \infty} \int_a^0 f(x)\,dx + \lim_{b \to \infty} \int_0^b f(x)\,dx. \quad (6.42)$$

If both limits exist, we may couple the two independent passages to $-\infty$ and ∞, and write

$$\int_{-\infty}^{\infty} f(x)dx = \lim_{r \to \infty} \int_{-r}^{r} f(x)dx. \tag{6.43}$$

We assume that the function $f(x)$ is a real rational function whose denominator is different from zero for all real x and is of degree at least two units higher than the degree of the numerator. Then the limits in (6.42) exist and we can start from (6.43). We consider the corresponding contour integral

$$\oint_C f(z)dz,$$

along a contour C consisting of the line along the x-axis from $-r$ to r and the semicircle Γ above (or below) the x-axis having this line as its diameter (Fig. 6.16). Then let $r \to \infty$. If $f(x)$ is an even function this can be used to evaluate

$$\int_0^{\infty} f(x)dx.$$

Let us see why this works. Since $f(x)$ is rational, $f(z)$ has finitely many poles in the upper half-plane, and if we choose r large enough, C encloses all these poles. Then by the residue theorem we have

$$\oint_C f(z)dz = \int_\Gamma f(z)dz + \int_{-r}^{r} f(x)dx = 2\pi i \sum \text{Res } f(z).$$

This gives

$$\int_{-r}^{r} f(x)dx = 2\pi i \sum \text{Res } f(z) - \int_\Gamma f(z)dz.$$

We next prove that $\int_\Gamma f(z)dz \to 0$ if $r \to \infty$. To this end, we set $z = re^{i\theta}$, then Γ is represented by $r = $ const, and as z ranges along Γ, θ ranges from 0 to π. Since

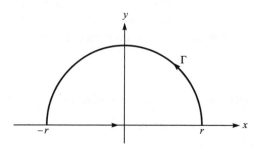

Figure 6.16.   Path of the contour integral.

the degree of the denominator of $f(z)$ is at least 2 units higher than the degree of the numerator, we have

$$|f(z)| < k/|z|^2 \qquad (|z| = r > r_0)$$

for sufficiently large constants k and r. By applying (6.24) we thus obtain

$$\left| \int_\Gamma f(z)dz \right| < \frac{k}{r^2}\pi r = \frac{k\pi}{r}.$$

Hence, as $r \to \infty$, the value of the integral over Γ approaches zero, and we obtain

$$\int_{-\infty}^{\infty} f(x)dx = 2\pi i \sum \text{Res } f(z). \qquad (6.44)$$

Example 6.28
Using (6.44), show that

$$\int_0^\infty \frac{dx}{1+x^4} = \frac{\pi}{2\sqrt{2}}.$$

Solution: $f(z) = 1/(1+z^4)$ has four simple poles at the points

$$z_1 = e^{\pi i/4}, \quad z_2 = e^{3\pi i/4}, \quad z_3 = e^{-3\pi i/4}, \quad z_4 = e^{-\pi i/4}.$$

The first two poles, z_1 and z_2, lie in the upper half-plane (Fig. 6.17) and we find, using L'Hospital's rule

$$\operatorname*{Res}_{z=z_1} f(z) = \left[\frac{1}{(1+z^4)'} \right]_{z=z_1} = \left[\frac{1}{4z^3} \right]_{z=z_1} = \frac{1}{4}e^{-3\pi i/4} = -\frac{1}{4}e^{\pi i/4},$$

$$\operatorname*{Res}_{z=z_2} f(z) = \left[\frac{1}{(1+z^4)'} \right]_{z=z_2} = \left[\frac{1}{4z^3} \right]_{z=z_2} = \frac{1}{4}e^{-9\pi i/4} = \frac{1}{4}e^{-\pi i/4},$$

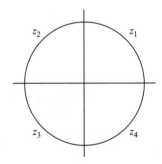

Figure 6.17.

then

$$\int_{-\infty}^{\infty} \frac{dx}{1+x^4} = \frac{2\pi i}{4}(-e^{\pi i/4} + e^{-\pi i/4}) = \pi \sin\frac{\pi}{4} = \frac{\pi}{\sqrt{2}}$$

and so

$$\int_0^{\infty} \frac{dx}{1+x^4} = \frac{1}{2}\int_{-\infty}^{\infty} \frac{dx}{1+x^4} = \frac{\pi}{2\sqrt{2}}.$$

Example 6.29

Show that

$$\int_{-\infty}^{\infty} \frac{x^2 dx}{(x^2+1)^2(x^2+2x+2)} = \frac{7\pi}{50}.$$

Solution: The poles of

$$f(z) = \frac{z^2}{(z^2+1)^2(z^2+2z+2)}$$

enclosed by the contour of Fig. 6.17 are $z = i$ of order 2 and $z = -1 + i$ of order 1. The residue at $z = i$ is

$$\lim_{z \to i} \frac{d}{dz}\left[(z-i)^2 \frac{z^2}{(z+i)^{21}(z-i)^2(z^2+2z+2)}\right] = \frac{9i-12}{100}.$$

The residue at $z = -1 + i$ is

$$\lim_{z \to -1+i} (z+1-i) \frac{z^2}{(z^2+1)^2(z+1-i)(z+1+i)} = \frac{3-4i}{25}.$$

Therefore

$$\int_{-\infty}^{\infty} \frac{x^2 dx}{(x^2+1)^2(x^2+2x+2)} = 2\pi i\left(\frac{9i-12}{100} + \frac{3-4i}{25}\right) = \frac{7\pi}{50}.$$

Integrals of the rational functions of $\sin\theta$ **and** $\cos\theta$ $\displaystyle\int_0^{2\pi} G(\sin\theta, \cos\theta)d\theta$

$G(\sin\theta, \cos\theta)$ is a real rational function of $\sin\theta$ and $\cos\theta$ finite on the interval $0 \le \theta \le 2\pi$. Let $z = e^{i\theta}$, then

$$dz = ie^{i\theta}d\theta, \quad \text{or} \quad d\theta = dz/iz, \quad \sin\theta = (z-z^{-1})/2i, \quad \cos\theta = (z+z^{-1})/2$$

and the given integrand becomes a rational function of z, say, $f(z)$. As θ ranges from 0 to 2π, the variable z ranges once around the unit circle $|z| = 1$ in the counterclockwise sense. The given integral takes the form

$$\oint_C f(z) \frac{dz}{iz},$$

the integration being taken in the counterclockwise sense around the unit circle.

Example 6.30
Evaluate

$$\int_0^{2\pi} \frac{d\theta}{3 - 2\cos\theta + \sin\theta}.$$

Solution: Let $z = e^{i\theta}$, then $dz = ie^{i\theta}d\theta$, or $d\theta = dz/iz$, and

$$\sin\theta = \frac{z - z^{-1}}{2i}, \quad \cos\theta = \frac{z + z^{-1}}{2},$$

then

$$\int_0^{2\pi} \frac{d\theta}{3 - 2\cos\theta + \sin\theta} = \oint_C \frac{2dz}{(1 - 2i)z^2 + 6iz - 1 - 2i},$$

where C is the circle of unit radius with its center at the origin (Fig. 6.18).
We need to find the poles of

$$\frac{1}{(1 - 2i)z^2 + 6iz - 1 - 2i}:$$

$$z = \frac{-6i \pm \sqrt{(6i)^2 - 4(1 - 2i)(-1 - 2i)}}{2(1 - 2i)}$$

$$= 2 - i, (2 - i)/5,$$

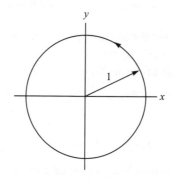

Figure 6.18.

only $(2 - i)/5$ lies inside C, and residue at this pole is

$$\lim_{z \to (2-i)/5} [z - (2 - i)/5] \left[\frac{2}{(1 - 2i)z^2 + 6iz - 1 - 2i} \right]$$

$$= \lim_{z \to (2-i)/5} \frac{2}{2(1 - 2i)z + 6i} = \frac{1}{2i} \quad \text{by L'Hospital's rule.}$$

Then

$$\int_0^{2\pi} \frac{d\theta}{3 - 2\cos\theta + \sin\theta} = \oint_C \frac{2dz}{(1 - 2i)z^2 + 6iz - 1 - 2i} = 2\pi i(1/2i) = \pi.$$

Fourier integrals of the form $\displaystyle\int_{-\infty}^{\infty} f(x) \begin{Bmatrix} \sin mx \\ \cos mx \end{Bmatrix} dx$

If $f(x)$ is a rational function satisfying the assumptions stated in connection with improper integrals of rational functions, then the above integrals may be evaluated in a similar way. Here we consider the corresponding integral

$$\oint_C f(z)e^{imz} dz$$

over the contour C as that in improper integrals of rational functions (Fig. 6.16), and obtain the formula

$$\int_{-\infty}^{\infty} f(x)e^{imx} dx = 2\pi i \sum \text{Res}[f(z)e^{imz}] \quad (m > 0), \tag{6.45}$$

where the sum consists of the residues of $f(z)e^{imz}$ at its poles in the upper half-plane. Equating the real and imaginary parts on each side of Eq. (6.45), we obtain

$$\int_{-\infty}^{\infty} f(x)\cos mx\, dx = -2\pi \sum \text{Im Res}[f(z)e^{imz}], \tag{6.46}$$

$$\int_{-\infty}^{\infty} f(x)\sin mx\, dx = 2\pi \sum \text{Re Res}[f(z)e^{imz}]. \tag{6.47}$$

To establish Eq. (6.45) we should now prove that the value of the integral over the semicircle Γ in Fig. 6.16 approaches zero as $r \to \infty$. This can be done as follows. Since Γ lies in the upper half-plane $y \geq 0$ and $m > 0$, it follows that

$$|e^{imz}| = |e^{imx}||e^{-my}| = e^{-my} \leq 1 \quad (y \geq 0, m > 0).$$

From this we obtain

$$|f(z)e^{imz}| = |f(z)| \leq |e^{imz}||f(z)| \quad (y \geq 0, m > 0),$$

which reduces our present problem to that of an improper integral of a rational function of this section, since $f(x)$ is a rational function satisfying the assumptions stated in connection these improper integrals. Continuing as before, we see that the value of the integral under consideration approaches zero as r approaches ∞, and Eq. (6.45) is established.

Example 6.31
Show that

$$\int_{-\infty}^{\infty} \frac{\cos mx}{k^2 + x^2}\, dx = \frac{\pi}{k}e^{-km}, \qquad \int_{-\infty}^{\infty} \frac{\sin mx}{k^2 + x^2}\, dx = 0 \qquad (m > 0, k > 0).$$

Solution: The function $f(z) = e^{imz}/(k^2 + z^2)$ has a simple pole at $z = ik$ which lies in the upper half-plane. The residue of $f(z)$ at $z = ik$ is

$$\operatorname*{Res}_{z=ik} \frac{e^{imz}}{k^2 + z^2} = \left[\frac{e^{imz}}{2z}\right]_{z=ik} = \frac{e^{-mk}}{2ik}.$$

Therefore

$$\int_{-\infty}^{\infty} \frac{e^{imx}}{k^2 + x^2}\, dx = 2\pi i\, \frac{e^{-mk}}{2ik} = \frac{\pi}{k}e^{-mk}$$

and this yields the above results.

Other types of real improper integrals

These are definite integrals

$$\int_{A}^{B} f(x)dx$$

whose integrand becomes infinite at a point a in the interval of integration, $\lim_{x \to a} |f(x)| = \infty$. This means that

$$\int_{A}^{B} f(x)dx = \lim_{\varepsilon \to 0} \int_{A}^{a-\varepsilon} f(x)dx + \lim_{\eta \to 0} \int_{a+\eta}^{B} f(x)dx,$$

where both ε and η approach zero independently and through positive values. It may happen that neither of these limits exists when $\varepsilon, \eta \to 0$ independently, but

$$\lim_{\varepsilon \to 0} \left[\int_{A}^{a-\varepsilon} f(x)dx + \int_{a+\varepsilon}^{B} f(x)dx\right]$$

exists; this is called Cauchy's principal value of the integral and is often written

$$\text{pr. v.} \int_{A}^{B} f(x)dx.$$

To evaluate improper integrals whose integrands have poles on the real axis, we can use a path which avoids these singularities by following small semicircles with centers at the singular points. We now illustrate the procedure with a simple example.

Example 6.32
Show that

$$\int_0^\infty \frac{\sin x}{x}\,dx = \frac{\pi}{2}.$$

Solution: The function $\sin(z)/z$ does not behave suitably at infinity. So we consider e^{iz}/z, which has a simple pole at $z = 0$, and integrate around the contour C or $ABDEFGA$ (Fig. 6.19). Since e^{iz}/z is analytic inside and on C, it follows from Cauchy's integral theorem that

$$\oint_C \frac{e^{iz}}{z}\,dz = 0$$

or

$$\int_{-R}^{-\varepsilon} \frac{e^{ix}}{x}\,dx + \int_{C_2} \frac{e^{iz}}{z}\,dz + \int_{\varepsilon}^{R} \frac{e^{ix}}{x}\,dx + \int_{C_1} \frac{e^{iz}}{z}\,dz = 0. \qquad (6.48)$$

We now prove that the value of the integral over large semicircle C_1 approaches zero as R approaches infinity. Setting $z = Re^{i\theta}$, we have $dz = iRe^{i\theta}d\theta$, $dz/z = id\theta$ and therefore

$$\left|\int_{C_1} \frac{e^{iz}}{z}\,dz\right| = \left|\int_0^\pi e^{iz}\,id\theta\right| \le \int_0^\pi |e^{iz}|\,d\theta.$$

In the integrand on the right,

$$|e^{iz}| = |e^{iR(\cos\theta + i\sin\theta)}| = |e^{iR\cos\theta}||e^{-R\sin\theta}| = e^{-R\sin\theta}.$$

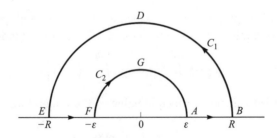

Figure 6.19.

By inserting this and using $\sin(\pi - \theta) = \sin\theta$ we obtain

$$\int_0^\pi |e^{iz}|\,d\theta = \int_0^\pi e^{-R\sin\theta}\,d\theta = 2\int_0^{\pi/2} e^{-R\sin\theta}\,d\theta$$

$$= 2\left[\int_0^\varepsilon e^{-R\sin\theta}\,d\theta + \int_\varepsilon^{\pi/2} e^{-R\sin\theta}\,d\theta\right],$$

where ε has any value between 0 and $\pi/2$. The absolute values of the integrands in the first and the last integrals on the right are at most equal to 1 and $e^{-R\sin\varepsilon}$, respectively, because the integrands are monotone decreasing functions of θ in the interval of integration. Consequently, the whole expression on the right is smaller than

$$2\left[\int_0^\varepsilon d\theta + e^{-R\sin\theta}\int_\varepsilon^{\pi/2} d\theta\right] = 2\left[\varepsilon + e^{-R\sin\theta}\left(\frac{\pi}{2} - \varepsilon\right)\right] < 2\varepsilon + \pi e^{-R\sin\varepsilon}.$$

Altogether

$$\left|\int_{C_1} \frac{e^{iz}}{z}\,dz\right| < 2\varepsilon + \pi e^{-R\sin\varepsilon}.$$

We first take ε arbitrarily small. Then, having fixed ε, the last term can be made as small as we please by choosing R sufficiently large. Hence the value of the integral along C_1 approaches 0 as $R \to \infty$.

We next prove that the value of the integral over the small semicircle C_2 approaches zero as $\varepsilon \to 0$. Let $z = \varepsilon e^{i\theta}$, then

$$\int_{C_2} \frac{e^{iz}}{z}\,dz = -\lim_{\varepsilon\to 0}\int_\pi^0 \frac{\exp(i\varepsilon e^{i\theta})}{\varepsilon e^{i\theta}} i\varepsilon e^{i\theta}\,d\theta = -\lim_{\varepsilon\to 0}\int_\pi^0 i\exp(i\varepsilon e^{i\theta})\,d\theta = \pi i$$

and Eq. (6.48) reduces to

$$\int_{-R}^{-\varepsilon} \frac{e^{ix}}{x}\,dx + \pi i + \int_\varepsilon^R \frac{e^{ix}}{x}\,dx = 0.$$

Replacing x by $-x$ in the first integral and combining with the last integral, we find

$$\int_\varepsilon^R \frac{e^{ix} - e^{-ix}}{x}\,dx + \pi i = 0.$$

Thus we have

$$2i\int_\varepsilon^R \frac{\sin x}{x}\,dx = \pi i.$$

Taking the limits $R \to \infty$ and $\varepsilon \to 0$

$$\int_0^\infty \frac{\sin x}{x} dx = \frac{\pi}{2}.$$

Problems

6.1. Given three complex numbers $z_1 = a + ib$, $z_2 = c + id$, and $z_3 = g + ih$, show that:

(a) $z_1 + z_2 = z_2 + z_1$ commutative law of addition;

(b) $z_1 + (z_2 + z_3) = (z_1 + z_2) + z_3$ associative law of addition;

(c) $z_1 z_2 = z_2 z_1$ commutative law of multiplication;

(d) $z_1 (z_2 z_3) = (z_1 z_2) z_3$ associative law of multiplication.

6.2. Given

$$z_1 = \frac{3 + 4i}{3 - 4i}, \quad z_2 = \left[\frac{1 + 2i}{1 - 3i}\right]^2$$

find their polar forms, complex conjugates, moduli, product, the quotient z_1/z_2.

6.3. The absolute value or modulus of a complex number $z = x + iy$ is defined as

$$|z| = \sqrt{zz^*} = \sqrt{x^2 + y^2}.$$

If $z_1, z_2, \ldots, z_m$ are complex numbers, show that the following hold:

(a) $|z_1 z_2| = |z_1||z_2|$ or $|z_1 z_2 \cdots z_m| = |z_1||z_2| \cdots |z_m|$.

(b) $|z_1/z_2| = |z_1|/|z_2|$ if $z_2 \neq 0$.

(c) $|z_1 + z_2| \leq |z_1| + |z_2|$.

(d) $|z_1 + z_2| \geq |z_1| - |z_2|$ or $|z_1 - z_2| \geq |z_1| - |z_2|$.

6.4 Find all roots of (a) $\sqrt[5]{-32}$, and (b) $\sqrt[3]{1 + i}$, and locate them in the complex plane.

6.5 Show, using De Moivre's theorem, that:

(a) $\cos 5\theta = 16 \cos^5 \theta - 20 \cos^3 \theta + 5 \cos \theta$;

(b) $\sin 5\theta = 5 \cos^4 \theta \sin \theta - 10 \cos^2 \theta \sin^3 \theta + \sin^5 \theta$.

6.6 Given $z = re^{i\theta}$, interpret $ze^{i\theta}$, where α is real geometrically.

6.7 Solve the quadratic equation $az^2 + bz + c = 0, a \neq 0$.

6.8 A point P moves in a counterclockwise direction around a circle of radius 1 with center at the origin in the z plane. If the mapping function is $w = z^2$, show that when P makes one complete revolution the image P' of P in the w plane makes three complete revolutions in a counterclockwise direction on a circle of radius 1 with center at the origin.

6.9 Show that $f(z) = \ln z$ has a branch point at $z = 0$.

6.10 Let $w = f(z) = (z^2 + 1)^{1/2}$, show that:

(a) $f(z)$ has branch points at $z = \pm I$.

(b) a complete circuit around both branch points produces no change in the branches of $f(z)$.

6.11 Apply the definition of limits to prove that:

$$\lim_{z \to 1} \frac{z^2 - 1}{z - 1} = 2.$$

6.12. Prove that:

(a) $f(z) = z^2$ is continuous at $z = z_0$, and

(b) $f(z) = \begin{cases} z^2, z \neq z_0 \\ 0, z = z_0 \end{cases}$ is discontinuous at $z = z_0$, where $z_0 \neq 0$.

6.13 Given $f(z) = z*$, show that $f'(i)$ does not exist.

6.14 Using the definition, find the derivative of $f(z) = z^3 - 2z$ at the point where:

(a) $z = z_0$, and (b) $z = -1$.

6.15. Show that f is an analytic function of z if it does not depend on $z*: f(z, z*) = f(z)$. In other words, $f(x, y) = f(x + iy)$, that is, x and y enter f only in the combination $x + iy$.

6.16. (a) Show that $u = y^3 - 3x^2y$ is harmonic.

(b) Find v such that $f(z) = u + iv$ is analytic.

6.17 (a) If $f(z) = u(x, y) + iv(x, y)$ is analytic in some region R of the z plane, show that the one-parameter families of curves $u(x, y) = C_1$ and $v(x, y) = C_2$ are orthogonal families.

(b) Illustrate (a) by using $f(z) = z^2$.

6.18 For each of the following functions locate and name the singularities in the finite z plane:

(a) $f(z) = \dfrac{z}{(z^2 + 4)^4}$;   (b) $f(z) = \dfrac{\sin \sqrt{z}}{\sqrt{z}}$;   (c) $f(z) = \sum_{n=0}^{\infty} \dfrac{1}{z^n n!}$.

6.19 (a) Locate and name all the singularities of

$$f(z) = \frac{z^8 + z^4 + 2}{(z - 1)^3 (3z + 2)^2}.$$

(b) Determine where $f(z)$ is analytic.

6.20 (a) Given $e^z = e^x(\cos y + i \sin y)$, show that $(d/dz)e^z = e^z$.

(b) Show that $e^{z_1} e^{z_2} = e^{z_1 + z_2}$.

(Hint: set $z_1 = x_1 + iy_1$ and $z_2 = x_2 + iy_2$ and apply the addition formulas for the sine and cosine.)

6.21 Show that: (a) $\ln e^z = z + 2n\pi i$, (b) $\ln z_1/z_2 = \ln z_1 - \ln z_2 + 2n\pi i$.

6.22 Find the values of: (a) $\ln i$, (b) $\ln (1 - i)$.

6.23 Evaluate $\int_C z* \, dz$ from $z = 0$ to $z = 4 + 2i$ along the curve C given by:

(a) $z = t^2 + it$;

(b) the line from $z = 0$ to $z = 2i$ and then the line from $z = 2i$ to $z = 4 + 2i$.

6.24 Evaluate $\oint_C dz/(z-a)^n$, $n = 2, 3, 4, \ldots$ where $z = a$ is inside the simple closed curve C.

6.25 If $f(z)$ is analytic in a simply-connected region R, and a and z are any two points in R, show that the integral

$$\int_a^z f(z)dz$$

is independent of the path in R joining a and z.

6.26 Let $f(z)$ be continuous in a simply-connected region R and let a and z be points in R. Prove that $F(z) = \int_a^z f(z')dz'$ is analytic in R, and $F'(z) = f(z)$.

6.27 Evaluate

(a) $\oint\limits_C \dfrac{\sin \pi z^2 + \cos \pi z^2}{(z-1)(z-2)} dz$

(b) $\oint\limits_C \dfrac{e^{2z}}{(z+1)^4} dz,$

where C is the circle $|z| = 1$.

6.28 Evaluate

$$\oint_C \frac{2\sin z^2}{(z-1)^4} dz,$$

where C is any simple closed path not passing through 1.

6.29 Show that the complex sequence

$$z_n = \frac{1}{n} - \frac{n^2 - 1}{n} i$$

diverges.

6.30 Find the region of convergence of the series $\sum_{n=1}^{\infty} (z+2)^{n+1}/(n+1)^3 4^n$.

6.31 Find the Maclaurin series of $f(z) = 1/(1+z^2)$.

6.32 Find the Taylor series of $f(z) = \sin z$ about $z = \pi/4$, and determine its circle of convergence. (Hint: $\sin z = \sin[a + (z-a)]$.)

6.33 Find the Laurent series about the indicated singularity for each of the following functions. Name the singularity in each case and give the region of convergence of each series.

(a) $(z-3)\sin\dfrac{1}{z+2}$, $z = -2$;

(b) $\dfrac{z}{(z+1)(z+2)}$, $z = -2$;

(c) $\dfrac{1}{z(z-3)^2}$, $z = 3$.

6.34 Expand $f(z) = 1/[(z+1)(z+3)]$ in a Laurent series valid for:
(a) $1 < |z| < 3$,   (b) $|z| > 3$,   (c) $0 < |z+1| < 2$.

6.35 Evaluate

$$\int_{-\infty}^{\infty} \frac{x^2 dx}{(x^2 + a^2)(x^2 + b^2)}, \qquad a > 0, b > 0.$$

6.36 Evaluate

(a)
$$\int_0^{2\pi} \frac{d\theta}{1 - 2p \cos \theta + p^2},$$

where p is a fixed number in the interval $0 < p < 1$;

(b)
$$\int_0^{2\pi} \frac{d\theta}{(5 - 3 \sin \theta)^2}.$$

6.37 Evaluate

$$\int_{-\infty}^{\infty} \frac{x \sin \pi x}{x^2 + 2x + 5} dx.$$

6.38 Show that:

(a)
$$\int_0^{\infty} \sin x^2 dx = \int_0^{\infty} \cos x^2 dx = \frac{1}{2} \sqrt{\frac{\pi}{2}};$$

(b)
$$\int_0^{\infty} \frac{x^{p-1}}{1 + x} dx = \frac{\pi}{\sin p\pi}, \qquad 0 < p < 1.$$

7

Special functions of mathematical physics

The functions discussed in this chapter arise as solutions of second-order differential equations which appear in special, rather than in general, physical problems. So these functions are usually known as the special functions of mathematical physics. We start with Legendre's equation (Adrien Marie Legendre, 1752–1833, French mathematician).

Legendre's equation

Legendre's differential equation

$$(1 - x^2)\frac{d^2y}{dx^2} - 2x\frac{dy}{dx} + \nu(\nu + 1)y = 0, \tag{7.1}$$

where ν is a positive constant, is of great importance in classical and quantum physics. The reader will see this equation in the study of central force motion in quantum mechanics. In general, Legendre's equation appears in problems in classical mechanics, electromagnetic theory, heat, and quantum mechanics, with spherical symmetry.

Dividing Eq. (7.1) by $1 - x^2$, we obtain the standard form

$$\frac{d^2y}{dx^2} - \frac{2x}{1 - x^2}\frac{dy}{dx} + \frac{\nu(\nu + 1)}{1 - x^2}y = 0.$$

We see that the coefficients of the resulting equation are analytic at $x = 0$, so the origin is an ordinary point and we may write the series solution in the form

$$y = \sum_{m=0}^{\infty} a_m x^m. \tag{7.2}$$

Substituting this and its derivatives into Eq. (7.1) and denoting the constant $\nu(\nu+1)$ by k we obtain

$$(1-x^2)\sum_{m=2}^{\infty}m(m-1)a_m x^{m-2} - 2x\sum_{m=1}^{\infty}ma_m x^{m-1} + k\sum_{m=0}^{\infty}a_m x^m = 0.$$

By writing the first term as two separate series we have

$$\sum_{m=2}^{\infty}m(m-1)a_m x^{m-2} - \sum_{m=2}^{\infty}m(m-1)a_m x^m - 2\sum_{m=1}^{\infty}ma_m x^m + k\sum_{m=0}^{\infty}a_m x^m = 0,$$

which can be written as:

$$\begin{aligned}
2\times 1a_2 + 3\times 2a_3 x + 4\times 3a_4 x^2 + \cdots &+ (s+2)(s+1)a_{s+2}x^s + \cdots \\
-2\times 1a_2 x^2 - \cdots\quad &- (s(s-1)a_s x^s - \cdots \\
-2\times 1a_1 x - 2\times 2a_2 x^2 - \cdots\quad &- 2sa_s x^s - \cdots \\
+ka_0\quad + ka_1 x\quad + ka_2 x^2 + \cdots\quad &+ ka_s x^s + \cdots = 0.
\end{aligned}$$

Since this must be an identity in x if Eq. (7.2) is to be a solution of Eq. (7.1), the sum of the coefficients of each power of x must be zero; remembering that $k = \nu(\nu+1)$ we thus have

$$2a_2 + \nu(\nu+1)a_0 = 0, \tag{7.3a}$$

$$6a3 + [-2 + v(v+1)]a_1 = 0, \tag{7.3b}$$

and in general, when $s = 2, 3, \ldots$,

$$(s+2)(s+1)a_{s+2} + [-s(s-1) - 2s + \nu(\nu+1)]a_s = 0. \tag{4.4}$$

The expression in square brackets [...] can be written

$$(\nu - s)(\nu + s + 1).$$

We thus obtain from Eq. (7.4)

$$a_{s+2} = -\frac{(\nu - s)(\nu + s + 1)}{(s+2)(s+1)}a_s \qquad (s = 0, 1, \ldots). \tag{7.5}$$

This is a recursion formula, giving each coefficient in terms of the one two places before it in the series, except for a_0 and a_1, which are left as arbitrary constants.

We find successively

$$a_2 = -\frac{\nu(\nu+1)}{2!}a_0, \qquad\qquad a_3 = -\frac{(\nu-1)(\nu+2)}{3!}a_1,$$

$$a_4 = -\frac{(\nu-2)(\nu+3)}{4\cdot 3}a_2, \qquad a_5 = -\frac{(\nu-3)(\nu+4)}{3!}a_3,$$

$$= \frac{(\nu-2)\nu(\nu+1)(\nu+3)}{4!}a_0, \qquad = \frac{(\nu-3)(\nu-1)(\nu+2)(\nu+4)}{5!}a_1,$$

etc. By inserting these values for the coefficients into Eq. (7.2) we obtain

$$y(x) = a_0 y_1(x) + a_1 y_2(x), \tag{7.6}$$

where

$$y_1(x) = 1 - \frac{\nu(\nu+1)}{2!}x^2 + \frac{(\nu-2)\nu(\nu+1)(\nu+3)}{4!}x^4 - + \cdots \tag{7.7a}$$

and

$$y_2(x) = x - \frac{(\nu-1)(\nu+2)}{3!}x^3 + \frac{(\nu-2)(\nu-1)(\nu+2)(\nu+4)}{5!}x^5 - + \cdots. \tag{7.7b}$$

These series converge for $|x| < 1$. Since Eq. (7.7a) contains even powers of x, and Eq. (7.7b) contains odd powers of x, the ratio y_1/y_2 is not a constant, and y_1 and y_2 are linearly independent solutions. Hence Eq. (7.6) is a general solution of Eq. (7.1) on the interval $-1 < x < 1$.

In many applications the parameter ν in Legendre's equation is a positive integer n. Then the right hand side of Eq. (7.5) is zero when $s = n$ and, therefore, $a_{n+2} = 0$ and $a_{n+4} = 0, \ldots$. Hence, if n is even, $y_1(x)$ reduces to a polynomial of degree n. If n is odd, the same is true with respect to $y_2(x)$. These polynomials, multiplied by some constants, are called Legendre polynomials. Since they are of great practical importance, we will consider them in some detail. For this purpose we rewrite Eq. (7.5) in the form

$$a_s = -\frac{(s+2)(s+1)}{(n-s)(n+s+1)}a_{s+2} \tag{7.8}$$

and then express all the non-vanishing coefficients in terms of the coefficient a_n of the highest power of x of the polynomial. The coefficient a_n is then arbitrary. It is customary to choose $a_n = 1$ when $n = 0$ and

$$a_n = \frac{(2n)!}{2^n(n!)^2} = \frac{1 \times 3 \times 5 \cdots (2n-1)}{n!}, \qquad n = 1, 2, \ldots, \tag{7.9}$$

the reason being that for this choice of a_n all those polynomials will have the value 1 when $x = 1$. We then obtain from Eqs. (7.8) and (7.9)

$$a_{n-2} = -\frac{n(n-1)}{2(2n-1)} a_n = -\frac{n(n-1)(2n)!}{2(2n-1)2^n(n!)^2}$$

$$= -\frac{n(n-1)2n(2n-1)(2n-2)!!}{2(2n-1)2^n n(n-1)!n(n-1)(n-2)!},$$

that is,

$$a_{n-2} = -\frac{(2n-2)!}{2^n(n-1)!(n-2)!}.$$

Similarly,

$$a_{n-4} = -\frac{(n-2)(n-3)}{4(2n-3)} a_{n-2} = \frac{(2n-4)!}{2^n 2!(n-2)!(n-4)!}$$

etc., and in general

$$a_{n-2m} = (-1)^m \frac{(2n-2m)!}{2^n m!(n-m)!(n-2m)!}. \tag{7.10}$$

The resulting solution of Legendre's equation is called the Legendre polynomial of degree n and is denoted by $P_n(x)$; from Eq. (7.10) we obtain

$$P_n(x) = \sum_{m=0}^{M} (-1)^m \frac{(2n-2m)!}{2^n m!(n-m)!(n-2m)!} x^{n-2m}$$

$$= \frac{(2n)!}{2^n(n!)^2} x^n - \frac{(2n-2)!}{2^n 1!(n-1)!(n-2)!} x^{n-2} + - \cdots, \tag{7.11}$$

where $M = n/2$ or $(n-1)/2$, whichever is an integer. In particular (Fig. 7.1)

$$P_0(x) = 1, \quad P_1(x) = x, \quad P_2(x) = \tfrac{1}{2}(3x^2 - 1), \quad P_3(x) = \tfrac{1}{2}(5x^3 - 3x),$$

$$P_4(x) = \tfrac{1}{8}(35x^4 - 30x^2 + 3), \quad P_5(x) = \tfrac{1}{8}(63x^5 - 70x^3 + 15x).$$

Rodrigues' formula for $P_n(x)$

The Legendre polynomials $P_n(x)$ are given by the formula

$$P_n(x) = \frac{1}{2^n n!} \frac{d^n}{dx^n} [(x^2 - 1)^n]. \tag{7.12}$$

We shall establish this result by actually carrying out the indicated differentiations, using the Leibnitz rule for nth derivative of a product, which we state below without proof:

299

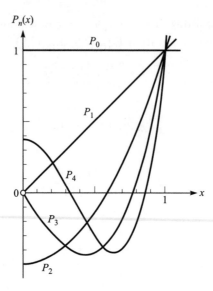

Figure 7.1.   Legendre polynomials.

If we write $D^n u$ as u_n and $D^n v$ as v_n, then

$$(uv)_n = uv_n + {}^nC_1 u_1 v_{n-1} + \cdots + {}^nC_r u_r v_{n-r} + \cdots + u_n v,$$

where $D = d/dx$ and ${}^nC_r$ is the binomial coefficient and is equal to $n!/[r!(n-r)!]$. We first notice that Eq. (7.12) holds for $n = 0, 1$. Then, write

$$z = (x^2 - 1)^n / 2^n n!$$

so that

$$(x^2 - 1)Dz = 2nxz. \tag{7.13}$$

Differentiating Eq. (7.13) $(n+1)$ times by the Leibnitz rule, we get

$$(1 - x^2)D^{n+2}z - 2xD^{n+1}z + n(n+1)D^n z = 0.$$

Writing $y = D^n z$, we then have:

(i) y is a polynomial.
(ii) The coefficient of x^n in $(x^2 - 1)^n$ is $(-1)^{n/2}\,{}^nC_{n/2}$ (n even) or 0 (n odd). Therefore the lowest power of x in $y(x)$ is x^0 (n even) or x^1 (n odd). It follows that

$$y_n(0) = 0 \qquad (n \text{ odd})$$

and

$$y_n(0) = \frac{1}{2^n n!}(-1)^{n/2}\,{}^nC_{n/2}n! = \frac{(-1)^{n/2}n!}{2^n[(n/2)!]^2} \qquad (n \text{ even}).$$

300

By Eq. (7.11) it follows that

$$y_n(0) = P_n(0) \quad \text{(all } n\text{)}.$$

(iii) $(1 - x^2)D^2y - 2xDy + n(n+1)y = 0$, which is Legendre's equation.

Hence Eq. (7.12) is true for all n.

The generating function for $P_n(x)$

One can prove that the polynomials $P_n(x)$ are the coefficients of z^n in the expansion of the function $\Phi(x, z) = (1 - 2xz + z^2)^{-1/2}$, with $|z| < 1$; that is,

$$\Phi(x, z) = (1 - 2xz + z^2)^{-1/2} = \sum_{n=0}^{\infty} P_n(x)z^n, \qquad |z| < 1. \tag{7.14}$$

$\Phi(x, z)$ is called the generating function for Legendre polynomials $P_n(x)$. We shall be concerned only with the case in which

$$x = \cos\theta \quad (-\pi < \theta \le \pi)$$

and then

$$z^2 - 2xz + 1 \equiv (z - e^{i\theta})(z - e^{i\theta}).$$

The expansion (7.14) is therefore possible when $|z| < 1$. To prove expansion (7.14) we have

$$\text{lhs} = 1 + \tfrac{1}{2}z(2x - 1) + \frac{1 \times 3}{2^2 \times 2!}z^2(2x - z)^2 + \cdots$$

$$+ \frac{1 \times 3 \cdots (2n - 1)}{2^n n!}z^n(2x - z)^n + \cdots.$$

The coefficient of z^n in this power series is

$$\frac{1 \times 3 \cdots (2n - 1)}{2^n n!}(2^n x^n) + \frac{1 \times 3 \cdots (2n - 3)}{2^{n-1}(n - 1)!}[-(n - 1)(2x)^{n-2}] + \cdots = P_n(x)$$

by Eq. (7.11). We can use Eq. (7.14) to find successive polynomials explicitly. Thus, differentiating Eq. (7.14) with respect to z so that

$$(x - z)(1 - 2xz + z^2)^{-3/2} = \sum_{n=1}^{\infty} nz^{n-1}P_n(x)$$

and using Eq. (7.14) again gives

$$(x - z)\left[P_0(x) + \sum_{n=1}^{\infty} P_n(x)z^n\right] = (1 - 2xz + z^2)\sum_{n=1}^{\infty} nz^{n-1}P_n(x). \tag{7.15}$$

301

Then expanding coefficients of z^n in Eq. (7.15) leads to the recurrence relation

$$(2n + 1)xP_n(x) = (n + 1)P_{n+1}(x) + nP_{n-1}(x). \tag{7.16}$$

This gives P_4, P_5, P_6, etc. very quickly in terms of P_0, P_1, and P_3.

Recurrence relations are very useful in simplifying work, helping in proofs or derivations. We list four more recurrence relations below without proofs or derivations:

$$xP'_n(x) - P'_{n-1}(x) = nP_n(x); \tag{7.16a}$$

$$P'_n(x) - xP'_{n-1}(x) = nP_{n-1}(x); \tag{7.16b}$$

$$(1 - x^2)P'_n(x) = nP_{n-1}(x) - nxP_n(x); \tag{7.16c}$$

$$(2n + 1)P_n(x) = P'_{n+1}(x) - P'_{n-1}(x). \tag{7.16d}$$

With the help of the recurrence formulas (7.16) and (7.16b), it is straightforward to establish the other three. Omitting the full details, which are left for the reader, these relations can be obtained as follows:

(*i*) differentiation of Eq. (7.16) with respect to x and the use of Eq. (7.16b) to eliminate $P'_{n+1}(x)$ leads to relation (7.16a);

(*ii*) the addition of Eqs. (7.16a) and (7.16b) immediately yields relation (7.16d);

(*iii*) the elimination of $P'_{n-1}(x)$ between Eqs. (7.16b) and (7.16a) gives relation (7.16c).

Example 7.1

The physical significance of expansion (7.14) is apparent in this simple example: find the potential V of a point charge at point P due to a charge $+q$ at Q.

Solution: Suppose the origin is at O (Fig. 7.2). Then

$$V_P = \frac{q}{R} = q(\rho^2 - 2r\rho\cos\theta + r^2)^{-1/2}.$$

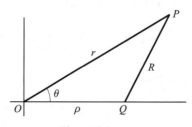

Figure 7.2.

Thus, if $r < \rho$

$$V_P = \frac{q}{\rho}(1 - 2z\cos\theta + z^2)^{-1/2}, \qquad z = r/\rho,$$

which gives

$$V_P = \frac{q}{\rho}\sum_{n=0}^{\infty}\left(\frac{r}{\rho}\right)^n P_n(\cos\theta) \qquad (r < \rho).$$

Similarly, when $r > \rho$, we get

$$V_P = \frac{q}{\rho}\sum_{n=0}^{\infty}\left(\frac{\rho}{r}\right)^{n+1} P_n(\cos\theta).$$

There are many problems in which it is essential that the Legendre polynomials be expressed in terms of θ, the colatitude angle of the spherical coordinate system. This can be done by replacing x by $\cos\theta$. But this will lead to expressions that are quite inconvenient because of the powers of $\cos\theta$ they contain. Fortunately, using the generating function provided by Eq. (7.14), we can derive more useful forms in which cosines of multiples of θ take the place of powers of $\cos\theta$. To do this, let us substitute

$$x = \cos\theta = (e^{i\theta} + e^{-i\theta})/2$$

into the generating function, which gives

$$[1 - z(e^{i\theta} + e^{-i\theta}) + z^2]^{-1/2} = [(1 - ze^{i\theta})(1 - ze^{-i\theta})]^{-1/2} = \sum_{n=0}^{\infty} P_n(\cos\theta)z^n.$$

Now by the binomial theorem, we have

$$(1 - ze^{i\theta})^{-1/2} = \sum_{n=0}^{\infty} a_n z^n e^{ni\theta}, \qquad (1 - ze^{-i\theta})^{-1/2} = \sum_{n=0}^{\infty} a_n z^n e^{-ni\theta},$$

where

$$a_n = \frac{1 \times 3 \times 5 \cdots (2n-1)}{2 \times 4 \times 6 \cdots (2n)}, \qquad n \geq 1, \qquad a_0 = 1. \tag{7.17}$$

To find the coefficient of z^n in the product of these two series, we need to form the Cauchy product of these two series. What is a Cauchy product of two series? We state it below for the reader who is in need of a review:

The Cauchy product of two infinite series, $\sum_{n=0}^{\infty} u_n(x)$ and $\sum_{n=0}^{\infty} v_n(x)$, is defined as the sum over n

$$\sum_{n=0}^{\infty} s_n(x) = \sum_{n=0}^{\infty}\sum_{k=0}^{n} u_k(x)v_{n-k}(x),$$

303

where $s_n(x)$ is given by

$$s_n(x) = \sum_{k=0}^{n} u_k(x)v_{n-k}(x) = u_0(x)v_n(x) + \cdots + u_n(x)v_0(x).$$

Now the Cauchy product for our two series is given by

$$\sum_{n=0}^{\infty}\sum_{k=0}^{n}\left(a_{n-k}z^{n-k}e^{(n-k)i\theta}\right)\left(a_k z^k e^{-ki\theta}\right) = \sum_{n=0}^{\infty}\left(z^n\sum_{k=0}^{n}a_k a_{n-k}e^{(n-2k)i\theta}\right). \tag{7.18}$$

In the inner sum, which is the sum of interest to us, it is straightforward to prove that, for $n \geq 1$, the terms corresponding to $k = j$ and $k = n - j$ are identical except that the exponents on e are of opposite sign. Hence these terms can be paired, and we have for the coefficient of z^n,

$$P_n(\cos\theta) = a_0 a_n(e^{ni\theta} + e^{-ni\theta}) + a_1 a_{n-1}(e^{(n-2)i\theta} + e^{-(n-2)i\theta}) + \cdots$$
$$= 2[a_0 a_n \cos n\theta + a_1 a_{n-1}\cos(n-2)\theta + \cdots]. \tag{7.19}$$

If n is odd, the number of terms is even and each has a place in one of the pairs. In this case, the last term in the sum is

$$a_{(n-1)/2}a_{(n+1)/2}\cos\theta.$$

If n is even, the number of terms is odd and the middle term is unpaired. In this case, the series (7.19) for $P_n(\cos\theta)$ ends with the constant term

$$a_{n/2}a_{n/2}.$$

Using Eq. (7.17) to compute values of the a_n, we find from the unit coefficient of z^0 in Eqs. (7.18) and (7.19), whether n is odd or even, the specific expressions

$$P_0(\cos\theta) = 1, \qquad P_1(\cos\theta) = \cos\theta, \qquad P_2(\cos\theta) = (3\cos 2\theta + 1)/4$$
$$P_3(\cos\theta) = (5\cos 3\theta + 3\cos\theta)/8$$
$$P_4(\cos\theta) = (35\cos 4\theta + 20\cos 2\theta + 9)/64$$
$$P_5(\cos\theta) = (63\cos 5\theta + 35\cos 3\theta + 30\cos\theta)/128$$
$$P_6(\cos\theta) = (231\cos 6\theta + 126\cos 4\theta + 105\cos 2\theta + 50)/512$$

$$\tag{7.20}$$

Orthogonality of Legendre polynomials

The set of Legendre polynomials $\{P_n(x)\}$ is orthogonal for $-1 \leq x \leq +1$. In particular we can show that

$$\int_{-1}^{+1} P_n(x)P_m(x)dx = \begin{cases} 2/(2n+1) & \text{if } m = n \\ 0 & \text{if } m \neq n \end{cases}. \tag{7.21}$$

(i) $m \neq n$:   Let us rewrite the Legendre equation (7.1) for $P_m(x)$ in the form

$$\frac{d}{dx}[(1-x^2)P'_m(x)] + m(m+1)P_m(x) = 0 \qquad (7.22)$$

and the one for $P_n(x)$

$$\frac{d}{dx}[(1-x^2)P'_n(x)] + n(n+1)P_n(x) = 0. \qquad (7.23)$$

We then multiply Eq. (7.22) by $P_n(x)$ and Eq. (7.23) by $P_m(x)$, and subtract to get

$$P_m\frac{d}{dx}[(1-x^2)P'_n] - P_n\frac{d}{dx}[(1-x^2)P'_m] + [n(n+1) - m(m+1)]P_mP_n = 0.$$

The first two terms in the last equation can be written as

$$\frac{d}{dx}[(1-x^2)(P_mP'_n - P_nP'_m)].$$

Combining this with the last equation we have

$$\frac{d}{dx}[(1-x^2)(P_mP'_n - P_nP'_m)] + [n(n+1) - m(m+1)]P_mP_n = 0.$$

Integrating the above equation between -1 and 1 we obtain

$$(1-x^2)(P_mP'_n - P_nP'_m)|_{-1}^{1} + [n(n+1) - m(m+1)]\int_{-1}^{1} P_m(x)P_n(x)dx = 0.$$

The integrated term is zero because $(1-x^2) = 0$ at $x = \pm 1$, and $P_m(x)$ and $P_n(x)$ are finite. The bracket in front of the integral is not zero since $m \neq n$. Therefore the integral must be zero and we have

$$\int_{-1}^{1} P_m(x)P_n(x)dx = 0, \qquad m \neq n.$$

(ii) $m = n$: We now use the recurrence relation (7.16a), namely

$$nP_n(x) = xP'_n(x) - P'_{n-1}(x).$$

Multiplying this recurrence relation by $P_n(x)$ and integrating between -1 and 1, we obtain

$$n\int_{-1}^{1} [P_n(x)]^2 dx = \int_{-1}^{1} xP_n(x)P'_n(x)dx - \int_{-1}^{1} P_n(x)P'_{n-1}(x)dx. \qquad (7.24)$$

The second integral on the right hand side is zero. (Why?) To evaluate the first integral on the right hand side, we integrate by parts

$$\int_{-1}^{1} xP_n(x)P'_n(x)dx = \frac{x}{2}[P_n(x)]^2|_{-1}^{1} - \frac{1}{2}\int_{-1}^{1} [P_n(x)]^2 dx = 1 - \frac{1}{2}\int_{-1}^{1} [P_n(x)]^2 dx.$$

Substituting these into Eq. (7.24) we obtain

$$n \int_{-1}^{1} [P_n(x)]^2 dx = 1 - \frac{1}{2} \int_{-1}^{1} [P_n(x)]^2 dx,$$

which can be simplified to

$$\int_{-1}^{1} [P_n(x)]^2 dx = \frac{2}{2n+1}.$$

Alternatively, we can use generating function

$$\frac{1}{\sqrt{1 - 2xz + z^2}} = \sum_{n=0}^{\infty} P_n(x)z^n.$$

We have on squaring both sides of this:

$$\frac{1}{1 - 2xz + z^2} = \sum_{m=0}^{\infty} \sum_{n=0}^{\infty} P_m(x) P_n(x) z^{m+n}.$$

Then by integrating from -1 to 1 we have

$$\int_{-1}^{1} \frac{dx}{1 - 2xz + z^2} = \sum_{m=0}^{\infty} \sum_{n=0}^{\infty} \left\{ \int_{-1}^{1} P_m(x) P_n(x) dx \right\} z^{m+n}.$$

Now

$$\int_{-1}^{1} \frac{dx}{1 - 2xz + z^2} = -\frac{1}{2z} \int_{-1}^{1} \frac{d(1 - 2xz + z^2)}{1 - 2xz + z^2} = -\frac{1}{2z} \ln(1 - 2xz + z^2)|_{-1}^{1}$$

and

$$\int_{-1}^{1} P_m(x) P_n(x) dx = 0, \qquad m \neq n.$$

Thus, we have

$$-\frac{1}{2z} \ln(1 - 2xz + z^2)|_{-1}^{1} = \sum_{n=0}^{\infty} \left\{ \int_{-1}^{1} P_n^2(x) dx \right\} z^{2n}$$

or

$$\frac{1}{z} \ln\left(\frac{1+z}{1-z}\right) = \sum_{n=0}^{\infty} \left\{ \int_{-1}^{1} P_n^2(x) dx \right\} z^{2n},$$

that is,

$$\sum_{n=0}^{\infty} \frac{2z^{2n}}{2n+1} = \sum_{n=0}^{\infty} \left\{ \int_{-1}^{1} P_n^2(x) dx \right\} z^{2n}.$$

Equating coefficients of z^{2n} we have as required $\int_{-1}^{1} P_n^2(x) dx = 2/(2n+1)$.

Since the Legendre polynomials form a complete orthogonal set on $(-1, 1)$, we can expand functions in Legendre series just as we expanded functions in Fourier series:

$$f(x) = \sum_{i=0}^{\infty} c_i P_i(x).$$

The coefficients c_i can be found by a method parallel to the one we used in finding the formulas for the coefficients in a Fourier series. We shall not pursue this line further.

There is a second solution of Legendre's equation. However, this solution is usually only required in practical applications in which $|x| > 1$ and we shall only briefly discuss it for such values of x. Now solutions of Legendre's equation relative to the regular singular point at infinity can be investigated by writing $x^2 = t$. With this substitution,

$$\frac{dy}{dx} = \frac{dy}{dt}\frac{dt}{dx} = 2t^{1/2}\frac{dy}{dt} \quad \text{and} \quad \frac{d^2y}{dx^2} = \frac{d}{dx}\left(\frac{dy}{dx}\right) = 2\frac{dy}{dx} + 4t\frac{d^2y}{dt^2},$$

and Legendre's equation becomes, after some simplifications,

$$t(1-t)\frac{d^2y}{dt^2} + \left(\frac{1}{2} - \frac{3}{2}t\right)\frac{dy}{dt} + \frac{\nu(\nu+1)}{4}y = 0.$$

This is the hypergeometric equation with $\alpha = -\nu/2$, $\beta = (1+\nu)/2$, and $\gamma = \frac{1}{2}$:

$$x(1-x)\frac{d^2y}{dx^2} + [\gamma - (\alpha + \beta + 1)x]\frac{dy}{dx} - \alpha\beta y = 0;$$

we shall not seek its solutions. The second solution of Legendre's equation is commonly denoted by $Q_\nu(x)$ and is called the Legendre function of the second kind of order ν. Thus the general solution of Legendre's equation (7.1) can be written

$$y = AP_\nu(x) + BQ_\nu(x),$$

A and B being arbitrary constants. $P_\nu(x)$ is called the Legendre function of the first kind of order ν and it reduces to the Legendre polynomial $P_n(x)$ when ν is an integer n.

The associated Legendre functions

These are the functions of integral order which are solutions of the associated Legendre equation

$$(1-x^2)y'' - 2xy' + \left\{n(n+1) - \frac{m^2}{1-x^2}\right\}y = 0 \qquad (7.25)$$

with $m^2 \leq n^2$.

We could solve Eq. (7.25) by series; but it is more useful to know how the solutions are related to Legendre polynomials, so we shall proceed in the following way. We write

$$y = (1 - x^2)^{m/2} u(x)$$

and substitute into Eq. (7.25) whence we get, after a little simplification,

$$(1 - x^2)u'' - 2(m+1)xu' + [n(n+1) - m(m+1)]u = 0. \tag{7.26}$$

For $m = 0$, this is a Legendre equation with solution $P_n(x)$. Now we differentiate Eq. (7.26) and get

$$(1 - x^2)(u')'' - 2[(m+1)+1]x(u')' + [n(n+1) - (m+1)(m+2)]u' = 0. \tag{7.27}$$

Note that Eq. (7.27) is just Eq. (7.26) with u' in place of u, and $(m+1)$ in place of m. Thus, if $P_n(x)$ is a solution of Eq. (7.26) with $m = 0$, $P_n'(x)$ is a solution of Eq. (7.26) with $m = 1$, $P_n''(x)$ is a solution with $m = 2$, and in general for integral $m, 0 \leq m \leq n$, $(d^m/dx^m)P_n(x)$ is a solution of Eq. (7.26). Then

$$y = (1 - x^2)^{m/2} \frac{d^m}{dx^m} P_n(x) \tag{7.28}$$

is a solution of the associated Legendre equation (7.25). The functions in Eq. (7.28) are called associated Legendre functions and are denoted by

$$P_n^m(x) = (1 - x^2)^{m/2} \frac{d^m}{dx^m} P_n(x). \tag{7.29}$$

Some authors include a factor $(-1)^m$ in the definition of $P_n^m(x)$.

A negative value of m in Eq. (7.25) does not change m^2, so a solution of Eq. (7.25) for positive m is also a solution for the corresponding negative m. Thus many references define $P_n^m(x)$ for $-n \leq m \leq n$ as equal to $P_n^{|m|}(x)$.

When we write $x = \cos\theta$, Eq. (7.25) becomes

$$\frac{1}{\sin\theta} \frac{d}{d\theta} \left(\sin\theta \frac{dy}{d\theta}\right) + \left\{n(n+1) - \frac{m^2}{\sin^2\theta}\right\} y = 0 \tag{7.30}$$

and Eq. (7.29) becomes

$$P_n^m(\cos\theta) = \sin^m\theta \frac{d^m}{d(\cos\theta)^m} \{P_n(\cos\theta)\}.$$

In particular

$$D^{-1} \text{ means } \int_1^x P_n(x)dx.$$

Orthogonality of associated Legendre functions

As in the case of Legendre polynomials, the associated Legendre functions $P_n^m(x)$ are orthogonal for $-1 \le x \le 1$ and in particular

$$\int_{-1}^{1} P_m^s(x) P_n^s(x) dx = \frac{(n+s)!}{(n-s)!} \delta_{mn}. \tag{7.31}$$

To prove this, let us write for simplicity

$$M = P_s^m(x), \quad \text{and} \quad N = P_n^s(x)$$

and from Eq. (7.25), the associated Legendre equation, we have

$$\frac{d}{dx}\left\{(1-x^2)\frac{dM}{dx}\right\} + \left\{m(m+1) - \frac{s^2}{1-x^2}\right\}M = 0 \tag{7.32}$$

and

$$\frac{d}{dx}\left\{(1-x^2)\frac{dN}{dx}\right\} + \left\{n(n+1) - \frac{s^2}{1-x^2}\right\}N = 0. \tag{7.33}$$

Multiplying Eq. (7.32) by N, Eq. (7.33) by M and subtracting, we get

$$M\frac{d}{dx}\left\{(1-x^2)\frac{dN}{dx}\right\} - N\frac{d}{dx}\left\{(1-x^2)\frac{dM}{dx}\right\} = \{m(m+1) - n(n+1)\}MN.$$

Integration between -1 and 1 gives

$$(m-n)(m+n-1)\int_{-1}^{1} MN dx = \int_{-1}^{1}\left[M\frac{d}{dx}\left\{(1-x^2)\frac{dN}{dx}\right\}\right.$$

$$\left. - N\frac{d}{dx}\left\{(1-x^2)\frac{dM}{dx}\right\}\right]dx. \tag{7.34}$$

Integration by parts gives

$$\int_{-1}^{1} M\frac{d}{dx}\{(1-x^2)N'\}dx = [MN'(1-x^2)]_{-1}^{1} - \int_{-1}^{1}(1-x^2)M'N'dx$$

$$= -\int_{-1}^{1}(1-x^2)M'N'dx.$$

Then integrating by parts once more, we obtain

$$\int_{-1}^{1} M \frac{d}{dx}\{(1-x^2)N'\}dx = -\int_{-1}^{1}(1-x^2)M'N'dx$$

$$= -[MN(1-x^2)]_{-1}^{1} + \int_{-1}^{1} N\frac{d}{dx}\{(1-x^2)M'\}dx$$

$$= \int_{-1}^{1} N\frac{d}{dx}\{(1-x^2)M'\}dx.$$

Substituting this in Eq. (7.34) we get

$$(m-n)(m+n-1)\int_{-1}^{1} MN dx = 0.$$

If $m \le n$, we have

$$\int_{-1}^{1} MN dx = \int_{-1}^{1} P_m^s(x)P_m^s(x)dx = 0 \qquad (m \ne n).$$

If $m = n$, let us write

$$P_n^s(x) = (1-x^2)^{s/2}\frac{d^s}{dx^s}P_n(x) = \frac{(1-x^2)^{s/2}}{2^n n!}\frac{d^{s+n}}{dx^{s+n}}\{(x^2-1)^n\}.$$

Hence

$$\int_{-1}^{1} P_n^s(x)P_n^s(x)dx = \frac{1}{2^{2n}(n!)^2}\int_{-1}^{1}(1-x^2)^s D^{n+s}\{(x^2-1)^n\}D^{n+s}$$

$$\times \{(x^2-1)^n\}dx, \qquad (D^k = d^k/dx^k).$$

Integration by parts gives

$$\frac{1}{2^{2n}(n!)^2}[(1-x^2)^s D^{n+s}\{(x^2-1)^n\}D^{n+s-1}\{(x^2-1)^n\}]_{-1}^{1}$$

$$-\frac{1}{2^{2n}(n!)^2}\int_{-1}^{1} D[(1-x^2)^s D^{n+s}\{(x^2-1)^n\}]D^{n+s-1}\{(x^2-1)^n\}dx.$$

The first term vanishes at both limits and we have

$$\int_{-1}^{1}\{P_n^s(x)\}^2 dx = \frac{-1}{2^{2n}(n!)^2}\int_{-1}^{1} D[(1-x^2)^s D^{n+s}\{(x^2-1)^n\}]D^{n+s-1}\{(x^2-1)^n\}dx.$$

$$(7.35)$$

We can continue to integrate Eq. (7.35) by parts and the first term continues to vanish since $D^p[(1-x^2)^s D^{n+s}\{(x^2-1)^n\}]$ contains the factor $(1-x^2)$ when $p < s$

and $D^{n+s-p}\{(x^2-1)^n\}$ contains it when $p \geq s$. After integrating $(n+s)$ times we find

$$\int_{-1}^{1} \{P_n^s(x)\}^2 dx = \frac{(-1)^{n+s}}{2^{2n}(n!)^2} \int_{-1}^{1} D^{n+s}[(1-x^2)^s D^{n+s}\{(x^2-1)^n\}](x^2-1)^n dx. \quad (7.36)$$

But $D^{n+s}\{(x^2-1)^n\}$ is a polynomial of degree $(n-s)$ so that $(1-x^2)^s D^{n+s}\{(x^2-1)^n\}$ is of degree $n-2+2s = n+s$. Hence the first factor in the integrand is a polynomial of degree zero. We can find this constant by examining the following:

$$D^{n+s}(x^{2n}) = 2n(2n-1)(2n-2)\cdots(n-+1)x^{n-s}.$$

Hence the highest power in $(1-x^2)^s D^{n+s}\{(x^2-1)^n\}$ is the term

$$(-1)^s 2n(2n-1)\cdots(n-s+1)x^{n+s},$$

so that

$$D^{n+s}[(1-x^2)^s D^{n+s}\{(x^2-1)^n\}] = (-1)^s(2n)! \frac{(n+s)!}{(n-s)!}.$$

Now Eq. (7.36) gives, by writing $x = \cos\theta$,

$$\int_{-1}^{1} P_n^s\{(x)\}^2 dx = \frac{(-1)^n}{2^{2n}(n!)^2} \int_{-1}^{1} (2n)! \frac{(n+s)!}{(n-s)!} (x^2-1)^n dx$$

$$= \frac{2}{2n+1} \frac{(n+s)!}{(n-s)!} \quad (7.37)$$

Hermite's equation

Hermite's equation is

$$y'' - 2xy' + 2\nu y = 0, \quad (7.38)$$

where $y' = dy/dx$. The reader will see this equation in quantum mechanics (when solving the Schrödinger equation for a linear harmonic potential function).

The origin $x = 0$ is an ordinary point and we may write the solution in the form

$$y = a_0 + a_1 x + a_2 x^2 + \cdots = \sum_{j=0}^{\infty} a_j x^j. \quad (7.39)$$

Differentiating the series term by term, we have

$$y' = \sum_{j=0}^{\infty} j a_j x^{j-1}, \qquad y'' = \sum_{j=0}^{\infty} (j+1)(j+2)a_{j+2} x^j.$$

311

Substituting these into Eq. (7.38) we obtain

$$\sum_{j=0}^{\infty} [(j+1)(j+2)a_{j+2} + 2(\nu - j)a_j]x^j = 0.$$

For a power series to vanish the coefficient of each power of x must be zero; this gives

$$(j+1)(j+2)a_{j+2} + 2(\nu - j)a_j = 0,$$

from which we obtain the recurrence relations

$$a_{j+2} = \frac{2(j - \nu)}{(j+1)(j+2)} a_j. \tag{7.40}$$

We obtain polynomial solutions of Eq. (7.38) when $\nu = n$, a positive integer. Then Eq. (7.40) gives

$$a_{n+2} = a_{n+4} = \cdots = 0.$$

For even n, Eq. (7.40) gives

$$a_2 = (-1)\frac{2n}{2!}a_0, \quad a_4 = (-1)^2 \frac{2^2(n-2)n}{4!}a_0, \quad a_6 = (-1)^3 \frac{2^3(n-4)(n-2)n}{6!}a_0$$

and generally

$$a_n = (-1)^{n/2} \frac{2^{n/2}n(n-2)\cdots 4 \times 2}{n!}a_0.$$

This solution is called a Hermite polynomial of degree n and is written $H_n(x)$. If we choose

$$a_0 = \frac{(-1)^{n/2}2^{n/2}n!}{n(n-2)\cdots 4 \times 2} = \frac{(-1)^{n/2}n!}{(n/2)!}$$

we can write

$$H_n(x) = (2x)^n - \frac{n(n-1)}{1!}(2x)^{n-2} + \frac{n(n-1)(n-2)(n-3)}{2!}(2x)^{n-4} + \cdots. \tag{7.41}$$

When n is odd the polynomial solution of Eq. (7.38) can still be written as Eq. (7.41) if we write

$$a_1 = \frac{(-1)^{(n-1)/2}2n!}{(n/2 - 1/2)!}.$$

In particular,

$$H_0(x) = 1, \quad H_1(x) = 2x, \quad H_3(x) = 4x^2 - 2, \quad H_3(x) = 8x^2 - 12x,$$

$$H_4(x) = 16x^4 - 48x^2 + 12, \quad H_5(x) = 32x^5 - 160x^3 + 120x, \ldots.$$

Rodrigues' formula for Hermite polynomials $H_n(x)$

The Hermite polynomials are also given by the formula

$$H_n(x) = (-1)^n e^{x^2} \frac{d^n}{dx^n} (e^{-x^2}). \tag{7.42}$$

To prove this formula, let us write $q = e^{-x^2}$. Then

$$Dq + 2xq = 0, \quad D = \frac{d}{dx}.$$

Differentiate this $(n + 1)$ times by the Leibnitz' rule giving

$$D^{n+2}q + 2xD^{n+1}q + 2(n+1)D^n q = 0.$$

Writing $y = (-1)^n D^n q$ gives

$$D^2 y + 2xDy + 2(n+1)y = 0 \tag{7.43}$$

substitute $u = e^{x^2} y$ then

$$Du = e^{x^2} \{2xy + Dy\}$$

and

$$D^2 u = e^{x^2} \{D^2 y + 4xDy + 4x^2 y + 2y\}.$$

Hence by Eq. (7.43) we get

$$D^2 u - 2xDu + 2nu = 0,$$

which indicates that

$$u = (-1)^n e^{x^2} D^n (e^{-x^2})$$

is a polynomial solution of Hermite's equation (7.38).

Recurrence relations for Hermite polynomials

Rodrigues' formula gives on differentiation

$$H_n'(x) = (-1)^n 2xe^{x^2} D^n (e^{-x^2}) + (-1)^n e^{x^2} D^{n+1}(e^{-x^2}).$$

that is,

$$H_n'(x) = 2xH_n(x) - H_{n+1}(x). \tag{7.44}$$

Eq. (7.44) gives on differentiation

$$H_n''(x) = 2H_n(x) + 2xH_n'(x) - H_{n+1}'(x).$$

Now $H_n(x)$ satisfies Hermite's equation

$$H_n''(x) - 2xH_n'(x) + 2nH_n(x) = 0.$$

Eliminating $H_n''(x)$ from the last two equations, we obtain

$$2xH_n'(x) - 2nH_n(x) = 2H_n(x) + 2xH_n'(x) - H_{n+1}'(x)$$

which reduces to

$$H_{n+1}'(x) = 2(n+1)H_n(x). \tag{7.45}$$

Replacing n by $n+1$ in Eq. (7.44), we have

$$H_{n+1}'(x) = 2xH_{n+1}(x) - H_{n+2}(x).$$

Combining this with Eq. (7.45) we obtain

$$H_{n+2}(x) = 2xH_{n+1}(x) - 2(n+1)H_n(x). \tag{7.46}$$

This will quickly give the higher polynomials.

Generating function for the $H_n(x)$

By using Rodrigues' formula we can also find a generating formula for the $H_n(x)$. This is

$$\Phi(x,t) = e^{2tx-t^2} = e^{\{x^2-(t-x)^2\}} = \sum_{n=0}^{\infty} \frac{H_n(x)}{n!} t^n. \tag{7.47}$$

Differentiating Eq. (7.47) n times with respect to t we get

$$e^{x^2} \frac{\partial^n}{\partial t^n} e^{-(t-x)^2} = e^{x^2}(-1)^n \frac{\partial^n}{\partial x^n} e^{-(t-x)^2} = \sum_{k=0}^{\infty} H_{n+k}(x) \frac{t^k}{k!}.$$

Put $t = 0$ in the last equation and we obtain Rodrigues' formula

$$H_n(x) = (-1)^n e^{x^2} \frac{d^n}{dx^n} (e^{-x^2}).$$

The orthogonal Hermite functions

These are defined by

$$F_n(x) = e^{-x^2/2} H_n(x); \tag{7.48}$$

from which we have

$$DF_n(x) = -xF_n(x) + e^{-x^2/2} H_n'(x),$$

$$D^2 F_n(x) = e^{-x^2/2} H_n''(x) - 2xe^{-x^2/2} H_n'(x) + x^2 e^{-x^2/2} H_n(x) - F_n(x)$$

$$= e^{-x^2/2} [H_n''(x) - 2xH_n'(x)] + x^2 F_n(x) - F_n(x),$$

314

but $H_n''(x) - 2xH_n'(x) = -2nH_n(x)$, so we can rewrite the last equation as

$$D^2 F_n(x) = e^{-x^2/2}[-2nH_n'(x)] + x^2 F_n(x) - F_n(x)$$
$$= -2nF_n(x) + x^2 F_n(x) - F_n(x),$$

which gives

$$D^2 F_n(x) - x^2 F_n(x) + (2n+1)F_n(x) = 0. \tag{7.49}$$

We can now show that the set $\{F_n(x)\}$ is orthogonal in the infinite range $-\infty < x < \infty$. Multiplying Eq. (7.49) by $F_m(x)$ we have

$$F_m(x)D^2 F_n(x) - x^2 F_n(x)F_m(x) + (2n+1)F_n(x)F_m(x) = 0.$$

Interchanging m and n gives

$$F_n(x)D^2 F_m(x) - x^2 F_m(x)F_n(x) + (2m+1)F_m(x)F_n(x) = 0.$$

Subtracting the last two equations from the previous one and then integrating from $-\infty$ to $+\infty$, we have

$$I_{n,m} = \int_{-\infty}^{\infty} F_n(x)F_m(x)dx = \frac{1}{2(n-m)}\int_{-\infty}^{\infty}(F_n'' F_m - F_m'' F_n)dx.$$

The integration by parts gives

$$2(n-m)I_{n,m} = \left[F_n' F_m - F_m' F_n\right]_{-\infty}^{\infty} - \int_{-\infty}^{\infty}(F_n' F_m' - F_m' F_n')dx.$$

Since the right hand side vanishes at both limits and if $m \neq m$, we have

$$I_{n,m} = \int_{-\infty}^{\infty} F_n(x)F_m(x)dx = 0. \tag{7.50}$$

When $n = m$ we can proceed as follows

$$I_{n,n} = \int_{-\infty}^{\infty} e^{-x^2} H_n(x)H_n(x)dx = \int_{-\infty}^{\infty} e^{x^2} D^n(e^{-x^2})D^m(e^{-x^2})dx.$$

Integration by parts, that is, $\int u\,dv = uv - \int v\,du$ with $u = e^{-x^2}D^n(e^{-x^2})$ and $v = D^{n-1}(e^{-x^2})$, gives

$$I_{n,n} = -\int_{-\infty}^{\infty}[2xe^{x^2} D^n(e^{-x^2}) + e^{x^2} D^{n+1}(e^{-x^2})]D^{n-1}(e^{-x^2})dx.$$

By using Eq. (7.43) which is true for $y = (-1)^n D^n q = (-1)^n D^n(e^{-x^2})$ we obtain

$$I_{n,n} = \int_{-\infty}^{\infty} 2ne^{x^2} D^{n-1}(e^{-x^2})D^{n-1}(e^{-x^2})dx = 2nI_{n-1,n-1}.$$

Since

$$I_{0,0} = \int_{-\infty}^{\infty} e^{-x^2} dx = \Gamma(1/2) = \sqrt{\pi},$$

we find that

$$I_{n,n} = \int_{-\infty}^{\infty} e^{-x^2} H_n(x)H_n(x)dx = 2^n n! \sqrt{\pi}. \tag{7.51}$$

We can also use the generating function for the Hermite polynomials:

$$e^{2tx-t^2} = \sum_{n=0}^{\infty} \frac{H_n(x)t^n}{n!}, \qquad e^{2sx-s^2} = \sum_{m=0}^{\infty} \frac{H_m(x)s^m}{m!}.$$

Multiplying these, we have

$$e^{2tx-t^2+2sx-s^2} = \sum_{m=0}^{\infty} \sum_{n=0}^{\infty} \frac{H_m(x)H_n(x)s^m t^n}{m!n!}.$$

Multiplying by e^{-x^2} and integrating from $-\infty$ to ∞ gives

$$\int_{-\infty}^{\infty} e^{-[(x+s+t)^2 - 2st]} dx = \sum_{m=0}^{\infty} \sum_{n=0}^{\infty} \frac{s^m t^n}{m!n!} \int_{-\infty}^{\infty} e^{-x^2} H_m(x)H_n(x)dx.$$

Now the left hand side is equal to

$$e^{2st} \int_{-\infty}^{\infty} e^{-(x+s+t)^2} dx = e^{2st} \int_{-\infty}^{\infty} e^{-u^2} du = e^{2st}\sqrt{\pi} = \sqrt{\pi} \sum_{m=0}^{\infty} \frac{2^m s^m t^m}{m!}.$$

By equating coefficients the required result follows.

It follows that the functions $(1/2^n n! \sqrt{n})^{1/2} e^{-x^2} H_n(x)$ form an orthonormal set. We shall assume it is complete.

Laguerre's equation

Laguerre's equation is

$$xD^2 y + (1-x)Dy + \nu y = 0. \tag{7.52}$$

This equation and its solutions (Laguerre functions) are of interest in quantum mechanics (e.g., the hydrogen problem). The origin $x = 0$ is a regular singular point and so we write

$$y(x) = \sum_{k=0}^{\infty} a_k x^{k+\rho}. \tag{7.53}$$

By substitution, Eq. (7.52) becomes

$$\sum_{k=0}^{\infty} [(k+\rho)^2 a_k x^{k+\rho-1} + (\nu - k + \rho)a_k x^k] = 0 \tag{7.54}$$

from which we find that the indicial equation is $\rho^2 = 0$. And then (7.54) reduces to

$$\sum_{k=0}^{\infty} [k^2 a_k x^{k-1} + (\nu - k)a_k x^k] = 0.$$

Changing $k - 1$ to k' in the first term, then renaming $k' = k$, we obtain

$$\sum_{k=0}^{\infty} \{(k + 1)^2 a_{k+1} + (\nu - k)a_k\}x^k = 0,$$

whence the recurrence relations are

$$a_{k+1} = \frac{k - \nu}{(k + 1)^2} a_k. \tag{7.55}$$

When ν is a positive integer n, the recurrence relations give $a_{k+1} = a_{k+2} = \cdots = 0$, and

$$a_1 = \frac{-n}{1^2} a_0, \qquad a_2 = \frac{-(n - 1)}{2^2} a_1 = \frac{(-1)^2(n - 1)n}{(1 \times 2)^2} a_0,$$

$$a_3 = \frac{-(n - 2)}{3^2} a_2 = \frac{(-1)^3(n - 2)(n - 1)n}{(1 \times 2 \times 3)^2} a_0, \quad \text{etc.}$$

In general

$$a_k = (-1)^k \frac{(n - k + 1)(n - k + 2) \cdots (n - 1)n}{(k!)^2} a_0. \tag{7.56}$$

We usually choose $a_0 = (-1)n!$, then the polynomial solution of Eq. (7.52) is given by

$$L_n(x) = (-1)^n \left\{ x^n - \frac{n^2}{1!} x^{n-1} + \frac{n^2(n - 1)^2}{2!} x^{n-2} - + \cdots + (-1)^n n! \right\}. \tag{7.57}$$

This is called the Laguerre polynomial of degree n. We list the first four Laguerre polynomials below:

$$L_0(x) = 1, \quad L_1(x) = 1 - x, \quad L_2(x) = 2 - 4x + x^2, \quad L_3(x) = 6 - 18x + 9x^2 - x^3.$$

The generating function for the Laguerre polynomials $L_n(x)$

This is given by

$$\Phi(x, z) = \frac{e^{-xz/(1-z)}}{1 - z} = \sum_{n=0}^{\infty} \frac{L_n(x)}{n!} z^n. \tag{7.58}$$

By writing the series for the exponential and collecting powers of z, you can verify the first few terms of the series. And it is also straightforward to show that

$$x\frac{\partial^2 \Phi}{\partial x^2} + (1-x)\frac{\partial \Phi}{\partial x} + z\frac{\partial \Phi}{\partial z} = 0.$$

Substituting the right hand side of Eq. (7.58), that is, $\Phi(x, z) = \sum_{n=0}^{\infty} [L_n(x)/n!]z^n$, into the last equation we see that the functions $L_n(x)$ satisfy Laguerre's equation. Thus we identify $\Phi(x, z)$ as the generating function for the Laguerre polynomials.

Now multiplying Eq. (7.58) by z^{-n-1} and integrating around the origin, we obtain

$$L_n(x) = \frac{n!}{2\pi i} \oint \frac{e^{-xz/(1-z)}}{(1-z)z^{n+1}} dz, \tag{7.59}$$

which is an integral representation of $L_n(x)$.

By differentiating the generating function in Eq. (7.58) with respect to x and z, we obtain the recurrence relations

$$\left.\begin{aligned} L_{n+1}(x) &= (2n+1-x)L_n(x) - n^2 L_{n-1}(x), \\ nL_{n-1}(x) &= nL'_{n-1}(x) - L'_n(x). \end{aligned}\right\} \tag{7.60}$$

Rodrigues' formula for the Laguerre polynomials $L_n(x)$

The Laguerre polynomials are also given by Rodrigues' formula

$$L_n(x) = e^x \frac{d^n}{dx^n}(x^n e^{-x}). \tag{7.61}$$

To prove this formula, let us go back to the integral representation of $L_n(x)$, Eq. (7.59). With the transformation

$$\frac{xz}{1-z} = s - x \quad \text{or} \quad z = \frac{s-x}{s},$$

Eq. (7.59) becomes

$$L_n(x) = \frac{n!e^x}{2\pi i} \oint \frac{s^n e^{-n}}{(s-x)^{n+1}} ds,$$

the new contour enclosing the point $s = x$ in the s plane. By Cauchy's integral formula (for derivatives) this reduces to

$$L_n(x) = e^x \frac{d^n}{dx^n}(x^n e^{-x}),$$

which is Rodrigues' formula.

Alternatively, we can differentiate Eq. (7.58) n times with respect to z and afterwards put $z=0$, and thus obtain

$$e^x \lim_{z \to 0} \frac{\partial^n}{\partial z^n}\left[(1-z)^{-1}\exp\left(\frac{-x}{1-z}\right)\right] = L_n(x).$$

But

$$\lim_{z \to 0} \frac{\partial^n}{\partial z^n} \left[(1-z)^{-1} \exp\left(\frac{-x}{1-z}\right) \right] = \frac{d^n}{dx^n} (x^n e^{-x}),$$

hence

$$L_n(x) = e^x \frac{d^n}{dx^n} (x^n e^{-x}).$$

The orthogonal Laguerre functions

The Laguerre polynomials, $L_n(x)$, do not by themselves form an orthogonal set. But the functions $e^{-x/2} L_n(x)$ are orthogonal in the interval $(0, \infty)$. For any two Laguerre polynomials $L_m(x)$ and $L_n(x)$ we have, from Laguerre's equation,

$$xL_m'' + (1-x)L_m' + mL_m = 0,$$

$$xL_n'' + (1-x)L_n' + mL_n = 0.$$

Multiplying these equations by $L_n(x)$ and $L_m(x)$ respectively and subtracting, we find

$$x[L_n L_m'' - L_m L_n''] + (1-x)[L_n L_m' - L_m L_n'] = (n-m)L_m L_n$$

or

$$\frac{d}{dx}[L_n L_m' - L_m L_n'] + \frac{1-x}{x}[L_n L_m' - L_m L_n'] = \frac{(n-m)L_m L_n}{x}.$$

Then multiplying by the integrating factor

$$\exp \int [(1-x)/x]dx = \exp(\ln x - x) = xe^{-x},$$

we have

$$\frac{d}{dx}\{xe^{-x}[L_n L_m' - L_m L_n']\} = (n-m)e^{-x}L_m L_n.$$

Integrating from 0 to ∞ gives

$$(n-m) \int_0^\infty e^{-x} L_m(x) L_n(x) dx = xe^{-x}[L_n L_m' - L_m L_n']|_0^\infty = 0.$$

Thus if $m \neq n$

$$\int_0^\infty e^{-x} L_m(x) L_n(x) dx = 0 \quad (m \neq n), \tag{7.62}$$

which proves the required result.

Alternatively, we can use Rodrigues' formula (7.61). If m is a positive integer,

$$\int_0^\infty e^{-x} x^m L_m(x) dx = \int_0^\infty x^m \frac{d^n}{dx^n}(x^n e^{-x}) dx = (-1)^m m! \int_0^\infty \frac{d^{n-m}}{dx^{n-m}}(x^n e^{-x}) dx,$$

$$\tag{7.63}$$

319

the last step resulting from integrating by parts m times. The integral on the right hand side is zero when $n > m$ and, since $L_n(x)$ is a polynomial of degree m in x, it follows that

$$\int_0^\infty e^{-x} L_m(x) L_n(x) dx = 0 \quad (m \neq n),$$

which is Eq. (7.62). The reader can also apply Eq. (7.63) to show that

$$\int_0^\infty e^{-x} \{L_n(x)\}^2 dx = (n!)^2. \tag{7.64}$$

Hence the functions $\{e^{-x/2} L_n(x)/n!\}$ form an orthonormal system.

The associated Laguerre polynomials $L_n^m(x)$

Differentiating Laguerre's equation (7.52) m times by the Leibnitz theorem we obtain

$$xD^{m+2}y + (m+1-x)D^{m+1}y + (n-m)D^m y = 0 \quad (\nu = n)$$

and writing $z = D^m y$ we obtain

$$xD^2 z + (m+1-x)Dz + (n-m)z = 0. \tag{7.65}$$

This is Laguerre's associated equation and it clearly possesses a polynomial solution

$$z = D^m L_n(x) \equiv L_n^m(x) \quad (m \leq n), \tag{7.66}$$

called the associated Laguerre polynomial of degree $(n-m)$. Using Rodrigues' formula for Laguerre polynomial $L_n(x)$, Eq. (7.61), we obtain

$$L_n^m(x) = \frac{d^m}{dx^m} L_n(x) = \frac{d^m}{dx^m} \left\{ e^x \frac{d^n}{dx^n} (x^n e^{-x}) \right\}. \tag{7.67}$$

This result is very useful in establishing further properties of the associated Laguerre polynomials. The first few polynomials are listed below:

$$L_0^0(x) = 1; \quad L_1^0(x) = 1 - x; \quad L_1^1(x) = -1;$$

$$L_2^0(x) = 2 - 4x + x^2; \quad L_2^1(x) = -4 + 2x; \quad L_2^2(x) = 2.$$

Generating function for the associated Laguerre polynomials

The Laguerre polynomial $L_n(x)$ can be generated by the function

$$\frac{1}{1-t} \exp\left(\frac{-xt}{1-t}\right) = \sum_{n=0}^\infty L_n(x) \frac{t^n}{n!}.$$

Differentiating this k times with respect to x, it is seen at once that

$$(-1)^k (1-t)^{-1} \left(\frac{t}{1-t}\right)^k \exp\left(\frac{-xt}{1-t}\right) = \sum_{\lambda=k}^{\infty} \frac{L_\lambda^k(x)}{\lambda!} t^\lambda. \tag{7.68}$$

Associated Laguerre function of integral order

A function of great importance in quantum mechanics is the associated Laguerre function that is defined as

$$G_n^m(x) = e^{-x/2} x^{(m-1)/2} L_n^m(x) \quad (m \le n). \tag{7.69}$$

It is significant largely because $|G_n^m(x)| \to 0$ as $x \to \infty$. It satisfies the differential equation

$$x^2 D^2 u + 2x Du + \left[\left(n - \frac{m-1}{2}\right) x - \frac{x^2}{4} - \frac{m^2-1}{4}\right] u = 0. \tag{7.70}$$

If we substitute $u = e^{-x/2} x^{(m-1)/2} z$ in this equation, it reduces to Laguerre's associated equation (7.65). Thus $u = G_n^m$ satisfies Eq. (7.70). You will meet this equation in quantum mechanics in the study of the hydrogen atom.

Certain integrals involving G_n^m are often used in quantum mechanics and they are of the form

$$I_{n,m} = \int_0^\infty e^{-x} x^{k-1} L_n^k(x) L_m^k(x) x^p \, dx,$$

where p is also an integer. We will not consider these here and instead refer the interested reader to the following book: *The Mathematics of Physics and Chemistry*, by Henry Margenau and George M. Murphy; D. Van Nostrand Co. Inc., New York, 1956.

Bessel's equation

The differential equation

$$x^2 y'' + xy' + (x^2 - \alpha^2) y = 0 \tag{7.71}$$

in which α is a real and positive constant, is known as Bessel's equation and its solutions are called Bessel functions. These functions were used by Bessel (Friedrich Wilhelm Bessel, 1784–1864, German mathematician and astronomer) extensively in a problem of dynamical astronomy. The importance of this equation and its solutions (Bessel functions) lies in the fact that they occur frequently in the boundary-value problems of mathematical physics and engineering

321

involving cylindrical symmetry (so Bessel functions are sometimes called cylindrical functions), and many others. There are whole books on Bessel functions.

The origin is a regular singular point, and all other values of x are ordinary points. At the origin we seek a series solution of the form

$$y(x) = \sum_{m=0}^{\infty} a_m x^{m+\rho} \quad (a_0 \neq 0). \tag{7.72}$$

Substituting this and its derivatives into Bessel's equation (7.71), we have

$$\sum_{m=0}^{\infty} (m+\rho)(m+\rho-1)a_m x^{m+\rho} + \sum_{m=0}^{\infty} (m+\rho)a_m x^{m+\rho}$$

$$+ \sum_{m=0}^{\infty} a_m x^{m+\rho+2} - \alpha^2 \sum_{m=0}^{\infty} a_m x^{m+\rho} = 0.$$

This will be an identity if and only if the coefficient of every power of x is zero. By equating the sum of the coefficients of $x^{k+\rho}$ to zero we find

$$\rho(\rho-1)a_0 + \rho a_0 - \alpha^2 a_0 = 0 \quad (k=0), \tag{7.73a}$$

$$(\rho-1)\rho a_1 + (\rho+1)a_1 - \alpha^2 a_1 = 0 \quad (k=1), \tag{7.73b}$$

$$(k+\rho)(k+\rho-1)a_k + (k+\rho)a_k + a_{k-2} - \alpha^2 a_k = 0 \quad (k=2,3,\ldots). \tag{7.73c}$$

From Eq. (7.73a) we obtain the indicial equation

$$\rho(\rho-1) + \rho - \alpha^2 = (\rho+\alpha)(\rho-\alpha) = 0.$$

The roots are $\rho = \pm\alpha$. We first determine a solution corresponding to the positive root. For $\rho = +\alpha$, Eq. (7.73b) yields $a_1 = 0$, and Eq. (7.73c) takes the form

$$(k+2\alpha)ka_k + a_{k-2} = 0, \quad \text{or} \quad a_k = \frac{-1}{k(k+2\alpha)}a_{k-2}, \tag{7.74}$$

which is a recurrence formula: since $a_1 = 0$ and $\alpha \geq 0$, it follows that $a_3 = 0, a_5 = 0, \ldots$, successively. If we set $k = 2m$ in Eq. (7.74), the recurrence formula becomes

$$a_{2m} = -\frac{1}{2^2 m(\alpha+m)}a_{2m-2}, \quad m = 1,2,\ldots \tag{7.75}$$

and we can determine the coefficients a_2, a_4, successively. We can rewrite a_{2m} in terms of a_0:

$$a_{2m} = \frac{(-1)^m}{2^{2m}m!(\alpha+m)\cdots(\alpha+2)(\alpha+1)}a_0.$$

Now a_{2m} is the coefficient of $x^{\alpha+2m}$ in the series (7.72) for y. Hence it would be convenient if a_{2m} contained the factor $2^{\alpha+2m}$ in its denominator instead of just 2^{2m}. To achieve this, we write

$$a_{2m} = \frac{(-1)^m}{2^{\alpha+2m}m!(\alpha+m)\cdots(\alpha+2)(\alpha+1)}(2^\alpha a_0).$$

Furthermore, the factors

$$(\alpha+m)\cdots(\alpha+2)(\alpha+1)$$

suggest a factorial. In fact, if α were an integer, a factorial could be created by multiplying numerator by $\alpha!$. However, since α is not necessarily an integer, we must use not $\alpha!$ but its generalization $\Gamma(\alpha+1)$ for this purpose. Then, except for the values

$$\alpha = -1, -2, -3, \ldots$$

for which $\Gamma(\alpha+1)$ is not defined, we can write

$$a_{2m} = \frac{(-1)^m}{2^{\alpha+2m}m!(\alpha+m)\cdots(\alpha+2)(\alpha+1)\Gamma(\alpha+1)}[2^\alpha\Gamma(\alpha+1)a_0].$$

Since the gamma function satisfies the recurrence relation $z\Gamma(z) = \Gamma(z+1)$, the expression for a_{2m} becomes finally

$$a_{2m} = \frac{(-1)^m}{2^{\alpha+2m}m!\Gamma(\alpha+m+1)}[2^\alpha\Gamma(\alpha+1)a_0].$$

Since a_0 is arbitrary, and since we are looking only for particular solutions, we choose

$$a_0 = \frac{1}{2^\alpha\Gamma(\alpha+1)},$$

so that

$$a_{2m} = \frac{(-1)^m}{2^{\alpha+2m}m!\Gamma(\alpha+m+1)}, \qquad a_{2m+1} = 0$$

and the series for y is, from Eq. (7.72),

$$y(x) = x^\alpha\left[\frac{1}{2^\alpha\Gamma(\alpha+1)} - \frac{x^2}{2^{\alpha+2}\Gamma(\alpha+2)} + \frac{x^4}{2^{\alpha+4}2!\Gamma(\alpha+3)} - +\cdots\right]$$

$$= \sum_{m=0}^{\infty}\frac{(-1)^m}{2^{\alpha+2m}m!\Gamma(\alpha+m+1)}x^{\alpha+2m}. \tag{7.76}$$

The function defined by this infinite series is known as the Bessel function of the first kind of order α and is denoted by the symbol $J_\alpha(x)$. Since Bessel's equation of order α has no finite singular points except the origin, the ratio test will show that the series for $J_\alpha(x)$ converges for all values of x if $\alpha \geq 0$.

When $\alpha = n$, an integer, solution (7.76) becomes, for $n \geq 0$

$$J_n(x) = x^n \sum_{m=0}^{\infty} \frac{(-1)^m x^{2m}}{2^{2m+n} m!(n+m)!}. \tag{7.76a}$$

The graphs of $J_0(x), J_1(x)$, and $J_2(x)$ are shown in Fig. 7.3. Their resemblance to the graphs of $\cos x$ and $\sin x$ is interesting (Problem 7.16 illustrates this for the first few terms). Fig. 7.3 also illustrates the important fact that for every value of α the equation $J_\alpha(x) = 0$ has infinitely many real roots.

With the second root $\rho = -\alpha$ of the indicial equation, the recurrence relation takes the form (from Eq. (7.73c))

$$a_k = \frac{-1}{k(k - 2\alpha)} a_{k-2}. \tag{7.77}$$

If α is not an integer, this leads to an independent second solution that can be written

$$J_{-\alpha}(x) = \sum_{m=0}^{\infty} \frac{(-1)^m}{m! \Gamma(-\alpha + m + 1)} (x/2)^{-\alpha+2m} \tag{7.78}$$

and the complete solution of Bessel's equation is then

$$y(x) = A J_\alpha(x) + B J_{-\alpha}(x), \tag{7.79}$$

where A and B are arbitrary constants.

When α is a positive integer n, it can be shown that the formal expression for $J_{-n}(x)$ is equal to $(-1)^n J_n(x)$. So $J_n(x)$ and $J_{-n}(x)$ are linearly dependent and Eq. (7.79) cannot be a general solution. In fact, if α is a positive integer, the recurrence

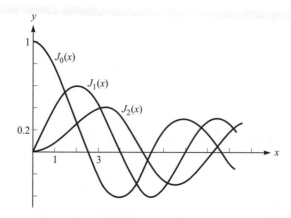

Figure 7.3. Bessel functions of the first kind.

relation (7.77) breaks down when $2\alpha = k$ and a second solution has to be found by other methods. There is a difficulty also when $\alpha = 0$, in which case the two roots of the indicial equation are equal; the second solution must also found by other methods. These will be discussed in next section.

The results of Problem 7.16 are a special case of an important general theorem which states that $J_\alpha(x)$ is expressible in finite terms by means of algebraic and trigonometrical functions of x whenever α is half of an odd integer. Further examples are

$$J_{3/2}(x) = \left(\frac{2}{\pi x}\right)^{1/2}\left(\frac{\sin x}{x} - \cos x\right),$$

$$J_{-5/2}(x) = \left(\frac{2}{\pi x}\right)^{1/2}\left\{\frac{3\sin x}{x} + \left(\frac{3}{x^2} - 1\right)\cos x\right\}.$$

The functions $J_{(n+1/2)}(x)$ and $J_{-(n+1/2)}(x)$, where n is a positive integer or zero, are called spherical Bessel functions; they have important applications in problems of wave motion in which spherical polar coordinates are appropriate.

Bessel functions of the second kind $Y_n(x)$

For integer $\alpha = n$, $J_n(x)$ and $J_{-n}(x)$ are linearly dependent and do not form a fundamental system. We shall now obtain a second independent solution, starting with the case $n = 0$. In this case Bessel's equation may be written

$$xy'' + y' + xy = 0, \tag{7.80}$$

the indicial equation (7.73a) now, with $\alpha = 0$, has the double root $\rho = 0$. Then we see from Eq. (7.33) that the desired solution must be of the form

$$y_2(x) = J_0(x)\ln x + \sum_{m=1}^{\infty} A_m x^m. \tag{7.81}$$

Next we substitute y_2 and its derivatives

$$y_2' = J_0'\ln x + \frac{J_0}{x} + \sum_{m=1}^{\infty} mA_m x^{m-1},$$

$$y_2'' = J_0''\ln x + \frac{2J_0'}{x} - \frac{J_0}{x^2} + \sum_{m=1}^{\infty} m(m-1)A_m x^{m-2}$$

into Eq. (7.80). Then the logarithmic terms disappear because J_0 is a solution of Eq. (7.80), the other two terms containing J_0 cancel, and we find

$$2J_0' + \sum_{m=1}^{\infty} m(m-1)A_m x^{m-1} + \sum_{m=1}^{\infty} mA_m x^{m-1} + \sum_{m=1}^{\infty} A_m x^{m+1} = 0.$$

From Eq. (7.76a) we obtain J_0' as

$$J_0'(x) = \sum_{m=1}^{\infty} \frac{(-1)^m 2mx^{2m-1}}{2^{2m}(m!)^2} = \sum_{m=1}^{\infty} \frac{(-1)^m x^{2m-1}}{2^{2m-1}m!(m-1)!}.$$

By inserting this series we have

$$\sum_{m=1}^{\infty} \frac{(-1)^m x^{2m-1}}{2^{2m-2}m!(m-1)!} + \sum_{m=1}^{\infty} m^2 A_m x^{m-1} + \sum_{m=1}^{\infty} A_m x^{m+1} = 0.$$

We first show that A_m with odd subscripts are all zero. The coefficient of the power x^0 is A_1 and so $A_1 = 0$. By equating the sum of the coefficients of the power x^{2s} to zero we obtain

$$(2s+1)^2 A_{2s+1} + A_{2s-1} = 0, \qquad s = 1, 2, \dots.$$

Since $A_1 = 0$, we thus obtain $A_3 = 0, A_5 = 0, \dots$, successively. We now equate the sum of the coefficients of x^{2s+1} to zero. For $s = 0$ this gives

$$-1 + 4A_2 = 0 \qquad \text{or} \qquad A_2 = 1/4.$$

For the other values of s we obtain

$$\frac{(-1)^{s+1}}{2^s(s+1)!s!} + (2s+2)^2 A_{2s+2} + A_{2s} = 0.$$

For $s = 1$ this yields

$$1/8 + 16A_4 + A_2 = 0 \quad \text{or} \quad A_4 = -3/128$$

and in general

$$A_{2m} = \frac{(-1)^{m-1}}{2^m(m!)^2}\left(1 + \frac{1}{2} + \frac{1}{3} + \dots + \frac{1}{m}\right), \qquad m = 1, 2, \dots. \qquad (7.82)$$

Using the short notation

$$h_m = 1 + \frac{1}{2} + \frac{1}{3} + \dots + \frac{1}{m}$$

and inserting Eq. (7.82) and $A_1 = A_3 = \dots = 0$ into Eq. (7.81) we obtain the result

$$y_2(x) = J_0(x) \ln x + \sum_{m=1}^{\infty} \frac{(-1)^{m-1} h_m}{2^{2m}(m!)^2} x^{2m}$$

$$= J_0(x) \ln x + \frac{1}{4}x^2 - \frac{3}{128}x^4 + - \dots. \qquad (7.83)$$

Since J_0 and y_2 are linearly independent functions, they form a fundamental system of Eq. (7.80). Of course, another fundamental system is obtained by replacing y_2 by an independent particular solution of the form $a(y_2 + bJ_0)$, where $a(\neq 0)$ and b are constants. It is customary to choose $a = 2/\pi$ and

$b = \gamma - \ln 2$, where $\gamma = 0.577\,215\,664\,90\ldots$ is the so-called Euler constant, which is defined as the limit of

$$1 + \frac{1}{2} + \cdots + \frac{1}{s} - \ln s$$

as s approaches infinity. The standard particular solution thus obtained is known as the Bessel function of the second kind of order zero or Neumann's function of order zero and is denoted by $Y_0(x)$:

$$Y_0(x) = \frac{2}{\pi} \left[J_0(x)\left(\ln\frac{x}{2} + \gamma\right) + \sum_{m=1}^{\infty} \frac{(-1)^{m-1}h_m}{2^{2m}(m!)^2} x^{2m} \right]. \tag{7.84}$$

If $\alpha = 1, 2, \ldots$, a second solution can be obtained by similar manipulations, starting from Eq. (7.35). It turns out that in this case also the solution contains a logarithmic term. So the second solution is unbounded near the origin and is useful in applications only for $x \neq 0$.

Note that the second solution is defined differently, depending on whether the order α is integral or not. To provide uniformity of formalism and numerical tabulation, it is desirable to adopt a form of the second solution that is valid for all values of the order. The common choice for the standard second solution defined for all α is given by the formula

$$Y_\alpha(x) = \frac{J_\alpha(x)\cos\alpha\pi - J_{-\alpha}(x)}{\sin\alpha\pi}, \quad Y_n(x) = \lim_{\alpha \to n} Y_\alpha(x). \tag{7.85}$$

This function is known as the Bessel function of the second kind of order α. It is also known as Neumann's function of order α and is denoted by $N_\alpha(x)$ (Carl Neumann 1832–1925, German mathematician and physicist). In G. N. Watson's *A Treatise on the Theory of Bessel Functions* (2nd ed. Cambridge University Press, Cambridge, 1944), it was called Weber's function and the notation $Y_\alpha(x)$ was used. It can be shown that

$$Y_{-n}(x) = (-1)^n Y_n(x).$$

We plot the first three $Y_n(x)$ in Fig. 7.4.

A general solution of Bessel's equation for all values of α can now be written:

$$y(x) = c_1 J_\alpha(x) + c_2 Y_\alpha(x).$$

In some applications it is convenient to use solutions of Bessel's equation that are complex for all values of x, so the following solutions were introduced

$$\left.\begin{array}{l} H_\alpha^{(1)}(x) = J_\alpha(x) + i Y_\alpha(x), \\ H_\alpha^{(2)}(x) = J_\alpha(x) - i Y_\alpha(x). \end{array}\right\} \tag{7.86}$$

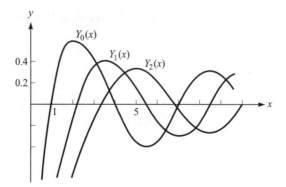

Figure 7.4.   Bessel functions of the second kind.

These linearly independent functions are known as Bessel functions of the third kind of order α or first and second Hankel functions of order α (Hermann Hankel, 1839–1873, German mathematician).

To illustrate how Bessel functions enter into the analysis of physical problems, we consider one example in classical physics: small oscillations of a hanging chain, which was first considered as early as 1732 by Daniel Bernoulli.

Hanging flexible chain

Fig. 7.5 shows a uniform heavy flexible chain of length l hanging vertically under its own weight. The x-axis is the position of stable equilibrium of the chain and its lowest end is at $x = 0$. We consider the problem of small oscillations in the vertical xy plane caused by small displacements from the stable equilibrium position. This is essentially the problem of the vibrating string which we discussed in Chapter 4, with two important differences: here, instead of being constant, the tension T at a given point of the chain is equal to the weight of the chain below that point, and now one end of the chain is free, whereas before both ends were fixed. The analysis of Chapter 4 generally holds. To derive an equation for y, consider an element dx, then Newton's second law gives

$$\left(T\frac{\partial y}{\partial x}\right)_2 - \left(T\frac{\partial y}{\partial x}\right)_1 = \rho dx \frac{\partial^2 y}{\partial t^2}$$

or

$$\rho dx \frac{\partial^2 y}{\partial t^2} = \frac{\partial}{\partial x}\left(T\frac{\partial y}{\partial x}\right)dx,$$

328

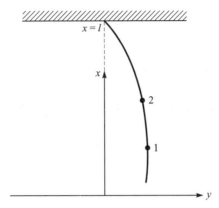

Figure 7.5. A flexible chain.

from which we obtain

$$\rho \frac{\partial^2 y}{\partial t^2} = \frac{\partial}{\partial x}\left(T \frac{\partial y}{\partial x}\right).$$

Now $T = \rho g x$. Substituting this into the above equation for y, we obtain

$$\frac{\partial^2 y}{\partial t^2} = g \frac{\partial y}{\partial x} + g x \frac{\partial^2 y}{\partial x^2},$$

where y is a function of two variables x and t. The first step in the solution is to separate the variables. Let us attempt a solution of the form $y(x, t) = u(x)f(t)$. Substitution of this into the partial differential equation yields two equations:

$$f''(t) + \omega^2 f(t) = 0, \quad xu''(x) + u'(x) + (\omega^2/g)u(x) = 0,$$

where ω^2 is the separation constant. The differential equation for $f(t)$ is ready for integration and the result is $f(t) = \cos(\omega t - \delta)$, with δ a phase constant. The differential equation for $u(x)$ is not in a recognizable form yet. To solve it, first change variables by putting

$$x = g z^2/4, \quad w(z) = u(x),$$

then the differential equation for $u(x)$ becomes Bessel's equation of order zero:

$$zw''(z) + w'(z) + \omega^2 zw(z) = 0.$$

Its general solution is

$$w(z) = A J_0(\omega z) + B Y_0(\omega z)$$

or

$$u(x) = A J_0\left(2\omega\sqrt{\frac{x}{g}}\right) + B Y_0\left(2\omega\sqrt{\frac{x}{g}}\right).$$

329

Since $Y_0(2\omega\sqrt{x/g}) \to -\infty$ as $x \to 0$, we are forced by physics to choose $B = 0$ and then

$$y(x, t) = AJ_0\left(2\omega\sqrt{\frac{x}{g}}\right)\cos(\omega t - \delta).$$

The upper end of the chain at $x = l$ is fixed, requiring that

$$J_0\left(2\omega\sqrt{\frac{l}{g}}\right) = 0.$$

The frequencies of the normal vibrations of the chain are given by

$$2\omega_n\sqrt{\frac{l}{g}} = \alpha_n,$$

where α_n are the roots of J_0. Some values of $J_0(x)$ and $J_1(x)$ are tabulated at the end of this chapter.

Generating function for $J_n(x)$

The function

$$\Phi(x, t) = e^{(x/2)(t-t^{-1})} = \sum_{n=-\infty}^{\infty} J_n(x)t^n \tag{7.87}$$

is called the generating function for Bessel functions of the first kind of integral order. It is very useful in obtaining properties of $J_n(x)$ for integral values of n which can then often be proved for all values of n.

To prove Eq. (7.87), let us consider the exponential functions $e^{xt/2}$ and $e^{-xt/2}$. The Laurent expansions for these two exponential functions about $t = 0$ are

$$e^{xt/2} = \sum_{k=0}^{\infty} \frac{(xt/2)^k}{k!}, \quad e^{-xt/2} = \sum_{m=0}^{\infty} \frac{(-xt/2)^k}{m!}.$$

Multiplying them together, we get

$$e^{x(t-t^{-1})/2} = \sum_{k=0}^{\infty}\sum_{m=0}^{\infty} \frac{(-1)^m}{k!m!}\left(\frac{x}{2}\right)^{k+m} t^{k-m}. \tag{7.88}$$

It is easy to recognize that the coefficient of the t^0 term which is made up of those terms with $k = m$ is just $J_0(x)$:

$$\sum_{k=0}^{\infty} \frac{(-1)^k}{2^{2k}(k!)^2} x^{2k} = J_0(x).$$

Similarly, the coefficient of the term t^n which is made up of those terms for which $k - m = n$ is just $J_n(x)$:

$$\sum_{k=0}^{\infty} \frac{(-1)^k}{(k+n)!k!2^{2k+n}} x^{2k+n} = J_n(x).$$

This shows clearly that the coefficients in the Laurent expansion (7.88) of the generating function are just the Bessel functions of integral order. Thus we have proved Eq. (7.87).

Bessel's integral representation

With the help of the generating function, we can express $J_n(x)$ in terms of a definite integral with a parameter. To do this, let $t = e^{i\theta}$ in the generating function, then

$$e^{x(t-t^{-1})/2} = e^{x(e^{i\theta}-e^{-i\theta})/2} = e^{ix\sin\theta}$$

$$= \cos(x\sin\theta) + i\sin(x\cos\theta).$$

Substituting this into Eq. (7.87) we obtain

$$\cos(x\sin\theta) + i\sin(x\cos\theta) = \sum_{n=-\infty}^{\infty} J_n(x)(\cos\theta + i\sin\theta)^n$$

$$= \sum_{-\infty}^{\infty} J_n(x)\cos n\theta + i\sum_{-\infty}^{\infty} J_n(x)\sin n\theta.$$

Since $J_{-n}(x) = (-1)^n J_n(x), \cos n\theta = \cos(-n\theta)$, and $\sin n\theta = -\sin(-n\theta)$, we have, upon equating the real and imaginary parts of the above equation,

$$\cos(x\sin\theta) = J_0(x) + 2\sum_{n=1}^{\infty} J_{2n}(x)\cos 2n\theta,$$

$$\sin(x\sin\theta) = 2\sum_{n=1}^{\infty} J_{2n-1}(x)\sin(2n-1)\theta.$$

It is interesting to note that these are the Fourier cosine and sine series of $\cos(x\sin\theta)$ and $\sin(x\sin\theta)$. Multiplying the first equation by $\cos k\theta$ and integrating from 0 to π, we obtain

$$\frac{1}{\pi}\int_0^{\pi} \cos k\theta \cos(x\sin\theta)d\theta = \begin{cases} J_k(x), & \text{if } k = 0, 2, 4, \dots \\ 0, & \text{if } k = 1, 3, 5, \dots \end{cases}.$$

Now multiplying the second equation by $\sin k\theta$ and integrating from 0 to π, we obtain

$$\frac{1}{\pi}\int_0^\pi \sin k\theta \sin(x\sin\theta)d\theta = \begin{cases} J_k(x), & \text{if } k = 1,3,5,\dots \\ 0, & \text{if } k = 0,2,4,\dots \end{cases}.$$

Adding these two together we obtain Bessel's integral representation

$$J_n(x) = \frac{1}{\pi}\int_0^\pi \cos(n\theta - x\sin\theta)d\theta, \quad n = \text{positive integer}. \tag{7.89}$$

Recurrence formulas for $J_n(x)$

Bessel functions of the first kind, $J_n(x)$, are the most useful, because they are bounded near the origin. And there exist some useful recurrence formulas between Bessel functions of different orders and their derivatives.

(1) $\quad J_{n+1}(x) = \dfrac{2n}{x}J_n(x) - J_{n-1}(x).$ $\hfill (7.90)$

Proof: Differentiating both sides of the generating function with respect to t, we obtain

$$e^{x(t-t^{-1})/2}\frac{x}{2}\left(1 + \frac{1}{t^2}\right) = \sum_{n=-\infty}^{\infty} nJ_n(x)t^{n-1}$$

or

$$\frac{x}{2}\left(1 + \frac{1}{t^2}\right)\sum_{n=-\infty}^{\infty} J_n(x)t^n = \sum_{n=-\infty}^{\infty} nJ_n(x)t^{n-1}.$$

This can be rewritten as

$$\frac{x}{2}\sum_{n=-\infty}^{\infty} J_n(x)t^n + \frac{x}{2}\sum_{n=-\infty}^{\infty} J_n(x)t^{n-2} = \sum_{n=-\infty}^{\infty} nJ_n(x)t^{n-1}$$

or

$$\frac{x}{2}\sum_{n=-\infty}^{\infty} J_n(x)t^n + \frac{x}{2}\sum_{n=-\infty}^{\infty} J_{n+2}(x)t^n = \sum_{n=-\infty}^{\infty} (n+1)J_{n+1}(x)t^n.$$

Equating coefficients of t^n on both sides, we obtain

$$\frac{x}{2}J_n(x) + \frac{x}{2}J_{n+2}(x) = (n+1)J_n(x).$$

Replacing n by $n-1$, we obtain the required result.

(2) $\quad xJ_n'(x) = nJ_n(x) - xJ_{n+1}(x).$ $\hfill (7.91)$

Proof:

$$J_n(x) = \sum_{k=0}^{\infty} \frac{(-1)^k}{k!\Gamma(n+k+1)2^{n+2k}} x^{n+2k}.$$

Differentiating both sides once, we obtain

$$J_n'(x) = \sum_{k=0}^{\infty} \frac{(n+2k)(-1)^k}{k!\Gamma(n+k+1)2^{n+2k}} x^{n+2k-1},$$

from which we have

$$xJ_n'(x) = nJ_n(x) + x\sum_{k=1}^{\infty} \frac{(-1)^k}{(k-1)!\Gamma(n+k+1)2^{n+2k-1}} x^{n+2k-1}.$$

Letting $k = m+1$ in the sum on the right hand side, we obtain

$$xJ_n'(x) = nJ_n(x) - x\sum_{m=0}^{\infty} \frac{(-1)^m}{m!\Gamma(n+m+2)2^{n+2m+1}} x^{n+2m+1}$$

$$= nJ_n(x) - xJ_{n+1}(x).$$

(3) $xJ_n'(x) = -nJ_n(x) + xJ_{n-1}(x).$ (7.92)

Proof: Differentiating both sides of the following equation with respect to x

$$x^n J_n(x) = \sum_{k=0}^{\infty} \frac{(-1)^k}{k!\Gamma(n+k+1)2^{n+2k}} x^{2n+2k},$$

we have

$$\frac{d}{dx}\{x^n J_n(x)\} = x^n J_n'(x) + nx^{n-1} J_n(x),$$

$$\frac{d}{dx}\sum_{k=0}^{\infty} \frac{(-1)^k x^{2n+2k}}{2^{n+2k}k!\Gamma(n+k+1)} = \sum_{k=0}^{\infty} \frac{(-1)^k x^{2n+2k-1}}{2^{n+2k-1}k!\Gamma(n+k)}$$

$$= x^n \sum_{k=0}^{\infty} \frac{(-1)^k x^{(n-1)+2k}}{2^{(n-1)+2k}k!\Gamma[(n-1)+k+1]}$$

$$= x^n J_{n-1}(x).$$

Equating these two results, we have

$$x^n J_n'(x) + nx^{n-1} J_n(x) = x^n J_{n-1}(x).$$

333

Canceling out the common factor x^{n-1}, we obtained the required result (7.92).

(4) $J_n'(x) = [J_{n-1}(x) - J_{n+1}(x)]/2.$ (7.93)

Proof: Adding (7.91) and (7.92) and dividing by $2x$, we obtain the required result (7.93).

If we subtract (7.91) from (7.92), $J_n'(x)$ is eliminated and we obtain

$$xJ_{n+1}(x) + xJ_{n-1}(x) = 2nJ_n(x)$$

which is Eq. (7.90).

These recurrence formulas (or important identities) are very useful. Here are some illustrative examples.

Example 7.2
Show that $J_0'(x) = J_{-1}(x) = -J_1(x)$.

Solution: From Eq. (7.93), we have

$$J_0'(x) = [J_{-1}(x) - J_1(x)]/2,$$

then using the fact that $J_{-n}(x) = (-1)^n J_n(x)$, we obtain the required results.

Example 7.3
Show that

$$J_3(x) = \left(\frac{8}{x^2} - 1\right)J_1(x) - \frac{4}{x}J_0(x).$$

Solution: Letting $n = 4$ in (7.90), we have

$$J_3(x) = \frac{4}{x}J_2(x) - J_1(x).$$

Similarly, for $J_2(x)$ we have

$$J_2(x) = \frac{2}{x}J_1(x) - J_0(x).$$

Substituting this into the expression for $J_3(x)$, we obtain the required result.

Example 7.4
Find $\int_0^t xJ_0(x)dx.$

Solution: Taking derivative of the quantity $xJ_1(x)$ with respect to x, we obtain

$$\frac{d}{dx}\{xJ_1(x)\} = J_1(x) + xJ_1'(x).$$

Then using Eq. (7.92) with $n = 1$, $xJ_1'(x) = -J_1(x) + xJ_0(x)$, we find

$$\frac{d}{dx}\{xJ_1(x)\} = J_1(x) + xJ_1'(x) = xJ_0(x),$$

thus,

$$\int_0^t xJ_0(x)dx = xJ_1(x)|_0^t = tJ_1(t).$$

Approximations to the Bessel functions

For very large or very small values of x we might be able to make some approximations to the Bessel functions of the first kind $J_n(x)$. By a rough argument, we can see that the Bessel functions behave something like a damped cosine function when the value of x is very large. To see this, let us go back to Bessel's equation (7.71)

$$x^2y'' + xy' + (x^2 - \alpha^2)y = 0$$

and rewrite it as

$$y'' + \frac{1}{x}y' + \left(1 - \frac{\alpha^2}{x^2}\right)y = 0.$$

If x is very large, let us drop the term α^2/x^2 and then the differential equation reduces to

$$y'' + \frac{1}{x}y' + y = 0.$$

Let $u = yx^{1/2}$, then $u' = y'x^{1/2} + \frac{1}{2}x^{-1/2}y$, and $u'' = y''x^{1/2} + x^{-1/2}y' - \frac{1}{4}x^{-3/2}y$. From u'' we have

$$y'' + \frac{1}{x}y' = x^{-1/2}u'' + \frac{1}{4x^2}y.$$

Adding y on both sides, we obtain

$$y'' + \frac{1}{x}y' + y = 0 = x^{-1/2}u'' + \frac{1}{4x^2}y + y,$$

$$x^{-1/2}u'' + \frac{1}{4x^2}y + y = 0$$

335

or

$$u'' + \left(\frac{1}{4x^2} + 1\right)x^{1/2}y = u'' + \left(\frac{1}{4x^2} + 1\right)u = 0,$$

the solution of which is

$$u = A\cos x + B\sin x.$$

Thus the approximate solution to Bessel's equation for very large values of x is

$$y = x^{-1/2}(A\cos x + B\sin x) = Cx^{-1/2}\cos(x + \beta).$$

A more rigorous argument leads to the following asymptotic formula

$$J_n(x) \approx \left(\frac{2}{\pi x}\right)^{1/2}\cos\left(x - \frac{\pi}{4} - \frac{n\pi}{2}\right). \tag{7.94}$$

For very small values of x (that is, near 0), by examining the solution itself and dropping all terms after the first, we find

$$J_n(x) \approx \frac{x^n}{2^n\Gamma(n+1)}. \tag{7.95}$$

Orthogonality of Bessel functions

Bessel functions enjoy a property which is called orthogonality and is of general importance in mathematical physics. If λ and μ are two different constants, we can show that under certain conditions

$$\int_0^1 xJ_n(\lambda x)J_n(\mu x)dx = 0.$$

Let us see what these conditions are. First, we can show that

$$\int_0^1 xJ_n(\lambda x)J_n(\mu x)dx = \frac{\mu J_n(\lambda)J_n'(\mu) - \lambda J_n(\mu)J_n'(\lambda)}{\lambda^2 - \mu^2}. \tag{7.96}$$

To show this, let us go back to Bessel's equation (7.71) and change the independent variable to λx, where λ is a constant, then the resulting equation is

$$x^2y'' + xy' + (\lambda^2x^2 - n^2)y = 0$$

and its general solution is $J_n(\lambda x)$. Now suppose we have two such equations, one for y_1 with constant λ, and one for y_2 with constant μ:

$$x^2y_1'' + xy_1' + (\lambda^2x^2 - n^2)y_1 = 0, \quad x^2y_2'' + xy_2' + (\mu^2x^2 - n^2)y_2 = 0.$$

Now multiplying the first equation by y_2, the second by y_1 and subtracting, we get

$$x^2[y_2y_1'' - y_1y_2''] + x[y_2y_1' - y_1y_2'] = (\mu^2 - \lambda^2)x^2y_1y_2.$$

336

Dividing by x we obtain

$$x \frac{d}{dx}[y_2 y_1' - y_1 y_2'] + [y_2 y_1' - y_1 y_2'] = (\mu^2 - \lambda^2) x y_1 y_2$$

or

$$\frac{d}{dx} \{ x[y_2 y_1' - y_1 y_2'] \} = (\mu^2 - \lambda^2) x y_1 y_2$$

and then integration gives

$$(\mu^2 - \lambda^2) \int x y_1 y_2 dx = x[y_2 y_1' - y_1 y_2'],$$

where we have omitted the constant of integration. Now $y_1 = J_n(\lambda x), y_2 = J_n(x)$, and if $\lambda \neq \mu$ we then have

$$\int x J_n(\lambda x) J_n(\mu x) dx = \frac{x[\lambda J_n(\mu x) J_n'(\lambda x) - \mu J_n(\lambda x) J_n'(\mu x)]}{\mu^2 - \lambda^2}.$$

Thus

$$\int_0^1 x J_n(\lambda x) J_n(\mu x) dx = \frac{\mu J_n(\lambda) J_n'(\mu) - \lambda J_n(\mu) J_n'(\lambda)}{\lambda^2 - \mu^2} \quad \text{q.e.d.}$$

Now letting $\mu \to \lambda$ and using L'Hospital's rule, we obtain

$$\int_0^1 x J_n^2(\lambda x) dx = \lim_{\mu \to \lambda} \frac{\lambda J_n'(\mu) J_n'(\lambda) - J_n(\lambda) J_n'(\mu) - \mu J_n(\lambda) J_n''(\mu)}{2\mu}$$

$$= \frac{\lambda J_n'^2(\lambda) - J_n(\lambda) J_n'(\lambda) - \lambda J_n(\lambda) J_n''(\lambda)}{2\lambda}.$$

But

$$\lambda^2 J_n''(\lambda) + \lambda J_n'(\lambda) + (\lambda^2 - n^2) J_n(\lambda) = 0.$$

Solving for $J_n''(\lambda)$ and substituting, we obtain

$$\int_0^1 x J_n^2(\lambda x) dx = \frac{1}{2} \left[J_n'^2(\lambda) + \left(1 - \frac{n^2}{\lambda^2} \right) J_n^2(x) \right]. \tag{7.97}$$

Furthermore, if λ and μ are any two different roots of the equation $R J_n(x) + S x J_n'(x) = 0$, where R and S are constant, we then have

$$R J_n(\lambda) + S \lambda J_n'(\lambda) = 0, \quad R J_n(\mu) + S \mu J_n'(\mu) = 0;$$

from these two equations we find, if $R \neq 0, S \neq 0$,

$$\mu J_n(\lambda) J_n'(\mu) - \lambda J_n(\mu) J_n'(\lambda) = 0$$

and then from Eq. (7.96) we obtain

$$\int_0^1 x J_n(\lambda x) J_n(\mu x) dx = 0. \tag{7.98}$$

Thus, the two functions $\sqrt{x} J_n(\lambda x)$ and $\sqrt{x} J_n(\mu x)$ are orthogonal in $(0, 1)$. We can also say that the two functions $J_n(\lambda x)$ and $J_n(\mu x)$ are orthogonal with respect to the weighted function x.

Eq. (7.98) is also easily proved if $R = 0$ and $S \neq 0$, or $R \neq 0$ but $S = 0$. In this case, λ and μ can be any two different roots of $J_n(x) = 0$ or $J_n'(x) = 0$.

Spherical Bessel functions

In physics we often meet the following equation

$$\frac{d}{dr} \left(r^2 \frac{dR}{dr} \right) + [k^2 r^2 - l(l+1)] R = 0, \qquad (l = 0, 1, 2, \ldots). \tag{7.99}$$

In fact, this is the radial equation of the wave and the Helmholtz partial differential equation in the spherical coordinate system (see Problem 7.22). If we let $x = kr$ and $y(x) = R(r)$, then Eq. (7.99) becomes

$$x^2 y'' + 2xy' + [x^2 - l(l+1)]y = 0 \qquad (l = 0, 1, 2, \ldots), \tag{7.100}$$

where $y' = dy/dx$. This equation almost matches Bessel's equation (7.71). Let us make the further substitution

$$y(x) = w(x)/\sqrt{x},$$

then we obtain

$$x^2 w'' + xw' + [x^2 - (l + \tfrac{1}{2})]w = 0 \qquad (l = 0, 1, 2, \ldots). \tag{7.101}$$

The reader should recognize this equation as Bessel's equation of order $l + \frac{1}{2}$. It follows that the solutions of Eq. (7.100) can be written in the form

$$y(x) = A \frac{J_{l+1/2}(x)}{\sqrt{x}} + B \frac{J_{-l-1/2}(x)}{\sqrt{x}}.$$

This leads us to define spherical Bessel functions $j_l(x) = C J_{l+E}(x)/\sqrt{x}$. The factor C is usually chosen to be $\sqrt{\pi/2}$ for a reason to be explained later:

$$j_l(x) = \sqrt{\pi/2x}\, J_{l+E}(x). \tag{7.102}$$

Similarly, we can define

$$n_l(x) = \sqrt{\pi/2x}\, N_{l+E}(x).$$

We can express $j_l(x)$ in terms of $j_0(x)$. To do this, let us go back to $J_n(x)$ and we find that

$$\frac{d}{dx}\{x^{-n}J_n(x)\} = -x^{-n}J_{n+1}(x), \quad \text{or} \quad J_{n+1}(x) = -x^n\frac{d}{dx}\{x^{-n}J_n(x)\}.$$

The proof is simple and straightforward:

$$\frac{d}{dx}\{x^{-n}J_n(x)\} = \frac{d}{dx}\sum_{k=0}^{\infty}\frac{(-1)^k x^{2k}}{2^{n+2k}k!\Gamma(n+k+1)}$$

$$= x^{-n}\sum_{k=0}^{\infty}\frac{(-1)^k x^{n+2k-1}}{2^{n+2k-1}(k-1)!\Gamma(n+k+1)}$$

$$= x^{-n}\sum_{k=0}^{\infty}\frac{(-1)^{k+1} x^{n+2k+1}}{2^{n+2k+1}k!\Gamma[(n+k+2]} = -x^{-n}J_{n+1}(x).$$

Now if we set $n = l + \frac{1}{2}$ and divide by $x^{l+3/2}$, we obtain

$$\frac{J_{l+3/2}(x)}{x^{l+3/2}} = -\frac{1}{x}\frac{d}{dx}\left[\frac{J_{l+1/2}(x)}{x^{l+1/2}}\right] \quad \text{or} \quad \frac{j_{l+1}(x)}{x^{l+1}} = -\frac{1}{x}\frac{d}{dx}\left[\frac{j_l(x)}{x^l}\right].$$

Starting with $l = 0$ and applying this formula l times, we obtain

$$j_l(x) = x^l\left(-\frac{1}{x}\frac{d}{dx}\right)^l j_0(x) \quad (l = 1, 2, 3, \ldots). \tag{7.103}$$

Once $j_0(x)$ has been chosen, all $j_l(x)$ are uniquely determined by Eq. (7.103).

Now let us go back to Eq. (7.102) and see why we chose the constant factor C to be $\sqrt{\pi/2}$. If we set $l = 0$ in Eq. (7.101), the resulting equation is

$$xy'' + 2y' + xy = 0.$$

Solving this equation by the power series method, the reader will find that functions $\sin(x)/x$ and $\cos(x)/x$ are among the solutions. It is customary to define

$$j_0(x) = \sin(x)/x.$$

Now by using Eq. (7.76), we find

$$J_{1/2}(x) = \sum_{k=0}^{\infty}\frac{(-1)^k(x/2)^{1/2+2k}}{k!\Gamma(k+3/2)}$$

$$= \frac{(x/2)^{1/2}}{(1/2)\sqrt{\pi}}\left(1 - \frac{x^2}{3!} + \frac{x^4}{5!} - \cdots\right) = \frac{(x/2)^{1/2}}{(1/2)\sqrt{\pi}}\frac{\sin x}{x} = \sqrt{\frac{2}{\pi x}}\sin x.$$

Comparing this with $j_0(x)$ shows that $j_0(x) = \sqrt{\pi/2x}J_{1/2}(x)$, and this explains the factor $\sqrt{\pi/2}$ chosen earlier.

339

Sturm–Liouville systems

A boundary-value problem having the form

$$\frac{d}{dx}\left[r(x)\frac{dy}{dx}\right] + [q(x) + \lambda p(x)]y = 0, \qquad a \le x \le b \qquad (7.104)$$

and satisfying boundary conditions of the form

$$k_1 y(a) + k_2 y'(a) = 0, \qquad l_1 y(b) + l_2 y'(b) = 0 \qquad (7.104a)$$

is called a Sturm–Liouville boundary-value problem; Eq. (7.104) is known as the Sturm–Liouville equation. Legendre's equation, Bessel's equation and many other important equations can be written in the form of (7.104).

Legendre's equation (7.1) can be written as

$$[(1 - x^2)y']' + \lambda y = 0, \qquad \lambda = \nu(\nu + 1);$$

we can then see it is a Sturm–Liouville equation with $r = 1 - x^2, q = 0$ and $p = 1$.

Then, how do Bessel functions fit into the Sturm–Liouville framework? $J(s)$ satisfies the Bessel equation (7.71)

$$s^2 \ddot{J}_n + s\dot{J}_n + (s^2 - n^2)J_n = 0, \qquad \dot{J}_n = dJ_n/ds. \qquad (7.71a)$$

We assume n is a positive integer and setting $s = \lambda x$, with λ a non-zero constant, we have

$$\frac{ds}{dx} = \lambda, \quad \dot{J}_n = \frac{dJ_n}{dx}\frac{dx}{ds} = \frac{1}{\lambda}\frac{dJ_n}{dx}, \quad \ddot{J}_n = \frac{d}{dx}\left(\frac{1}{\lambda}\frac{dJ_n}{dx}\right)\frac{dx}{ds} = \frac{1}{\lambda^2}\frac{d^2 J_n}{dx^2}$$

and Eq. (7.71a) becomes

$$x^2 J_n''(\lambda x) + x J_n'(\lambda x) + (\lambda^2 x^2 - n^2)J_n(\lambda x) = 0, \quad J_n' = dJ_n/dx$$

or

$$x J_n''(\lambda x) + J_n'(\lambda x) + (\lambda^2 x - n^2/x)J_n(\lambda x) = 0,$$

which can be written as

$$[x J_n'(\lambda x)]' + \left(-\frac{n^2}{x} + \lambda^2 x\right)J_n(\lambda x) = 0.$$

It is easy to see that for each fixed n this is a Sturm–Liouville equation (7.104), with $r(x) = x, q(x) = -n^2/x, p(x) = x$, and with the parameter λ now written as λ^2.

For the Sturm–Liouville system (7.104) and (7.104a), a non-trivial solution exists in general only for a particular set of values of the parameter λ. These values are called the eigenvalues of the system. If $r(x)$ and $q(x)$ are real, the eigenvalues are real. The corresponding solutions are called eigenfunctions of the system. In general there is one eigenfunction to each eigenvalue. This is the non-degenerate case. In the degenerate case, more than one eigenfunction may correspond to the same eigenvalue. The eigenfunctions form an orthogonal set with respect to the density function $p(x)$ which is generally ≥ 0. Thus by suitable normalization the set of functions can be made an orthonormal set with respect to $p(x)$ in $a \leq x \leq b$. We now proceed to prove these two general claims.

Property 1
If $r(x)$ and $q(x)$ are real, the eigenvalues of a Sturm–Liouville system are real.

We start with the Sturm–Liouville equation (7.104) and the boundary conditions (7.104a):

$$\frac{d}{dx}\left[r(x)\frac{dy}{dx}\right] + [q(x) + \lambda p(x)]y = 0, \qquad a \leq x \leq b,$$

$$k_1 y(a) + k_2 y'(a) = 0, \quad l_1 y(b) + l_2 y'(b) = 0,$$

and assume that $r(x), q(x), p(x), k_1, k_2, l_1$, and l_2 are all real, but λ and y may be complex. Now take the complex conjugates

$$\frac{d}{dx}\left[r(x)\frac{d\bar{y}}{dx}\right] + [q(x) + \bar{\lambda}p(x)]\bar{y} = 0, \tag{7.105}$$

$$k_1 \bar{y}(a) + k_2 \bar{y}'(a) = 0, l_1 \bar{y}(b) + l_2 \bar{y}'(b) = 0, \tag{7.105a}$$

where $\bar{y}$ and $\bar{\lambda}$ are the complex conjugates of y and λ, respectively.

Multiplying (7.104) by $\bar{y}$, (7.105) by y, and subtracting, we obtain after simplifying

$$\frac{d}{dx}[r(x)(y\bar{y}' - \bar{y}y')] = (\lambda - \bar{\lambda})p(x)y\bar{y}.$$

Integrating from a to b, and using the boundary conditions (7.104a) and (7.105a), we then obtain

$$(\lambda - \bar{\lambda})\int_a^b p(x)|y'|^2 dx = r(x)(y\bar{y}' - \bar{y}y')|_a^b = 0.$$

Since $p(x) \geq 0$ in $a \leq x \leq b$, the integral on the left is positive and therefore $\lambda = \bar{\lambda}$, that is, λ is real.

Property 2
The eigenfunctions corresponding to two different eigenvalues are orthogonal with respect to $p(x)$ in $a \leq x \leq b$.

If y_1 and y_2 are eigenfunctions corresponding to the two different eigenvalues λ_1, λ_2, respectively,

$$\frac{d}{dx}\left[r(x)\frac{dy_1}{dx}\right] + [q(x) + \lambda_1 p(x)]y_1 = 0, \qquad a \le x \le b, \tag{7.106}$$

$$k_1 y_1(a) + k_2 y_1'(a) = 0, \quad l_1 y_1(b) + l_2 y_1'(b) = 0; \tag{7.106a}$$

$$\frac{d}{dx}\left[r(x)\frac{dy_2}{dx}\right] + [q(x) + \lambda_2 p(x)]y_2 = 0, \qquad a \le x \le b, \tag{7.107}$$

$$k_1 y_2(a) + k_2 y_2'(a) = 0, \quad l_1 y_2(b) + l_2 y_2'(b) = 0. \tag{7.107a}$$

Multiplying (7.106) by y_2 and (7.107) by y_1, then subtracting, we obtain

$$\frac{d}{dx}[r(x)(y_1 y_2' - y_2 y_1')] = (\lambda - \bar{\lambda})p(x)y_1 y_2.$$

Integrating from a to b, and using (7.106a) and (7.107a), we obtain

$$(\lambda_1 - \lambda_2)\int_a^b p(x)y_1 y_2 dx = r(x)(y_1 y_2' - y_2 y_1')|_a^b = 0.$$

Since $\lambda_1 \ne \lambda_2$ we have the required result; that is,

$$\int_a^b p(x)y_1 y_2 dx = 0.$$

We can normalize these eigenfunctions to make them an orthonormal set, and so we can expand a given function in a series of these orthonormal eigenfunctions.

We have shown that Legendre's equation is a Sturm–Liouville equation with $r(x) = 1 - x, q = 0$ and $p = 1$. Since $r = 0$ when $x = \pm 1$, no boundary conditions are needed to form a Sturm–Liouville problem on the interval $-1 \le x \le 1$. The numbers $\lambda_n = n(n+1)$ are eigenvalues with $n = 0, 1, 2, 3, \ldots$. The corresponding eigenfunctions are $y_n = P_n(x)$. Property 2 tells us that

$$\int_{-1}^1 P_n(x)P_m(x)dx = 0 \qquad n \ne m.$$

For Bessel functions we saw that

$$[xJ_n'(\lambda x)]' + \left(-\frac{n^2}{x} + \lambda^2 x\right)J_n(\lambda x) = 0$$

is a Sturm–Liouville equation (7.104), with $r(x) = x, q(x) = -n^2/x, p(x) = x$, and with the parameter λ now written as λ^2. Typically, we want to solve this equation on an interval $0 \leq x \leq b$ subject to

$$J_n(\lambda b) = 0.$$

which limits the selection of λ. Property 2 then tells us that

$$\int_0^b x J_n(\lambda_k x) J_n(\lambda_l x) dx = 0, \qquad k \neq l.$$

Problems

7.1 Using Eq. (7.11), show that $P_n(-x) = (-1)^n P_n(x)$ and $P_n{}'(-x) = (-1)^{n+1} P_n'(x)$.

7.2 Find $P_0(x), P_1(x), P_2(x), P_3(x)$, and $P_4(x)$ from Rodrigues' formula (7.12). Compare your results with Eq. (7.11).

7.3 Establish the recurrence formula (7.16b) by manipulating Rodrigues' formula.

7.4 Prove that $P_5'(x) = 9P_4(x) + 5P_2(x) + P_0(x)$.
Hint: Use the recurrence relation (7.16d).

7.5 Let P and Q be two points in space (Fig. 7.6). Using Eq. (7.14), show that

$$\frac{1}{r} = \frac{1}{\sqrt{r_1^2 + r_2^2 - 2r_1 r_2 \cos\theta}}$$

$$= \frac{1}{r_2}\left[P_0 + P_1(\cos\theta)\frac{r_1}{r_2} + P_2(\cos\theta)\left(\frac{r_1}{r_2}\right)^2 + \cdots \right].$$

7.6 What is $P_n(1)$? What is $P_n(-1)$?

7.7 Obtain the associated Legendre functions: (a) $P_2^1(x)$, (b) $P_3^2(x)$, (c) $P_2^3(x)$.

7.8 Verify that $P_3^2(x)$ is a solution of Legendre's associated equation (7.25) for $m = 2, n = 3$.

7.9 Verify the orthogonality conditions (7.31) for the functions $P_2^1(x)$ and $P_3^1(x)$.

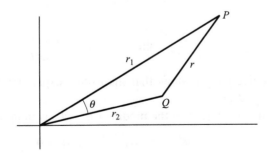

Figure 7.6.

7.10 Verify Eq. (7.37) for the function $P_2^1(x)$.

7.11 Show that

$$\frac{d^{n-m}}{dx^{n-m}}(x^2-1)^n = \frac{(n-m)!}{(n+m)!}(x^2-1)^m \frac{d^{n+m}}{dx^{n+m}}(x^2-1)^m$$

Hint: Write $(x^2-1)^n = (x-1)^n(x+1)^n$ and find the derivatives by Leibnitz's rule.

7.12 Use the generating function for the Hermite polynomials to find:
(a) $H_0(x)$; (b) $H_1(x)$; (c) $H_2(x)$; (d) $H_3(x)$.

7.13 Verify that the generating function Φ satisfies the identity

$$\frac{\partial^2 \Phi}{\partial x^2} - 2x \frac{\partial \Phi}{\partial x} + 2t \frac{\partial \Phi}{\partial t} = 0.$$

Show that the functions $H_n(x)$ in Eq. (7.47) satisfy Eq. (7.38).

7.14 Given the differential equation $y'' + (\varepsilon - x^2)y = 0$, find the possible values of ε (eigenvalues) such that the solution $y(x)$ of the given differential equation tends to zero as $x \to \pm\infty$. For these values of ε, find the eigenfunctions $y(x)$.

7.15 In Eq. (7.58), write the series for the exponential and collect powers of z to verify the first few terms of the series. Verify the identity

$$x\frac{\partial^2 \Phi}{\partial x^2} + (1-x)\frac{\partial \Phi}{\partial x} + z\frac{\partial \Phi}{\partial z} = 0.$$

Substituting the series (7.58) into this identity, show that the functions $L_n(x)$ in Eq. (7.58) satisfy Laguerre's equation.

7.16 Show that

$$J_0(x) = 1 - \frac{x^2}{2^2(1!)^2} + \frac{x^4}{2^4(2!)^2} - \frac{x^6}{2^6(3!)^2} + - \cdots,$$

$$J_1(x) = \frac{x}{2} - \frac{x^3}{2^3 1!2!} + \frac{x^5}{2^5 2!3!} - \frac{x^7}{2^7 3!4!} + - \cdots.$$

7.17 Show that

$$J_{1/2}(x) = \left(\frac{2}{\pi x}\right)^{1/2} \sin x, \quad J_{-1/2}(x) = \left(\frac{2}{\pi x}\right)^{1/2} \cos x.$$

7.18 If n is a positive integer, show that the formal expression for $J_{-n}(x)$ gives $J_{-n}(x) = (-1)^n J_n(x)$.

7.19 Find the general solution to the modified Bessel's equation

$$x^2 y'' + xy' + (x^2 s^2 - \alpha^2)y = 0$$

which differs from Bessel's equation only in that sx takes the place of x.

(Hint: Reduce the given equation to Bessel's equation first.)

7.20 The lengthening simple pendulum: Consider a small mass m suspended by a string of length l. If its length is increased at a steady rate r as it swings back and forth freely in a vertical plane, find the equation of motion and the solution for small oscillations.

7.21 Evaluate the integrals:

$$(a) \int x^n J_{n-1}(x)\,dx; \qquad (b) \int x^{-n} J_{n+1}(x)\,dx; \qquad (c) \int x^{-1} J_1(x)\,dx.$$

7.22 In quantum mechanics, the three-dimensional Schrödinger equation is

$$i\hbar \frac{\partial \psi(\mathbf{r}, t)}{\partial t} = -\frac{\hbar^2}{2m} \nabla^2 \psi(\mathbf{r}, t) + V\psi(\mathbf{r}, t), \qquad i = \sqrt{-1}, \hbar = h/2\pi.$$

(a) When the potential V is independent of time, we can write $\psi(\mathbf{r}, t) = u(\mathbf{r})T(t)$. Show that in this case the Schrödinger equation reduces to

$$-\frac{\hbar^2}{2m} \nabla^2 u(\mathbf{r}) + Vu(\mathbf{r}) = Eu(\mathbf{r}),$$

a time-independent equation along with $T(t) = e^{-iEt/\hbar}$, where E is a separation constant.

(b) Show that, in spherical coordinates, the time-independent Schrödinger equation takes the form

$$-\frac{\hbar^2}{2m}\left[\frac{1}{r^2}\frac{\partial}{\partial r}\left(r^2 \frac{\partial u}{\partial r}\right) + \frac{1}{r^2 \sin\theta}\frac{\partial}{\partial \theta}\left(\sin\theta \frac{\partial u}{\partial \theta}\right) + \frac{1}{r^2 \sin^2\theta}\frac{\partial^2 u}{\partial \phi^2}\right] + V(r)u = Eu,$$

then use separation of variables, $u(r, \theta, \phi) = R(r)Y(\theta, \phi)$, to split it into two equations, with α as a new separation constant:

$$-\frac{\hbar^2}{2m r^2}\frac{d}{dr}\left(r^2 \frac{dR}{dr}\right) + \left(V + \frac{\alpha}{r^2}\right)R = ER,$$

$$-\frac{\hbar^2}{2m \sin\theta}\frac{\partial}{\partial \theta}\left(\sin\theta \frac{\partial Y}{\partial \theta}\right) - \frac{\hbar^2}{2m \sin^2\theta}\frac{\partial^2 Y}{\partial \phi^2} = \alpha Y.$$

It is straightforward to see that the radial equation is in the form of Eq. (7.99). Continuing the separation process by putting $Y(\theta, \phi) = \Theta(\theta)\Phi(\theta)$, the angular equation can be separated further into two equations, with β as separation constant:

$$-\frac{\hbar^2}{2m}\frac{1}{\Phi}\frac{d^2\Phi}{d\phi^2} = \beta,$$

$$-\frac{\hbar^2}{2m}\sin\theta\frac{d}{d\theta}\left(\sin\theta \frac{d\Theta}{d\theta}\right) - \alpha \sin^2\theta\Theta + \beta\Theta = 0.$$

The first equation is ready for integration. Do you recognize the second equation in θ as Legendre's equation? (Compare it with Eq. (7.30).) If you are unsure, try to simplify it by putting $\gamma = 2m\alpha/\hbar$, $\mu = (2m\beta/\hbar)^{1/2}$, and you will obtain

$$\sin\theta\frac{d}{d\theta}\left(\sin\theta\frac{d\Theta}{d\theta}\right) + (\gamma\sin^2\theta - \mu^2)\Theta = 0$$

or

$$\frac{1}{\sin\theta}\frac{d}{d\theta}\left(\sin\theta\frac{d\Theta}{d\theta}\right) + \left(\gamma - \frac{\mu^2}{\sin^2\theta}\right)\Theta = 0,$$

which more closely resembles Eq. (7.30).

7.23 Consider the differential equation

$$y'' + R(x)y' + [Q(x) + \lambda P(x)]y = 0.$$

Show that it can be put into the form of the Sturm–Liouville equation (7.104) with

$$r(x) = e^{\int R(x)dx}, \quad q(x) = Q(x)e^{\int R(x)dx}, \quad \text{and} \quad p(x) = P(x)e^{\int R(x)dx}.$$

7.24. (a) Show that the system $y'' + \lambda y = 0, y(0) = 0, y(1) = 0$ is a Sturm–Liouville system.

(b) Find the eigenvalues and eigenfunctions of the system.

(c) Prove that the eigenfunctions are orthogonal on the interval $0 \le x \le 1$.

(d) Find the corresponding set of normalized eigenfunctions, and expand the function $f(x) = 1$ in a series of these orthonormal functions.

8

The calculus of variations

The calculus of variations, in its present form, provides a powerful method for the treatment of variational principles in physics and has become increasingly important in the development of modern physics. It is originated as a study of certain extremum (maximum and minimum) problems not treatable by elementary calculus. To see this more precisely let us consider the following integral whose integrand is a function of x, y, and of the first derivative $y'(x) = dy/dx$:

$$I = \int_{x_1}^{x_2} f\{y(x), y'(x); x\} dx, \tag{8.1}$$

where the semicolon in f separates the independent variable x from the dependent variable $y(x)$ and its derivative $y'(x)$. For what function $y(x)$ is the value of the integral I a maximum or a minimum? This is the basic problem of the calculus of variations.

The quantity f depends on the functional form of the dependent variable $y(x)$ and is called the functional which is considered as given, the limits of integration are also given. It is also understood that $y = y_1$ at $x = x_1$, $y = y_2$ at $x = x_2$. In contrast with the simple extreme-value problem of differential calculus, the function $y(x)$ is not known here, but is to be varied until an extreme value of the integral I is found. By this we mean that if $y(x)$ is a curve which gives to I a minimum value, then any neighboring curve will make I increase.

We can make the definition of a neighboring curve clear by giving $y(x)$ a parametric representation:

$$y(\varepsilon, x) = y(0, x) + \varepsilon \eta(x), \tag{8.2}$$

where $\eta(x)$ is an arbitrary function which has a continuous first derivative and ε is a small arbitrary parameter. In order for the curve (8.2) to pass through (x_1, y_1) and (x_2, y_2), we require that $\eta(x_1) = \eta(x_2) = 0$ (see Fig. 8.1). Now the integral I

347

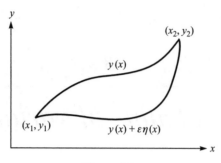

Figure 8.1.

also becomes a function of the parameter ε

$$I(\varepsilon) = \int_{x_1}^{x_2} f\{y(\varepsilon, x), y'(\varepsilon, x); x\}dx. \tag{8.3}$$

We then require that $y(x) = y(0, x)$ makes the integral I an extreme, that is, the integral $I(\varepsilon)$ has an extreme value for $\varepsilon = 0$:

$$I(\varepsilon) = \int_{x_1}^{x_2} f\{y(\varepsilon, x), y'(\varepsilon, x); x\}dx = \text{extremum for } \varepsilon = 0.$$

This gives us a very simple method of determining the extreme value of the integral I. The necessary condition is

$$\left.\frac{dI}{d\varepsilon}\right|_{\varepsilon=0} = 0 \tag{8.4}$$

for all functions $\eta(x)$. The sufficient conditions are quite involved and we shall not pursue them. The interested reader is referred to mathematical texts on the calculus of variations.

The problem of the extreme-value of an integral occurs very often in geometry and physics. The simplest example is provided by the problem of determining the shortest curve (or distance) between two given points. In a plane, this is the straight line. But if the two given points lie on a given arbitrary surface, then the analytic equation of this curve, which is called a geodesic, is found by solution of the above extreme-value problem.

The Euler–Lagrange equation

In order to find the required curve $y(x)$ we carry out the indicated differentiation in the extremum condition (8.4):

$$\frac{\partial I}{\partial \varepsilon} = \frac{\partial}{\partial \varepsilon} \int_{x_1}^{x_2} f\{y(\varepsilon, x), y'(\varepsilon, x); x\} dx$$

$$= \int_{x_1}^{x_2} \left(\frac{\partial f}{\partial y} \frac{\partial y}{\partial \varepsilon} + \frac{\partial f}{\partial y'} \frac{\partial y'}{\partial \varepsilon} \right) dx, \tag{8.5}$$

where we have employed the fact that the limits of integration are fixed, so the differential operation affects only the integrand. From Eq. (8.2) we have

$$\frac{\partial y}{\partial \varepsilon} = \eta(x) \quad \text{and} \quad \frac{\partial y'}{\partial \varepsilon} = \frac{d\eta}{dx}.$$

Substituting these into Eq. (8.5) we obtain

$$\frac{\partial I}{\partial \varepsilon} = \int_{x_1}^{x_2} \left(\frac{\partial f}{\partial y} \eta(x) + \frac{\partial f}{\partial y'} \frac{d\eta}{dx} \right) dx. \tag{8.6}$$

Using integration by parts, the second term on the right hand side becomes

$$\int_{x_1}^{x_2} \frac{\partial f}{\partial y'} \frac{d\eta}{dx} dx = \frac{\partial f}{\partial y'} \eta(x) \Big|_{x_1}^{x_2} - \int_{x_1}^{x_2} \frac{d}{dx} \left(\frac{\partial f}{\partial y'} \right) \eta(x) dx.$$

The integrated term on the right hand side vanishes because $\eta(x_1) = \eta(x_2) = 0$ and Eq. (8.6) becomes

$$\frac{\partial I}{\partial \varepsilon} = \int_{x_1}^{x_2} \left(\frac{\partial f}{\partial y} \frac{\partial y}{\partial \varepsilon} - \frac{d}{dx} \left(\frac{\partial f}{\partial y'} \right) \frac{\partial y}{\partial \varepsilon} \right) dx$$

$$= \int_{x_1}^{x_2} \left(\frac{\partial f}{\partial y} - \frac{d}{dx} \left(\frac{\partial f}{\partial y'} \right) \right) \eta(x) dx. \tag{8.7}$$

Note that $\partial f/\partial y$ and $\partial f/\partial y'$ are still functions of ε. However, when $\varepsilon = 0, y(\varepsilon, x) = y(x)$ and the dependence on ε disappears.

Then $(\partial I/\partial \varepsilon)|_{\varepsilon=0}$ vanishes, and since $\eta(x)$ is an arbitrary function, the integrand in Eq. (8.7) must vanish for $\varepsilon = 0$:

$$\frac{d}{dx} \frac{\partial f}{\partial y'} - \frac{\partial f}{\partial y} = 0. \tag{8.8}$$

Eq. (8.8) is known as the Euler–Lagrange equation; it is a necessary but not sufficient condition that the integral I have an extreme value. Thus, the solution of the Euler–Lagrange equation may not yield the minimizing curve. Ordinarily we must verify whether or not this solution yields the curve that actually minimizes the integral, but frequently physical or geometrical considerations enable us to tell whether the curve so obtained makes the integral a minimum or a maximum. The Euler–Lagrange equation can be written in the form (Problem 8.2)

$$\frac{d}{dx} \left(f - y' \frac{\partial f}{\partial y'} \right) - \frac{\partial f}{\partial x} = 0. \tag{8.8a}$$

This is often called the second form of the Euler–Lagrange equation. If f does not involve x explicitly, it can be integrated to yield

$$f - y' \frac{\partial f}{\partial y'} = c, \tag{8.8b}$$

where c is an integration constant.

The Euler–Lagrange equation can be extended to the case in which f is a functional of several dependent variables:

$$f = f\{y_1(x), y_1'(x), y_2(x), y_2'(x), \ldots; x\}.$$

Then, in analogy with Eq. (8.2), we now have

$$y_i(\varepsilon, x) = y_i(0, x) + \varepsilon \eta_i(x), \qquad i = 1, 2, \ldots, n.$$

The development proceeds in an exactly analogous manner, with the result

$$\frac{\partial I}{\partial \varepsilon} = \int_{x_1}^{x_2} \left(\frac{\partial f}{\partial y_i} - \frac{d}{dx} \left(\frac{\partial f}{\partial y_i'} \right) \right) \eta_i(x) dx.$$

Since the individual variations, that is, the $\eta_i(x)$, are all independent, the vanishing of the above equation when evaluated at $\varepsilon = 0$ requires the separate vanishing of each expression in the brackets:

$$\frac{d}{dx} \frac{\partial f}{\partial y_i'} - \frac{\partial f}{\partial y_i} = 0, \qquad i = 1, 2, \ldots, n. \tag{8.9}$$

Example 8.1

The brachistochrone problem: Historically, the brachistochrone problem was the first to be treated by the method of the calculus of variations (first solved by Johann Bernoulli in 1696). As shown in Fig. 8.2, a particle is constrained to move in a gravitational field starting at rest from some point P_1 to some lower

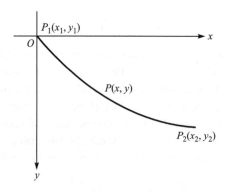

Figure 8.2

point P_2. Find the shape of the path such that the particle goes from P_1 to P_2 in the least time. (The word brachistochrone was derived from the Greek *brachistos* (shortest) and *chronos* (time).)

Solution: If O and P are not very far apart, the gravitational field is constant, and if we ignore the possibility of friction, then the total energy of the particle is conserved:

$$0 + mgy_1 = \frac{1}{2}m\left(\frac{ds}{dt}\right)^2 + mg(y_1 - y),$$

where the left hand side is the sum of the kinetic energy and the potential energy of the particle at point P_1, and the right hand side refers to point $P(x, y)$. Solving for ds/dt:

$$ds/dt = \sqrt{2gy}.$$

Thus the time required for the particle to move from P_1 to P_2 is

$$t = \int_{P_1}^{P_2} dt = \int_{P_1}^{P_2} \frac{ds}{\sqrt{2gy}}.$$

The line element ds can be expressed as

$$ds = \sqrt{dx^2 + dy^2} = \sqrt{1 + y'^2}dx, \qquad y' = dy/dx;$$

thus, we have

$$t = \int_{P_1}^{P_2} dt = \int_{P_1}^{P_2} \frac{ds}{\sqrt{2gy}} = \frac{1}{\sqrt{2g}} \int_0^{x_2} \frac{\sqrt{1 + y'^2}}{\sqrt{y}} dx.$$

We now apply the Euler–Lagrange equation to find the shape of the path for the particle to go from P_1 to P_2 in the least time. The constant does not affect the final equation and the functional f may be identified as

$$f = \sqrt{1 + y'^2}/\sqrt{y},$$

which does not involve x explicitly. Using Problem 8.2(b), we find

$$f - y'\frac{\partial f}{\partial y'} = \frac{\sqrt{1 + y'^2}}{\sqrt{y}} - y'\left[\frac{y'}{\sqrt{1 + y'^2}\sqrt{y}}\right] = c,$$

which simplifies to

$$\sqrt{1 + y'^2}\sqrt{y} = 1/c.$$

Letting $1/c = \sqrt{a}$ and solving for y' gives

$$y' = \frac{dy}{dx} = \sqrt{\frac{a-y}{y}},$$

and solving for dx and integrating we obtain

$$\int dx = \int \sqrt{\frac{y}{a-y}}\, dy.$$

We then let

$$y = a\sin^2\theta = \frac{a}{2}(1 - \cos 2\theta)$$

which leads to

$$x = 2a\int \sin^2\theta\, d\theta = a\int (1 - \cos 2\theta)d\theta = \frac{a}{2}(2\theta - \sin 2\theta) + k.$$

Thus the parametric equation of the path is given by

$$x = b(1 - \cos\phi), \qquad y = b(\phi - \sin\phi) + k,$$

where $b = a/2$, $\phi = 2\theta$. The path passes through the origin so we have $k = 0$ and

$$x = b(1 - \cos\phi), \qquad y = b(\phi - \sin\phi).$$

The constant b is determined from the condition that the particle passes through $P_2(x_2, y_2)$.

The required path is a cycloid and is the path of a fixed point P' on a circle of radius b as it rolls along the x-axis (Fig. 8.3).

A line that represents the shortest path between any two points on some surface is called a geodesic. On a flat surface, the geodesic is a straight line. It is easy to show that, on a sphere, the geodesic is a great circle; we leave this as an exercise for the reader (Problem 8.3).

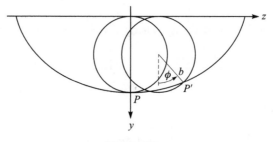

Figure 8.3.

Variational problems with constraints

In certain problems we seek a minimum or maximum value of the integral (8.1)

$$I = \int_{x_1}^{x_2} f\{y(x), y'(x); x\} dx \tag{8.1}$$

subject to the condition that another integral

$$J = \int_{x_1}^{x_2} g\{y(x), y'(x); x\} dx \tag{8.10}$$

has a known constant value. A simple problem of this sort is the problem of determining the curve of a given perimeter which encloses the largest area, or finding the shape of a chain of fixed length which minimizes the potential energy.

In this case we can use the method of Lagrange multipliers which is based on the following theorem:

> *The problem of the stationary value of $F(x, y)$ subject to the condition $G(x, y) = const.$ is equivalent to the problem of stationary values, without constraint, of $F + \lambda G$ for some constant λ, provided either $\partial G/\partial x$ or $\partial G/\partial y$ does not vanish at the critical point.*

The constant λ is called a Lagrange multiplier and the method is known as the method of Lagrange multipliers. To see the ideas behind this theorem, let us assume that $G(x, y) = 0$ defines y as a unique function of x, say, $y = g(x)$, having a continuous derivative $g'(x)$. Then

$$F(x, y) = F[x, g(x)]$$

and its maximum or minimum can be found by setting the derivative with respect to x equal to zero:

$$\frac{\partial F}{\partial x} + \frac{\partial F}{\partial y}\frac{dy}{dx} = 0 \qquad \text{or} \qquad F_x + F_y g'(x) = 0. \tag{8.11}$$

We also have

$$G[x, g(x)] = 0,$$

from which we find

$$\frac{\partial G}{\partial x} + \frac{\partial G}{\partial y}\frac{dy}{dx} = 0 \qquad \text{or} \qquad G_x + G_y g'(x) = 0. \tag{8.12}$$

Eliminating $g'(x)$ between Eq. (8.11) and Eq. (8.12) we obtain

$$F_x - (F_y/G_y)G_x = 0, \tag{8.13}$$

353

provided $G_y = \partial G/\partial y \neq 0$. Defining $\lambda = -F_y/G_y$ or

$$F_y + \lambda G_y = \frac{\partial F}{\partial y} + \lambda \frac{\partial G}{\partial y} = 0, \qquad (8.14)$$

Eq. (8.13) becomes

$$F_x + \lambda G_x = \frac{\partial F}{\partial x} + \lambda \frac{\partial G}{\partial x} = 0. \qquad (8.15)$$

If we define

$$H(x, y) = F(x, y) + \lambda G(x, y),$$

then Eqs. (8.14) and (8.15) become

$$\partial H(x, y)/\partial x = 0, \qquad H(x, y)/\partial y = 0,$$

and this is the basic idea behind the method of Lagrange multipliers.

It is natural to attempt to solve the problem I = minimum subject to the condition J = constant by the method of Lagrange multipliers. We construct the integral

$$I + \lambda J = \int_{x_1}^{x_2} [F(y, y'; x) + \lambda G(y, y'; x)] dx$$

and consider its free extremum. This implies that the function $y(x)$ that makes the value of the integral an extremum must satisfy the equation

$$\frac{d}{dx} \frac{\partial(F + \lambda G)}{\partial y'} \frac{\partial(F + \lambda G)}{\partial y} = 0 \qquad (8.16)$$

or

$$\left[\frac{d}{dx}\left(\frac{\partial F}{\partial y'}\right) - \frac{\partial F}{\partial y}\right] + \lambda \left[\frac{d}{dx}\left(\frac{\partial G}{\partial y'}\right) - \frac{\partial G}{\partial y}\right] = 0. \qquad (8.16a)$$

Example 8.2
Isoperimetric problem: Find that curve C having the given perimeter l that encloses the largest area.

Solution: The area bounded by C can be expressed as

$$A = \frac{1}{2} \int_C (x\,dy - y\,dx) = \frac{1}{2} \int_C (xy' - y) dx$$

and the length of the curve C is

$$s = \int_C \sqrt{1 + y'^2}\,dx = l.$$

Then the function H is

$$H = \int_C [\tfrac{1}{2}(xy' - y) + \lambda\sqrt{1 + y'^2}]dx$$

and the Euler–Lagrange equation gives

$$\frac{d}{dx}\left(\frac{1}{2}x + \frac{\lambda y'}{\sqrt{1 + y'^2}}\right) + \frac{1}{2} = 0$$

or

$$\frac{\lambda y'}{\sqrt{1 + y'^2}} = -x + c_1.$$

Solving for y', we get

$$y' = \frac{dy}{dx} = \pm\frac{x - c_1}{\sqrt{\lambda^2 - (x - c_1)^2}},$$

which on integrating gives

$$y - c_2 = \pm\sqrt{\lambda^2 - (x - c_1)^2}$$

or

$$(x - c_1)^2 + (y - c_2)^2 = \lambda^2, \quad \text{a circle.}$$

Hamilton's principle and Lagrange's equation of motion

One of the most important applications of the calculus of variations is in classical mechanics. In this case, the functional f in Eq. (8.1) is taken to be the Lagrangian L of a dynamical system. For a conservative system, the Lagrangian L is defined as the difference of kinetic and potential energies of the system:

$$L = T - V,$$

where time t is the independent variable and the generalized coordinates $q_i(t)$ are the dependent variables. What do we mean by generalized coordinates? Any convenient set of parameters or quantities that can be used to specify the configuration (or state) of the system can be assumed to be generalized coordinates; therefore they need not be geometrical quantities, such as distances or angles. In suitable circumstances, for example, they could be electric currents.

Eq. (8.1) now takes the form that is known as the action (or the action integral)

$$I = \int_{t_1}^{t_2} L(q_i(t), \dot{q}_i(t); t)dt, \qquad \dot{q} = dq/dt \qquad (8.17)$$

and Eq. (8.4) becomes

$$\delta I = \frac{\partial I}{\partial \varepsilon}\bigg|_{\varepsilon=0} d\varepsilon = \delta \int_{t_1}^{t_2} L(q_i(t), \qquad \dot{q}_i(t); t)dt = 0, \qquad (8.18)$$

where $q_i(t)$, and hence $\dot{q}_i(t)$, is to be varied subject to $\delta q_i(t_1) = \delta q_i(t_2) = 0$. Equation (8.18) is a mathematical statement of Hamilton's principle of classical mechanics. In this variational approach to mechanics, the Lagrangian L is given, and $q_i(t)$ taken on the prescribed values at t_1 and t_2, but may be arbitrarily varied for values of t between t_1 and t_2.

In words, Hamilton's principle states that for a conservative dynamical system, the motion of the system from its position in configuration space at time t_1 to its position at time t_2 follows a path for which the action integral (8.17) has a stationary value. The resulting Euler–Lagrange equations are known as the Lagrange equations of motion:

$$\frac{d}{dt}\frac{\partial L}{\partial \dot{q}_i} - \frac{\partial L}{\partial q_i} = 0. \qquad (8.19)$$

These Lagrange equations can be derived from Newton's equations of motion (that is, the second law written in differential equation form) and Newton's equations can be derived from Lagrange's equations. Thus they are 'equivalent.' However, Hamilton's principle can be applied to a wide range of physical phenomena, particularly those involving fields, with which Newton's equations are not usually associated. Therefore, Hamilton's principle is considered to be more fundamental than Newton's equations and is often introduced as a basic postulate from which various formulations of classical dynamics are derived.

Example 8.3
Electric oscillations: As an illustration of the generality of Lagrangian dynamics, we consider its application to an *LC* circuit (inductive–capacitive circuit) as shown in Fig. 8.4. At some instant of time the charge on the capacitor C is $Q(t)$ and the current flowing through the inductor is $I(t) = \dot{Q}(t)$. The voltage drop around the

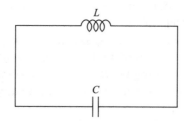

Figure 8.4.   *LC* circuit.

circuit is, according to Kirchhoff's law

$$L\frac{dI}{dt} + \frac{1}{C}\int I(t)dt = 0$$

or in terms of Q

$$L\ddot{Q} + \frac{1}{C}Q = 0.$$

This equation is of exactly the same form as that for a simple mechanical oscillator:

$$m\ddot{x} + kx = 0.$$

If the electric circuit also contains a resistor R, Kirchhoff's law then gives

$$L\ddot{Q} + R\dot{Q} + \frac{1}{C}Q = 0,$$

which is of exactly the same form as that for a damped oscillator

$$m\ddot{x} + b\dot{x} + kx = 0,$$

where b is the damping constant.

By comparing the corresponding terms in these equations, an analogy between mechanical and electric quantities can be established:

x	displacement	Q	charge (generalized coordinate)
$\dot{x}$	velocity	$\dot{Q} = I$	electric current
m	mass	L	inductance
$1/k$	$k =$ spring constant	C	capacitance
b	damping constant	R	electric resistance
$\frac{1}{2}m\dot{x}^2$	kinetic energy	$\frac{1}{2}L\dot{Q}^2$	energy stored in inductance
$\frac{1}{2}mx^2$	potential energy	$\frac{1}{2}Q^2/C$	energy stored in capacitance

If we recognize in the beginning that the charge Q in the circuit plays the role of a generalized coordinate, and $T = \frac{1}{2}L\dot{Q}^2$ and $V = \frac{1}{2}Q^2/C$, then the Langrangian L of the system is

$$L = T - V = \tfrac{1}{2}L\dot{Q}^2 - \tfrac{1}{2}Q^2/C$$

and the Lagrange equation gives

$$L\ddot{Q} + \frac{1}{C}Q = 0,$$

the same equation as given by Kirchhoff's law.

357

Example 8.4

A bead of mass m slides freely on a frictionless wire of radius b that rotates in a horizontal plane about a point on the circular wire with a constant angular velocity w. Show that the bead oscillates as a pendulum of length $l = g/w^2$.

Solution: The circular wire rotates in the xy plane about the point O, as shown in Fig. 8.5. The rotation is in the counterclockwise direction, C is the center of the circular wire, and the angles θ and ϕ are as indicated. The wire rotates with an angular velocity w, so $\phi = wt$. Now the coordinates x and y of the bead are given by

$$x = b\cos wt + b\cos(\theta + wt),$$

$$y = b\sin wt + b\sin(\theta + wt),$$

and the generalized coordinate is θ. The potential energy of the bead (in a horizontal plane) can be taken to be zero, while its kinetic energy is

$$T = \tfrac{1}{2}m(\dot{x}^2 + \dot{y}^2) = \tfrac{1}{2}mb^2[w^2 + (\dot{\theta} + w)^2 + 2w(\dot{\theta} + w)\cos\theta],$$

which is also the Lagrangian of the bead. Inserting this into Lagrange's equation

$$\frac{d}{d\theta}\left(\frac{\partial L}{\partial\dot{\theta}}\right) - \frac{\partial L}{\partial\theta} = 0$$

we obtain, after some simplifications,

$$\ddot{\theta} + w^2\sin\theta = 0.$$

Comparing this equation with Lagrange's equation for a simple pendulum of length l

$$\ddot{\theta} + (g/l)\sin\theta = 0$$

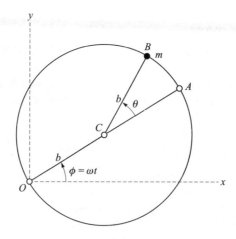

Figure 8.5.

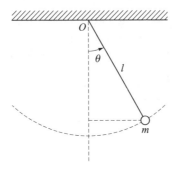

Figure 8.6.

(Fig. 8.6) we see that the bead oscillates about the line OA like a pendulum of length $l = g/\omega^2$.

Rayleigh–Ritz method

Hamilton's principle views the motion of a dynamical system as a whole and involves a search for the path in configuration space that yields a stationary value for the action integral (8.17):

$$\delta I = \delta \int_{t_1}^{t_2} L(q_i(t), \dot{q}_i(t); t)dt = 0, \qquad (8.18)$$

with $\delta q_i(t_1) = \delta q_i(t_2) = 0$. Ordinarily it is used as a variational method to obtain Lagrange's and Hamilton's equations of motion, so we do not often think of it as a computational tool. But in other areas of physics variational formulations are used in a much more active way. For example, the variational method for determining the approximate ground-state energies in quantum mechanics is very well known. We now use the Rayleigh–Ritz method to illustrate that Hamilton's principle can be used as computational device in classical mechanics. The Rayleigh–Ritz method is a procedure for obtaining approximate solutions of problems expressed in variational form directly from the variational equation.

The Lagrangian is a function of the generalized coordinates qs and their time derivatives $\dot{q}$s. The basic idea of the approximation method is to guess a solution for the qs that depends on time and a number of parameters. The parameters are then adjusted so that Hamilton's principle is satisfied. The Rayleigh–Ritz method takes a special form for the trial solution. A complete set of functions $\{f_i(t)\}$ is chosen and the solution is assumed to be a linear combination of a finite number of these functions. The coefficients in this linear combination are the parameters that are chosen to satisfy Hamilton's principle (8.18). Since the variations of the qs must vanish at the endpoints of the integral, the variations of the parameter must be so chosen that this condition is satisfied.

To summarize, suppose a given system can be described by the action integral

$$I = \int_{t_1}^{t_2} L(q_i(t), \dot{q}_i(t); t) dt, \qquad \dot{q} = dq/dt.$$

The Rayleigh–Ritz method requires the selection of a trial solution, ideally in the form

$$q = \sum_{i=1}^{n} a_i f_i(t), \tag{8.20}$$

which satisfies the appropriate conditions at both the initial and final times, and where as are undetermined constant coefficients and the fs are arbitrarily chosen functions. This trial solution is substituted into the action integral I and integration is performed so that we obtain an expression for the integral I in terms of the coefficients. The integral I is then made 'stationary' with respect to the assumed solution by requiring that

$$\frac{\partial I}{\partial a_i} = 0 \tag{8.21}$$

after which the resulting set of n simultaneous equations is solved for the values of the coefficients a_i. To illustrate this method, we apply it to two simple examples.

Example 8.5
A simple harmonic oscillator consists of a mass M attached to a spring of force constant k. As a trial function we take the displacement x as a function t in the form

$$x(t) = \sum_{n=1}^{\infty} A_n \sin n\omega t.$$

For the boundary conditions we have $x = 0, t = 0$, and $x = 0, t = 2\pi/\omega$. Then the potential energy and the kinetic energy are given by, respectively,

$$V = \tfrac{1}{2}kx^2 = \tfrac{1}{2}k \sum_{n=1}^{\infty} \sum_{m=1}^{\infty} A_n A_m \sin n\omega t \sin m\omega t,$$

$$T = \tfrac{1}{2}M\dot{x}^2 = \tfrac{1}{2}M\omega^2 \sum_{n=1}^{\infty} \sum_{m=1}^{\infty} A_n A_m nm \cos n\omega t \cos m\omega t.$$

The action I has the form

$$I = \int_{0}^{2\pi/\omega} L dt = \int_{0}^{2\pi/\omega} (T - V) dt = \frac{\pi}{2\omega} \sum_{n=1}^{\infty} (kA_n^2 - Mn^2 A_n^2 \omega^2).$$

In order to satisfy Hamilton's principle we must choose the values of A_n so as to make I an extremum:

$$\frac{dI}{dA_n} = (k - n^2\omega^2 M)A_n = 0.$$

The solution that meets the physics of the problem is

$$A_1 = 0, \quad \omega^2 = k/M; \quad \text{or} \quad \tau = (2\pi/\omega)^{1/2} = 2\pi(M/k)^{1/2},$$

$$A_n = 0, \quad \text{for} \quad n = 2, 3, \text{etc.}$$

Example 8.6

As a second example, we consider a bead of mass M sliding freely along a wire shaped in the form of a parabola along the vertical axis and of the form $y = ax^2$. In this case, we have

$$L = T - V = \tfrac{1}{2}M(\dot{x}^2 + \dot{y}^2) - Mgy = \tfrac{1}{2}M(1 + 4a^2x^2)\dot{x}^2 - Mgy.$$

We assume

$$x = A \sin \omega t$$

to be an approximate value for the displacement x, and then the action integral becomes

$$I = \int_0^{2\pi/\omega} L dt = \int_0^{2\pi/\omega} (T - V)dt = A^2 \left\{ \frac{\omega^2(1 + a^2 A^2)}{2} - ga \right\} \frac{M\pi}{\omega}.$$

The extremum condition, $dI/dA = 0$, gives an approximate ω:

$$\omega = \frac{\sqrt{2ga}}{1 + a^2 A^2},$$

and the approximate period is

$$\tau = \frac{2\pi(1 + a^2 A^2)}{\sqrt{2ga}}.$$

The Rayleigh–Ritz method discussed in this section is a special case of the general Rayleigh–Ritz methods that are designed for finding approximate solutions of boundary-value problems by use of varitional principles, for example, the eigenvalues and eigenfunctions of the Sturm–Liouville systems.

Hamilton's principle and canonical equations of motion

Newton first formulated classical mechanics in the seventeenth century and it is known as Newtonian mechanics. The essential physics involved in Newtonian mechanics is contained in Newton's three laws of motion, with the second law serving as the equation of motion. Classical mechanics has since been reformulated in a few different forms: the Lagrange, the Hamilton, and the Hamilton–Jacobi formalisms, to name just a few.

The essential physics of Lagrangian dynamics is contained in the Lagrange function L of the dynamical system and Lagrange's equations (the equations of motion). The Lagrangian L is defined in terms of independent generalized coordinates q_i and the corresponding generalized velocity $\dot{q}_i$. In Hamiltonian dynamics, we describe the state of a system by Hamilton's function (or the Hamiltonian) H defined in terms of the generalized coordinates q_i and the corresponding generalized momenta p_i, and the equations of motion are given by Hamilton's equations or canonical equations

$$\dot{q}_i = \frac{\partial H}{\partial p_i}, \qquad \dot{p}_i = -\frac{\partial H}{\partial q_i}, \qquad i = 1, 2, \ldots, n. \tag{8.22}$$

Hamilton's equations of motion can be derived from Hamilton's principle. Before doing so, we have to define the generalized momentum and the Hamiltonian. The generalized momentum p_i corresponding to q_i is defined as

$$p_i = \frac{\partial L}{\partial q_i} \tag{8.23}$$

and the Hamiltonian of the system is defined by

$$H = \sum_i p_i \dot{q}_i - L. \tag{8.24}$$

Even though $\dot{q}_i$ explicitly appears in the defining expression (8.24), H is a function of the generalized coordinates q_i, the generalized momenta p_i, and the time t, because the defining expression (8.23) can be solved explicitly for the $\dot{q}_i$s in terms of p_i, q_i, and t. The qs and ps are now treated the same: $H = H(q_i, p_i, t)$. Just as with the configuration space spanned by the n independent qs, we can imagine a space of $2n$ dimensions spanned by the $2n$ variables $q_1, q_2, \ldots, q_n, p_1, p_2, \ldots, p_n$. Such a space is called phase space, and is particularly useful in both statistical mechanics and the study of non-linear oscillations. The evolution of a representative point in this space is determined by Hamilton's equations.

We are ready to deduce Hamilton's equation from Hamilton's principle. The original Hamilton's principle refers to paths in configuration space, so in order to extend the principle to phase space, we must modify it such that the integrand of the action I is a function of both the generalized coordinates and momenta and their derivatives. The action I can then be evaluated over the paths of the system point in phase space. To do this, first we solve Eq. (8.24) for L

$$L = \sum_i p_i \dot{q}_i - H$$

and then substitute L into Eq. (8.18) and we obtain

$$\delta I = \delta \int_{t_1}^{t_2} \left(\sum_i p_i \dot{q}_i - H(p, q, t) \right) dt = 0, \tag{8.25}$$

where $q_I(t)$ is still varied subject to $\delta q_i(t_1) = \delta q_i(t_2) = 0$, but p_i is varied without such end-point restrictions.

Carrying out the variation, we obtain

$$\int_{t_1}^{t_2} \sum_i \left(p_i \delta \dot{q}_i + \dot{q}_i \delta p_i - \frac{\partial H}{\partial q_i} \delta q_i - \frac{\partial H}{\partial p_i} \delta p_i \right) dt = 0, \tag{8.26}$$

where the $\delta \dot{q}$s are related to the δqs by the relation

$$\delta \dot{q}_i = \frac{d}{dt} \delta q_i. \tag{8.27}$$

Now we integrate the term $p_i \delta \dot{q}_i dt$ by parts. Using Eq. (8.27) and the endpoint conditions on δq_i, we find that

$$\int_{t_1}^{t_2} \sum_i p_i \delta \dot{q}_i dt = \int_{t_1}^{t_2} \sum_i p_i \frac{d}{dt} \delta q_i dt$$

$$= \int_{t_1}^{t_2} \sum_i \frac{d}{dt} p_i \delta q_i dt - \int_{t_1}^{t_2} \sum_i \dot{p}_i \delta q_i dt$$

$$= p_i \delta q_i \Big|_{t_1}^{t_2} - \int_{t_1}^{t_2} \sum_i \dot{p}_i \delta q_i dt$$

$$= - \int_{t_1}^{t_2} \sum_i \dot{p}_i \delta q_i dt.$$

Substituting this back into Eq. (8.26), we obtain

$$\int_{t_1}^{t_2} \sum_i \left[\left(\dot{q}_i - \frac{\partial H}{\partial p_i} \right) \delta p_i - \left(\dot{p}_i + \frac{\partial H}{\partial q_i} \right) \delta q_i \right] dt = 0. \tag{8.28}$$

Since we view Hamilton's principle as a variational principle in phase space, both the δqs and the δps are arbitrary, the coefficients of δq_i and δp_i in Eq. (8.28) must vanish separately, which results in the $2n$ Hamilton's equations (8.22).

Example 8.7
Obtain Hamilton's equations of motion for a one-dimensional harmonic oscillator.

Solution: We have

$$T = \tfrac{1}{2}m\dot{x}^2, \qquad V = \tfrac{1}{2}Kx^2,$$

$$p = \frac{\partial L}{\partial \dot{x}} = \frac{\partial T}{\partial \dot{x}} = m\dot{x}, \qquad \dot{x} = \frac{p}{m}.$$

Hence

$$H = p\dot{x} - L = T + V = \frac{1}{2m}p^2 + \frac{1}{2}Kx^2.$$

Hamilton's equations

$$\dot{x} = \frac{\partial H}{\partial p}, \qquad \dot{p} = -\frac{\partial H}{\partial x}$$

then read

$$\dot{x} = \frac{p}{m}, \qquad \dot{p} = -Kx.$$

Using the first equation, the second can be written

$$\frac{d}{dt}(m\dot{x}) = -Kx \quad \text{or} \quad m\ddot{x} + Kx = 0$$

which is the familiar equation of the harmonic oscillator.

The modified Hamilton's principle and the Hamilton–Jacobi equation

The Hamilton–Jacobi equation is the cornerstone of a general method of integrating equations of motion. Before the advent of modern quantum theory, Bohr's atomic theory was treated in terms of Hamilton–Jacobi theory. It also plays an important role in optics as well as in canonical perturbation theory. In classical mechanics books, the Hamilton–Jacobi equation is often obtained via canonical transformations. We want to show that the Hamilton–Jacobi equation can also be obtained directly from Hamilton's principle, or, a modified Hamilton's principle.

In formulating Hamilton's principle, we have considered the action

$$I = \int_{t_1}^{t_2} L(q_i(t), \dot{q}_i(t); t)dt, \qquad \dot{q} = dq/dt,$$

taken along a path between two given positions $q_i(t_1)$ and $q_i(t_2)$ which the dynamical system occupies at given instants t_1 and t_2. In varying the action, we compare the values of the action for neighboring paths with fixed ends, that is, with $\delta q_i(t_1) = \delta q_i(t_2) = 0$. Only one of these paths corresponds to the true dynamical path for which the action has its extremum value.

We now consider another aspect of the concept of action, by regarding I as a quantity characterizing the motion along the true path, and comparing the value of I for paths having a common beginning at $q_i(t_1)$, but passing through different points at time t_2. In other words we consider the action I for the true path as a function of the coordinates at the upper limit of integration:

$$I = I(q_i, t),$$

where q_i are the coordinates of the final position of the system, and t is the instant when this position is reached.

If $q_i(t_2)$ are the coordinates of the final position of the system reached at time t_2, the coordinates of a point near the point $q_i(t_2)$ can be written as $q_i(t_1) + \delta q_i$, where δq_i is a small quantity. The action for the trajectory bringing the system to the point $q_i(t_1) + \delta q_i$ differs from the action for the trajectory bringing the system to the point $q_i(t_2)$ by the quantity

$$\delta I = \int_{t_1}^{t_2} \left[\frac{\partial L}{\partial q_i} \delta q_i + \frac{\partial L}{\partial \dot{q}_i} \delta \dot{q}_i \right] dt, \tag{8.29}$$

where δq_i is the difference between the values of q_i taken for both paths at the same instant t; similarly, $\delta \dot{q}_i$ is the difference between the values of $\dot{q}_i$ at the instant t.

We now integrate the second term on the right hand side of Eq. (8.25) by parts:

$$\int_{t_1}^{t_2} \frac{\partial L}{\partial \dot{q}_i} \delta \dot{q}_i dt = \frac{\partial L}{\partial \dot{q}_i} \delta q_i - \int_{t_1}^{t_2} \frac{d}{dt} \left(\frac{\partial L}{\partial \dot{q}_i} \right) \delta q_i dt$$

$$= p_i \delta q_i - \int_{t_1}^{t_2} \frac{d}{dt} \left(\frac{\partial L}{\partial \dot{q}_i} \right) \delta q_i dt, \tag{8.30}$$

where we have used the fact that the starting points of both paths coincide, hence $\delta q_i(t_1) = 0$; the quantity $\delta q_i(t_2)$ is now written as just δq_i. Substituting Eq. (8.30) into Eq. (8.29), we obtain

$$\delta I = \sum_i p_i \delta q_i + \int_{t_1}^{t_2} \sum_i \left[\frac{\partial L}{\partial q_i} - \frac{d}{dt} \left(\frac{\partial L}{\partial \dot{q}_i} \right) \right] \delta q_i dt. \tag{8.31}$$

Since the true path satisfies Lagrange's equations of motion, the integrand and, consequently, the integral itself vanish. We have thus obtained the following value for the increment of the action I due to the change in the coordinates of the final position of the system by δq_i (at a constant time of motion):

$$\delta I = \sum_i p_i \delta q_i, \tag{8.32}$$

from which it follows that

$$\frac{\partial I}{\partial q_i} = p_i, \tag{8.33}$$

that is, the partial derivatives of the action with respect to the generalized coordinates equal the corresponding generalized momenta.

The action I may similarly be regarded as an explicit function of time, by considering paths starting from a given point $q_i(1)$ at a given instant t_1, ending at a given point $q_i(2)$ at various times $t_2 = t$:

$$I = I(q_i, t).$$

Then the total time derivative of I is

$$\frac{dI}{dt} = \frac{\partial I}{\partial t} + \sum_i \frac{\partial I}{\partial q_i} \dot{q}_i = \frac{\partial I}{\partial t} + \sum_i p_i \dot{q}_i. \tag{8.34}$$

From the definition of the action, we have $dI/dt = L$. Substituting this into Eq. (8.34), we obtain

$$\frac{\partial I}{\partial t} = L - \sum_i p_i \dot{q}_i = -H$$

or

$$\frac{\partial I}{\partial t} + H(q_i, p_i, t) = 0. \tag{8.35}$$

Replacing the momenta p_i in the Hamiltonian H by $\partial I/\partial q_i$ as given by Eq. (8.33), we obtain the Hamilton–Jacobi equation

$$H(q_i, \partial I/\partial q_i, t) + \frac{\partial I}{\partial t} = 0. \tag{8.36}$$

For a conservative system with stationary constraints, the time is not contained explicitly in Hamiltonian H, and $H = E$ (the total energy of the system). Consequently, according to Eq. (8.35), the dependence of action I on time t is expressed by the term $-Et$. Therefore, the action breaks up into two terms, one of which depends only on q_i, and the other only on t:

$$I(q_i, t) = I_o(q_i) - Et. \tag{8.37}$$

The function $I_o(q_i)$ is sometimes called the contracted action, and the Hamilton–Jacobi equation (8.36) reduces to

$$H(q_i, \partial I_o/\partial q_i) = E. \tag{8.38}$$

Example 8.8
To illustrate the method of Hamilton–Jacobi, let us consider the motion of an electron of charge $-e$ revolving about an atomic nucleus of charge Ze (Fig. 8.7). As the mass M of the nucleus is much greater than the mass m of the electron, we may consider the nucleus to remain stationary without making any very appreciable error. This is a central force motion and so its motion lies entirely in one plane (see *Classical Mechanics*, by Tai L. Chow, John Wiley, 1995). Employing

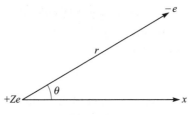

Figure 8.7.

polar coordinates r and θ in the plane of motion to specify the position of the electron relative to the nucleus, the kinetic and potential energies are, respectively,

$$T = \frac{1}{2}m(\dot{r}^2 + r^2\dot{\theta}^2), \qquad V = -\frac{Ze^2}{r}.$$

Then

$$L = T - V = \frac{1}{2}m(\dot{r}^2 + r^2\dot{\theta}^2) + \frac{Ze^2}{r}$$

and

$$p_r = \frac{\partial L}{\partial \dot{r}} = m\dot{r}p_\theta \qquad p_\theta = \frac{\partial L}{\partial \dot{\theta}} = mr^2\dot{\theta}.$$

The Hamiltonian H is

$$H = \frac{1}{2m}\left(p_r^2 + \frac{p_\theta^2}{r^2}\right) - \frac{Ze^2}{r}.$$

Replacing p_r and p_θ in the Hamiltonian by $\partial I/\partial r$ and $\partial I/\partial \theta$, respectively, we obtain, by Eq. (8.36), the Hamilton–Jacobi equation

$$\frac{1}{2m}\left[\left(\frac{\partial I}{\partial r}\right)^2 + \frac{1}{r^2}\left(\frac{\partial I}{\partial \theta}\right)^2\right] - \frac{Ze^2}{r} + \frac{\partial I}{\partial t} = 0.$$

Variational problems with several independent variables

The functional f in Eq. (8.1) contains only one independent variable, but very often f may contain several independent variables. Let us now extend the theory to this case of several independent variables:

$$I = \iiint_V f\{u, u_x, u_y, u_z; x, y, z\}\, dx\,dy\,dz, \tag{8.39}$$

where V is assumed to be a bounded volume in space with prescribed values of $u(x, y, z)$ at its boundary S; $u_x = \partial u/\partial x$, and so on. Now, the variational problem

367

is to find the function $u(x, y, z)$ for which I is stationary with respect to small changes in the functional form $u(x, y, z)$.

Generalizing Eq. (8.2), we now let

$$u(x, y, z, \varepsilon) = u(x, y, z, 0) + \varepsilon\eta(x, y, z), \tag{8.40}$$

where $\eta(x, y, z)$ is an arbitrary well-behaved (that is, differentiable) function which vanishes at the boundary S. Then we have, from Eq. (8.40),

$$u_x(x, y, z, \varepsilon) = u_x(x, y, z, 0) + \varepsilon\eta_x,$$

and similar expressions for u_y, u_z; and

$$\left.\frac{\partial I}{\partial \varepsilon}\right|_{\varepsilon=0} = \iiint_V \left(\frac{\partial f}{\partial u}\eta + \frac{\partial f}{\partial u_x}\eta_x + \frac{\partial f}{\partial u_y}\eta_y + \frac{\partial f}{\partial u_z}\eta_z \right) dxdydz = 0.$$

We next integrate each of the terms $(\partial f/\partial u_i)\eta_i$ using 'integration by parts' and the integrated terms vanish at the boundary as required. After some simplifications, we finally obtain

$$\iiint_V \left\{ \frac{\partial f}{\partial u} - \frac{\partial}{\partial x}\frac{\partial f}{\partial u_x} - \frac{\partial}{\partial y}\frac{\partial f}{\partial u_y} - \frac{\partial}{\partial z}\frac{\partial f}{\partial u_z} \right\}\eta(x, y, z)dxdydz = 0.$$

Again, since $\eta(x, y, z)$ is arbitrary, the term in the braces may be set equal to zero, and we obtain the Euler–Lagrange equation:

$$\frac{\partial f}{\partial u} - \frac{\partial}{\partial x}\frac{\partial f}{\partial u_x} - \frac{\partial}{\partial y}\frac{\partial f}{\partial u_y} - \frac{\partial}{\partial z}\frac{\partial f}{\partial u_z} = 0. \tag{8.41}$$

Note that in Eq. (8.41) $\partial/\partial x$ is a partial derivative, in that y and z are constant. But $\partial/\partial x$ is also a total derivative in that it acts on implicit x dependence and on explicit x dependence:

$$\frac{\partial}{\partial x}\frac{\partial f}{\partial u_x} = \frac{\partial^2 f}{\partial x \partial u_x} + \frac{\partial^2 f}{\partial u \partial u_x}u_x + \frac{\partial^2 f}{\partial u_x^2}u_x + \frac{\partial^2 f}{\partial u_y \partial u_x}u_{xy} + \frac{\partial^2 f}{\partial u_z \partial u_x}u_{xz}. \tag{8.42}$$

Example 8.9

The Schrödinger wave equation. The equations of motion of classical mechanics are the Euler–Lagrange differential equations of Hamilton's principle. Similarly, the Schrödinger equation, the basic equation of quantum mechanics, is also a Euler–Lagrange differential equation of a variational principle the form of which is, in the case of a system of N particles, the following

$$\delta \int L d\tau = 0, \tag{8.43}$$

with

$$L = \sum_{i=1}^{N} \frac{\hbar^2}{2m_i} \left(\frac{\partial \psi^*}{\partial x_i} \frac{\partial \psi}{\partial x_i} + \frac{\partial \psi^*}{\partial y_i} \frac{\partial \psi}{\partial y_i} + \frac{\partial \psi^*}{\partial z_i} \frac{\partial \psi}{\partial z_i} \right) + V \psi^* \psi \qquad (8.44)$$

and the constraint

$$\int \psi^* \psi d\tau = 1, \qquad (8.45)$$

where m_i is the mass of particle I, V is the potential energy of the system, and $d\tau$ is a volume element of the $3N$-dimensional space.

Condition (8.45) can be taken into consideration by introducing a Lagrangian multiplier $-E$:

$$\delta \int (L - E\psi^* \psi) d\tau = 0. \qquad (8.46)$$

Performing the variation we obtain the Schrödinger equation for a system of N particles

$$\sum_{i=1}^{N} \frac{\hbar^2}{2m_i} \nabla_i^2 \psi + (E - V)\psi = 0, \qquad (8.47)$$

where ∇_i^2 is the Laplace operator relating to particle i. Can you see that E is the energy parameter of the system? If we use the Hamiltonian operator $\hat{H}$, Eq. (8.47) can be written as

$$\hat{H}\psi = E\psi. \qquad (8.48)$$

From this we obtain for E

$$E = \frac{\int \psi^* H \psi d\tau}{\int \psi^* \psi d\tau}. \qquad (8.49)$$

Through partial integration we obtain

$$\int L d\tau = \int \psi^* H \psi d\tau$$

and thus the variational principle can be formulated in another way: $\delta \int \psi^* (H - E)\psi d\tau = 0$.

Problems

8.1 As a simple practice of using varied paths and the extremum condition, we consider the simple function $y(x) = x$ and the neighboring paths

$y(\varepsilon, x) = x + \varepsilon \sin x$. Draw these paths in the xy plane between the limits $x = 0$ and $x = 2\pi$ for $\varepsilon = 0$ for two different non-vanishing values of ε. If the integral $I(\varepsilon)$ is given by

$$I(\varepsilon) = \int_0^{2\pi} (dy/dx)^2 dx,$$

show that the value of $I(\varepsilon)$ is always greater than $I(0)$, no matter what value of ε (positive or negative) is chosen. This is just condition (8.4).

8.2 (a) Show that the Euler–Lagrange equation can be written in the form

$$\frac{d}{dx}\left(f - y'\frac{\partial f}{\partial y'}\right) - \frac{\partial f}{\partial x} = 0.$$

This is often called the second form of the Euler–Lagrange equation.

(b) If f does not involve x explicitly, show that the Euler–Lagrange equation can be integrated to yield

$$f - y'\frac{\partial f}{\partial y'} = c,$$

where c is an integration constant.

8.3 As shown in Fig. 8.8, a curve C joining points (x_1, y_1) and (x_2, y_2) is revolved about the x-axis. Find the shape of the curve such that the surface thus generated is a minimum.

8.4 A geodesic is a line that represents the shortest distance between two points. Find the geodesic on the surface of a sphere.

8.5 Show that the geodesic on the surface of a right circular cylinder is a helix.

8.6 Find the shape of a heavy chain which minimizes the potential energy while the length of the chain is constant.

8.7 A wedge of mass M and angle α slides freely on a horizontal plane. A particle of mass m moves freely on the wedge. Determine the motion of the particle as well as that of the wedge (Fig. 8.9).

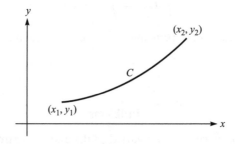

Figure 8.8.

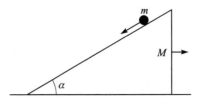

Figure 8.9.

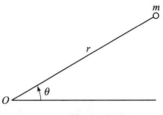

Figure 8.10.

8.8 Use the Rayleigh–Ritz method to analyze the forced oscillations of a harmonic oscillation:

$$m\ddot{x} + kx = F_0 \sin \omega t.$$

8.9 A particle of mass m is attracted to a fixed point O by an inverse square force $F_r = -k/r^2$ (Fig. 8.10). Find the canonical equations of motion.

8.10 Set up the Hamilton–Jacobi equation for the simple harmonic oscillator.

9

The Laplace transformation

The Laplace transformation method is generally useful for obtaining solutions of linear differential equations (both ordinary and partial). It enables us to reduce a differential equation to an algebraic equation, thus avoiding going to the trouble of finding the general solution and then evaluating the arbitrary constants. This procedure or technique can be extended to systems of equations and to integral equations, and it often yields results more readily than other techniques. In this chapter we shall first define the Laplace transformation, then evaluate the transformation for some elementary functions, and finally apply it to solve some simple physical problems.

Definition of the Lapace transform

The Laplace transform $L[f(x)]$ of a function $f(x)$ is defined by the integral

$$L[f(x)] = \int_0^\infty e^{-px} f(x)dx = F(p), \qquad (9.1)$$

whenever this integral exists. The integral in Eq. (9.1) is a function of the parameter p and we denote it by $F(p)$. The function $F(p)$ is called the Laplace transform of $f(x)$. We may also look upon Eq. (9.1) as a definition of a Laplace transform operator L which tranforms $f(x)$ in to $F(p)$. The operator L is linear, since from Eq. (9.1) we have

$$L[c_1 f(x) + c_2 g(x)] = \int_0^\infty e^{-px} \{c_1 f(x) + c_2 g(x)\}dx$$

$$= c_1 \int_0^\infty e^{-px} f(x)dx + c_2 \int_0^\infty e^{-px} g(x)dx$$

$$= c_1 L[f(x)] + c_2 L[g(x)],$$

372

where c_1 and c_2 are arbitrary constants and $g(x)$ is an arbitrary function defined for $x > 0$.

The inverse Laplace transform of $F(p)$ is a function $f(x)$ such that $L[f(x)] = F(p)$. We denote the operation of taking an inverse Laplace transform by L^{-1}:

$$L^{-1}[F(p)] = f(x).\qquad(9.2)$$

That is, we operate algebraically with the operators L and L^{-1}, bringing them from one side of an equation to the other side just as we would in writing $ax = b$ implies $x = a^{-1}b$. To illustrate the calculation of a Laplace transform, let us consider the following simple example.

Example 9.1
Find $L[e^{ax}]$, where a is a constant.

Solution: The transform is

$$L[e^{ax}] = \int_0^\infty e^{-px}e^{ax}\,dx = \int_0^\infty e^{-(p-a)x}\,dx.$$

For $p \le a$, the exponent on e is positive or zero and the integral diverges. For $p > a$, the integral converges:

$$L[e^{ax}] = \int_0^\infty e^{-px}e^{ax}\,dx = \int_0^\infty e^{-(p-a)x}\,dx = \frac{e^{-(p-a)x}}{-(p-a)}\bigg|_0^\infty = \frac{1}{p-a}.$$

This example enables us to investigate the existence of Eq. (9.1) for a general function $f(x)$.

Existence of Laplace transforms

We can prove that:

(1) if $f(x)$ is piecewise continuous on every finite interval $0 \le x \le X$, and
(2) if we can find constants M and a such that $|f(x)| \le Me^{ax}$ for $x \ge X$,

then $L[f(x)]$ exists for $p > a$. A function $f(x)$ which satisfies condition (2) is said to be of *exponential order* as $x \to \infty$; this is mathematician's jargon!

These are sufficient conditions on $f(x)$ under which we can guarantee the existence of $L[f(x)]$. Under these conditions the integral converges for $p > a$:

$$\left|\int_0^X f(x)e^{-px}\,dx\right| \le \int_0^X |f(x)|e^{-px}\,dx \le \int_0^X Me^{ax}e^{-px}\,dx$$

$$\le M\int_0^\infty e^{-(p-a)x}\,dx = \frac{M}{p-a}.$$

This establishes not only the convergence but the absolute convergence of the integral defining $L[f(x)]$. Note that $M/(p-a)$ tends to zero as $p \to \infty$. This shows that

$$\lim_{p \to \infty} F(p) = 0 \tag{9.3}$$

for all functions $F(p) = L[f(x)]$ such that $f(x)$ satisfies the foregoing conditions (1) and (2). It follows that if $\lim_{p \to \infty} F(p) \neq 0$, $F(p)$ cannot be the Laplace transform of any function $f(x)$.

It is obvious that functions of exponential order play a dominant role in the use of Laplace transforms. One simple way of determining whether or not a specified function is of exponential order is the following one: if a constant b exists such that

$$\lim_{x \to \infty} \left[e^{-bx} |f(x)| \right] \tag{9.4}$$

exists, the function $f(x)$ is of exponential order (of the order of e^{-bx}). To see this, let the value of the above limit be $K \neq 0$. Then, when x is large enough, $|e^{-bx} f(x)|$ can be made as close to K as possible, so certainly

$$|e^{-bx} f(x)| < 2K.$$

Thus, for sufficiently large x,

$$|f(x)| < 2K e^{bx}$$

or

$$|f(x)| < M e^{bx}, \quad \text{with} \quad M = 2K.$$

On the other hand, if

$$\lim_{x \to \infty} [e^{-cx} |f(x)|] = \infty \tag{9.5}$$

for every fixed c, the function $f(x)$ is not of exponential order. To see this, let us assume that b exists such that

$$|f(x)| < M e^{bx} \quad \text{for} \quad x \geq X$$

from which it follows that

$$|e^{-2bx} f(x)| < M e^{-bx}.$$

Then the choice of $c = 2b$ would give us $|e^{-cx} f(x)| < M e^{-bx}$, and $e^{-cx} f(x) \to 0$ as $x \to \infty$ which contradicts Eq. (9.5).

Example 9.2
Show that x^3 is of exponential order as $x \to \infty$.

Solution: We have to check whether or not

$$\lim_{x \to \infty} \left(e^{-bx} x^3 \right) = \lim_{x \to \infty} \frac{x^3}{e^{bx}}$$

exists. Now if $b > 0$, then L'Hospital's rule gives

$$\lim_{x \to \infty} \left(e^{-bx} x^3 \right) = \lim_{x \to \infty} \frac{x^3}{e^{bx}} = \lim_{x \to \infty} \frac{3x^2}{be^{bx}} = \lim_{x \to \infty} \frac{6x}{b^2 e^{bx}} = \lim_{x \to \infty} \frac{6}{b^3 e^{bx}} = 0.$$

Therefore x^3 is of exponential order as $x \to \infty$.

Laplace transforms of some elementary functions

Using the definition (9.1) we now obtain the transforms of polynomials, exponential and trigonometric functions.

(1) $f(x) = 1$ for $x > 0$.

By definition, we have

$$L[1] = \int_0^\infty e^{-px} dx = \frac{1}{p}, \qquad p > 0.$$

(2) $f(x) = x^n$, where n is a positive integer.

By definition, we have

$$L[x^n] = \int_0^\infty e^{-px} x^n dx.$$

Using integration by parts:

$$\int uv' dx = uv - \int vu' dx$$

with

$$u = x^n, \quad dv = v' dx = e^{-px} dx = -(1/p) d(e^{-px}), \quad v = -(1/p) e^{-px},$$

we obtain

$$\int_0^\infty e^{-px} x^n dx = \left[\frac{-x^n e^{-px}}{p} \right]_0^\infty + \frac{n}{p} \int_0^\infty e^{-px} x^{n-1} dx.$$

For $p > 0$ and $n > 0$, the first term on the right hand side of the above equation is zero, and so we have

$$\int_0^\infty e^{-px} x^n dx = \frac{n}{p} \int_0^\infty e^{-px} x^{n-1} dx$$

or

$$L[x^n] = \frac{n}{p} L[x^{n-1}]$$

from which we may obtain for $n > 1$

$$L[x^{n-1}] = \frac{n-1}{p} L[x^{n-2}].$$

Iteration of this process yields

$$L[x^n] = \frac{n(n-1)(n-2) \cdots 2 \cdot 1}{p^n} L[x^0].$$

By (1) above we have

$$L[x^0] = L[1] = 1/p.$$

Hence we finally have

$$L[x^n] = \frac{n!}{p^{n+1}}, \qquad p > 0.$$

(3) $f(x) = e^{ax}$, where a is a real constant.

$$L[e^{ax}] = \int_0^\infty e^{-px} e^{ax} dx = \frac{1}{p-a},$$

where $p > a$ for convegence. (For details, see Example 9.1.)

(4) $f(x) = \sin ax$, where a is a real constant.

$$L[\sin ax] = \int_0^\infty e^{-px} \sin ax dx.$$

Using

$$\int uv' dx = uv - \int vu' dx \quad \text{with} \quad u = e^{-px}, \quad dv = -d(\cos ax)/a,$$

and

$$\int e^{mx} \sin nx dx = \frac{e^{mx}(m \sin nx - n \cos nx)}{n^2 + m^2}$$

(you can obtain this simply by using integration by parts twice) we obtain

$$L[\sin ax] = \int_0^\infty e^{-px} \sin ax dx = \left[\frac{e^{-px}(-p \sin ax - a \cos ax)}{p^2 + a^2} \right]_0^\infty.$$

Since p is positive, $e^{-px} \to 0$ as $x \to \infty$, but $\sin ax$ and $\cos ax$ are bounded as $x \to \infty$, so we obtain

$$L[\sin ax] = 0 - \frac{1(0-a)}{p^2 + a^2} = \frac{a}{p^2 + a^2}, \qquad p > 0.$$

376

(5) $f(x) = \cos ax$, where a is a real constant.

Using the result

$$\int e^{mx} \cos nx\, dx = \frac{e^{mx}(m\cos nx + n\sin mx)}{n^2 + m^2},$$

we obtain

$$L[\cos ax] = \int_0^\infty e^{-px} \cos ax\, dx = \frac{p}{p^2 + a^2}, \qquad p > 0.$$

(6) $f(x) = \sinh ax$, where a is a real constant.

Using the linearity property of the Laplace transform operator L, we obtain

$$L[\cosh ax] = L\left[\frac{e^{ax} + e^{-ax}}{2}\right] = \frac{1}{2}L[e^{ax}] + \frac{1}{2}L[e^{-ax}]$$

$$= \frac{1}{2}\left(\frac{1}{p-a} + \frac{1}{p+a}\right) = \frac{p}{p^2 - a^2}.$$

(7) $f(x) = x^k$, where $k > -1$.

By definition we have

$$L[x^k] = \int_0^\infty e^{-px} x^k\, dx.$$

Let $px = u$, then $dx = p^{-1}du$, $x^k = u^k/p^k$, and so

$$L[x^k] = \int_0^\infty e^{-px} x^k\, dx = \frac{1}{p^{k+1}} \int_0^\infty u^k e^{-u}\, du = \frac{\Gamma(k+1)}{p^{k+1}}.$$

Note that the integral defining the gamma function converges if and only if $k > -1$.

The following example illustrates the calculation of inverse Laplace transforms which is equally important in solving differential equations.

Example 9.3

Find

(a) $L^{-1}\left[\dfrac{5}{p+2}\right]$,   (b) $L^{-1}\left[\dfrac{1}{p^s}\right]$, $s > 0$.

Solution:

(a) $L^{-1}\left[\dfrac{5}{p+2}\right] = 5L^{-1}\left[\dfrac{1}{p+2}\right]$.

Recall $L[e^{ax}] = 1/(p - a)$, hence $L^{-1}[1/(p - a)] = e^{ax}$. It follows that

$$L^{-1}\left[\frac{5}{p + 2}\right] = 5L^{-1}\left[\frac{1}{p + 2}\right] = 5e^{-2x}.$$

(b) Recall

$$L[x^k] = \int_0^\infty e^{-px} x^k dx = \frac{1}{p^{k+1}} \int_0^\infty u^k e^{-u} du = \frac{\Gamma(k + 1)}{p^{k+1}}.$$

From this we have

$$L\left[\frac{x^k}{\Gamma(k + 1)}\right] = \frac{1}{p^{k+1}},$$

hence

$$L^{-1}\left[\frac{1}{p^{k+1}}\right] = \frac{x^k}{\Gamma(k + 1)}.$$

If we now let $k + 1 = s$, then

$$L^{-1}\left[\frac{1}{p^s}\right] = \frac{x^{s-1}}{\Gamma(s)}.$$

Shifting (or translation) theorems

In practical applications, we often meet functions multiplied by exponential factors. If we know the Laplace transform of a function, then multiplying it by an exponential factor does not require a new computation as shown by the following theorem.

The first shifting theorem

If $L[f(x)] = F(p)$, $p > b$; then $L[e^{at}f(x)] = F(p - a)$, $p > a + b$.

Note that $F(p - a)$ denotes the function $F(p)$ 'shifted' a units to the right. Hence the theorem is called the shifting theorem.

The proof is simple and straightforward. By definition (9.1) we have

$$L[f(x)] = \int_0^\infty e^{-px} f(x) dx = F(p).$$

Then

$$L[e^{ax}f(x)] = \int_0^\infty e^{-px}\{e^{ax}f(x)\}dx = \int_0^\infty e^{-(p-a)x}f(x)dx = F(p - a).$$

The following examples illustrate the use of this theorem.

Example 9.4
Show that:

(a) $L[e^{-ax}x^n] = \dfrac{n!}{(p+a)^{n+1}}, \qquad p > -a;$

(b) $L[e^{-ax}\sin bx] = \dfrac{b}{(p+a)^2 + b^2}, \qquad p > -a.$

Solution: (a) Recall

$$L[x^n] = n!/p^{n+1}, \qquad p > 0;$$

the shifting theorem then gives

$$L[e^{-ax}x^n] = \frac{n!}{(p+a)^{n+1}}, \qquad p > -a.$$

(b) Since

$$L[\sin ax] = \frac{a}{p^2 + a^2},$$

it follows from the shifting theorem that

$$L[e^{-ax}\sin bx] = \frac{b}{(p+a)^2 + b^2}, \qquad p > -a.$$

Because of the relationship between Laplace transforms and inverse Laplace transforms, any theorem involving Laplace transforms will have a corresponding theorem involving inverse Lapace transforms. Thus

$$\text{If } L^{-1}[F(p)] = f(x), \quad \text{then} \quad L^{-1}[F(p-a)] = e^{ax}f(x).$$

The second shifting theorem

This second shifting theorem involves the shifting x variable and states that

Given $L[f(x)] = F(p)$, where $f(x) = 0$ for $x < 0$; and if $g(x) = f(x-a)$, then

$$L[g(x)] = e^{-ap}L[f(x)].$$

To prove this theorem, let us start with

$$F(p) = L[f(x)] = \int_0^\infty e^{-px}f(x)dx$$

from which it follows that

$$e^{-ap}F(p) = e^{-ap}L[f(x)] = \int_0^\infty e^{-p(x+a)}f(x)dx.$$

Let $u = x + a$, then

$$e^{-ap}F(p) = \int_0^\infty e^{-p(x+a)}f(x)dx = \int_0^\infty e^{-pu}f(u-a)du$$

$$= \int_0^a e^{-pu}0\,du + \int_a^\infty e^{-pu}f(u-a)du$$

$$= \int_0^\infty e^{-pu}g(u)du = L[g(u)].$$

Example 9.5
Show that given

$$f(x) = \begin{cases} x & \text{for} \quad x \geq 0 \\ 0 & \text{for} \quad x < 0 \end{cases},$$

and if

$$g(x) = \begin{cases} 0, & \text{for} \quad x < 5 \\ x - 5, & \text{for} \quad x \geq 5 \end{cases}$$

then

$$L[g(x)] = e^{-5p}/p^2.$$

Solution: We first notice that

$$g(x) = f(x - 5).$$

Then the second shifting theorem gives

$$L[g(x)] = e^{-5p}L[x] = e^{-5p}/p^2.$$

The unit step function

It is often possible to express various discontinuous functions in terms of the unit step function, which is defined as

$$U(x - a) = \begin{cases} 0 & x < a \\ 1 & x \geq a \end{cases}.$$

Sometimes it is convenient to state the second shifting theorem in terms of the unit step function:

If $f(x) = 0$ for $x < 0$ and $L[f(x)] = F(p)$, then

$$L[U(x - a)f(x - a)] = e^{-ap}F(p).$$

380

The proof is straightforward:

$$L[U(x-a)f(x-a)] = \int_0^\infty e^{-px}U(x-a)f(x-)dx$$

$$= \int_0^a e^{-px}0\,dx + \int_a^\infty e^{-px}f(x-a)dx.$$

Let $x - a = u$, then

$$L[U(x-a)f(x-a)] = \int_a^\infty e^{-px}f(x-a)dx$$

$$= \int_a^\infty e^{-p(u+a)}f(u)du = e^{-ap}\int_a^\infty e^{-pu}f(u)du = e^{-ap}F(p).$$

The corresponding theorem involving inverse Laplace transforms can be stated as

If $f(x) = 0$ for $x < 0$ and $L^{-1}[F(p)] = f(x)$, then
$$L^{-1}[e^{-ap}F(p)] = U(x-a)f(x-a).$$

Laplace transform of a periodic function

If $f(x)$ is a periodic function of period $P > 0$, that is, if $f(x+P) = f(x)$, then

$$L[f(x)] = \frac{1}{1 - e^{-pP}}\int_0^P e^{-px}f(x)dx.$$

To prove this, we assume that the Laplace transform of $f(x)$ exists:

$$L[f(x)] = \int_0^\infty e^{-px}f(x)dx = \int_0^P e^{-px}f(x)dx + \int_P^{2P} e^{-px}f(x)dx$$

$$+ \int_{2P}^{3P} e^{-px}f(x)dx + \cdots.$$

On the right hand side, let $x = u + P$ in the second integral, $x = u + 2P$ in the third integral, and so on, we then have

$$L[f(x)] = \int_0^P e^{-px}f(x)dx + \int_0^P e^{-p(u+P)}f(u+P)du$$

$$+ \int_0^P e^{-p(u+2P)}f(u+2P)du + \cdots.$$

381

But $f(u+P) = f(u)$, $f(u+2P) = f(u)$, etc. Also, let us replace the dummy variable u by x, then the above equation becomes

$$L[f(x)] = \int_0^P e^{-px}f(x)dx + \int_0^P e^{-p(x+P)}f(x)dx + \int_0^P e^{-p(x+2P)}f(x)dx + \cdots$$

$$= \int_0^P e^{-px}f(x)dx + e^{-pP}\int_0^P e^{-px}f(x)dx + e^{-2pP}\int_0^P e^{-px}f(x)dx + \cdots$$

$$= (1 + e^{-pP} + e^{-2pP} + \cdots)\int_0^P e^{-px}f(x)dx$$

$$= \frac{1}{1-e^{-pP}}\int_0^P e^{-px}f(x)dx.$$

Laplace transforms of derivatives

If $f(x)$ is a continuous for $x \geq 0$, and $f'(x)$ is piecewise continuous in every finite interval $0 \leq x \leq k$, and if $|f(x)| \leq Me^{bx}$ (that is, $f(x)$ is of exponential order), then

$$L[f'(x)] = pL[f(x)] - f(0), \quad p > b.$$

We may employ integration by parts to prove this result:

$$\int u dv = uv - \int v du \quad \text{with} \quad u = e^{-px}, \quad \text{and} \quad dv = f'(x)dx;$$

$$L[f'(x)] = \int_0^\infty e^{-px}f'(x)dx = [e^{-px}f(x)]_0^\infty - \int_0^\infty (-p)e^{-px}f(x)dx.$$

Since $|f(x)| \leq Me^{bx}$ for sufficiently large x, then $|f(x)e^{-px}| \leq Me^{(b-p)}$ for sufficiently large x. If $p > b$, then $Me^{(b-p)} \to 0$ as $x \to \infty$; and $e^{-px}f(x) \to 0$ as $x \to \infty$. Next, $f(x)$ is continuous at $x = 0$, and so $e^{-px}f(x) \to f(0)$ as $x \to 0$. Thus, the desired result follows:

$$L[f'(x)] = pL[f(x)] - f(0), \quad p > b.$$

This result can be extended as follows:

If $f(x)$ is such that $f^{(n-1)}(x)$ is continuous and $f^{(n)}(x)$ piecewise continuous in every interval $0 \leq x \leq k$ and furthermore, if $f(x)$, $f'(x)$, ..., $f^{(n)}(x)$ are of exponential order for $0 > k$, then

$$L[f^{(n)}(x)] = p^n L[f(x)] - p^{n-1}f(0) - p^{n-2}f'(0) - \cdots - f^{(n-1)}(0).$$

Example 9.6

Solve the initial value problem:

$y'' + y = 0; y(0) = y'(0) = 0$, and $f(t) = 0$ for $t < 0$ but $f(t) = 1$ for $t \geq 0$.

Solution: Note that $y' = dy/dt$. We know how to solve this simple differential equation, but as an illustration we now solve it using Laplace transforms. Taking both sides of the equation we obtain

$$L[y''] + L[y] = L[1], \quad (L[f] = L[1]).$$

Now

$$L[y''] = pL[y'] - y'(0) = p\{pL[y] - y(0)\} - y'(0)$$
$$= p^2 L[y] - py(0) - y'(0)$$
$$= p^2 L[y]$$

and

$$L[1] = 1/p.$$

The transformed equation then becomes

$$p^2 L[y] + L[y] = 1/p$$

or

$$L[y] = \frac{1}{p(p^2 + 1)} = \frac{1}{p} - \frac{p}{p^2 + 1},$$

therefore

$$y = L^{-1}\left[\frac{1}{p}\right] - L^{-1}\left[\frac{p}{p^2 + 1}\right].$$

We find from Eqs. (9.6) and (9.10) that

$$L^{-1}\left[\frac{1}{p}\right] = 1 \quad \text{and} \quad L^{-1}\left[\frac{p}{p^2 + 1}\right] = \cos t.$$

Thus, the solution of the initial problem is

$$y = 1 - \cos t \quad \text{for} \quad t \ge 0, \qquad y = 0 \quad \text{for} \quad t < 0.$$

Laplace transforms of functions defined by integrals

If $g(x) = \int_0^x f(u)\,du$, and if $L[f(x)] = F(p)$, then $L[g(x)] = F(p)/p$.

Similarly, if $L^{-1}[F(p)] = f(x)$, then $L^{-1}[F(p)/p] = g(x)$.
It is easy to prove this. If $g(x) = \int_0^x f(u)\,du$, then $g(0) = 0, g'(x) = f(x)$. Taking Laplace transform, we obtain

$$L[g'(x)] = L[f(x)]$$

but

$$L[g'(x)] = pL[g(x)] - g(0) = pL[g(x)]$$

and so

$$pL[g(x)] = L[f(x)], \quad \text{or} \quad L[g(x)] = \frac{1}{p}L[f(x)] = \frac{F(p)}{p}.$$

From this we have

$$L^{-1}[F(p)/p] = g(x).$$

Example 9.7
If $g(x) = \int_0^u \sin au \, du$, then

$$L[g(x)] = L\left[\int_0^u \sin au \, du\right] = \frac{1}{p}L[\sin au] = \frac{a}{p(p^2 + a^2)}.$$

A note on integral transformations

A Laplace transform is one of the integral transformations. The integral transformation $T[f(x)]$ of a function $f(x)$ is defined by the integral equation

$$T[f(x)] = \int_a^b f(x)K(p, x)dx = F(p), \tag{9.6}$$

where $K(p, x)$, a known function of p and x, is called the kernel of the transformation. In the application of integral transformations to the solution of boundary-value problems, we have so far made use of five different kernels:

Laplace transform: $K(p, x) = e^{-px}$, and $a = 0, b = \infty$:

$$L[f(x)] = \int_0^\infty e^{-px}f(x)dx = (p).$$

Fourier sine and cosine transforms: $K(p, x) = \sin px$ or $\cos px$, and
$a = 0, b = \infty$:

$$F[f(x)] = \int_0^\infty f(x)\left\{\begin{array}{c} \sin(px) \\ \cos(px) \end{array}\right. dx = F(p).$$

Complex Fourier transform: $K(p, x) = e^{ipx}$, and $a = -\infty, b = -\infty$:

$$F[f(x)] = \int_{-\infty}^\infty e^{ipx}f(x)dx = F(p).$$

Hankel transform: $K(p, x) = xJ_n(px), a = 0, b = \infty$, where $J_n(px)$ is the Bessel function of the first kind of order n:

$$H[f(x)] = \int_0^\infty f(x)xJ_n(x)dx = F(p).$$

Mellin transform: $K(p, x) = x^{p-1}$, and $a = 0, b = \infty$:

$$M[f(x)] = \int_0^\infty f(x)x^{p-1}dx = F(p).$$

The Laplace transform has been the subject of this chapter, and the Fouier transform was treated in Chapter 4. It is beyond the scope of this book to include Hankel and Mellin transformations.

Problems

9.1　Show that:
　　(a) e^{t^2} is not of exponential order as $x \to \infty$.
　　(b) $\sin e^{t^2}$ is of exponential order as $x \to \infty$.

9.2　Show that:

　　(a) $L[\sinh ax] = \dfrac{a}{p^2 - a^2}$,　　$p > 0$.

　　(b) $L[3x^4 - 2x^{3/2} + 6] = \dfrac{72}{p^5} - \dfrac{3\sqrt{\pi}}{2p^{5/2}} + \dfrac{6}{p}$.

　　(c) $L[\sin x \cos x] = 1/(p^2 + 4)$.

　　(d) If

$$f(x) = \begin{cases} x, & 0 < x < 4 \\ 5, & x > 4 \end{cases},$$

　　then

$$L[f(x)] = \frac{1}{p^2} + \frac{e^{-4p}}{p} - \frac{e^{-4p}}{p^2}.$$

9.4　Show that $L[U(x - a)] = e^{-ap}/p, \; p > 0$.

9.5　Find the Laplace transform of $H(x)$, where

$$H(x) = \begin{cases} x, & 0 < x < 4 \\ 5, & x > 4 \end{cases}.$$

9.5　Let $f(x)$ be the rectified sine wave of period $P = 2\pi$:

$$f(x) = \begin{cases} \sin x, & 0 < x < \pi \\ 0, & \pi \le x < 2\pi \end{cases}.$$

Find the Laplace transform of $f(x)$.

9.6   Find

$$L^{-1}\left[\frac{15}{p^2 + 4p + 13}\right].$$

9.7   Prove that if $f'(x)$ is continuous and $f''(x)$ is piecewise continuous in every finite interval $0 \leq x \leq k$ and if $f(x)$ and $f'(x)$ are of exponential order for $x > k$, then

$$L[f'''(x)] = p^2 L[f(x)] - pf(0) - f'(0).$$

(Hint: Use (9.19) with $f'(x)$ in place of $f(x)$ and $f''(x)$ in place of $f'(x)$.)

9.8   Solve the initial problem $y''(t) + \beta^2 y(t) = A \sin \omega t$; $y(0) = 1$, $y'(0) = 0$.

9.9   Solve the initial problem $y'''(t) - y'(t) = \sin t$ subject to

$$y(0) = 2, \qquad y(0) = 0, \qquad y''(0) = 1.$$

9.10   Solve the linear simultaneous differential equation with constant coefficients

$$y'' + 2y - x = 0,$$

$$x'' + 2x - y = 0,$$

subject to $x(0) = 2$, $y(0) = 0$, and $x'(0) = y'(0) = 0$, where x and y are the dependent variables and t is the independent variable.

9.11   Find

$$L\left[\int_0^\infty \cos au \, du\right].$$

9.12.   Prove that if $L[f(x)] = F(p)$ then

$$L[f(ax)] = \frac{1}{a}F\left(\frac{p}{a}\right).$$

Similarly if $L^{-1}[F(p)] = f(x)$ then

$$L^{-1}\left[F\left(\frac{p}{a}\right)\right] = af(ax).$$

10

Partial differential equations

We have met some partial differential equations in previous chapters. In this chapter we will study some elementary methods of solving partial differential equations which occur frequently in physics and in engineering. In general, the solution of partial differential equations presents a much more difficult problem than the solution of ordinary differential equations. A complete discussion of the general theory of partial differential equations is well beyond the scope of this book. We therefore limit ourselves to a few solvable partial differential equations that are of physical interest.

Any equation that contains an unknown function of two or more variables and its partial derivatives with respect to these variables is called a partial differential equation, the order of the equation being equal to the order of the highest partial derivatives present. For example, the equations

$$3y^2 \frac{\partial u}{\partial x} + \frac{\partial u}{\partial y} = 2u, \quad \frac{\partial^2 u}{\partial x \partial y} = 2x - y$$

are typical partial differential equations of the first and second orders, respectively, x and y being independent variables and $u(x, y)$ the function to be found. These two equations are linear, because both u and its derivatives occur only to the first order and products of u and its derivatives are absent. We shall not consider non-linear partial differential equations.

We have seen that the general solution of an ordinary differential equation contains arbitrary constants equal in number to the order of the equation. But the general solution of a partial differential equation contains arbitrary functions (equal in number to the order of the equation). After the particular choice of the arbitrary functions is made, the general solution becomes a particular solution.

The problem of finding the solution of a given differential equation subject to given initial conditions is called a boundary-value problem or an initial-value

387

problem. We have seen already that such problems often lead to eigenvalue problems.

Linear second-order partial differential equations

Many physical processes can be described to some degree of accuracy by linear second-order partial differential equations. For simplicity, we shall restrict our discussion to the second-order linear partial differential equation in two independent variables, which has the general form

$$A\frac{\partial^2 u}{\partial x^2} + B\frac{\partial^2 u}{\partial x \partial y} + C\frac{\partial^2 u}{\partial y^2} + D\frac{\partial u}{\partial x} + E\frac{\partial u}{\partial y} + Fu = G, \tag{10.1}$$

where $A, B, C, \ldots, G$ may be dependent on variables x and y.

If G is a zero function, then Eq. (10.1) is called homogeneous; otherwise it is said to be non-homogeneous. If $u_1, u_2, \ldots, u_n$ are solutions of a linear homogeneous partial differential equation, then $c_1 u_1 + c_2 u_2 + \cdots + c_n u_n$ is also a solution, where $c_1, c_2, \ldots$ are constants. This is known as the superposition principle; it does not apply to non-linear equations. The general solution of a linear non-homogeneous partial differential equation is obtained by adding a particular solution of the non-homogeneous equation to the general solution of the homogeneous equation.

The homogeneous form of Eq. (10.1) resembles the equation of a general conic:

$$ax^2 + bxy + cy^2 + dx + ey + f = 0.$$

We thus say that Eq. (10.1) is of

$$\left.\begin{array}{c} \text{elliptic} \\ \text{hyperbolic} \\ \text{parabolic} \end{array}\right\} \text{type} \qquad \text{when} \left\{ \begin{array}{c} B^2 - 4AC < 0 \\ B^2 - 4AC > 0. \\ B^2 - 4AC = 0 \end{array}\right.$$

For example, according to this classification the two-dimensional Laplace equation

$$\frac{\partial^2 u}{\partial x^2} + \frac{\partial^2 u}{\partial y^2} = 0$$

is of elliptic type ($A = C = 1, B = D = E = F = G = 0$), and the equation

$$\frac{\partial^2 u}{\partial x^2} - \alpha^2 \frac{\partial^2 u}{\partial y^2} = 0 \quad (\alpha \text{ is a real constant})$$

is of hyperbolic type. Similarly, the equation

$$\frac{\partial^2 u}{\partial x^2} - \alpha \frac{\partial u}{\partial y} = 0 \quad (\alpha \text{ is a real constant})$$

is of parabolic type.

We now list some important linear second-order partial differential equations that are of physical interest and we have seen already:

(1) Laplace's equation:

$$\nabla^2 u = 0, \tag{10.2}$$

where ∇^2 is the Laplacian operator. The function u may be the electrostatic potential in a charge-free region. It may be the gravitational potential in a region containing no matter or the velocity potential for an incompressible fluid with no sources or sinks.

(2) Poisson's equation:

$$\nabla^2 u = \rho(x, y, z), \tag{10.3}$$

where the function $\rho(x, y, z)$ is called the source density. For example, if u represents the electrostatic potential in a region containing charges, then ρ is proportional to the electrical charge density. Similarly, for the gravitational potential case, ρ is proportional to the mass density in the region.

(3) Wave equation:

$$\nabla^2 u = \frac{1}{v^2} \frac{\partial^2 u}{\partial t^2}, \tag{10.4}$$

transverse vibrations of a string, longitudinal vibrations of a beam, or propagation of an electromagnetic wave all obey this same type of equation. For a vibrating string, u represents the displacement from equilibrium of the string; for a vibrating beam, u is the longitudinal displacement from the equilibrium. Similarly, for an electromagnetic wave, u may be a component of electric field $\mathbf{E}$ or magnetic field $\mathbf{H}$.

(4) Heat conduction equation:

$$\frac{\partial u}{\partial t} = \alpha \nabla^2 u, \tag{10.5}$$

where u is the temperature in a solid at time t. The constant α is called the diffusivity and is related to the thermal conductivity, the specific heat capacity, and the mass density of the object. Eq. (10.5) can also be used as a diffusion equation: u is then the concentration of a diffusing substance.

It is obvious that Eqs. (10.2)–(10.5) all are homogeneous linear equations with constant coefficients.

Example 10.1

Laplace's equation: arises in almost all branches of analysis. A simple example can be found from the motion of an incompressible fluid. Its velocity $\mathbf{v}(x, y, z, t)$ and the fluid density $\rho(x, y, z, t)$ must satisfy the equation of continuity:

$$\frac{\partial \rho}{\partial t} + \nabla \cdot (\rho \mathbf{v}) = 0.$$

If ρ is constant we then have

$$\nabla \cdot \mathbf{v} = 0.$$

If, furthermore, the motion is irrotational, the velocity vector can be expressed as the gradient of a scalar function V:

$$\mathbf{v} = -\nabla V,$$

and the equation of continuity becomes Laplace's equation:

$$\nabla \cdot \mathbf{v} = \nabla \cdot (-\nabla V) = 0, \quad \text{or} \quad \nabla^2 V = 0.$$

The scalar function V is called the velocity potential.

Example 10.2

Poisson's equation: The electrostatic field provides a good example of Poisson's equation. The electric force between any two charges q and q' in a homogeneous isotropic medium is given by Coulomb's law

$$\mathbf{F} = C\frac{qq'}{r^2}\hat{r},$$

where r is the distance between the charges, and $\hat{r}$ is a unit vector in the direction of the force. The constant C determines the system of units, which is not of interest to us; thus we leave C as it is.

An electric field $\mathbf{E}$ is said to exist in a region if a stationary charge q' in that region experiences a force $\mathbf{F}$:

$$\mathbf{E} = \lim_{q' \to 0}(\mathbf{F}/q').$$

The $\lim_{q' \to 0}$ guarantees that the test charge q' will not alter the charge distribution that existed prior to the introduction of the test charge q'. From this definition and Coulomb's law we find that the electric field at a point r distant from a point charge is given by

$$\mathbf{E} = C\frac{q}{r^2}\hat{r}.$$

Taking the curl on both sides we get

$$\nabla \times \mathbf{E} = 0,$$

which shows that the electrostatic field is a conservative field. Hence a potential function ϕ exists such that

$$\mathbf{E} = -\nabla\phi.$$

Taking the divergence of both sides

$$\nabla \cdot (\nabla\phi) = -\nabla \cdot \mathbf{E}$$

or

$$\nabla^2 \phi = -\nabla \cdot \mathbf{E}.$$

$\nabla \cdot \mathbf{E}$ is given by Gauss' law. To see this, consider a volume τ containing a total charge q. Let $d\mathbf{s}$ be an element of the surface S which bounds the volume τ. Then

$$\iint_S \mathbf{E} \cdot d\mathbf{s} = Cq \iint_S \frac{\hat{r} \cdot d\mathbf{s}}{r^2}.$$

The quantity $\hat{r} \cdot d\mathbf{s}$ is the projection of the element area $d\mathbf{s}$ on a plane perpendicular to $\mathbf{r}$. This projected area divided by r^2 is the solid angle subtended by $d\mathbf{s}$, which is written $d\Omega$. Thus, we have

$$\iint_S \mathbf{E} \cdot d\mathbf{s} = Cq \iint_S \frac{\hat{r} \cdot d\mathbf{s}}{r^2} = Cq \iint_S d\Omega = 4\pi Cq.$$

If we write q as

$$q = \iiint_\tau \rho dV,$$

where ρ is the charge density, then

$$\iint_S \mathbf{E} \cdot d\mathbf{s} = 4\pi C \iiint_\tau \rho dV.$$

But (by the divergence theorem)

$$\iint_S \mathbf{E} \cdot d\mathbf{s} = \iiint_\tau \nabla \cdot \mathbf{E} dV.$$

Substituting this into the previous equation, we obtain

$$\iiint_\tau \nabla \cdot \mathbf{E} dV = 4\pi C \iiint_\tau \rho dV$$

or

$$\iiint_\tau (\nabla \cdot \mathbf{E} - 4\pi C\rho) dV = 0.$$

This equation must be valid for all volumes, that is, for any choice of the volume τ. Thus, we have Gauss' law in differential form:

$$\nabla \cdot \mathbf{E} = 4\pi C\rho.$$

Substituting this into the equation $\nabla^2 \phi = -\nabla \cdot \mathbf{E}$, we get

$$\nabla^2 \phi = -4\pi C\rho,$$

which is Poisson's equation. In the Gaussian system of units, $C = 1$; in the SI system of units, $C = 1/4\pi\varepsilon_0$, where the constant ε_0 is known as the permittivity of free space. If we use SI units, then

$$\nabla^2 \phi = -\rho/\varepsilon_0.$$

In the particular case of zero charge density it reduces to Laplace's equation,

$$\nabla^2 \phi = 0.$$

In the following sections, we shall consider a number of problems to illustrate some useful methods of solving linear partial differential equations. There are many methods by which homogeneous linear equations with constant coefficients can be solved. The following are commonly used in the applications.

(1) General solutions: In this method we first find the general solution and then that particular solution which satisfies the boundary conditions. It is always satisfying from the point of view of a mathematician to be able to find general solutions of partial differential equations; however, general solutions are difficult to find and such solutions are sometimes of little value when given boundary conditions are to be imposed on the solution. To overcome this difficulty it is best to find a less general type of solution which is satisfied by the type of boundary conditions to be imposed. This is the method of separation of variables.

(2) Separation of variables: The method of separation of variables makes use of the principle of superposition in building up a linear combination of individual solutions to form a solution satisfying the boundary conditions. The basic approach of this method in attempting to solve a differential equation (in, say, two dependent variables x and y) is to write the dependent variable $u(x, y)$ as a product of functions of the separate variables $u(x, y) = X(x) Y(y)$. In many cases the partial differential equation reduces to ordinary differential equations for X and Y.

(3) Laplace transform method: We first obtain the Laplace transform of the partial differential equation and the associated boundary conditions with respect to one of the independent variables, and then solve the resulting equation for the Laplace transform of the required solution which can be found by taking the inverse Laplace transform.

Solutions of Laplace's equation: separation of variables

(1) Laplace's equation in two dimensions (x, y): If the potential ϕ is a function of only two rectangular coordinates, Laplace's equation reads

$$\frac{\partial^2 \phi}{\partial x^2} + \frac{\partial^2 \phi}{\partial y^2} = 0.$$

It is possible to obtain the general solution to this equation by means of a transformation to a new set of independent variables:

$$\xi = x + iy, \quad \eta = x - iy,$$

where I is the unit imaginary number. In terms of these we have

$$\frac{\partial}{\partial x} = \frac{\partial}{\partial \xi}\frac{\partial \xi}{\partial x} + \frac{\partial}{\partial \eta}\frac{\partial \eta}{\partial x} = \frac{\partial}{\partial \xi} + \frac{\partial}{\partial \eta},$$

$$\frac{\partial^2}{\partial x^2} = \frac{\partial}{\partial x}\left(\frac{\partial}{\partial \xi} + \frac{\partial}{\partial \eta}\right)$$

$$= \frac{\partial}{\partial \xi}\left(\frac{\partial}{\partial \xi} + \frac{\partial}{\partial \eta}\right)\frac{\partial \xi}{\partial x} + \frac{\partial}{\partial \eta}\left(\frac{\partial}{\partial \xi} + \frac{\partial}{\partial \eta}\right)\frac{\partial \eta}{\partial x}$$

$$= \frac{\partial^2}{\partial \xi^2} + 2\frac{\partial}{\partial \xi}\frac{\partial}{\partial \eta} + \frac{\partial^2}{\partial \eta^2}.$$

Similarly, we have

$$\frac{\partial^2}{\partial y^2} = -\frac{\partial^2}{\partial \xi^2} + 2\frac{\partial}{\partial \xi}\frac{\partial}{\partial \eta} - \frac{\partial^2}{\partial \eta^2}$$

and Laplace's equation now reads

$$\nabla^2 \phi = 4\frac{\partial^2 \phi}{\partial \xi \partial \eta} = 0.$$

Clearly, a very general solution to this equation is

$$\phi = f_1(\xi) + f_2(\eta) = f_1(x + iy) + f_2(x - iy),$$

where f_1 and f_2 are arbitrary functions which are twice differentiable. However, it is a somewhat difficult matter to choose the functions f_1 and f_2 such that the equation is, for example, satisfied inside a square region defined by the lines $x = 0, x = a, y = 0, y = b$ and such that ϕ takes prescribed values on the boundary of this region. For many problems the method of separation of variables is more satisfactory. Let us apply this method to Laplace's equation in three dimensions.

(2) Laplace's equation in three dimensions (x, y, z): Now we have

$$\frac{\partial^2 \phi}{\partial x^2} + \frac{\partial^2 \phi}{\partial y^2} + \frac{\partial^2 \phi}{\partial z^2} = 0. \tag{10.6}$$

We make the assumption, justifiable by its success, that $\phi(x, y, z)$ may be written as the product

$$\phi(x, y, z) = X(x)Y(y)Z(z).$$

Substitution of this into Eq. (10.6) yields, after division by ϕ,

$$\frac{1}{X}\frac{d^2X}{dx^2} + \frac{1}{Y}\frac{d^2Y}{dy^2} = -\frac{1}{Z}\frac{d^2Z}{dz^2}. \tag{10.7}$$

393

The left hand side of Eq. (10.7) is a function of x and y, while the right hand side is a function of z alone. If Eq. (10.7) is to have a solution at all, each side of the equation must be equal to the same constant, say k_3^2. Then Eq. (10.7) leads to

$$\frac{d^2 Z}{dz^2} + k_3^2 Z = 0, \tag{10.8}$$

$$\frac{1}{X}\frac{d^2 X}{dx^2} = -\frac{1}{Y}\frac{d^2 Y}{dy^2} + k_3^2. \tag{10.9}$$

The left hand side of Eq. (10.9) is a function of x only, while the right hand side is a function of y only. Thus, each side of the equation must be equal to a constant, say k_1^2. Therefore

$$\frac{d^2 X}{dx^2} + k_1^2 X = 0, \tag{10.10}$$

$$\frac{d^2 Y}{dy^2} + k_2^2 Y = 0, \tag{10.11}$$

where

$$k_2^2 = k_1^2 - k_3^2.$$

The solution of Eq. (10.10) is of the form

$$X(x) = a(k_1)e^{k_1 x}, \quad k_1 \neq 0, \quad -\infty < k_1 < \infty$$

or

$$X(x) = a(k_1)e^{k_1 x} + a'(k_1)e^{-k_1 x}, \quad k_1 \neq 0, \quad 0 < k_1 < \infty. \tag{10.12}$$

Similarly, the solutions of Eqs. (10.11) and (10.8) are of the forms

$$Y(y) = b(k_2)e^{k_2 y} + b'(k_2)e^{-k_2 y}, \quad k_2 \neq 0, \quad 0 < k_2 < \infty, \tag{10.13}$$

$$Z(z) = c(k_3)e^{k_3 z} + c'(k_3)e^{-k_3 z}, \quad k_3 \neq 0, \quad 0 < k_3 < \infty. \tag{10.14}$$

Hence

$$\phi = [a(k_1)e^{k_1 x} + a'(k_1)e^{-k_1 x}][b(k_2)e^{k_2 y} + b'(k_2)e^{-k_2 y}][c(k_3)e^{k_3 z} + c'(k_3)e^{-k_3 z}],$$

and the general solution of Eq. (10.6) is obtained by integrating the above equation over all the permissible values of the k_i $(i = 1, 2, 3)$.

In the special case when $k_i = 0$ $(i = 1, 2, 3)$, Eqs. (10.8), (10.10), and (10.11) have solutions of the form

$$X_i(x_i) = a_i x_i + b_i,$$

where $x_1 = x$, and $X_1 = X$ etc.

Let us now apply the above result to a simple problem in electrostatics: that of finding the potential ϕ at a point P a distance h from a uniformly charged infinite plane in a dielectric of permittivity ε. Let σ be the charge per unit area of the plane, and take the origin of the coordinates in the plane and the x-axis perpendicular to the plane. It is evident that ϕ is a function of x only. There are two types of solutions, namely:

$$\phi(x) = a(k_1)e^{k_1 x} + a'(k_1)e^{-k_1 x},$$

$$\phi(x) = a_1 x + b_1;$$

the boundary conditions will eliminate the unwanted one. The first boundary condition is that the plane is an equipotential, that is, $\phi(0) = \text{constant}$, and the second condition is that $E = -\partial\phi/\partial x = \sigma/2\varepsilon$. Clearly, only the second type of solution satisfies both the boundary conditions. Hence $b_1 = \phi(0), a_1 = -\sigma/2\varepsilon$, and the solution is

$$\phi(x) = -\frac{\sigma}{2\varepsilon}x + \phi(0).$$

(3) Laplace's equation in cylindrical coordinates (ρ, φ, z): The cylindrical coordinates are shown in Fig. 10.1, where

$$\left.\begin{aligned} x &= \rho\cos\varphi \\ y &= \rho\sin\varphi \\ z &= z \end{aligned}\right\} \quad \text{or} \quad \left\{\begin{aligned} \rho^2 &= x^2 + y^2 \\ \varphi &= \tan^{-1}(y/x) \\ z &= z. \end{aligned}\right.$$

Laplace's equation now reads

$$\nabla^2\phi(\rho, \varphi, z) = \frac{1}{\rho}\frac{\partial}{\partial\rho}\left(\rho\frac{\partial\phi}{\partial\rho}\right) + \frac{1}{\rho^2}\frac{\partial^2\phi}{\partial\varphi^2} + \frac{\partial^2\phi}{\partial z^2} = 0. \tag{10.15}$$

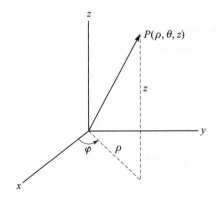

Figure 10.1.   Cylindrical coordinates.

395

We assume that

$$\phi(\rho, \varphi, z) = R(\rho)\Phi(\varphi)Z(z).$$ (10.16)

Substitution into Eq. (10.15) yields, after division by ϕ,

$$\frac{1}{\rho R}\frac{d}{d\rho}\left(\rho\frac{dR}{d\rho}\right) + \frac{1}{\rho^2 \Phi}\frac{d^2\Phi}{d\varphi^2} = -\frac{1}{Z}\frac{d^2Z}{dz^2}.$$ (10.17)

Clearly, both sides of Eq. (10.17) must be equal to a constant, say $-k^2$. Then

$$\frac{1}{Z}\frac{d^2Z}{dz^2} = k^2 \quad \text{or} \quad \frac{d^2Z}{dz^2} - k^2Z = 0$$ (10.18)

and

$$\frac{1}{\rho R}\frac{d}{d\rho}\left(\rho\frac{dR}{d\rho}\right) + \frac{1}{\rho^2 \Phi}\frac{d^2\Phi}{d\varphi^2} = -k^2$$

or

$$\frac{\rho}{R}\frac{d}{d\rho}\left(\rho\frac{dR}{d\rho}\right) + k^2\rho^2 = -\frac{1}{\Phi}\frac{d^2\Phi}{d\varphi^2}.$$

Both sides of this last equation must be equal to a constant, say α^2. Hence

$$\frac{d^2\Phi}{d\varphi^2} + \alpha^2\Phi = 0,$$ (10.19)

$$\frac{1}{R}\frac{d}{d\rho}\left(\rho\frac{dR}{d\rho}\right) + \left(k^2 - \frac{\alpha^2}{\rho^2}\right)R = 0.$$ (10.20)

Equation (10.18) has for solutions

$$Z(z) = \begin{cases} c(k)e^{kz} + c'(k)e^{-kz}, & k \neq 0, \quad 0 < k < \infty, \\ c_1 z + c_2, & k = 0, \end{cases}$$ (10.21)

where c and c' are arbitrary functions of k and c_1 and c_2 are arbitrary constants. Equation (10.19) has solutions of the form

$$\Phi(\varphi) = \begin{cases} a(\alpha)e^{i\alpha\varphi}, & \alpha \neq 0, \quad -\infty < \alpha < \infty, \\ b\varphi + b', & \alpha = 0. \end{cases}$$

That the potential must be single-valued requires that $\Phi(\varphi) = \Phi(\varphi + 2n\pi)$, where n is an integer. It follows from this that α must be an integer or zero and that $b = 0$. Then the solution $\Phi(\varphi)$ becomes

$$\Phi(\varphi) = \begin{cases} a(\alpha)e^{i\alpha\varphi} + a'(\alpha)e^{-i\alpha\varphi}, & \alpha \neq 0, \quad \alpha = \text{integer}, \\ b', & \alpha = 0. \end{cases}$$ (10.22)

In the special case $k = 0$, Eq. (10.20) has solutions of the form

$$R(\rho) = \begin{cases} d(\alpha)\rho^\alpha + d'(\alpha)\rho^{-\alpha}, & \alpha \neq 0, \\ f \ln \rho + g, & \alpha = 0. \end{cases} \tag{10.23}$$

When $k \neq 0$, a simple change of variable can put Eq. (10.20) in the form of Bessel's equation. Let $x = k\rho$, then $dx = kd\rho$ and Eq. (10.20) becomes

$$\frac{d^2 R}{dx^2} + \frac{1}{x}\frac{dR}{dx} + \left(1 - \frac{\alpha^2}{x^2}\right)R = 0, \tag{10.24}$$

the well-known Bessel's equation (Eq. (7.71)). As shown in Chapter 7, $R(x)$ can be written as

$$R(x) = AJ_\alpha(x) + BJ_{-\alpha}(x), \tag{10.25}$$

where A and B are constants, and $J_\alpha(x)$ is the Bessel function of the first kind. When α is not an integer, J_α and $J_{-\alpha}$ are independent. But when α is an integer, $J_{-\alpha}(x) = (-1)^n J_\alpha(x)$, thus J_α and $J_{-\alpha}$ are linearly dependent, and Eq. (10.25) cannot be a general solution. In this case the general solution is given by

$$R(x) = A_1 J_\alpha(x) + B_1 Y_\alpha(x), \tag{10.26}$$

where A_1 and B_2 are constants; $Y_\alpha(x)$ is the Bessel function of the second kind of order α or Neumann's function of order $\alpha N_\alpha(x)$.

The general solution of Eq. (10.20) when $k \neq 0$ is therefore

$$R(\rho) = p(\alpha)J_\alpha(k\rho) + q(\alpha)Y_\alpha(k\rho), \tag{10.27}$$

where p and q are arbitrary functions of α. Then these functions are also solutions:

$$H_\alpha^{(1)}(k\rho) = J_\alpha(k\rho) + iY_\alpha(k\rho), \quad H_\alpha^{(2)}(k\rho) = J_\alpha(k\rho) - iY_\alpha(k\rho).$$

These are the Hankel functions of the first and second kinds of order α, respectively.

The functions J_α, Y_α (or N_α), and $H_\alpha^{(1)}$, and $H_\alpha^{(2)}$ which satisfy Eq. (10.20) are known as cylindrical functions of integral order α and are denoted by $Z_\alpha(k\rho)$, which is not the same as $Z(z)$. The solution of Laplace's equation (10.15) can now be written

$$\phi(\rho, \varphi, z) \begin{cases} = (c_1 z + b)(f \ln \rho + g), & k = 0, \quad \alpha = 0, \\ = (c_1 z + b)[d(\alpha)\rho^\alpha + d'(\alpha)\rho^{-\alpha}][a(\alpha)e^{i\alpha\varphi} + a'(\alpha)e^{-i\alpha\varphi}], \\ & k = 0, \quad \alpha \neq 0, \\ = [c(k)e^{kz} + c'(k)e^{-kz}]Z_0(k\rho), & k \neq 0, \quad \alpha = 0, \\ = [c(k)e^{kz} + c'(k)e^{-kz}]Z_\alpha(k\rho)[a(\alpha)e^{i\alpha\varphi} + a'(\alpha)e^{-i\alpha\varphi}], & k \neq 0, \quad \alpha \neq 0. \end{cases}$$

Let us now apply the solutions of Laplace's equation in cylindrical coordinates to an infinitely long cylindrical conductor with radius l and charge per unit length λ. We want to find the potential at a point P a distance $\rho > l$ from the axis of the cylindrical. Take the origin of the coordinates on the axis of the cylinder that is taken to be the z-axis. The surface of the cylinder is an equipotential:

$$\phi(l) = \text{const.} \quad \text{for } r = l \text{ and all } \varphi \text{ and } z.$$

The secondary boundary condition is that

$$E = -\partial \phi/\partial \rho = \lambda/2\pi l \varepsilon \quad \text{for } r = l \text{ and all } \varphi \text{ and } z.$$

Of the four types of solutions to Laplace's equation in cylindrical coordinates listed above only the first can satisfy these two boundary conditions. Thus

$$\phi(\rho) = b(f \ln \rho + g) = -\frac{\lambda}{2\pi\varepsilon} \ln \frac{\rho}{l} + \phi(a).$$

(4) Laplace's equation in spherical coordinates (r, θ, φ): The spherical coordinates are shown in Fig. 10.2, where

$$x = r \sin \theta \cos \varphi,$$

$$y = r \sin \theta \sin \varphi,$$

$$z = r \cos \varphi.$$

Laplace's equation now reads

$$\nabla^2 \phi(r, \theta, \varphi) = \frac{1}{r} \frac{\partial}{\partial r} \left(r^2 \frac{\partial \phi}{\partial r} \right) + \frac{1}{r^2 \sin \theta} \frac{\partial}{\partial \theta} \left(\sin \theta \frac{\partial \phi}{\partial \theta} \right)$$

$$+ \frac{1}{r^2 \sin^2 \theta} \frac{\partial^2 \phi}{\partial \varphi^2} = 0.$$

(10.28)

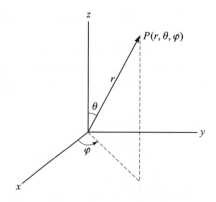

Figure 10.2.   Spherical coordinates.

Again, assume that

$$\phi(r,\theta,\varphi) = R(r)\Theta(\theta)\Phi(\varphi). \tag{10.29}$$

Substituting into Eq. (10.28) and dividing by ϕ we obtain

$$\frac{\sin^2\theta}{R}\frac{d}{dr}\left(r^2\frac{dR}{dr}\right) + \frac{\sin\theta}{\Theta}\frac{d}{d\theta}\left(\sin\theta\frac{d\Theta}{d\theta}\right) = -\frac{1}{\Phi}\frac{d^2\Phi}{d\varphi^2}.$$

For a solution, both sides of this last equation must be equal to a constant, say m^2. Then we have two equations

$$\frac{d^2\Phi}{d\varphi^2} + m^2\Phi = 0, \tag{10.30}$$

$$\frac{\sin^2\theta}{R}\frac{d}{dr}\left(r^2\frac{dR}{dr}\right) + \frac{\sin\theta}{\Theta}\frac{d}{d\theta}\left(\sin\theta\frac{d\Theta}{d\theta}\right) = m^2;$$

the last equation can be rewritten as

$$\frac{1}{\Theta\sin\theta}\frac{d}{d\theta}\left(\sin\theta\frac{d\Theta}{d\theta}\right) - \frac{m^2}{\sin^2\theta} = -\frac{1}{R}\frac{d}{dr}\left(r^2\frac{dR}{dr}\right).$$

Again, both sides of the last equation must be equal to a constant, say $-\beta$. This yields two equations

$$\frac{1}{R}\frac{d}{dr}\left(r^2\frac{dR}{dr}\right) = \beta, \tag{10.31}$$

$$\frac{1}{\Theta\sin\theta}\frac{d}{d\theta}\left(\sin\theta\frac{d\Theta}{d\theta}\right) - \frac{m^2}{\sin^2\theta} = -\beta.$$

By a simple substitution: $x = \cos\theta$, we can put the last equation in a more familiar form:

$$\frac{d}{dx}\left[(1-x^2)\frac{dP}{dx}\right] + \left(\beta - \frac{m^2}{1-x^2}\right)P = 0 \tag{10.32}$$

or

$$(1-x^2)\frac{d^2P}{dx^2} - 2x\frac{dP}{dx} + \left[\beta - \frac{m^2}{1-x^2}\right]P = 0, \tag{10.32a}$$

where we have set $P(x) = \Theta(\theta)$.

You may have already noticed that Eq. (10.32) is very similar to Eq. (10.25), the associated Legendre equation. Let us take a close look at this resemblance. In Eq. (10.32), the points $x = \pm 1$ are regular singular points of the equation. Let us first study the behavior of the solution near point $x = 1$; it is convenient to

399

bring this regular singular point to the origin, so we make the substitution $u = 1 - x$, $U(u) = P(x)$. Then Eq. (10.32) becomes

$$\frac{d}{du}\left[u(2-u)\frac{dU}{du}\right] + \left[\beta - \frac{m^2}{u(2-u)}\right]U = 0.$$

When we solve this equation by a power series: $U = \sum_{n=0}^{\infty} a_n u^{n+\rho}$, we find that the indicial equation leads to the values $\pm m/2$ for ρ. For the point $x = -1$, we make the substitution $v = 1 + x$, and then solve the resulting differential equation by the power series method; we find that the indicial equation leads to the same values $\pm m/2$ for ρ.

Let us first consider the value $+m/2, m \geq 0$. The above considerations lead us to assume

$$P(x) = (1-x)^{m/2}(1+x)^{m/2}y(x) = (1-x^2)^{m/2}y(x), \qquad m \geq 0$$

as the solution of Eq. (10.32). Substituting this into Eq. (10.32) we find

$$(1-x^2)\frac{d^2y}{dx^2} - 2(m+1)x\frac{dy}{dx} + [\beta - m(m+1)]y = 0.$$

Solving this equation by a power series

$$y(x) = \sum_{n=0}^{\infty} c_n x^{n+\delta},$$

we find that the indicial equation is $\delta(\delta - 1) = 0$. Thus the solution can be written

$$y(x) = \sum_{n \text{ even}} c_n x^n + \sum_{n \text{ odd}} c_n x^n.$$

The recursion formula is

$$c_{n+2} = \frac{(n+m)(n+m+1) - \beta}{(n+1)(n+2)}c_n.$$

Now consider the convergence of the series. By the ratio test,

$$R_n = \left|\frac{c_n x^n}{c_{n-2}x^{n-2}}\right| = \left|\frac{(n+m)(n+m+1) - \beta}{(n+1)(n+2)}\right| \cdot |x|^2.$$

The series converges for $|x| < 1$, whatever the finite value of β may be. For $|x| = 1$, the ratio test is inconclusive. However, the integral test yields

$$\int_M \frac{(t+m)(t+m+1) - \beta}{(t+1)(t+2)}dt = \int_M \frac{(t+m)(t+m+1)}{(t+1)(t+2)}dt - \int_M \frac{\beta}{(t+1)(t+2)}dt$$

and since

$$\int_M \frac{(t+m)(t+m+1)}{(t+1)(t+2)}dt \to \infty \quad \text{as} \quad M \to \infty,$$

400

the series diverges for $|x| = 1$. A solution which converges for all x can be obtained if either the even or odd series is terminated at the term in x^j. This may be done by setting β equal to

$$\beta = (j+m)(j+m+1) = l(l+1).$$

On substituting this into Eq. (10.32a), the resulting equation is

$$(1-x^2)\frac{d^2P}{dx^2} - 2x\frac{dP}{dx} + \left[l(l+1) - \frac{m^2}{1-x^2}\right]P = 0,$$

which is identical to Eq. (7.25). Special solutions were studied there: they were written in the form $P_l^m(x)$ and are known as the associated Legendre functions of the first kind of degree l and order m, where l and m, take on the values $l = 0, 1, 2, \ldots,$ and $m = 0, 1, 2, \ldots, l$. The general solution of Eq. (10.32) for $m \geq 0$ is therefore

$$P(x) = \Theta(\theta) = a_l P_l^m(x). \tag{10.33}$$

The second solution of Eq. (10.32) is given by the associated Legendre function of the second kind of degree l and order m: $Q_l^m(x)$. However, only the associated Legendre function of the first kind remains finite over the range $-1 \leq x \leq 1$ (or $0 \leq \theta \leq 2\pi$).

Equation (10.31) for $R(r)$ becomes

$$\frac{d}{dr}\left(r^2\frac{dR}{dr}\right) - l(l+1)R = 0. \tag{10.31a}$$

When $l \neq 0$, its solution is

$$R(r) = b(l)r^l + b'(l)r^{-l-1}, \tag{10.34}$$

and when $l = 0$, its solution is

$$R(r) = cr^{-1} + d. \tag{10.35}$$

The solution of Eq. (10.30) is

$$\Phi = \begin{cases} f(m)e^{im\varphi} + f'(l)e^{-im\varphi}, & m \neq 0, \quad \text{positive integer,} \\ g, & m = 0. \end{cases} \tag{10.36}$$

The solution of Laplace's equation (10.28) is therefore given by

$$\phi(r,\theta,\varphi) = \begin{cases} [br^l + b'r^{-l-1}]P_l^m(\cos\theta)[fe^{im\varphi} + f'e^{-im\varphi}], & l \neq 0, \quad m \neq 0, \\ [br^l + b'r^{-l-1}]P_l(\cos\theta), & l \neq 0, \quad m = 0, \quad (10.37) \\ [cr^{-1} + d]P_0(\cos\theta), & l = 0, \quad m = 0, \end{cases}$$

where $P_l = P_l^0$.

We now illustrate the usefulness of the above result for an electrostatic problem having spherical symmetry. Consider a conducting spherical shell of radius a and charge σ per unit area. The problem is to find the potential $\phi(r, \theta, \varphi)$ at a point P a distance $r > a$ from the center of shell. Take the origin of coordinates to be at the center of the shell. As the surface of the shell is an equipotential, we have the first boundary condition

$$\phi(r) = \text{constant} = \phi(a) \quad \text{for } r = a \quad \text{and all } \theta \text{ and } \varphi. \tag{10.38}$$

The second boundary condition is that

$$\phi \to 0 \quad \text{for } r \to \infty \text{ and all } \theta \text{ and } \varphi. \tag{10.39}$$

Of the three types of solutions (10.37) only the last can satisfy the boundary conditions. Thus

$$\phi(r, \theta, \varphi) = (cr^{-1} + d)P_0(\cos\theta). \tag{10.40}$$

Now $P_0(\cos\theta) = 1$, and from Eq. (10.38) we have

$$\phi(a) = ca^{-1} + d.$$

But the boundary condition (10.39) requires that $d = 0$. Thus $\phi(a) = ca^{-1}$, or $c = a\phi(a)$, and Eq. (10.40) reduces to

$$\phi(r) = \frac{a\phi(a)}{r}. \tag{10.41}$$

Now

$$\phi(a)/a = E(a) = Q/4\pi a^2\varepsilon,$$

where ε is the permittivity of the dielectric in which the shell is embedded, $Q = 4\pi a^2\sigma$. Thus $\phi(a) = a\sigma/\varepsilon$, and Eq. (10.41) becomes

$$\phi(r) = \frac{\sigma a^2}{\varepsilon r}. \tag{10.42}$$

Solutions of the wave equation: separation of variables

We now use the method of separation of variables to solve the wave equation

$$\frac{\partial^2 u(x, t)}{\partial x^2} = v^{-2}\frac{\partial^2 u(x, t)}{\partial t^2}, \tag{10.43}$$

subject to the following boundary conditions:

$$u(0, t) = u(l, t) = 0, \quad t \geq 0, \tag{10.44}$$

$$u(x, 0) = f(t), \quad 0 \leq x \leq l, \tag{10.45}$$

402

and

$$\frac{\partial u(x,t)}{\partial t}\bigg|_{t=0} = g(x), \quad 0 \le x \le l, \tag{10.46}$$

where f and g are given functions.

Assuming that the solution of Eq. (10.43) may be written as a product

$$u(x,t) = X(x)T(t), \tag{10.47}$$

then substituting into Eq. (10.43) and dividing by XT we obtain

$$\frac{1}{X}\frac{d^2X}{dx^2} = \frac{1}{v^2T}\frac{d^2T}{dt^2}.$$

Both sides of this last equation must be equal to a constant, say $-b^2/v^2$. Then we have two equations

$$\frac{1}{X}\frac{d^2X}{dx^2} = -\frac{b^2}{v^2}, \tag{10.48}$$

$$\frac{1}{T}\frac{d^2T}{dt^2} = -b^2. \tag{10.49}$$

The solutions of these equations are periodic, and it is more convenient to write them in terms of trigonometric functions

$$X(x) = A\sin\frac{bx}{v} + B\cos\frac{bx}{v}, \quad T(t) = C\sin bt + D\cos bt, \tag{10.50}$$

where A, B, C, and D are arbitrary constants, to be fixed by the boundary conditions. Equation (10.47) then becomes

$$u(x,t) = \left(A\sin\frac{bx}{v} + B\cos\frac{bx}{v}\right)(C\sin bt + D\cos bt). \tag{10.51}$$

The boundary condition $u(0,t) = 0(t > 0)$ gives

$$0 = B(C\sin bt + D\cos bt)$$

for all t, which implies

$$B = 0. \tag{10.52}$$

Next, from the boundary condition $u(l,t) = 0(t > 0)$ we have

$$0 = A\sin\frac{bl}{v}(C\sin bt + D\cos bt).$$

Note that $B = 0$ would make $u = 0$. However, the last equation can be satisfied for all t when

$$\sin\frac{bl}{v} = 0,$$

which implies

$$b = \frac{n\pi v}{l}, \quad n = 1, 2, 3, \ldots. \tag{10.53}$$

Note that n cannot be equal to zero, because it would make $b = 0$, which in turn would make $u = 0$.

Substituting Eq. (10.53) into Eq. (10.51) we have

$$u_n(x, t) = \sin \frac{n\pi x}{l} \left(C_n \sin \frac{n\pi v t}{l} + D_n \cos \frac{n\pi v t}{l} \right), \quad n = 1, 2, 3, \ldots \tag{10.54}$$

We see that there is an infinite set of discrete values of b and that to each value of b there corresponds a particular solution. Any linear combination of these particular solutions is also a solution:

$$u_n(x, t) = \sum_{n=1}^{\infty} \sin \frac{n\pi x}{l} \left(C_n \sin \frac{n\pi v t}{l} + D_n \cos \frac{n\pi v t}{l} \right). \tag{10.55}$$

The constants C_n and D_n are fixed by the boundary conditions (10.45) and (10.46).

Application of boundary condition (10.45) yields

$$f(x) = \sum_{n=1}^{\infty} D_n \sin \frac{n\pi x}{l}. \tag{10.56}$$

Similarly, application of boundary condition (10.46) gives

$$g(x) = \frac{\pi v}{l} \sum_{n=1}^{\infty} n C_n \sin \frac{n\pi x}{l}. \tag{10.57}$$

The coefficients C_n and D_n may then be determined by the Fourier series method:

$$D_n = \frac{2}{l} \int_0^l f(x) \sin \frac{n\pi x}{l} dx, \quad C_n = \frac{2}{n\pi v} \int_0^l g(x) \sin \frac{n\pi x}{l} dx. \tag{10.58}$$

We can use the method of separation of variable to solve the heat conduction equation. We shall leave this as a home work problem.

In the following sections, we shall consider two more methods for the solution of linear partial differential equations: the method of Green's functions, and the method of the Laplace transformation which was used in Chapter 9 for the solution of ordinary linear differential equations with constant coefficients.

Solution of Poisson's equation. Green's functions

The Green's function approach to boundary-value problems is a very powerful technique. The field at a point caused by a source can be considered to be the total effect due to each "unit" (or elementary portion) of the source. If $G(x; x')$ is the field at a point x due to a unit point source at x', then the total field at x due to a distributed source $\rho(x')$ is the integral of $G\rho$ over the range of x' occupied by the source. The function $G(x; x')$ is the well-known Green's function. We now apply this technique to solve Poisson's equation for electric potential ϕ (Example 10.2)

$$\nabla^2\phi(\mathbf{r}) = -\frac{1}{\varepsilon}\rho(\mathbf{r}), \qquad (10.59)$$

where ρ is the charge density and ε the permittivity of the medium, both are given.

By definition, Green's function $G(\mathbf{r}; \mathbf{r}')$ is the solution of

$$\nabla^2 G(\mathbf{r}; \mathbf{r}') = \delta(\mathbf{r} - \mathbf{r}'), \qquad (10.60)$$

where $\delta(\mathbf{r} - \mathbf{r}')$ is the Dirac delta function.

Now, multiplying Eq. (10.60) by ϕ and Eq. (10.59) by G, and then subtracting, we find

$$\phi(\mathbf{r})\nabla^2 G(\mathbf{r}; \mathbf{r}') - G(\mathbf{r}; \mathbf{r}')\nabla^2\phi(\mathbf{r}) = \phi(\mathbf{r})\delta(\mathbf{r} - \mathbf{r}') + \frac{1}{\varepsilon}G(\mathbf{r}; \mathbf{r}')\rho(\mathbf{r});$$

and on interchanging $\mathbf{r}$ and $\mathbf{r}'$,

$$\phi(\mathbf{r}')\nabla'^2 G(\mathbf{r}'; \mathbf{r}) - G(\mathbf{r}'; \mathbf{r})\nabla'^2\phi(\mathbf{r}') = \phi(\mathbf{r}')\delta(\mathbf{r}' - \mathbf{r}) + \frac{1}{\varepsilon}G(\mathbf{r}'; \mathbf{r})\rho(\mathbf{r}')$$

or

$$\phi(\mathbf{r}')\delta(\mathbf{r}' - \mathbf{r}) = \phi(\mathbf{r}')\nabla'^2 G(\mathbf{r}'; \mathbf{r}) - G(\mathbf{r}'; \mathbf{r})\nabla'^2\phi(\mathbf{r}') - \frac{1}{\varepsilon}G(\mathbf{r}'; \mathbf{r})\rho(\mathbf{r}'), \qquad (10.61)$$

the prime on ∇ indicates that differentiation is with respect to the primed co-ordinates. Integrating this last equation over all $\mathbf{r}'$ within and on the surface S' which encloses all sources (charges) yields

$$\phi(\mathbf{r}) = -\frac{1}{\varepsilon}\int G(\mathbf{r}; \mathbf{r}')\rho(\mathbf{r}')d\mathbf{r}'$$

$$+ \int [\phi(\mathbf{r}')\nabla'^2 G(\mathbf{r}; \mathbf{r}') - G(\mathbf{r}; \mathbf{r}')\nabla'^2\phi(\mathbf{r}')]d\mathbf{r}', \qquad (10.62)$$

where we have used the property of the delta function

$$\int_{-\infty}^{+\infty} f(\mathbf{r}')\delta(\mathbf{r} - \mathbf{r}')d\mathbf{r}' = f(\mathbf{r}).$$

We now use Green's theorem

$$\iiint (f\nabla'^2\psi - \psi\nabla'^2 f)d\tau' = \iint (f\nabla'\psi - \psi\nabla'f)\cdot d\mathbf{S}$$

to transform the second term on the right hand side of Eq. (10.62) and obtain

$$\phi(\mathbf{r}) = -\frac{1}{\varepsilon} \int G(\mathbf{r};\mathbf{r}')\rho(\mathbf{r}')d\mathbf{r}'$$

$$+ \int [\phi(\mathbf{r}')\nabla' G(\mathbf{r};\mathbf{r}') - G(\mathbf{r};\mathbf{r}')\nabla'\phi(\mathbf{r}')] \cdot d\mathbf{S}' \qquad (10.63)$$

or

$$\phi(\mathbf{r}) = -\frac{1}{\varepsilon} \int G(\mathbf{r};\mathbf{r}')\rho(\mathbf{r}')d\mathbf{r}'$$

$$+ \int \left[\phi(\mathbf{r}')\frac{\partial}{\partial n'} G(\mathbf{r};\mathbf{r}') - G(\mathbf{r};\mathbf{r}')\frac{\partial}{\partial n'}\phi(\mathbf{r}') \right] \cdot d\mathbf{S}', \qquad (10.64)$$

where $\mathbf{n}'$ is the outward normal to dS'. The Green's function $G(\mathbf{r};\mathbf{r}')$ can be found from Eq. (10.60) subject to the appropriate boundary conditions.

If the potential ϕ vanishes on the surface S' or $\partial\phi/\partial n'$ vanishes, Eq. (10.64) reduces to

$$\phi(\mathbf{r}) = -\frac{1}{\varepsilon} \int G(\mathbf{r};\mathbf{r}')\rho(\mathbf{r}')d\mathbf{r}'. \qquad (10.65)$$

On the other hand, if the surface S' encloses no charge, then Poisson's equation reduces to Laplace's equation and Eq. (10.64) reduces to

$$\phi(\mathbf{r}) = \int \left[\phi(\mathbf{r}')\frac{\partial}{\partial n'} G(\mathbf{r};\mathbf{r}') - G(\mathbf{r};\mathbf{r}')\frac{\partial}{\partial n'}\phi(\mathbf{r}') \right] \cdot d\mathbf{S}'. \qquad (10.66)$$

The potential at a field point $\mathbf{r}$ due to a point charge q located at the point $\mathbf{r}'$ is

$$\phi(\mathbf{r}) = \frac{1}{4\pi\varepsilon} \frac{q}{|\mathbf{r} - \mathbf{r}'|}.$$

Now

$$\nabla^2 \left(\frac{1}{|\mathbf{r} - \mathbf{r}'|} \right) = -4\pi\delta(\mathbf{r} - \mathbf{r}')$$

(the proof is left as an exercise for the reader) and it follows that the Green's function $G(\mathbf{r};\mathbf{r}')$ in this case is equal

$$G(\mathbf{r};\mathbf{r}') = \frac{1}{4\pi\varepsilon} \frac{1}{|\mathbf{r} - \mathbf{r}'|}.$$

If the medium is bounded, the Green's function can be obtained by direct solution of Eq. (10.60) subject to the appropriate boundary conditions.

To illustrate the procedure of the Green's function technique, let us consider a simple example that can easily be solved by other methods. Consider two grounded parallel conducting plates of infinite extent: if the electric charge density ρ between the two plates is given, find the electric potential distribution ϕ between the plates. The electric potential distribution ϕ is described by solving Poisson's equation

$$\nabla^2 \phi = -\rho/\varepsilon$$

subject to the boundary conditions

(1) $\phi(0) = 0$,
(2) $\phi(1) = 0$.

We take the coordinates shown in Fig. 10.3. Poisson's equation reduces to the simple form

$$\frac{d^2\phi}{dx^2} = -\frac{\rho}{\varepsilon}. \tag{10.67}$$

Instead of using the general result (10.64), it is more convenient to proceed directly. Multiplying Eq. (10.67) by $G(x; x')$ and integrating, we obtain

$$\int_0^1 G \frac{d^2\phi}{dx^2} dx = -\int_0^1 \frac{\rho(x)G}{\varepsilon} dx. \tag{10.68}$$

Then using integration by parts gives

$$\int_0^1 G \frac{d^2\phi}{dx^2} dx = G(x; x') \frac{d\phi(x)}{dx} \Big|_0^1 - \int_0^1 \frac{dG}{dx} \frac{d\phi}{dx} dx$$

and using integration by parts again on the right hand side, we obtain

$$-\int_0^1 G \frac{d^2\phi}{dx^2} dx = -G(x; x') \frac{d\phi(x)}{dx} \Big|_0^1 + \left[\frac{dG}{dx} \phi \Big|_0^1 - \int_0^1 \phi \frac{d^2G}{dx^2} dx \right]$$

$$= G(0; x') \frac{d\phi(0)}{dx} - G(1; x') \frac{d\phi(1)}{dx} - \int_0^1 \phi \frac{d^2G}{dx^2} dx.$$

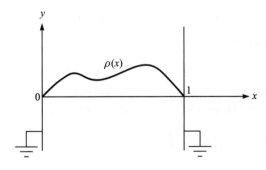

Figure 10.3.

Substituting this into Eq. (10.68) we obtain

$$G(0;x')\frac{d\phi(0)}{dx} - G(1;x')\frac{d\phi(1)}{dx} - \int_0^1 \phi\frac{d^2G}{dx^2}\,dx = \int_0^1 \frac{G(x;x')\rho(x)}{\varepsilon}\,dx$$

or

$$\int_0^1 \phi\frac{d^2G}{dx^2}\,dx = G(1;x')\frac{d\phi(1)}{dx} - G(0;x')\frac{d\phi(0)}{dx} - \int_0^1 \frac{G(x;x')\rho(x)}{\varepsilon}\,dx. \quad (10.69)$$

We must now choose a Green's function which satisfies the following equation and the boundary conditions:

$$\frac{d^2G}{dx^2} = -\delta(x-x'), \quad G(0;x') = G(1;x') = 0. \quad (10.70)$$

Combining these with Eq. (10.69) we find the solution to be

$$\phi(x') = \int_0^1 \frac{1}{\varepsilon}\rho(x)G(x;x')\,dx. \quad (10.71)$$

It remains to find $G(x;x')$. By integration, we obtain from Eq. (10.70)

$$\frac{dG}{dx} = -\int \delta(x-x')\,dx + a = -U(x-x') + a,$$

where U is the unit step function and a is an integration constant to be determined later. Integrating once we get

$$G(x;x') = -\int U(x-x')\,dx + ax + b = -(x-x')U(x-x') + ax + b.$$

Imposing the boundary conditions on this general solution yields two equations:

$$G(0;x') = x'U(-x') + a\cdot 0 + b = 0 + 0 + b = 0,$$

$$G(1;x') = -(1-x')U(1-x') + a + b = 0.$$

From these we find

$$a = (1-x')U(1-x'), \quad b = 0$$

and the Green's function is

$$G(x;x') = -(x-x')U(x-x') + (1-x')x. \quad (10.72)$$

This gives the response at x' due to a unit source at x. Interchanging x and x' in Eqs. (10.70) and (10.71) we find the solution of Eq. (10.67) to be

$$\phi(x) = \int_0^1 \frac{1}{\varepsilon}\rho(x')G(x';x)\,dx' = \int_0^1 \frac{1}{\varepsilon}\rho(x')[-(x'-x)U(x'-x) + (1-x)x']\,dx'.$$

$$(10.73)$$

408

Note that the Green's function in the last equation can be written in the form

$$G(x; x') = \begin{cases} (1-x)x & x < x' \\ (1-x)x' & x > x' \end{cases}.$$

Laplace transform solutions of boundary-value problems

Laplace and Fourier transforms are useful in solving a variety of partial differential equations, the choice of the appropriate transforms depends on the type of boundary conditions imposed on the problem. To illustrate the use of the Lapace transforms in solving boundary-value problems, we solve the following equation:

$$\frac{\partial u}{\partial t} = 2\frac{\partial^2 u}{\partial x^2}, \tag{10.74}$$

$$u(0, t) = u(3, t) = 0, \qquad u(x, 0) = 10\sin 2\pi x - 6\sin 4\pi x. \tag{10.75}$$

Taking the Laplace transform of Eq. (10.74) with respect to t gives

$$L\left[\frac{\partial u}{\partial t}\right] = 2L\left[\frac{\partial^2 u}{\partial x^2}\right].$$

Now

$$L\left[\frac{\partial u}{\partial t}\right] = pL(u) - u(x, 0)$$

and

$$L\left[\frac{\partial^2 u}{\partial x^2}\right] = \int_0^\infty e^{-pt}\frac{\partial^2 u}{\partial x^2}dt = \frac{\partial^2}{\partial x^2}\int_0^\infty e^{-pt}u(x, t)dt = \frac{\partial^2}{\partial x^2}L[u].$$

Here $\partial^2/\partial x^2$ and $\int_0^\infty \cdots dt$ are interchangeable because x and t are independent. For convenience, let

$$U = U(x, p) = L[u(x, t)] = \int_0^\infty e^{-pt}u(x, t)dt.$$

We then have

$$pU - u(x, 0) = 2\frac{d^2 U}{dx^2},$$

from which we obtain, on using the given condition (10.75),

$$\frac{d^2 U}{dx^2} - \frac{1}{2}pU = 3\sin 4\pi x - 5\sin 2\pi x. \tag{10.76}$$

Now think of this as a differential equation in terms of x, with p as a parameter. Then taking the Laplace transform of the given conditions $u(0, t) = u(3, t) = 0$, we have

$$L[u(0, t)] = 0, \quad L[u(3, t)] = 0$$

or

$$U(0, p) = 0, \quad U(3, p) = 0.$$

These are the boundary conditions on $U(x, p)$. Solving Eq. (10.76) subject to these conditions we find

$$U(x, p) = \frac{5 \sin 2\pi x}{p + 16\pi^2} - \frac{3 \sin 4\pi x}{p + 64\pi^2}.$$

The solution to Eq. (10.74) can now be obtained by taking the inverse Laplace transform

$$u(x, t) = L^{-1}[U(x, p)] = 5e^{-16\pi^2 t} \sin 2\pi x - 3e^{-64\pi^2} \sin 4\pi x.$$

The Fourier transform method was used in Chapter 4 for the solution of ordinary linear ordinary differential equations with constant coefficients. It can be extended to solve a variety of partial differential equations. However, we shall not discuss this here. Also, there are other methods for the solution of linear partial differential equations. In general, it is a difficult task to solve partial differential equations analytically, and very often a numerical method is the best way of obtaining a solution that satisfies given boundary conditions.

Problems

10.1 (a) Show that $y(x, t) = F(2x + 5t) + G(2x - 5t)$ is a general solution of

$$4 \frac{\partial^2 y}{\partial t^2} = 25 \frac{\partial^2 y}{\partial x^2}.$$

(b) Find a particular solution satisfying the conditions

$$y(0, t) = y(\pi, t) = 0, \quad y(x, 0) = \sin 2x, \quad y'(x, 0) = 0.$$

10.2. State the nature of each of the following equations (that is, whether elliptic, parabolic, or hyperbolic)

(a) $\dfrac{\partial^2 y}{\partial t^2} + \alpha \dfrac{\partial^2 y}{\partial x^2} = 0,$ (b) $x \dfrac{\partial^2 u}{\partial x^2} + y \dfrac{\partial^2 u}{\partial y^2} + 3y^2 \dfrac{\partial u}{\partial x}.$

10.3 The electromagnetic wave equation: Classical electromagnetic theory was worked out experimentally in bits and pieces by Coulomb, Oersted, Ampere, Faraday and many others, but the man who put it all together and built it into the compact and consistent theory it is today was James Clerk Maxwell.

His work led to the understanding of electromagnetic radiation, of which light is a special case.

Given the four Maxwell equations

$$\nabla \cdot \mathbf{E} = \rho/\varepsilon_0, \qquad \text{(Gauss' law)},$$

$$\nabla \times \mathbf{B} = \mu_0(\mathbf{j} + \varepsilon_0 \partial \mathbf{E}/\partial t) \quad \text{(Ampere's law)},$$

$$\nabla \cdot \mathbf{B} = 0 \qquad \text{(Gauss' law)},$$

$$\nabla \times \mathbf{E} = -\partial \mathbf{B}/\partial t \qquad \text{(Faraday's law)},$$

where $\mathbf{B}$ is the magnetic induction, $\mathbf{j} = \rho \mathbf{v}$ is the current density, and μ_0 is the permeability of the medium, show that:

(a) the electric field and the magnetic induction can be expressed as

$$\mathbf{E} = -\nabla \phi - \partial \mathbf{A}/\partial t, \qquad \mathbf{B} = \nabla \times \mathbf{A},$$

where $\mathbf{A}$ is called the vector potential, and ϕ the scalar potential. It should be noted that $\mathbf{E}$ and $\mathbf{B}$ are invariant under the following transformations:

$$\mathbf{A}' = \mathbf{A} + \nabla \chi, \qquad \phi' = \phi - \partial \phi/\partial t$$

in which χ is an arbitrary real function. That is, both $(\mathbf{A}', \phi)$, and $(\mathbf{A}', \phi')$ yield the same $\mathbf{E}$ and $\mathbf{B}$. Any condition which, for computational convenience, restricts the form of $\mathbf{A}$ and ϕ is said to define a gauge. Thus the above transformation is called a gauge transformation and χ is called a gauge parameter.

(b) If we impose the so-called Lorentz gauge condition on $\mathbf{A}$ and ϕ:

$$\nabla \cdot \mathbf{A} + \mu_0 \varepsilon_0 (\partial \phi/\partial t) = 0,$$

then both $\mathbf{A}$ and ϕ satisfy the following wave equations:

$$\nabla^2 \mathbf{A} - \mu_0 \varepsilon_0 \frac{\partial^2 \mathbf{A}}{\partial t^2} = -\mu_0 \mathbf{j},$$

$$\nabla^2 \phi - \mu_0 \varepsilon_0 \frac{\partial^2 \phi}{\partial t^2} = -\rho/\varepsilon_0.$$

10.4 Given Gauss' law $\iint_S \mathbf{E} \cdot d\mathbf{s} = q/\varepsilon$, find the electric field produced by a charged plane of infinite extension is given by $E = \sigma/\varepsilon$, where σ is the charge per unit area of the plane.

10.5 Consider an infinitely long uncharged conducting cylinder of radius l placed in an originally uniform electric field E_0 directed at right angles to the axis of the cylinder. Find the potential at a point $\rho(> l)$ from the axis of the cylinder. The boundary conditions are:

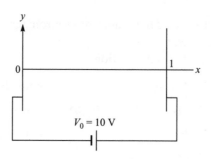

Figure 10.4.

$$\phi(\rho, \varphi) = \begin{cases} -E_0\rho \cos \varphi = -E_0x & \text{for} \quad \rho \to \infty, \\ 0 & \text{for} \quad \rho = l, \end{cases}$$

where the x-axis has been taken in the direction of the uniform field E_0.

10.6 Obtain the solution of the heat conduction equation

$$\frac{\partial^2 u(x, t)}{\partial x^2} = \frac{1}{\alpha} \frac{\partial u(x, t)}{\partial t}$$

which satisfies the boundary conditions
(1) $u(0, t) = u(l, t) = 0$, $t \geq 0$, (2) $u(x, 0) = f(x), 0 \leq x$, where $f(x)$ is a given function and l is a constant.

10.7 If a battery is connected to the plates as shown in Fig. 10.4, and if the charge density distribution between the two plates is still given by $\rho(x)$, find the potential distribution between the plates.

10.8 Find the Green's function that satisfies the equation

$$\frac{d^2 G}{dx^2} = \delta(x - x')$$

and the boundary conditions $G = 0$ when $x = 0$ and G remains bounded when x approaches infinity. (This Green's function is the potential due to a surface charge $-\varepsilon$ per unit area on a plane of infinite extent located at $x = x'$ in a dielectric medium of permittivity ε when a grounded conducting plane of infinite extent is located at $x = 0$.)

10.9 Solve by Laplace transforms the boundary-value problem

$$\frac{\partial^2 u}{\partial x^2} = \frac{1}{K} \frac{\partial u}{\partial t} \quad \text{for} \quad x > 0, \ t > 0,$$

given that $u = u_0$ (a constant) on $x = 0$ for $t > 0$, and $u = 0$ for $x > 0, t = 0$.

11

Simple linear integral equations

In previous chapters we have met equations in which the unknown functions appear under an integral sign. Such equations are called integral equations. Fourier and Laplace transforms are important integral equations, In Chapter 4, by introducing the method of Green's function we were led in a natural way to reformulate the problem in terms of integral equations. Integral equations have become one of the very useful and sometimes indispensable mathematical tools of theoretical physics and engineering.

Classification of linear integral equations

In this chapter we shall confine our attention to linear integral equations. Linear integral equations can be divided into two major groups:

(1) If the unknown function occurs only under the integral sign, the integral equation is said to be of the first kind. Integral equations having the unknown function both inside and outside the integral sign are of the second kind.

(2) If the limits of integration are constants, the equation is called a Fredholm integral equation. If one limit is variable, it is a Volterra equation.

These four kinds of linear integral equations can be written as follows:

$$f(x) = \int_a^b K(x,t)u(t)dt \qquad \text{Fredholm equation of the first kind;} \quad (11.1)$$

$$u(x) = f(x) + \lambda \int_a^b K(x,t)u(t)dt \quad \text{Fredholm equation of the second kind;}$$

$$(11.2)$$

413

$$f(x) = \int_a^x K(x, t)u(t)dt \qquad \text{Volterra equation of the first kind;} \quad (11.3)$$

$$u(x) = f(x) + \lambda \int_a^x K(x, t)u(t)dt \quad \text{Volterra equation of the second kind.}$$

$$(11.4)$$

In each case $u(t)$ is the unknown function, $K(x, t)$ and $f(x)$ are assumed to be known. $K(x, t)$ is called the kernel or nucleus of the integral equation. λ is a parameter, which often plays the role of an eigenvalue. The equation is said to be homogeneous if $f(x) = 0$.

If one or both of the limits of integration are infinite, or the kernel $K(x, t)$ becomes infinite in the range of integration, the equation is said to be singular; special techniques are required for its solution.

The general linear integral equation may be written as

$$h(x)u(x) = f(x) + \lambda \int_a^b K(x, t)u(t)dt. \qquad (11.5)$$

If $h(x) = 0$, we have a Fredholm equation of the first kind; if $h(x) = 1$, we have a Fredholm equation of the second kind. We have a Volterra equation when the upper limit is x.

It is beyond the scope of this book to present the purely mathematical general theory of these various types of equations. After a general discussion of a few methods of solution, we will illustrate them with some simple examples. We will then show with a few examples from physical problems how to convert differential equations into integral equations.

Some methods of solution

Separable kernel

When the two variables x and t which appear in the kernel $K(x, t)$ are separable, the problem of solving a Fredholm equation can be reduced to that of solving a system of algebraic equations, a much easier task. When the kernel $K(x, t)$ can be written as

$$K(x, t) = \sum_{i=1}^n g_i(x)h_i(t), \qquad (11.6)$$

where $g(x)$ is a function of x only and $h(t)$ a function of t only, it is said to be degenerate. Putting Eq. (11.6) into Eq. (11.2), we obtain

$$u(x) = f(x) + \lambda \sum_{i=1}^n \int_a^b g_i(x)h_i(t)u(t)dt.$$

414

Note that $g(x)$ is a constant as far as the t integration is concerned, hence it may be taken outside the integral sign and we have

$$u(x) = f(x) + \lambda \sum_{i=1}^{n} g_i(x) \int_a^b h_i(t)u(t)dt. \tag{11.7}$$

Now

$$\int_a^b h_i(t)u(t)dt = C_i \ (= \text{const.}). \tag{11.8}$$

Substituting this into Eq. (11.7) and solving for $u(t)$, we obtain

$$u(t) = f(x) + \lambda C \sum_{i=1}^{n} g_i(x). \tag{11.9}$$

The value of C_i may now be obtained by substituting Eq. (11.9) into Eq. (11.8). The solution is only valid for certain values of λ, and we call these the eigenvalues of the integral equation. The homogeneous equation has non-trivial solutions only if λ is one of these eigenvalues; these solutions are called eigenfunctions of the kernel (operator) K.

Example 11.1
As an example of this method, we consider the following equation:

$$u(x) = x + \lambda \int_0^1 (xt^2 + x^2t)u(t)dt. \tag{11.10}$$

This is a Fredholm equation of the second kind, with $f(x) = x$ and $K(x,t) = xt^2 + x^2t$. If we define

$$\alpha = \int_0^1 t^2 u(t)dt, \quad \beta = \int_0^1 tu(t)dt, \tag{11.11}$$

then Eq. (11.10) becomes

$$u(x) = x + \lambda(\alpha x + \beta x^2). \tag{11.12}$$

To determine A and B, we put Eq. (11.12) back into Eq. (11.11) and obtain

$$\alpha = \tfrac{1}{4} + \tfrac{1}{4}\lambda\alpha + \tfrac{1}{5}\lambda\beta, \quad \beta = \tfrac{1}{3} + \tfrac{1}{3}\lambda\alpha + \tfrac{1}{4}\lambda\beta. \tag{11.13}$$

Solving this for α and β we find

$$\alpha = \frac{60 + \lambda}{240 - 120\lambda - \lambda^2}, \quad \beta = \frac{80}{240 - 120\lambda - \lambda^2},$$

and the final solution is

$$u(t) = \frac{(240 - 60\lambda)x + 80\lambda x^2}{240 - 120\lambda - \lambda^2}.$$

The solution blows up when $\lambda = 117.96$ or $\lambda = 2.04$. These are the eigenvalues of the integral equation.

Fredholm found that if: (1) $f(x)$ is continuous, (2) $K(x, t)$ is piecewise continuous, (3) the integrals $\iint K^2(x, t)dxdt$, $\int f^2(t)dt$ exist, and (4) the integrals $\iint K^2(x, t)dt$ and $\int K^2(t, x)dt$ are bounded, then the following theorems apply:

(a) Either the inhomogeneous equation

$$u(x) = f(x) + \lambda \int_a^b K(x, t)u(t)dt$$

has a unique solution for any function $f(x)$ (λ is not an eigenvalue), or the homogeneous equation

$$u(x) = \lambda \int_a^b K(x, t)u(t)dt$$

has at least one non-trivial solution corresponding to a particular value of λ. In this case, λ is an eigenvalue and the solution is an eigenfunction.

(b) If λ is an eigenvalue, then λ is also an eigenvalue of the transposed equation

$$u(x) = \lambda \int_a^b K(t, x)u(t)dt;$$

and, if λ is not an eigenvalue, then λ is also not an eigenvalue of the transposed equation

$$u(x) = f(x) + \lambda \int_a^b K(t, x)u(t)dt.$$

(c) If λ is an eigenvalue, the inhomogeneous equation has a solution if, and only if,

$$\int_a^b u(x)f(x)dx = 0$$

for every function $f(x)$.

We refer the readers who are interested in the proof of these theorems to the book by R. Courant and D. Hilbert (*Methods of Mathematical Physics*, Vol. 1, Wiley, 1961).

Neumann series solutions

This method is due largely to Neumann, Liouville, and Volterra. In this method we solve the Fredholm equation (11.2)

$$u(x) = f(x) + \lambda \int_a^b K(x, t)u(t)dt$$

by iteration or successive approximations, and begin with the approximation

$$u(x) \approx u_0(x) \approx f(x).$$

This approximation is equivalent to saying that the constant λ or the integral is small. We then put this crude choice into the integral equation (11.2) under the integral sign to obtain a second approximation:

$$u_1(x) = f(x) + \lambda \int_a^b K(x,t)f(t)dt$$

and the process is then repeated and we obtain

$$u_2(x) = f(x) + \lambda \int_a^b K(x,t)f(t)dt + \lambda^2 \int_a^b \int_a^b K(x,t)K(t,t')f(t')dt'dt.$$

We can continue iterating this process, and the resulting series is known as the Neumann series, or Neumann solution:

$$u(x) = f(x) + \lambda \int_a^b K(x,t)f(t)dt + \lambda^2 \int_a^b \int_a^b K(x,t)K(t,t')f(t')dt'dt + \cdots.$$

This series can be written formally as

$$u_n(x) = \sum_{i=1}^n \lambda^i \varphi_i(x), \tag{11.14}$$

where

$$\varphi_0(x) = u_0(x) = f(x),$$

$$\varphi_1(x) = \int_a^b K(x,t_1)f(t_1)dt_1,$$

$$\varphi_2(x) = \int_a^b \int_a^b K(x,t_1)K(t_1,t_2)f(t_2)dt_1dt_2,$$

$$\vdots$$

$$\varphi_n(x) = \int_a^b \int_a^b \cdots \int_a^b K(x,t_1)K(t_1,t_2)\cdots K(t_{n-1},t_n)f(t_n)dt_1dt_2\cdots dt_n.$$

$$\left.\right\} \tag{11.15}$$

The series (11.14) will converge for sufficiently small λ, when the kernel $K(x,t)$ is bounded. This can be checked with the Cauchy ratio test (Problem 11.4).

Example 11.2
Use the Neumann method to solve the integral equation

$$u(x) = f(x) + \frac{1}{2}\int_{-1}^1 K(x,t)u(t)dt, \tag{11.16}$$

417

where

$$f(x) = x, \qquad K(x, t) = t - x.$$

Solution: We begin with

$$u_0(x) = f(x) = x.$$

Then

$$u_1(x) = x + \frac{1}{2} \int_{-1}^{1} (t - x) t \, dt = x + \frac{1}{3}.$$

Putting $u_1(x)$ into Eq. (11.16) under the integral sign, we obtain

$$u_2(x) = x + \frac{1}{2} \int_{-1}^{1} (t - x) \left(t + \frac{1}{3} \right) dt = x + \frac{1}{3} - \frac{x}{3}.$$

Repeating this process of substituting back into Eq. (11.16) once more, we obtain

$$u_3(x) = x + \frac{1}{3} - \frac{x}{3} - \frac{1}{3^2}.$$

We can improve the approximation by iterating the process, and the convergence of the resulting series (solution) can be checked out with the ratio test.

The Neumann method is also applicable to the Volterra equation, as shown by the following example.

Example 11.3
Use the Neumann method to solve the Volterra equation

$$u(x) = 1 + \lambda \int_0^x u(t) \, dt.$$

Solution: We begin with the zeroth approximation $u_0(x) = 1$. Then

$$u_1(x) = 1 + \lambda \int_0^x u_0(t) \, dt = 1 + \lambda \int_0^x dt = 1 + \lambda x.$$

This gives

$$u_2(x) = 1 + \lambda \int_0^x u_1(t) \, dt = 1 + \lambda \int_0^x (1 + \lambda t) \, dt = 1 + \lambda x + \frac{1}{2} \lambda^2 x^2;$$

similarly,

$$u_3(x) = 1 + \lambda \int_0^x \left(1 + \lambda t + \frac{1}{2} \lambda^2 t^2 \right) dt = 1 + \lambda t + \frac{1}{2} \lambda^2 t^2 + \frac{1}{3!} \lambda^3 x^3.$$

By induction

$$u_n(x) = \sum_{k=1}^{n} \frac{1}{k!} \lambda^k x^k.$$

When $n \to \infty$, $u_n(x)$ approaches

$$u(x) = e^{\lambda x}.$$

Transformation of an integral equation into a differential equation

Sometimes the Volterra integral equation can be transformed into an ordinary differential equation which may be easier to solve than the original integral equation, as shown by the following example.

Example 11.4

Consider the Volterra integral equation $u(x) = 2x + 4 \int_0^x (t - x)u(t)dt$. Before we transform it into a differential equation, let us recall the following very useful formula: if

$$I(\alpha) = \int_{a(\alpha)}^{b(\alpha)} f(x, \alpha)dx,$$

where a and b are continuous and at least once differentiable functions of α, then

$$\frac{dI(\alpha)}{d\alpha} = f(b, \alpha)\frac{db}{d\alpha} - f(a, \alpha)\frac{da}{d\alpha} + \int_a^b \frac{\partial f(x, \alpha)}{\partial \alpha}dx.$$

With the help of this formula, we obtain

$$\frac{d}{dx}u(x) = 2 + 4\left[\{(t - x)u(t)\}_{t=x} - \int_0^x u(t)dt\right]$$

$$= 2 - 4\int_0^x u(t)dt.$$

Differentiating again we obtain

$$\frac{d^2u(x)}{dx^2} = -4u(x).$$

This is a differentiation equation equivalent to the original integral equation, but its solution is much easier to find:

$$u(x) = A\cos 2x + B\sin 2x,$$

where A and B are integration constants. To determine their values, we put the solution back into the original integral equation under the integral sign, and then integration gives $A = 0$ and $B = 1$. Thus the solution of the original integral equation is

$$u(x) = \sin 2x.$$

Laplace transform solution

The Volterra integral equation can sometime be solved with the help of the Laplace transformation and the convolution theorem. Before we consider the Laplace transform solution, let us review the convolution theorem. If $f_1(x)$ and $f_2(x)$ are two arbitrary functions, we define their convolution (*faltung* in German) to be

$$g(x) = \int_{-\infty}^{\infty} f_1(y) f_2(x - y) dy.$$

Its Laplace transform is

$$L[g(x)] = L[f_1(x)] L[f_2(x)].$$

We now consider the Volterra equation

$$u(x) = f(x) + \lambda \int_0^x K(x, t) u(t) dt$$

$$= f(x) + \lambda \int_0^x g(x - t) u(t) dt, \tag{11.17}$$

where $K(x - t) = g(x - t)$, a so-called displacement kernel. Taking the Laplace transformation and using the convolution theorem, we obtain

$$L\left[\int_0^x g(x - t) u(t) dt\right] = L[g(x - t)] L[u(t)] = G(p) U(p),$$

where $U(p) = L[u(t)] = \int_0^\infty e^{-pt} u(t) dt$, and similarly for $G(p)$. Thus, taking the Laplace transformation of Eq. (11.17), we obtain

$$U(p) = F(p) + \lambda G(p) U(p)$$

or

$$U(p) = \frac{F(p)}{1 - \lambda G(p)}.$$

Inverting this we obtain $u(t)$:

$$u(t) = L^{-1}\left[\frac{F(p)}{1 - \lambda G(p)}\right].$$

420

Fourier transform solution

If the kernel is a displacement kernel and if the limits are $-\infty$ and $+\infty$, we can use Fourier transforms. Consider a Fredholm equation of the second kind

$$u(x) = f(x) + \lambda \int_{-\infty}^{\infty} K(x - t)u(t)dt. \tag{11.18}$$

Taking Fourier transforms (indicated by overbars)

$$\frac{1}{\sqrt{2\pi}} \int_{-\infty}^{\infty} dx f(x)e^{-ipx} = \bar{f}(p), \text{etc.},$$

and using the convolution theorem

$$\int_{-\infty}^{\infty} f(t)g(x - t)dt = \int_{-\infty}^{\infty} \bar{f}(y)\bar{g}(y)e^{-iyx}dy,$$

we obtain the transform of our integral equation (11.18):

$$\bar{u}(p) = \bar{f}(p) + \lambda \bar{K}(p)\bar{u}(p).$$

Solving for $\bar{u}(p)$ we obtain

$$\bar{u}(p) = \frac{\bar{f}(p)}{1 - \lambda \bar{K}(p)}.$$

If we can invert this equation, we can solve the original integral equation:

$$u(x) = \frac{1}{\sqrt{2\pi}} \int_{-\infty}^{\infty} \frac{f(t)e^{-ixt}}{1 - \sqrt{2\pi}\lambda K(t)}. \tag{11.19}$$

The Schmidt–Hilbert method of solution

In many physical problems, the kernel may be symmetric. In such cases, the integral equation may be solved by a method quite different from any of those in the preceding section. This method, devised by Schmidt and Hilbert, is based on considering the eigenfunctions and eigenvalues of the homogeneous integral equation.

A kernel $K(x, t)$ is said to be symmetric if $K(x, t) = K(t, x)$ and Hermitian if $K(x, y) = K^*(t, x)$. We shall limit our discussion to such kernels.

(*a*) The homogeneous Fredholm equation

$$u(x) = \lambda \int_a^b K(x, t)u(t)dt.$$

A Hermitian kernel has at least one eigenvalue and it may have an infinite number. The proof will be omitted and we refer interested readers to the book by Courant and Hibert mentioned earlier (Chapter 3).

The eigenvalues of a Hermitian kernel are real, and eigenfunctions belonging to different eigenvalues are orthogonal; two functions $f(x)$ and $g(x)$ are said to be orthogonal if

$$\int f^*(x)g(x)dx = 0.$$

To prove the reality of the eigenvalue, we multiply the homogeneous Fredholm equation by $u^*(x)$, then integrating with respect to x, we obtain

$$\int_a^b u^*(x)u(x)dx = \lambda \int_a^b \int_a^b K(x,t)u^*(x)u(t)dtdx. \tag{11.20}$$

Now, multiplying the complex conjugate of the Fredholm equation by $u(x)$ and then integrating with respect to x, we get

$$\int_a^b u^*(x)u(x)dx = \lambda^* \int_a^b \int_a^b K^*(x,t)u^*(t)u(x)dtdx.$$

Interchanging x and t on the right hand side of the last equation and remembering that the kernel is Hermitian $K^*(t,x) = K(x,t)$, we obtain

$$\int_a^b u^*(x)u(x)dx = \lambda^* \int_a^b \int_a^b K(x,t)u(t)u^*(x)dtdx.$$

Comparing this equation with Eq. (11.2), we see that $\lambda = \lambda^*$, that is, λ is real.

We now prove the orthogonality. Let λ_i, λ_j be two different eigenvalues and $u_i(x), u_j(x)$, the corresponding eigenfunctions. Then we have

$$u_i(x) = \lambda_i \int_a^b K(x,t)u_i(t)dt, \qquad u_j(x) = \lambda_j \int_a^b K(x,t)u_j(t)dt.$$

Now multiplying the first equation by $\lambda_j u_j(x)$, the second by $\lambda_i u_i(x)$, and then integrating with respect to x, we obtain

$$\lambda_j \int_a^b u_i(x)u_j(x)dx = \lambda_i\lambda_j \int_a^b \int_a^b K(x,t)u_i(t)u_j(x)dtdx,$$

$$\lambda_i \int_a^b u_i(x)u_j(x)dx = \lambda_i\lambda_j \int_a^b \int_a^b K(x,t)u_j(t)u_i(x)dtdx. \tag{11.21}$$

Now we interchange x and t on the right hand side of the last integral and because of the symmetry of the kernel, we have

$$\lambda_i \int_a^b u_i(x)u_j(x)dx = \lambda_i\lambda_j \int_a^b \int_a^b K(x,t)u_i(t)u_j(x)dtdx. \tag{11.22}$$

Subtracting Eq. (11.21) from Eq. (11.22), we obtain

$$(\lambda_i - \lambda_j) \int_a^b u_i(x)u_j(x)dx = 0. \tag{11.23}$$

Since $\lambda_i \neq \lambda_j$, it follows that

$$\int_a^b u_i(x)u_j(x)dx = 0. \tag{11.24}$$

Such functions may always be nomalized. We will assume that this has been done and so the solutions of the homogeneous Fredholm equation form a complete orthonomal set:

$$\int_a^b u_i(x)u_j(x)dx = \delta_{ij}. \tag{11.25}$$

Arbitrary functions of x, including the kernel for fixed t, may be expanded in terms of the eigenfunctions

$$K(x,t) = \sum C_i u_i(x). \tag{11.26}$$

Now substituting Eq. (11.26) into the original Fredholm equation, we have

$$u_j(t) = \lambda_j \int_a^b K(t,x)u_j(x)dx = \lambda_j \int_a^b K(x,t)u_j(x)dx$$

$$= \lambda_j \sum_i \int_a^b C_i u_i(x)u_j(x)dx = \lambda_j \sum_i C_i \delta_{ij} = \lambda_j C_j$$

or

$$C_i = u_i(t)/\lambda_i$$

and for our homogeneous Fredholm equation of the second kind the kernel may be expressed in terms of the eigenfunctions and eigenvalues as

$$K(x,t) = \sum_{n=1}^\infty \frac{u_n(x)u_n(t)}{\lambda_n}. \tag{11.27}$$

The Schmidt–Hilbert theory does not solve the homogeneous integral equation; its main function is to establish the properties of the eigenvalues (reality) and eigenfunctions (orthogonality and completeness). The solutions of the homogeneous integral equation come from the preceding section on methods of solution.

(b) Solution of the inhomogeneous equation

$$u(x) = f(x) + \lambda \int_a^b K(x,t)u(t)dt. \tag{11.28}$$

We assume that we have found the eigenfunctions of the homogeneous equation by the methods of the preceding section, and we denote them by $u_i(x)$. We may now expand both $u(x)$ and $f(x)$ in terms of $u_i(x)$, which forms an orthonormal complete set.

$$u(x) = \sum_{n=1}^{\infty} \alpha_n u_n(x), \quad f(x) = \sum_{n=1}^{\infty} \beta_n u_n(x). \tag{11.29}$$

Substituting Eq. (11.29) into Eq. (11.28), we obtain

$$\sum_{n=1}^{n} \alpha_n u_n(x) = \sum_{n=1}^{n} \beta_n u_n(x) + \lambda \int_a^b K(x,t) \sum_{n=1}^{n} \alpha_n u_n(t) dt$$

$$= \sum_{n=1}^{n} \beta_n u_n(x) + \lambda \sum_{n=1}^{\infty} \alpha_n \frac{u_m(x)}{\lambda_m} \int_a^b u_m(t) u_n(t) dt,$$

$$= \sum_{n=1}^{n} \beta_n u_n(x) + \lambda \sum_{n=1}^{\infty} \alpha_n \frac{u_m(x)}{\lambda_m} \delta_{nm},$$

from which it follows that

$$\sum_{n=1}^{n} \alpha_n u_n(x) = \sum_{n=1}^{n} \beta_n u_n(x) + \lambda \sum_{n=1}^{\infty} \frac{\alpha_n u_n(x)}{\lambda_n}. \tag{11.30}$$

Multiplying by $u_i(x)$ and then integrating with respect to x from a to b, we obtain

$$\alpha_n = \beta_n + \lambda \alpha_n / \lambda_n, \tag{11.31}$$

which can be solved for α_n in terms of β_n:

$$\alpha_n = \frac{\lambda_n}{\lambda_n - \lambda} \beta_n, \tag{11.32}$$

where β_n is given by

$$\beta_n = \int_a^b f(t) u_n(t) dt. \tag{11.33}$$

Finally, our solution is given by

$$u(x) = f(x) + \lambda \sum_{n=1}^{\infty} \frac{\alpha_n u_n(x)}{\lambda_n}$$

$$= f(x) + \lambda \sum_{n=1}^{\infty} \frac{\beta_n}{\lambda_n - \lambda} u_n(x), \tag{11.34}$$

where β_n is given by Eq. (11.33), and $\lambda_i \neq \lambda$.

When λ for the inhomogeneous equation is equal to one of the eigenvalues, λ_k, of the kernel, our solution (11.31) blows up. Let us return to Eq. (11.31) and see what happens to α_k:

$$\alpha_k = \beta_k + \lambda_k \alpha_k / \lambda_k = \beta_k + \alpha_k.$$

Clearly, $\beta_k = 0$, and α_k is no longer determined by β_k. But we have, according to Eq. (11.33),

$$\int_a^b f(t)u_k(t)dt = \beta_k = 0, \tag{11.35}$$

that is, $f(x)$ is orthogonal to the eigenfunction $u_k(x)$. Thus if $\lambda = \lambda_k$, the inhomogeneous equation has a solution only if $f(x)$ is orthogonal to the corresponding eigenfunction $u_k(x)$. The general solution of the equation is then

$$u(x) = f(x) + \alpha_k u_k(x) + \alpha_k \sum_{n=1'}^{\infty}{}' \frac{\int_a^b f(t)u_n(t)dt}{\lambda_n - \lambda_k} u_n(x), \tag{11.36}$$

where the prime on the summation sign means that the term $n = k$ is to be omitted from the sum. In Eq. (11.36) the α_k remains as an undetermined constant.

Relation between differential and integral equations

We have shown how an integral equation can be transformed into a differential equation that may be easier to solve than the original integral equation. We now show how to transform a differential equation into an integral equation. After we become familiar with the relation between differential and integral equations, we may state the physical problem in either form at will. Let us consider a linear second-order differential equation

$$x'' + A(t)x' + B(t)x = g(t), \tag{11.37}$$

with the initial condition

$$x(a) = x_0, \quad x'(a) = x_0'.$$

Integrating Eq. (11.37), we obtain

$$x' = -\int_a^t Ax' dt - \int_a^t Bx dt + \int_a^t g dt + C_1.$$

The initial conditions require that $C_1 = x_0'$. We next integrate the first integral on the right hand side by parts and obtain

$$x' = -Ax - \int_a^t (B - A')x dt + \int_a^t g dt + A(a)x_0 + x_0'.$$

Integrating again, we get

$$x = -\int_a^t Ax dt - \int_a^t \int_a^t [B(y) - A'(y)]x(y)dy dt$$
$$+ \int_a^t \int_a^t g(y)dy dt + [A(a)x_0 + x_0'](t - a) + x_0.$$

425

Then using the relation

$$\int_a^t \int_a^t f(y)\,dy\,dt = \int_a^t (t-y)f(y)\,dy,$$

we can rewrite the last equation as

$$x(t) = -\int_a^t \left[A(y) + (t-y)\{B(y) - A'(y)\}\right]x(y)\,dy$$

$$+ \int_a^t (t-y)g(y)\,dy + [A(a)x_0 + x_0'](t-a) + x_0, \qquad (11.38)$$

which can be put into the form of a Volterra equation of the second kind

$$x(t) = f(t) + \int_a^t K(t,y)x(y)\,dy, \qquad (11.39)$$

with

$$K(t,y) = (y-t)[B(y) - A'(y)] - A(y), \qquad (11.39a)$$

$$f(t) = \int_0^t (t-y)g(y)\,dy + [A(a)x_0 + x_0'](t-a) + x_0. \qquad (11.39b)$$

Use of integral equations

We have learned how linear integral equations of the more common types may be solved. We now show some uses of integral equations in physics; that is, we are going to state some physical problems in integral equation form. In 1823, Abel made one of the earliest applications of integral equations to a physical problem. Let us take a brief look at this old problem in mechanics.

Abel's integral equation

Consider a particle of mass m falling along a smooth curve in a vertical plane, the yz plane, under the influence of gravity, which acts in the negative z direction. Conservation of energy gives

$$\frac{1}{2}m(\dot{z}^2 + \dot{y}^2) + mgz = E,$$

where $\dot{z} = dz/dt$, and $\dot{y} = dy/dt$. If the shape of the curve is given by $y = F(z)$, we can write $\dot{y} = (dF/dz)\dot{z}$. Substituting this into the energy conservation equation and solving for $\dot{z}$, we obtain

$$\dot{z} = \frac{\sqrt{2E/m - 2gz}}{\sqrt{1 + (dF/dz)^2}} = \frac{\sqrt{E/mg - z}}{u(z)}, \qquad (11.40)$$

where

$$u(z) = \sqrt{1 + (dF/dz)^2/2g}.$$

If $\dot{z} = 0$ and $z = z_0$ at $t = 0$, then $E/mg = z_0$ and Eq. (11.40) becomes

$$\dot{z} = \sqrt{z_0 - z}/u(z).$$

Solving for time t, we obtain

$$t = -\int_z^{z_0} \frac{u(z)}{\sqrt{z_0 - z}}\, dz = \int_{z_0}^z \frac{u(z)}{\sqrt{z_0 - z}}\, dz,$$

where z is the height the particle reaches at time t.

Classical simple harmonic oscillator

Consider a linear oscillator

$$\ddot{x} + \omega^2 x = 0, \quad \text{with} \quad x(0) = 0, \quad \dot{x}(0) = 1.$$

We can transform this differential equation into an integral equation. Comparing with Eq. (11.37), we have

$$A(t) = 0, \quad B(t) = \omega^2, \quad \text{and} \quad g(t) = 0.$$

Substituting these into Eq. (11.38) (or (11.39), (11.39a), and (11.39b)), we obtain the integral equation

$$x(t) = t + \omega^2 \int_0^t (y - t)x(y)\,dy,$$

which is equivalent to the original differential equation plus the initial conditions.

Quantum simple harmonic oscillator

The Schrödinger equation for the energy eigenstates of the one-dimensional simple harmonic oscillator is

$$-\frac{\hbar^2}{2m}\frac{d^2\psi}{dx^2} + \frac{1}{2}m\omega^2 x^2\psi = E\psi. \tag{11.41}$$

Changing to the dimensionless variable $y = \sqrt{m\omega/\hbar}x$, Eq. (11.41) reduces to a simpler form:

$$\frac{d^2\psi}{dy^2} + (\alpha^2 - y^2)\psi = 0, \tag{11.42}$$

where $\alpha = \sqrt{2E/\hbar\omega}$. Taking the Fourier transform of Eq. (11.42), we obtain

$$\frac{d^2 g(k)}{dk^2} + (\alpha^2 - k^2)g(k) = 0, \tag{11.43}$$

427

where

$$g(k) = \frac{1}{\sqrt{2\pi}} \int_{-\infty}^{\infty} \psi(y) e^{iky} dy \tag{11.44}$$

and we also assume that ψ and ψ' vanish as $y \to \pm\infty$.

Eq. (11.43) is formally identical to Eq. (11.42). Since quantities such as the total probability and the expectation value of the potential energy must be remain finite for finite E, we should expect $g(k), dg(k)/dk \to 0$ as $k \to \pm\infty$. Thus g and ψ differ at most by a normalization constant

$$g(k) = c\psi(k).$$

It follows that ψ satisfies the integral equation

$$c\psi(k) = \frac{1}{\sqrt{2\pi}} \int_{-\infty}^{\infty} \psi(y) e^{iky} dy. \tag{11.45}$$

The constant c may be determined by substituting $c\psi$ on the right hand side:

$$c^2\psi(k) = \frac{1}{2\pi} \int_{-\infty}^{\infty} \int_{-\infty}^{\infty} \psi(z) e^{izy} e^{iky} dz dy$$

$$= \int_{-\infty}^{\infty} \psi(z) \delta(z+k) dz$$

$$= \psi(-k).$$

Recall that ψ may be simultaneously chosen to be a parity eigenstate $\psi(-x) = \pm\psi(x)$. We see that eigenstates of even parity require $c^2 = 1$, or $c = \pm 1$; and for eigenstates of odd parity we have $c^2 = -1$, or $c = \pm i$.

We shall leave the solution of Eq. (11.45), which can be approached in several ways, as an exercise for the reader.

Problems

11.1 Solve the following integral equations:

(a) $u(x) = \frac{1}{2} - x + \int_0^1 u(t) dt$;

(b) $u(x) = \lambda \int_0^1 u(t) dt$;

(c) $u(x) = x + \lambda \int_0^1 u(t) dt$.

11.2 Solve the Fredholm equation of the second kind

$$f(x) = u(x) + \lambda \int_a^b K(x,t) u(t) dt,$$

where $f(x) = \cosh x$, $K(x,t) = xt$.

11.3 The homogeneous Fredholm equation

$$u(x) = \lambda \int_0^{\pi/2} \sin x \sin t u(t)dt$$

only has a solution for a particular value of λ. Find the value of λ and the solution corresponding to this value of λ.

11.4 Solve homogeneous Fredholm equation $u(x) = \lambda \int_{-1}^1 (t + x)u(t)dt$. Find the values of λ and the corresponding solutions.

11.5 Check the convergence of the Neumann series (11.14) by the Cauchy ratio test.

11.6 Transform the following differential equations into integral equations:

(a) $\dfrac{dx}{dt} - x = 0$ with $x = 1$ when $t = 0$;

(b) $\dfrac{d^2x}{dt^2} + \dfrac{dx}{dt} + x = 1$ with $x = 0$, $\dfrac{dx}{dt} = 1$ when $t = 0$.

11.7 By using the Laplace transformation and the convolution theorem solve the equation

$$u(x) = x + \int_0^x \sin(x - t)u(t)dt.$$

11.8 Given the Fredholm integral equation

$$e^{-x^2} = \int_{-\infty}^{\infty} e^{-(x-t)^2} u(t)dt,$$

apply the Fouurier convolution technique to solve it for $u(t)$.

11.9 Find the solution of the Fredholm equation

$$u(x) = x + \lambda \int_0^1 (x + t)u(t)dt$$

by the Schmidt–Hilbert method for λ not equal to an eigenvalue. Show that there are no solutions when λ is an eigenvalue.

12

Elements of group theory

Group theory did not find a use in physics until the advent of modern quantum mechanics in 1925. In recent years group theory has been applied to many branches of physics and physical chemistry, notably to problems of molecules, atoms and atomic nuclei. Mostly recently, group theory has been being applied in the search for a pattern of 'family' relationships between elementary particles. Mathematicians are generally more interested in the abstract theory of groups, but the representation theory of groups of direct use in a large variety of physical problems is more useful to physicists. In this chapter, we shall give an elementary introduction to the theory of groups, which will be needed for understanding the representation theory.

Definition of a group (group axioms)

A group is a set of distinct elements for which a law of 'combination' is well defined. Hence, before we give 'group' a formal definition, we must first define what kind of 'elements' do we mean. Any collection of objects, quantities or operators form a set, and each individual object, quantity or operator is called an element of the set.

A group is a set of elements $A, B, C, \ldots$, finite or infinite in number, with a rule for combining any two of them to form a 'product', subject to the following four conditions:

(1) The product of any two group elements must be a group element; that is, if A and B are members of the group, then so is the product AB.

(2) The law of composition of the group elements is associative; that is, if A, B, and C are members of the group, then $(AB)C = A(BC)$.

(3) There exists a unit group element E, called the identity, such that $EA = AE = A$ for every member of the group.

430

(4) Every element has a unique inverse, A^{-1}, such that $AA^{-1} = A^{-1}A = E$.

The use of the word 'product' in the above definition requires comment. The law of combination is commonly referred as 'multiplication', and so the result of a combination of elements is referred to as a 'product'. However, the law of combination may be ordinary addition as in the group consisting of the set of all integers (positive, negative, and zero). Here $AB = A + B$, 'zero' is the identity, and $A^{-1} = (-A)$. The word 'product' is meant to symbolize a broad meaning of 'multiplication' in group theory, as will become clearer from the examples below.

A group with a finite number of elements is called a finite group; and the number of elements (in a finite group) is the order of the group.

A group containing an infinite number of elements is called an infinite group. An infinite group may be either discrete or continuous. If the number of the elements in an infinite group is denumerably infinite, the group is discrete; if the number of elements is non-denumerably infinite, the group is continuous.

A group is called Abelian (or commutative) if for every pair of elements A, B in the group, $AB = BA$. In general, groups are not Abelian and so it is necessary to preserve carefully the order of the factors in a group 'product'.

A subgroup is any subset of the elements of a group that by themselves satisfy the group axioms with the same law of combination.

Now let us consider some examples of groups.

Example 12.1

The real numbers 1 and -1 form a group of order two, under multiplication. The identity element is 1; and the inverse is $1/x$, where x stands for 1 or -1.

Example 12.2

The set of all integers (positive, negative, and zero) forms a discrete infinite group under addition. The identity element is zero; the inverse of each element is its negative. The group axioms are satisfied:

(1) is satisfied because the sum of any two integers (including any integer with itself) is always another integer.
(2) is satisfied because the associative law of addition $A + (B + C) = (A + B) + C$ is true for integers.
(3) is satisfied because the addition of 0 to any integer does not alter it.
(4) is satisfied because the addition of the inverse of an integer to the integer itself always gives 0, the identity element of our group: $A + (-A) = 0$.

Obviously, the group is Abelian since $A + B = B + A$. We denote this group by S_1.

431

The same set of all integers does not form a group under multiplication. Why? Because the inverses of integers are not integers and so they are not members of the set.

Example 12.3

The set of all rational numbers (p/q, with $q \neq 0$) forms a continuous infinite group under addition. It is an Abelian group, and we denote it by S_2. The identity element is 0; and the inverse of a given element is its negative.

Example 12.4

The set of all complex numbers ($z = x + iy$) forms an infinite group under addition. It is an Abelian group and we denote it by S_3. The identity element is 0; and the inverse of a given element is its negative (that is, $-z$ is the inverse of z).

The set of elements in S_1 is a subset of elements in S_2, and the set of elements in S_2 is a subset of elements in S_3. Furthermore, each of these sets forms a group under addition, thus S_1 is a subgroup of S_2, and S_2 a subgroup of S_3. Obviously S_1 is also a subgroup of S_3.

Example 12.5

The three matrices

$$\tilde{A} = \begin{pmatrix} 1 & 0 \\ 0 & 1 \end{pmatrix}, \quad \tilde{B} = \begin{pmatrix} 0 & 1 \\ -1 & -1 \end{pmatrix}, \quad \tilde{C} = \begin{pmatrix} -1 & -1 \\ 1 & 0 \end{pmatrix}$$

form an Abelian group of order three under matrix multiplication. The identity element is the unit matrix, $E = \tilde{A}$. The inverse of a given matrix is the inverse matrix of the given matrix:

$$\tilde{A}^{-1} = \begin{pmatrix} 1 & 0 \\ 0 & 1 \end{pmatrix} = \tilde{A}, \quad \tilde{B}^{-1} = \begin{pmatrix} -1 & -1 \\ 1 & 0 \end{pmatrix} = \tilde{C}, \quad \tilde{C}^{-1} = \begin{pmatrix} 0 & 1 \\ -1 & -1 \end{pmatrix} = \tilde{B}.$$

It is straightforward to check that all the four group axioms are satisfied. We leave this to the reader.

Example 12.6

The three permutation operations on three objects a, b, c

$$[1\ 2\ 3], [2\ 3\ 1], [3\ 1\ 2]$$

form an Abelian group of order three with sequential performance as the law of combination.

The operation $[1\ 2\ 3]$ means we put the object a first, object b second, and object c third. And two elements are multiplied by performing first the operation on the right, then the operation on the left. For example

$$[2\ 3\ 1][3\ 1\ 2]abc = [2\ 3\ 1]cab = abc.$$

Thus two operations performed sequentially are equivalent to the operation $[1\ 2\ 3]$:

$$[2\ 3\ 1][3\ 1\ 2] = [1\ 2\ 3].$$

similarly

$$[3\ 1\ 2][2\ 3\ 1]abc = [3\ 1\ 2]bca = abc,$$

that is,

$$[3\ 1\ 2][2\ 3\ 1] = [1\ 2\ 3].$$

This law of combination is commutative. What is the identity element of this group? And the inverse of a given element? We leave the reader to answer these questions. The group illustrated by this example is known as a cyclic group of order 3, C_3.

It can be shown that the set of all permutations of three objects

$$[1\ 2\ 3],\ [2\ 3\ 1],\ [3\ 1\ 2],\ [1\ 3\ 2],\ [3\ 2\ 1],\ [2\ 1\ 3]$$

forms a non-Abelian group of order six denoted by S_3. It is called the symmetric group of three objects. Note that C_3 is a subgroup of S_3.

Cyclic groups

We now revisit the cyclic groups. The elements of a cyclic group can be expressed as power of a single element A, say, as $A, A^2, A^3, \ldots, A^{p-1}, A^p = E$; p is the smallest integer for which $A^p = E$ and is the order of the group. The inverse of A^k is A^{p-k}, that is, an element of the set. It is straightforward to check that all group axioms are satisfied. We leave this to the reader. It is obvious that cyclic groups are Abelian since $A^k A = A A^k$ $(k < p)$.

Example 12.7
The complex numbers $1, i, -1, -i$ form a cyclic group of order 3. In this case, $A = i$ and $p = 3$: i^n, $n = 0, 1, 2, 3$. These group elements may be interpreted as successive $90°$ rotations in the complex plane $(0, \pi/2, \pi$, and $3\pi/2)$. Consequently, they can be represented by four 2×2 matrices. We shall come back to this later.

Example 12.8
We now consider a second example of cyclic groups: the group of rotations of an equilateral triangle in its plane about an axis passing through its center that brings

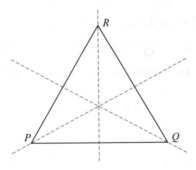

Figure 12.1.

it onto itself. This group contains three elements (see Fig. 12.1):

$E(=0°)$:   the identity; triangle is left alone;

$A(=120°)$:   the triangle is rotated through 120° counterclockwise, which sends P to Q, Q to R, and R to P;

$B(=240°)$:   the triangle is rotated through 240° counterclockwise, which sends P to R, R to Q, and Q to P;

$C(=360°)$:   the triangle is rotated through 360° counterclockwise, which sends P back to P, Q back to Q and R back to R.

Notice that $C = E$. Thus there are only three elements represented by E, A, and B. This set forms a group of order three under addition. The reader can check that all four group axioms are satisfied. It is also obvious that operation B is equivalent to performing operation A twice (240° = 120° + 120°), and the operation C corresponds to performing A three times. Thus the elements of the group may be expressed as the power of the single element A as E, A, A^2, A^3 ($=E$): that is, it is a cyclic group of order three, and is generated by the element A.

The cyclic group considered in Example 12.8 is a special case of groups of transformations (rotations, reflection, translations, permutations, etc.), the groups of particular interest to physicists. A transformation that leaves a physical system invariant is called a symmetry transformation of the system. The set of all symmetry transformations of a system is a group, as illustrated by this example.

Group multiplication table

A group of order n has n^2 products. Once the products of all ordered pairs of elements are specified the structure of a group is uniquely determined. It is sometimes convenient to arrange these products in a square array called a group multiplication table. Such a table is indicated schematically in Table 12.1. The element that appears at the intersection of the row labeled A and the column labeled B is the product AB, (in the table A^2 means AA, etc). It should be noted that all the

Table 12.1. Group multiplication table

	E	A	B	C	...
E	E	A	B	C	...
A	A	A^2	AB	AC	...
B	B	BA	B^2	BC	...
C	C	CA	CB	C^2	...
⋮	⋮	⋮	⋮	⋮	

Table 12.2.

	E	X	Y
E	E	X	Y
X^{-1}	X^{-1}	E	$X^{-1}Y$
Y^{-1}	Y^{-1}	$Y^{-1}X$	E

elements in each row or column of the group multiplication must be distinct: that is, each element appears once and only once in each row or column. This can be proved easily: if the same element appeared twice in a given row, the row labeled A say, then there would be two distinct elements C and D such that $AC = AD$. If we multiply the equation by A^{-1} on the left, then we would have $A^{-1}AC = A^{-1}AD$, or $EC = ED$. This cannot be true unless $C = D$, in contradiction to our hypothesis that C and D are distinct. Similarly, we can prove that all the elements in any column must be distinct.

As a simple practice, consider the group C_3 of Example 12.6 and label the elements as follows

$$[1\ 2\ 3] \rightarrow E, \quad [2\ 3\ 1] \rightarrow X, \quad [3\ 1\ 2] \rightarrow Y.$$

If we label the columns of the table with the elements E, X, Y and the rows with their respective inverses, E, X^{-1}, Y^{-1}, the group multiplication table then takes the form shown in Table 12.2.

Isomorphic groups

Two groups are isomorphic to each other if the elements of one group can be put in one-to-one correspondence with the elements of the other so that the corresponding elements multiply in the same way. Thus if the elements $A, B, C, \ldots$ of the group G

435

Table 12.3.

	1	i	-1	i			E	A	B	C
1	1	i	-1	i		E	E	A	B	C
i	i	-1	$-i$	1	or	A	A	B	C	E
-1	-1	$-i$	1	i		B	B	C	E	A
$-i$	$-i$	1	i	-1		C	C	E	A	B

correspond respectively to the elements $A', B', C', \ldots$ of G', then the equation $AB = C$ implies that $A'B' = C'$, etc., and vice versa. Two isomorphic groups have the same multiplication tables except for the labels attached to the group elements. Obviously, two isomorphic groups must have the same order.

Groups that are isomorphic and so have the same multiplication table are the same or identical, from an abstract point of view. That is why the concept of isomorphism is a key concept to physicists. Diverse groups of operators that act on diverse sets of objects have the same multiplication table; there is only one abstract group. This is where the value and beauty of the group theoretical method lie; the same abstract algebraic results may be applied in making predictions about a wide variety physical objects.

The isomorphism of groups is a special instance of homomorphism, which allows many-to-one correspondence.

Example 12.9
Consider the groups of Problems 12.2 and 12.4. The group G of Problem 12.2 consists of the four elements $E = 1, A = i, B = -1, C = -i$ with *ordinary multiplication* as the rule of combination. The group multiplication table has the form shown in Table 12.3. The group G' of Problem 12.4 consists of the following four elements, with *matrix multiplication* as the rule of combination

$$E' = \begin{pmatrix} 1 & 0 \\ 0 & 1 \end{pmatrix}, \quad A' = \begin{pmatrix} 0 & 1 \\ -1 & 0 \end{pmatrix}, \quad B' = \begin{pmatrix} -1 & 0 \\ 0 & -1 \end{pmatrix}, \quad C' = \begin{pmatrix} 0 & -1 \\ 1 & 0 \end{pmatrix}.$$

It is straightforward to check that the group multiplication table of group G' has the form of Table 12.4. Comparing Tables 12.3 and 12.4 we can see that they have precisely the same structure. The two groups are therefore isomorphic.

Example 12.10
We stated earlier that diverse groups of operators that act on diverse sets of objects have the same multiplication table; there is only one abstract group. To illustrate this, we consider, for simplicity, an abstract group of order two, G_2: that

Table 12.4.

	E'	A'	B'	C'
E'	E'	A'	B'	C'
A'	A'	B'	C'	E'
B'	B'	C'	E'	A'
C'	C'	E'	E'	B'

is, we make no *a priori* assumption about the significance of the two elements of our group. One of them must be the identity E, and we call the other X. Thus we have

$$E^2 = E, \quad EX = XE = E.$$

Since each element appears once and only once in each row and column, the group multiplication table takes the form:

	E	X
E	E	X
X	X	E

We next consider some groups of operators that are isomorphic to G_2. First, consider the following two transformations of three-dimensional space into itself:

(1) the transformation E', which leaves each point in its place, and
(2) the transformation R, which maps the point (x, y, z) into the point $(-x, -y, -z)$. Evidently, $R^2 = RR$ (the transformation R followed by R) will bring each point back to its original position. Thus we have $(E')^2 = E'$, $RE' = E'R = RE' = R$, $R^2 = E'$; and the group multiplication table has the same form as G_2: that is, the group formed by the set of the two operations E' and R is isomorphic to G_2.

We now associate with the two operations E' and R two operators $\hat{O}_{E'}$ and $\hat{O}_R$, which act on real- or complex-valued functions of the spatial coordinates (x, y, z), $\psi(x, y, z)$, with the following effects:

$$\hat{O}_{E'}\psi(x, y, z) = \psi(x, y, z), \quad \hat{O}_R\psi(x, y, z) = \psi(-x, -y, -z).$$

From these we see that

$$(\hat{O}_{E'})^2 = \hat{O}_{E'}, \quad \hat{O}_{E'}\hat{O}_R = \hat{O}_R\hat{O}_{E'} = \hat{O}_R, \quad (\hat{O}_R)^2 = \hat{O}_R.$$

Obviously these two operators form a group that is isomorphic to G_2. These two groups (formed by the elements E', R, and the elements $\hat{O}_{E'}$ and $\hat{O}_R$, respectively) are the two representations of the abstract group G_2. These two simple examples cannot illustrate the value and beauty of the group theoretical method, but they do serve to illustrate the key concept of isomorphism.

Group of permutations and Cayley's theorem

In Example 12.6 we examined briefly the group of permutations of three objects. We now come back to the general case of n objects $(1, 2, \ldots, n)$ placed in n boxes (or places) labeled $\alpha_1, \alpha_2, \ldots, \alpha_n$. This group, denoted by S_n, is called the symmetric group on n objects. It is of order $n!$ How do we know? The first object may be put in any of n boxes, and the second object may then be put in any of $n - 1$ boxes, and so forth: $n(n - 1)(n - 2) \times \cdots \times 3 \times 2 \times 1 = n!$.

We now define, following common practice, a permutation symbol P

$$P = \begin{pmatrix} 1 & 2 & 3 & \cdots & n \\ \alpha_1 & \alpha_2 & \alpha_3 & \cdots & \alpha_n \end{pmatrix}, \tag{12.1}$$

which shifts the object in box 1 to box α_1, the object in box 2 to box α_2, and so forth, where $\alpha_1 \alpha_2 \cdots \alpha_n$ is some arrangement of the numbers $1, 2, 3, \ldots, n$. The old notation in Example 12.6 can now be written as

$$[2\ 3\ 1] = \begin{pmatrix} 1 & 2 & 3 \\ 2 & 3 & 1 \end{pmatrix}.$$

For n objects there are $n!$ permutations or arrangements, each of which may be written in the form (12.1). Taking a specific example of three objects, we have

$$P_1 = \begin{pmatrix} 1 & 2 & 3 \\ 1 & 2 & 3 \end{pmatrix}, \quad P_2 = \begin{pmatrix} 1 & 2 & 3 \\ 2 & 3 & 1 \end{pmatrix}, \quad P_3 = \begin{pmatrix} 1 & 2 & 3 \\ 1 & 3 & 2 \end{pmatrix},$$

$$P_4 = \begin{pmatrix} 1 & 2 & 3 \\ 2 & 1 & 3 \end{pmatrix}, \quad P_5 = \begin{pmatrix} 1 & 2 & 3 \\ 3 & 2 & 1 \end{pmatrix}, \quad P_6 = \begin{pmatrix} 1 & 2 & 3 \\ 3 & 1 & 2 \end{pmatrix}.$$

For the product of two permutations $P_i P_j$ $(i, j = 1, 2, \ldots, 6)$, we first perform the one on the right, P_j, and then the one on the left, P_i. Thus

$$P_3 P_6 = \begin{pmatrix} 1 & 2 & 3 \\ 1 & 3 & 2 \end{pmatrix} \begin{pmatrix} 1 & 2 & 3 \\ 3 & 1 & 2 \end{pmatrix} = \begin{pmatrix} 1 & 2 & 3 \\ 2 & 1 & 3 \end{pmatrix} = P_4.$$

To the reader who has difficulty seeing this result, let us explain. Consider the first column. We first perform P_6, so that 1 is replaced by 3, we then perform P_3 and 3 is replaced by 2. So by the combined action 1 is replaced by 2 and we have the first column

$$\begin{pmatrix} 1 & \cdots & \cdots \\ 2 & \cdots & \cdots \end{pmatrix}.$$

We leave the other two columns to be completed by the reader.

Each element of a group has an inverse. Thus, for each permutation P_i there is P_i^{-1}, the inverse of P_i. We can use the property $P_i P_i^{-1} = P_1$ to find P_i^{-1}. Let us find P_6^{-1}:

$$P_6^{-1} = \begin{pmatrix} 3 & 1 & 2 \\ 1 & 2 & 3 \end{pmatrix} = \begin{pmatrix} 1 & 2 & 3 \\ 2 & 3 & 1 \end{pmatrix} = P_2.$$

It is straightforward to check that

$$P_6 P_6^{-1} = P_6 P_2 = \begin{pmatrix} 1 & 2 & 3 \\ 3 & 1 & 2 \end{pmatrix}\begin{pmatrix} 1 & 2 & 3 \\ 2 & 3 & 1 \end{pmatrix} = \begin{pmatrix} 1 & 2 & 3 \\ 1 & 2 & 3 \end{pmatrix} = P_1.$$

The reader can verify that our group S_3 is generated by the elements P_2 and P_3, while P_1 serves as the identity. This means that the other three distinct elements can be expressed as distinct multiplicative combinations of P_2 and P_3:

$$P_4 = P_2^2 P_3, \quad P_5 = P_2 P_3, \quad P_6 = P_2^2.$$

The symmetric group S_n plays an important role in the study of finite groups. Every finite group of order n is isomorphic to a subgroup of the permutation group S_n. This is known as Cayley's theorem. For a proof of this theorem the interested reader is referred to an advanced text on group theory.

In physics, these permutation groups are of considerable importance in the quantum mechanics of identical particles, where, if we interchange any two or more these particles, the resulting configuration is indistinguishable from the original one. Various quantities must be invariant under interchange or permutation of the particles. Details of the consequences of this invariant property may be found in most first-year graduate textbooks on quantum mechanics that cover the application of group theory to quantum mechanics.

Subgroups and cosets

A subset of a group G, which is itself a group, is called a subgroup of G. This idea was introduced earlier. And we also saw that C_3, a cyclic group of order 3, is a subgroup of S_3, a symmetric group of order 6. We note that the order of C_3 is a factor of the order of S_3. In fact, we will show that, in general,

> the order of a subgroup is a factor of the order of the full group
> (that is, the group from which the subgroup is derived).

This can be proved as follows. Let G be a group of order n with elements $g_1(=E)$, $g_2, \ldots, g_n$. Let H, of order m, be a subgroup of G with elements $h_1(=E)$, $h_2, \ldots, h_m$. Now form the set gh_k $(0 \le k \le m)$, where g is any element of G not in H. This collection of elements is called the left-coset of H with respect to g (the *left*-coset, because g is at the left of h_k).

If such an element g does not exist, then $H = G$, and the theorem holds trivially. If g does exist, than the elements gh_k are all different. Otherwise, we would have $gh_k = gh_\ell$, or $h_k = h_\ell$, which contradicts our assumption that H is a group. Moreover, the elements gh_k are not elements of H. Otherwise, $gh_k = h_j$, and we have

$$g = h_j/h_k.$$

This implies that g is an element of H, which contradicts our assumption that g does not belong to H.

This left-coset of H does not form a group because it does not contain the identity element $(g_1 = h_1 = E)$. If it did form a group, it would require for some h_j such that $gh_j = E$ or, equivalently, $g = h_j^{-1}$. This requires g to be an element of H. Again this is contrary to assumption that g does not belong to H.

Now every element g in G but not in H belongs to some coset gH. Thus G is a union of H and a number of non-overlapping cosets, each having m different elements. The order of G is therefore divisible by m. This proves that the order of a subgroup is a factor of the order of the full group. The ratio n/m is the index of H in G.

It is straightforward to prove that a group of order p, where p is a prime number, has no subgroup. It could be a cyclic group generated by an element a of period p.

Conjugate classes and invariant subgroups

Another way of dividing a group into subsets is to use the concept of classes. Let a, b, and u be any three elements of a group, and if

$$b = u^{-1}au,$$

b is said to be the transform of a by the element u; a and b are conjugate (or equivalent) to each other. It is straightforward to prove that conjugate has the following three properties:

(1) Every element is conjugate with itself (reflexivity). Allowing u to be the identity element E, then we have $a = E^{-1}aE$.

(2) If a is conjugate to b, then b is conjugate to a (symmetry). If $a = u^{-1}bu$, then $b = uau^{-1} = (u^{-1})^{-1}a(u^{-1})$, where u^{-1} is an element of G if u is.

(3) If a is conjugate with both b and c, then b and c are conjugate with each other (transitivity). If $a = u^{-1}bu$ and $b = v^{-1}cv$, then $a = u^{-1}v^{-1}cvu = (vu)^{-1}c(vu)$, where u and v belong to G so that vu is also an element of G.

We now divide our group up into subsets, such that all elements in any subset are conjugate to each other. These subsets are called classes of our group.

Example 12.11
The symmetric group S_3 has the following six distinct elements:

$$P_1 = E, \ P_2, \ P_3, \ P_4 = P_2^2 P_3, \ P_5 = P_2 P_3, \ P_6 = P_2^2,$$

which can be separated into three conjugate classes:

$$\{P_1\}; \quad \{P_2, P_6\}; \quad \{P_3, P_4, P_5\}.$$

We now state some simple facts about classes without proofs:

(a) The identity element always forms a class by itself.
(b) Each element of an Abelian group forms a class by itself.
(c) All elements of a class have the same period.

Starting from a subgroup H of a group G, we can form a set of elements $uh^{-1}u$ for each u belong to G. This set of elements can be seen to be itself a group. It is a subgroup of G and is isomorphic to H. It is said to be a conjugate subgroup to H in G. It may happen, for some subgroup H, that for all u belonging to G, the sets H and uhu^{-1} are identical. H is then an invariant or self-conjugate subgroup of G.

Example 12.12
Let us revisit S_3 of Example 12.11, taking it as our group $G = S_3$. Consider the subgroup $H = C_3 = \{P_1, P_2, P_6\}$. The following relation holds

$$P_2 \begin{pmatrix} P_1 \\ P_2 \\ P_6 \end{pmatrix} P_2^{-1} = P_1 P_2 \begin{pmatrix} P_1 \\ P_2 \\ P_6 \end{pmatrix} P_2^{-1} P_1^{-1} = (P_1 P_2) \begin{pmatrix} P_1 \\ P_2 \\ P_6 \end{pmatrix} (P_1 P_2)^{-1}$$

$$= P_1^2 P_2 \begin{pmatrix} P_1 \\ P_2 \\ P_6 \end{pmatrix} (P_1^2 P_2)^{-1} = \begin{pmatrix} P_1 \\ P_6 \\ P_2 \end{pmatrix}.$$

Hence $H = C_3 = \{P_1, P_2, P_6\}$ is an invariant subgroup of S_3.

Group representations

In previous sections we have seen some examples of groups which are isomorphic with matrix groups. Physicists have found that the representation of group elements by matrices is a very powerful technique. It is beyond the scope of this text to make a full study of the representation of groups; in this section we shall make a brief study of this important subject of the matrix representations of groups.

If to every element of a group G, $g_1, g_2, g_3, \ldots$, we can associate a non-singular square matrix $D(g_1), D(g_2), D(g_3), \ldots$, in such a way that

$$g_i g_j = g_k \qquad \text{implies} \qquad D(g_i) D(g_j) = D(g_k), \qquad (12.2)$$

then these matrices themselves form a group G', which is either isomorphic or homomorphic to G. The set of such non-singular square matrices is called a representation of group G. If the matrices are $n \times n$, we have an n-dimensional representation; that is, the order of the matrix is the dimension (or order) of the representation D_n. One trivial example of such a representation is the unit matrix associated with every element of the group. As shown in Example 12.9, the four matrices of Problem 12.4 form a two-dimensional representation of the group G of Problem 12.2.

If there is one-to-one correspondence between each element of G and the matrix representation group G', the two groups are isomorphic, and the representation is said to be faithful (or true). If one matrix D represents more than one group element of G, the group G is homomorphic to the matrix representation group G' and the representation is said to be unfaithful.

Now suppose a representation of a group G has been found which consists of matrices $D = D(g_1), D(g_2), D(g_3), \ldots, D(g_p)$, each matrix being of dimension n. We can form another representation D' by a similarity transformation

$$D'(g) = S^{-1} D(g) S, \qquad (12.3)$$

S being a non-singular matrix, then

$$\begin{aligned}
D'(g_i) D'(g_j) &= S^{-1} D(g_i) S S^{-1} D(g_j) S \\
&= S^{-1} D(g_i) D(g_j) S \\
&= S^{-1} D(g_i g_j) S \\
&= D'(g_i g_j).
\end{aligned}$$

In general, representations related in this way by a similarity transformation are regarded as being equivalent. However, the forms of the individual matrices in the two equivalent representations will be quite different. With this freedom in the choice of the forms of the matrices it is important to look for some quantity that is an invariant for a given transformation. This is found in considering the traces of the matrices of the representation group because the trace of a matrix is invariant under a similarity transformation. It is often possible to bring, by a similarity transformation, each matrix in the representation group into a diagonal form

$$S^{-1}DS = \begin{pmatrix} D^{(1)} & 0 \\ 0 & D^{(2)} \end{pmatrix},$$

(12.4)

where $D^{(1)}$ is of order $m, m < n$ and $D^{(2)}$ is of order $n - m$. Under these conditions, the original representation is said to be reducible to $D^{(1)}$ and $D^{(2)}$. We may write this result as

$$D = D^{(1)} \oplus D^{(2)}$$

(12.5)

and say that D has been decomposed into the two smaller representation $D^{(1)}$ and $D^{(2)}$; D is often called the direct sum of $D^{(1)}$ and $D^{(2)}$.

A representation $D(g)$ is called irreducible if it is not of the form (12.4) and cannot be put into this form by a similarity transformation. Irreducible representations are the simplest representations, all others may be built up from them, that is, they play the role of 'building blocks' for the study of group representation.

In general, a given group has many representations, and it is always possible to find a unitary representation – one whose matrices are unitary. Unitary matrices can be diagonalized, and the eigenvalues can serve for the description or classification of quantum states. Hence unitary representations play an especially important role in quantum mechanics.

The task of finding all the irreducible representations of a group is usually very laborious. Fortunately, for most physical applications, it is sufficient to know only the traces of the matrices forming the representation, for the trace of a matrix is invariant under a similarity transformation. Thus, the trace can be used to identify or characterize our representation, and so it is called the character in group theory. A further simplification is provided by the fact that the character of every element in a class is identical, since elements in the same class are related to each other by a similarity transformation. If we know all the characters of one element from every class of the group, we have all of the information concerning the group that is usually needed. Hence characters play an important part in the theory of group representations. However, this topic and others related to whether a given representation of a group can be reduced to one of smaller dimensions are beyond the scope of this book. There are several important theorems of representation theory, which we now state without proof.

(1) A matrix that commutes with all matrices of an irreducible representation of a group is a multiple of the unit matrix (perhaps null). That is, if matrix A commutes with $D(g)$ which is irreducible,

$$D(g)A = AD(g)$$

for all g in our group, then A is a multiple of the unit matrix.
(2) A representation of a group is irreducible if and only if the only matrices to commute with all matrices are multiple of the unit matrix.

Both theorems (1) and (2) are corollaries of Schur's lemma.
(3) Schur's lemma: Let $D^{(1)}$ and $D^{(2)}$ be two irreducible representations of (a group G) dimensionality n and n', if there exists a matrix A such that

$$AD^{(1)}(g) = D^{(2)}(g)A \qquad \text{for all } g \text{ in the group } G$$

then for $n \neq n'$, $A = 0$; for $n = n'$, either $A = 0$ or A is a non-singular matrix and $D^{(1)}$ and $D^{(2)}$ are equivalent representations under the similarity transformation generated by A.
(4) Orthogonality theorem: If G is a group of order h and $D^{(1)}$ and $D^{(2)}$ are any two inequivalent irreducible (unitary) representations, of dimensions d_1 and d_2, respectively, then

$$\sum_g [D^{(i)}_{\alpha\beta}(g)]^* D^{(j)}_{\gamma\delta}(g) = \frac{h}{d_1} \delta_{ij}\delta_{\alpha\gamma}\delta_{\beta\delta},$$

where $D^{(i)}(g)$ is a matrix, and $D^{(i)}_{\alpha\beta}(g)$ is a typical matrix element. The sum runs over all g in G.

Some special groups

Many physical systems possess symmetry properties that always lead to certain quantity being invariant. For example, translational symmetry (or spatial homogeneity) leads to the conservation of linear momentum for a closed system, and rotational symmetry (or isotropy of space) leads to the conservation of angular momentum. Group theory is most appropriate for the study of symmetry. In this section we consider the geometrical symmetries. This provides more illustrations of the group concepts and leads to some special groups.

Let us first review some symmetry operations. A plane of symmetry is a plane in the system such that each point on one side of the plane is the mirror image of a corresponding point on the other side. If the system takes up an identical position on rotation through a certain angle about an axis, that axis is called an axis of symmetry. A center of inversion is a point such that the system is invariant under the operation $\mathbf{r} \rightarrow -\mathbf{r}$, where $\mathbf{r}$ is the position vector of any point in the system referred to the inversion center. If the system takes up an identical position after a rotation followed by an inversion, the system possesses a rotation–inversion center.

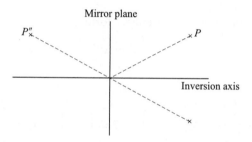

Figure 12.2.

Some symmetry operations are equivalent. As shown in Fig. 12.2, a two-fold inversion axis is equivalent to a mirror plane perpendicular to the axis.

There are two different ways of looking at a rotation, as shown in Fig. 12.3. According to the so-called active view, the system (the body) undergoes a rotation through an angle θ, say, in the clockwise direction about the x_3-axis. In the passive view, this is equivalent to a rotation of the coordinate system through the same angle but in the counterclockwise sense. The relation between the new and old coordinates of any point in the body is the same in both cases:

$$\left.\begin{aligned} x_1' &= x_1 \cos\theta + x_2 \sin\theta, \\ x_2' &= -x_1 \sin\theta + x_2 \cos\theta, \\ x_3' &= x_3, \end{aligned}\right\} \tag{12.6}$$

where the prime quantities represent the new coordinates.

A general rotation, reflection, or inversion can be represented by a linear transformation of the form

$$\left.\begin{aligned} x_1' &= \alpha_{11}x_1 + \alpha_{12}x_2 + \alpha_{13}x_3, \\ x_2' &= \alpha_{21}x_1 + \alpha_{22}x_2 + \alpha_{23}x_3, \\ x_3' &= \alpha_{31}x_1 + \alpha_{32}x_2 + \alpha_{33}x_3. \end{aligned}\right\} \tag{12.7}$$

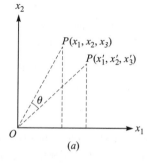
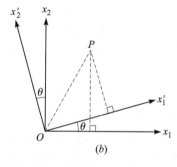

Figure 12.3. (a) Active view of rotation; (b) passive view of rotation.

Equation (12.7) can be written in matrix form

$$\tilde{x}' = \tilde{\lambda}\tilde{x} \tag{12.8}$$

with

$$\tilde{\lambda} = \begin{pmatrix} \lambda_{11} & \lambda_{12} & \lambda_{13} \\ \lambda_{21} & \lambda_{22} & \lambda_{23} \\ \lambda_{31} & \lambda_{32} & \lambda_{33} \end{pmatrix}, \quad \tilde{x} = \begin{pmatrix} x_1 \\ x_2 \\ x_3 \end{pmatrix}, \quad \tilde{x}' = \begin{pmatrix} x_1' \\ x_2' \\ x_3' \end{pmatrix}.$$

The matrix $\tilde{\lambda}$ is an orthogonal matrix and the value of its determinant is ± 1. The '-1' value corresponds to an operation involving an odd number of reflections. For Eq. (12.6) the matrix $\tilde{\lambda}$ has the form

$$\tilde{\lambda} = \begin{pmatrix} \cos\theta & \sin\theta & 0 \\ -\sin\theta & \cos\theta & 0 \\ 0 & 0 & 1 \end{pmatrix}. \tag{12.6a}$$

For a rotation, an inversion about an axis, or a reflection in a plane through the origin, the distance of a point from the origin remains unchanged:

$$r^2 = x_1^2 + x_2^2 + x_3^2 = x_1'^2 + x_2'^2 + x_3'^2. \tag{12.9}$$

The symmetry group D_2, D_3

Let us now examine two simple examples of symmetry and groups. The first one is on twofold symmetry axes. Our system consists of six particles: two identical particles A located at $\pm a$ on the x-axis, two particles, B at $\pm b$ on the y-axis, and two particles C at $\pm c$ on the z-axis. These particles could be the atoms of a molecule or part of a crystal. Each axis is a twofold symmetry axis. Clearly, the identity or unit operator (no rotation) will leave the system unchanged. What rotations can be carried out that will leave our system invariant? A certain combination of rotations of π radians about the three coordinate axes will do it. The orthogonal matrices that represent rotations about the three coordinate axes can be set up in a similar manner as was done for Eq. (12.6a), and they are

$$\tilde{\alpha}(\pi) = \begin{pmatrix} 1 & 0 & 0 \\ 0 & -1 & 0 \\ 0 & 0 & -1 \end{pmatrix}, \quad \tilde{\beta}(\pi) = \begin{pmatrix} -1 & 0 & 0 \\ 0 & 1 & 0 \\ 0 & 0 & -1 \end{pmatrix}, \quad \tilde{\gamma}(\pi) = \begin{pmatrix} -1 & 0 & 0 \\ 0 & -1 & 0 \\ 0 & 0 & 1 \end{pmatrix},$$

where $\tilde{\alpha}$ is the rotational matrix about the x-axis, and $\tilde{\beta}$ and $\tilde{\gamma}$ are the rotational matrices about y- and z-axes, respectively. Of course, the identity operator is a unit matrix

446

$$\tilde{E} = \begin{pmatrix} 1 & 0 & 0 \\ 0 & 1 & 0 \\ 0 & 0 & 1 \end{pmatrix}.$$

These four elements form an Abelian group with the group multiplication table shown in Table 12.5. It is easy to check this group table by matrix multiplication. Or you can check it by analyzing the operations themselves, a tedious task. This demonstrates the power of mathematics: when the system becomes too complex for a direct physical interpretation, the usefulness of mathematics shows.

Table 12.5.

	$\tilde{E}$	$\tilde{\alpha}$	$\tilde{\beta}$	$\tilde{\gamma}$
$\tilde{E}$	$\tilde{E}$	$\tilde{\alpha}$	$\tilde{\beta}$	$\tilde{\gamma}$
$\tilde{\alpha}$	$\tilde{\alpha}$	$\tilde{E}$	$\tilde{\gamma}$	$\tilde{\beta}$
$\tilde{\beta}$	$\tilde{\beta}$	$\tilde{\gamma}$	$\tilde{E}$	$\tilde{\alpha}$
$\tilde{\gamma}$	$\tilde{\gamma}$	$\tilde{\beta}$	$\tilde{\alpha}$	$\tilde{E}$

This symmetry group is usually labeled D_2, a dihedral group with a twofold symmetry axis. A dihedral group D_n with an n-fold symmetry axis has n axes with an angular separation of $2\pi/n$ radians and is very useful in crystallographic study.

We next consider an example of threefold symmetry axes. To this end, let us re-visit Example 12.8. Rotations of the triangle of $0°$, $120°$, $240°$, and $360°$ leave the triangle invariant. Rotation of the triangle of $0°$ means no rotation, the triangle is left unchanged; this is represented by a unit matrix (the identity element). The other two orthogonal rotational matrices can be set up easily:

$$\tilde{A} = R_z(120°) = \begin{pmatrix} -1/2 & -\sqrt{3}/2 \\ \sqrt{3}/2 & -1/2 \end{pmatrix},$$

$$\tilde{B} = R_z(240°) = \begin{pmatrix} -1/2 & \sqrt{3}/2 \\ -\sqrt{3}/2 & -1/2 \end{pmatrix},$$

and

$$\tilde{E} = R_z(0) = \begin{pmatrix} 1 & 0 \\ 0 & 1 \end{pmatrix}.$$

We notice that $\tilde{C} = R_z(360°) = \tilde{E}$. The set of the three elements $(\tilde{E}, \tilde{A}, \tilde{B})$ forms a cyclic group C_3 with the group multiplication table shown in Table 12.6. The z-

Table 12.6.

	$\tilde{E}$	$\tilde{A}$	$\tilde{B}$
$\tilde{E}$	$\tilde{E}$	$\tilde{A}$	$\tilde{B}$
$\tilde{A}$	$\tilde{A}$	$\tilde{B}$	$\tilde{E}$
B	$\tilde{B}$	$\tilde{E}$	$\tilde{A}$

axis is a threefold symmetry axis. There are three additional axes of symmetry in the xy plane: each corner and the geometric center O defining an axis; each of these is a twofold symmetry axis (Fig. 12.4). Now let us consider reflection operations. The following successive operations will bring the equilateral angle onto itself (that is, be invariant):

$\tilde{E}$ the identity; triangle is left unchanged;
$\tilde{A}$ triangle is rotated through 120° clockwise;
$\tilde{B}$ triangle is rotated through 240° clockwise;
$\tilde{C}$ triangle is reflected about axis OR (or the y-axis);
$\tilde{D}$ triangle is reflected about axis OQ;
$\tilde{F}$ triangle is reflected about axis OP.

Now the reflection about axis OR is just a rotation of 180° about axis OR, thus

$$\tilde{C} = R_{OR}(180°) = \begin{pmatrix} -1 & 0 \\ 0 & 1 \end{pmatrix}.$$

Next, we notice that reflection about axis OQ is equivalent to a rotation of 240° about the z-axis followed by a reflection of the x-axis (Fig. 12.5):

$$\tilde{D} = R_{OQ}(180°) = \tilde{C}\tilde{B} = \begin{pmatrix} -1 & 0 \\ 0 & 1 \end{pmatrix}\begin{pmatrix} -1/2 & \sqrt{3}/2 \\ -\sqrt{3}/2 & -1/2 \end{pmatrix} = \begin{pmatrix} 1/2 & -\sqrt{3}/2 \\ -\sqrt{3}/2 & -1/2 \end{pmatrix}.$$

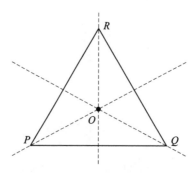

Figure 12.4.

448

Table 12.7.

	$\tilde{E}$	$\tilde{A}$	$\tilde{B}$	$\tilde{C}$	$\tilde{D}$	$\tilde{F}$
$\tilde{E}$	$\tilde{E}$	$\tilde{A}$	$\tilde{B}$	$\tilde{C}$	$\tilde{D}$	$\tilde{F}$
$\tilde{A}$	$\tilde{A}$	$\tilde{B}$	$\tilde{E}$	$\tilde{D}$	$\tilde{F}$	$\tilde{C}$
$\tilde{B}$	$\tilde{B}$	$\tilde{E}$	$\tilde{A}$	$\tilde{F}$	$\tilde{C}$	$\tilde{D}$
$\tilde{C}$	$\tilde{C}$	$\tilde{F}$	$\tilde{D}$	$\tilde{E}$	$\tilde{B}$	$\tilde{A}$
$\tilde{D}$	$\tilde{D}$	$\tilde{C}$	$\tilde{F}$	$\tilde{A}$	$\tilde{E}$	$\tilde{B}$
$\tilde{F}$	$\tilde{F}$	$\tilde{D}$	$\tilde{C}$	$\tilde{B}$	$\tilde{A}$	$\tilde{E}$

Similarly, reflection about axis OP is equivalent to a rotation of $180°$ followed by a reflection of the x-axis:

$$\tilde{F} = R_{OP}(180°) = \tilde{C}\tilde{A} = \begin{pmatrix} 1/2 & \sqrt{3}/2 \\ \sqrt{3}/2 & -1/2 \end{pmatrix}.$$

The group multiplication table is shown in Table 12.7. We have constructed a six-element non-Abelian group and a 2×2 irreducible matrix representation of it. Our group is known as D_3 in crystallography, the dihedral group with a threefold axis of symmetry.

One-dimensional unitrary group $U(1)$

We now consider groups with an infinite number of elements. The group element will contain one or more parameters that vary continuously over some range so they are also known as continuous groups. In Example 12.7, we saw that the complex numbers $(1, i, -1, -i)$ form a cyclic group of order 3. These group elements may be interpreted as successive $90°$ rotations in the complex plane $(0, \pi/2, \pi, 3\pi/2)$, and so they may be written as $e^{i\varphi}$ with $\varphi = 0, \pi/2, \pi, 3\pi/2$. If φ is allowed to vary continuously over the range $[0, 2\pi]$, then we will have, instead of a four-member cyclic group, a continuous group with multiplication for the composition rule. It is straightforward to check that the four group axioms are all

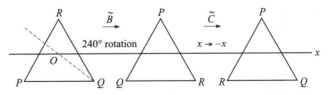

Figure 12.5.

449

met. In quantum mechanics, $e^{i\varphi}$ is a complex phase factor of a wave function, which we denote by $U(\varphi)$. Obviously, $U(0)$ is an identity element. Next,

$$U(\varphi)U(\varphi') = e^{i(\varphi+\varphi')} = U(\varphi + \varphi'),$$

and $U(\varphi + \varphi')$ is an element of the group. There is an inverse: $U^{-1}(\varphi) = U(-\varphi)$, since

$$U(\varphi)U(-\varphi) = U(-\varphi)U(\varphi) = U(0) = E$$

for any φ. The associative law is satisfied:

$$[U(\varphi_1)U(\varphi_2)]U(\varphi_3) = e^{i(\varphi_1+\varphi_2)}e^{i\varphi_3} = e^{i(\varphi_1+\varphi_2+\varphi_3)} = e^{i\varphi_1}e^{i(\varphi_2+\varphi_3)}$$

$$= U(\varphi_1)[U(\varphi_2)U(\varphi_3)].$$

This group is a one-dimensional unitary group; it is called $U(1)$. Each element is characterized by a continuous parameter φ, $0 \le \varphi \le 2\pi$; φ can take on an infinite number of values. Moreover, the elements are differentiable:

$$dU = U(\varphi + d\varphi) - U(\varphi) = e^{i(\varphi+d\varphi)} - e^{i\varphi}$$

$$= e^{i\varphi}(1 + id\varphi) - e^{i\varphi} = ie^{i\varphi}d\varphi = iUd\varphi$$

or

$$dU/d\varphi = iU.$$

Infinite groups whose elements are differentiable functions of their parameters are called Lie groups. The differentiability of group elements allows us to develop the concept of the generator. Furthermore, instead of studying the whole group, we can study the group elements in the neighborhood of the identity element. Thus Lie groups are of particular interest. Let us take a brief look at a few more Lie groups.

Orthogonal groups SO(2) and SO(3)

The rotations in an n-dimensional Euclidean space form a group, called $O(n)$. The group elements can be represented by $n \times n$ orthogonal matrices, each with $n(n-1)/2$ independent elements (Problem 12.12). If the determinant of O is set to be $+1$ (rotation only, no reflection), then the group is often labeled $SO(n)$. The label O_n^+ is also often used.

The elements of $SO(2)$ are familiar; they are the rotations in a plane, say the xy plane:

$$\begin{pmatrix} x' \\ y' \end{pmatrix} = \tilde{R}\begin{pmatrix} x \\ y \end{pmatrix} = \begin{pmatrix} \cos\theta & \sin\theta \\ -\sin\theta & \cos\theta \end{pmatrix}\begin{pmatrix} x \\ y \end{pmatrix}.$$

This group has one parameter: the angle θ. As we stated earlier, groups enter physics because we can carry out transformations on physical systems and the physical systems often are invariant under the transformations. Here $x^2 + y^2$ is left invariant.

We now introduce the concept of a generator and show that rotations of $SO(2)$ are generated by a special 2×2 matrix $\tilde{\sigma}_2$, where

$$\tilde{\sigma}_2 = \begin{pmatrix} 0 & -i \\ i & 0 \end{pmatrix}.$$

Using the Euler identity, $e^{i\theta} = \cos\theta + i\sin\theta$, we can express the 2×2 rotation matrices $R(\theta)$ in exponential form:

$$\tilde{R}(\theta) = \begin{pmatrix} \cos\theta & \sin\theta \\ -\sin\theta & \cos\theta \end{pmatrix} = \tilde{I}_2 \cos\theta + i\tilde{\sigma}_2 \sin\theta = e^{i\theta\tilde{\sigma}_2},$$

where $\tilde{I}_2$ is a 2×2 unit matrix. From the exponential form we see that multiplication is equivalent to addition of the arguments. The rotations close to the identity element have small angles $\theta \cong 0$. We call $\tilde{\sigma}_2$ the generator of rotations for $SO(2)$.

It has been shown that any element g of a Lie group can be written in the form

$$g(\theta_1, \theta_2, \ldots, \theta_n) = \exp\left(\sum_{i=1}^{n} i\theta_i F_i\right).$$

For n parameters there are n of the quantities F_i, and they are called the generators of the Lie group.

Note that we can get $\tilde{\sigma}_2$ from the rotation matrix $\tilde{R}(\theta)$ by differentiation at the identity of $SO(2)$, that is, $\theta \cong 0$. This suggests that we may find the generators of other groups in a similar manner.

For $n = 3$ there are three independent parameters, and the set of 3×3 orthogonal matrices with determinant $+1$ also forms a group, the $SO(3)$, its general member may be expressed in terms of the Euler angle rotation

$$R(\alpha, \beta, \gamma) = R_{z'}(0, 0, \alpha) R_y(0, \beta, 0) R_z(0, 0, \gamma),$$

where R_z is a rotation about the z-axis by an angle γ, R_y a rotation about the y-axis by an angle β, and $R_{z'}$ a rotation about the z'-axis (the new z-axis) by an angle α. This sequence can perform a general rotation. The separate rotations can be written as

$$\tilde{R}_y(\beta) = \begin{pmatrix} \cos\beta & 0 & -\sin\beta \\ 0 & 1 & 0 \\ \sin\beta & 0 & \cos\beta \end{pmatrix}, \quad \tilde{R}_z(\gamma) = \begin{pmatrix} \cos\gamma & \sin\gamma & 0 \\ -\sin\gamma & \cos\gamma & 0 \\ 0 & 0 & 1 \end{pmatrix},$$

$$\tilde{R}_x(\theta) = \begin{pmatrix} 1 & 0 & 0 \\ 0 & \cos\theta & \sin\theta \\ 0 & -\sin\theta & \cos\theta \end{pmatrix}.$$

The $SO(3)$ rotations leave $x^2 + y^2 + z^2$ invariant.

The rotations $R_z(\gamma)$ form a group, called the group R_z, which is an Abelian subgroup of $SO(3)$. To find the generator of this group, let us take the following differentiation

$$-i\, d\tilde{R}_z(\gamma)/d\gamma\big|_{\gamma=0} = \begin{pmatrix} 0 & -i & 0 \\ i & 0 & 0 \\ 0 & 0 & 0 \end{pmatrix} \equiv \tilde{S}_z,$$

where the insertion of i is to make $\tilde{S}_z$ Hermitian. The rotation $R_z(\delta\gamma)$ through an infinitesimal angle $\delta\gamma$ can be written in terms of $\tilde{S}_z$:

$$R_z(\delta\gamma) = \tilde{I}_3 + \frac{dR_z(\gamma)}{d\gamma}\bigg|_{\gamma=0}\delta\gamma + O((\delta\gamma)^2) = \tilde{I}_3 + i\delta\gamma\tilde{S}_z.$$

A finite rotation $R(\gamma)$ may be constructed from successive infinitesimal rotations

$$R_z(\delta\gamma_1 + \delta\gamma_2) = (\tilde{I}_3 + i\delta\gamma_1\tilde{S}_z)(\tilde{I}_3 + i\delta\gamma_2\tilde{S}_z).$$

Now let $(\delta\gamma = \gamma/N$ for N rotations, with $N \to \infty$, then

$$R_z(\gamma) = \lim_{N\to\infty}\left[\tilde{I}_3 + (i\gamma/N)\tilde{S}_z\right]^N = \exp(i\tilde{S}_z),$$

which identifies $\tilde{S}_z$ as the generator of the rotation group R_z. Similarly, we can find the generators of the subgroups of rotations about the x-axis and the y-axis.

The SU(n) groups

The $n \times n$ unitary matrices $\tilde{U}$ also form a group, the $U(n)$ group. If there is the additional restriction that the determinant of the matrices be $+1$, we have the special unitary or unitary unimodular group, $SU(n)$. Each $n \times n$ unitary matrix has $n^2 - 1$ independent parameters (Problem 12.14).

For $n = 2$ we have $SU(2)$ and possible ways to parameterize the matrix U are

$$\tilde{U} = \begin{pmatrix} a & b \\ -b^* & a^* \end{pmatrix},$$

where a, b are arbitrary complex numbers and $|a|^2 + |b|^2 = 1$. These parameters are often called the Cayley–Klein parameters, and were first introduced by Cayley and Klein in connection with problems of rotation in classical mechanics.

Now let us write our unitary matrix in exponential form:

$$\tilde{U} = e^{i\tilde{H}},$$

where $\tilde{H}$ is a Hermitian matrix. It is easy to show that $e^{i\tilde{H}}$ is unitary:

$$(e^{i\tilde{H}})^+ (e^{i\tilde{H}}) = e^{-i\tilde{H}^+} e^{i\tilde{H}} = e^{i(\tilde{H} - \tilde{H}^+)} = 1.$$

This implies that any $n \times n$ unitary matrix can be written in exponential form with a particularly selected set of n^2 Hermitian $n \times n$ matrices, $\tilde{H}_j$

$$\tilde{U} = \exp\left(i \sum_{j=1}^{n^2} \theta_j \tilde{H}_j \right),$$

where the θ_j are real parameters. The $n^2 \tilde{H}_j$ are the generators of the group $U(n)$. To specialize to $SU(n)$ we need to meet the restriction $\det \tilde{U} = 1$. To impose this restriction we need to use the identity

$$\det e^{\tilde{A}} = e^{\text{Tr} A}$$

for any square matrix $\tilde{A}$. The proof is left as homework (Problem 12.15). Thus the condition $\det \tilde{U} = 1$ requires $\text{Tr} \tilde{H} = 0$ for every $\tilde{H}$. Accordingly, the generators of $SU(n)$ are any set of $n \times n$ traceless Hermitian matrices.

For $n = 2$, $SU(n)$ reduces to $SU(2)$, which describes rotations in two-dimensional complex space. The determinant is $+1$. There are three continuous parameters $(2^2 - 1 = 3)$. We have expressed these as Cayley–Klein parameters. The orthogonal group $SO(3)$, determinant $+1$, describes rotations in ordinary three-dimensional space and leaves $x^2 + y^2 + z^2$ invariant. There are also three independent parameters. The rotation interpretations and the equality of numbers of independent parameters suggest these two groups may be isomorphic or homomorphic. The correspondence between these groups has been proved to be two-to-one. Thus $SU(2)$ and $SO(3)$ are isomorphic. It is beyond the scope of this book to reproduce the proof here.

The $SU(2)$ group has found various applications in particle physics. For example, we can think of the proton (p) and neutron (n) as two states of the same particle, a nucleon N, and use the electric charge as a label. It is also useful to imagine a particle space, called the strong isospin space, where the nucleon state points in some direction, as shown in Fig. 12.6. If (or assuming that) the theory that describes nucleon interactions is invariant under rotations in strong isospin space, then we may try to put the proton and the neutron as states of a spin-like doublet, or $SU(2)$ doublet. Other hadrons (strong-interacting particles) can also be classified as states in $SU(2)$ multiplets. Physicists do not have a deep understanding of why the Standard Model (of Elementary Particles) has an $SU(2)$ internal symmetry.

For $n = 3$ there are eight independent parameters $(3^2 - 1 = 8)$, and we have $SU(3)$, which is very useful in describing the color symmetry.

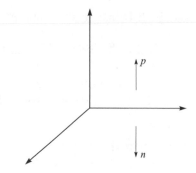

Figure 12.6.   The strong isospin space.

Homogeneous Lorentz group

Before we describe the homogeneous Lorentz group, we need to know the Lorentz transformation. This will bring us back to the origin of special theory of relativity. In classical mechanics, time is absolute and the Galilean transformation (the principle of Newtonian relativity) asserts that all inertial frames are equivalent for describing the laws of classical mechanics. But physicists in the nineteenth century found that electromagnetic theory did not seem to obey the principle of Newtonian relativity. Classical electromagnetic theory is summarized in Maxwell's equations, and one of the consequences of Maxwell's equations is that the speed of light (electromagnetic waves) is independent of the motion of the source. However, under the Galilean transformation, in a frame of reference moving uniformly with respect to the light source the light wave is no longer spherical and the speed of light is also different. Hence, for electromagnetic phenomena, inertial frames are not equivalent and Maxwell's equations are not invariant under Galilean transformation. A number of experiments were proposed to resolve this conflict. After the Michelson–Morley experiment failed to detect ether, physicists finally accepted that Maxwell's equations are correct and have the same form in all inertial frames. There had to be some transformation other than the Galilean transformation that would make both electromagnetic theory and classical mechanical invariant.

   This desired new transformation is the Lorentz transformation, worked out by H. Lorentz. But it was not until 1905 that Einstein realized its full implications and took the epoch-making step involved. In his paper, 'On the Electrodynamics of Moving Bodies' (*The Principle of Relativity*, Dover, New York, 1952), he developed the Special Theory of Relativity from two fundamental postulates, which are rephrased as follows:

(1) The laws of physics are the same in all inertial frame. No preferred inertial frame exists.

(2) The speed of light in free space is the same in all inertial frames and is independent of the motion of the source (the emitting body).

These postulates are often called Einstein's principle of relativity, and they radically revised our concepts of space and time. Newton's laws of motion abolish the concept of absolute space, because according to the laws of motion there is no absolute standard of rest. The non-existence of absolute rest means that we cannot give an event an absolute position in space. This in turn means that space is not absolute. This disturbed Newton, who insisted that there must be some absolute standard of rest for motion, remote stars or the ether system. Absolute space was finally abolished in its Maxwellian role as the ether. Then absolute time was abolished by Einstein's special relativity. We can see this by sending a pulse of light from one place to another. Since the speed of light is just the distance it has traveled divided by the time it has taken, in Newtonian theory, different observers would measure different speeds for the light because time is absolute. Now in relativity, all observers agree on the speed of light, but they do not agree on the distance the light has traveled. So they cannot agree on the time it has taken. That is, time is no longer absolute.

We now come to the Lorentz transformation, and suggest that the reader to consult books on special relativity for its derivation. For two inertial frames with their corresponding axes parallel and the relative velocity v along the $x_1(=x)$ axis, the Lorentz transformation has the form:

$$x_1' = \gamma(x_1 + i\beta x_4),$$
$$x_2' = x_2,$$
$$x_3' = x_3,$$
$$x_4' = \gamma(x_4 - i\beta x_1),$$

where $x_4 = ict$, $\beta = v/c$, and $\gamma = 1/\sqrt{1-\beta^2}$. We will drop the two directions perpendicular to the motion in the following discussion.

For an infinitesimal relative velocity δv, the Lorentz transformation reduces to

$$x_1' = x_1 + i\delta\beta x_4,$$
$$x_4' = x_4 - i\delta\beta x_1,$$

where $\delta\beta = \delta v/c$, $\gamma = 1/\sqrt{1-(\delta\beta)^2} \approx 1$. In matrix form we have

$$\begin{pmatrix} x_1' \\ x_4' \end{pmatrix} = \begin{pmatrix} 1 & i\delta\beta \\ -i\delta\beta & 1 \end{pmatrix} \begin{pmatrix} x_1 \\ x_4 \end{pmatrix}.$$

455

We can express the transformation matrix in exponential form:

$$\begin{pmatrix} 1 & i\delta\beta \\ -i\delta\beta & 1 \end{pmatrix} = \begin{pmatrix} 1 & 0 \\ 0 & 1 \end{pmatrix} + \delta\beta \begin{pmatrix} 0 & i \\ -i & 0 \end{pmatrix} = \tilde{I} + \delta\beta\tilde{\sigma},$$

where

$$\tilde{I} = \begin{pmatrix} 1 & 0 \\ 0 & 1 \end{pmatrix}, \quad \tilde{\sigma} = \begin{pmatrix} 0 & i \\ -i & 0 \end{pmatrix}.$$

Note that $\tilde{\sigma}$ is the negative of the Pauli spin matrix $\tilde{\sigma}_2$. Now we have

$$\begin{pmatrix} x_1' \\ x_4' \end{pmatrix} = (\tilde{I} + \delta\beta\tilde{\sigma}) \begin{pmatrix} x_1 \\ x_4 \end{pmatrix}.$$

We can generate a finite transformation by repeating the infinitesimal transformation N times with $N\delta\beta = \theta$:

$$\begin{pmatrix} x_1' \\ x_4' \end{pmatrix} = \left(\tilde{I} + \frac{\theta\tilde{\sigma}}{N} \right)^N \begin{pmatrix} x_1 \\ x_4 \end{pmatrix}.$$

In the limit as $N \to \infty$,

$$\lim_{N\to\infty} \left(\tilde{I} + \frac{\theta\tilde{\sigma}}{N} \right)^N = e^{\theta\tilde{\sigma}}.$$

Now we can expand the exponential in a Maclaurin series:

$$e^{\theta\tilde{\sigma}} = \tilde{I} + \theta\tilde{\sigma} + (\theta\tilde{\sigma})^2/2! + (\theta\tilde{\sigma})^3/3! + \cdots$$

and, noting that $\tilde{\sigma}^2 = 1$ and

$$\sinh\theta = \theta + \theta^3/3! + \theta^5/5! + \theta^7/7! + \cdots,$$

$$\cosh\theta = 1 + \theta^2/2! + \theta^4/4! + \theta^6/6! + \cdots,$$

we finally obtain

$$e^{\theta\tilde{\sigma}} = \tilde{I}\cosh\theta + \tilde{\sigma}\sinh\theta.$$

Our finite Lorentz transformation then takes the form:

$$\begin{pmatrix} x_1' \\ x_2' \end{pmatrix} = \begin{pmatrix} \cosh\theta & i\sinh\theta \\ -i\sinh\theta & \cosh\theta \end{pmatrix} \begin{pmatrix} x_1 \\ x_2 \end{pmatrix},$$

and $\tilde{\sigma}$ is the generator of the representations of our Lorentz transformation. The transformation

$$\begin{pmatrix} \cosh\theta & i\sinh\theta \\ -i\sinh\theta & \cosh\theta \end{pmatrix}$$

can be interpreted as the rotation matrix in the complex $x_4 x_1$ plane (Problem 12.16).

It is straightforward to generalize the above discussion to the general case where the relative velocity is in an arbitrary direction. The transformation matrix will be a 4×4 matrix, instead of a 2×2 matrix one. For this general case, we have to take x_2- and x_3-axes into consideration.

Problems

12.1.   Show that
(a) the unit element (the identity) in a group is unique, and
(b) the inverse of each group element is unique.

12.2.   Show that the set of complex numbers $1, i, -1$, and $-i$ form a group of order four under multiplication.

12.3.   Show that the set of all rational numbers, the set of all real numbers, and the set of all complex numbers form infinite Abelian groups under addition.

12.4.   Show that the four matrices

$$\tilde{A} = \begin{pmatrix} 1 & 0 \\ 0 & 1 \end{pmatrix}, \quad \tilde{B} = \begin{pmatrix} 0 & 1 \\ -1 & 0 \end{pmatrix}, \quad \tilde{C} = \begin{pmatrix} -1 & 0 \\ 0 & -1 \end{pmatrix}, \quad \tilde{D} = \begin{pmatrix} 0 & -1 \\ 1 & 0 \end{pmatrix}$$

form an Abelian group of order four under multiplication.

12.5.   Show that the set of all permutations of three objects

$$[1\ 2\ 3],\ [2\ 3\ 1],\ [3\ 1\ 2],\ [1\ 3\ 2],\ [3\ 2\ 1],\ [2\ 1\ 3]$$

forms a non-Abelian group of order six, with sequential performance as the law of combination.

12.6.   Given two elements A and B subject to the relations $A^2 = B^2 = E$ (the identity), show that:
(a) $AB \neq BA$, and
(b) the set of six elements E, A, B, A^2, AB, BA form a group.

12.7.   Show that the set of elements $1, A, A^2, \ldots, A^{n-1}, A^n = 1$, where $A = e^{2\pi i/n}$ forms a cyclic group of order n under multiplication.

12.8.   Consider the rotations of a line about the z-axis through the angles $\pi/2, \pi, 3\pi/2$, and 2π in the xy plane. This is a finite set of four elements, the four operations of rotating through $\pi/2, \pi, 3\pi/2$, and 2π. Show that this set of elements forms a group of order four under addition.

12.9.   Construct the group multiplication table for the group of Problem 12.2.

12.10. Consider the possible rearrangement of two objects. The operation E_p leaves each object in its place, and the operation I_p interchanges the two objects. Show that the two operations form a group that is isomorphic to G_2.

Next, we associate with the two operations two operators $\hat{O}_{E_p}$ and $\hat{O}_{I_p}$, which act on the real or complex function $f(x_1, y_1, z_1; x_2, y_2, z_2)$ with the following effects:

$$\hat{O}_{E_p} f = f, \quad \hat{O}_{I_p} f(x_1, y_1, z_1; x_2, y_2, z_2) = f(x_2, y_2, z_2; x_1, y_1, z_1).$$

Show that the two operators form a group that is isomorphic to G_2.

12.11. Verify that the multiplication table of S_3 has the form:

	P_1	P_2	P_3	P_4	P_5	P_6
P_1	P_1	P_2	P_3	P_4	P_5	P_6
P_2	P_2	P_1	P_6	P_5	P_6	P_4
P_3	P_3	P_4	P_5	P_6	P_2	P_1
P_4	P_4	P_5	P_3	P_1	P_6	P_2
P_5	P_5	P_3	P_4	P_2	P_1	P_6
P_6	P_6	P_2	P_1	P_3	P_4	P_5

12.12. Show that an $n \times n$ orthogonal matrix has $n(n-1)/2$ independent elements.

12.13. Show that the 2×2 matrix σ_2 can be obtained from the rotation matrix $R(\sigma)$ by differentiation at the identity of $SO(2)$, that is, $\theta = 0$.

12.14. Show that an $n \times n$ unitary matrix has $n^2 - 1$ independent parameters.

12.15. Show that $\det e^{\tilde{A}} = e^{\text{Tr}\,\tilde{A}}$ where $\tilde{A}$ is any square matrix.

12.16. Show that the Lorentz transformation

$$x_1' = \gamma(x_1 + i\beta x_4),$$
$$x_2' = x_2,$$
$$x_3' = x_3,$$
$$x_4' = \gamma(x_4 - i\beta x_1)$$

corresponds to an imaginary rotation in the $x_4 x_1$ plane. (A detailed discussion of this can be found in the book *Classical Mechanics*, by Tai L. Chow, John Wiley, 1995.)

13

Numerical methods

Very few of the mathematical problems which arise in physical sciences and engineering can be solved analytically. Therefore, a simple, perhaps crude, technique giving the desired values within specified limits of tolerance is often to be preferred. We do not give a full coverage of numerical analysis in this chapter; but some methods for numerically carrying out the processes of interpolation, finding roots of equations, integration, and solving ordinary differential equations will be presented.

Interpolation

In the eighteenth century Euler was probably the first person to use the interpolation technique to construct planetary elliptical orbits from a set of observed positions of the planets. We discuss here one of the most common interpolation techniques: the polynomial interpolation. Suppose we have a set of observed or measured data (x_0, y_0), $(x_1, y_1), \ldots, (x_n, y_n)$, how do we represent them by a smooth curve of the form $y = f(x)$? For analytical convenience, this smooth curve is usually assumed to be polynomial:

$$f(x) = a_0 + a_1 x^1 + a_2 x^2 + \cdots + a_n x^n \tag{13.1}$$

and we use the given points to evaluate the coefficients $a_0, a_1, \ldots, a_n$:

$$\left. \begin{aligned} f(x_0) &= a_0 + a_1 x_0 + a_2 x_0^2 + \cdots + a_n x_0^n = y_0, \\ f(x_1) &= a_0 + a_1 x_1 + a_2 x_1^2 + \cdots + a_n x_1^n = y_1, \\ &\vdots \\ f(x_n) &= a_0 + a_1 x_n + a_2 x_n^2 + \cdots + a_n x_n^n = y_n. \end{aligned} \right\} \tag{13.2}$$

This provides $n + 1$ equations to solve for the $n + 1$ coefficients $a_0, a_1, \ldots, a_n$. However, straightforward evaluation of coefficients in the way outlined above is rather tedious, as shown in Problem 13.1, hence many shortcuts have been devised, though we will not discuss these here because of limited space.

Finding roots of equations

A solution of an equation $f(x) = 0$ is sometimes called a root, where $f(x)$ is a real continuous function. If $f(x)$ is sufficiently complicated that a direct solution may not be possible, we can seek approximate solutions. In this section we will sketch some simple methods for determining the approximate solutions of algebraic and transcendental equations. A polynomial equation is an algebraic equation. An equation that is not reducible to an algebraic equation is called transcendental. Thus, $\tan x - x = 0$ and $e^x + 2\cos x = 0$ are transcendental equations.

Graphical methods

The approximate solution of the equation

$$f(x) = 0 \tag{13.3}$$

can be found by graphing the function $y = f(x)$ and reading from the graph the values of x for which $y = 0$. The graphing procedure can often be simplified by first rewriting Eq. (13.3) in the form

$$g(x) = h(x) \tag{13.4}$$

and then graphing $y = g(x)$ and $y = h(x)$. The x values of the intersection points of the two curves gives the approximate values of the roots of Eq. (13.4). As an example, consider the equation

$$f(x) = x^3 - 146.25x - 682.5 = 0,$$

we can graph

$$y = x^3 - 146.25x - 682.5$$

to find its roots. But it is simpler to graph the two curves

$$y = x^3 \text{ (a cubic)}$$

and

$$y = 146.25x + 682.5 \text{ (a straight line)}.$$

See Fig. 13.1.

There is one drawback of graphical methods: that is, they require plotting curves on a large scale to obtain a high degree of accuracy. To avoid this, methods of successive approximations (or simple iterative methods) have been devised, and we shall sketch a couple of these in the following sections.

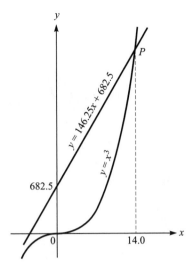

Figure 13.1.

Method of linear interpolation (method of false position)

Make an initial guess of the root of Eq. (13.3), say x_0, located between x_1 and x_2, and in the interval (x_1, x_2) the graph of $y = f(x)$ has the appearance as shown in Fig. 13.2. The straight line connecting P_1 and P_2 cuts the x-axis at point x_3, which is usually closer to x_0 than either x_1 or x_2. From similar triangles

$$\frac{x_3 - x_1}{-f(x_1)} = \frac{x_2 - x_1}{f(x_2)},$$

and solving for x_3 we get

$$x_3 = \frac{x_1 f(x_2) - x_2 f(x_1)}{f(x_2) - f(x_1)}.$$

Now the straight line connecting the points P_3 and P_2 intersects the x-axis at point x_4, which is a closer approximation to x_0 than x_3. By repeating this process we obtain a sequence of values $x_3, x_4, \ldots, x_n$ that generally converges to the root of the equation.

The iterative method described above can be simplified if we rewrite Eq. (13.3) in the form of Eq. (13.4). If the roots of

$$g(x) = c \tag{13.5}$$

can be determined for every real c, then we can start the iterative process as follows. Let x_1 be an approximate value of the root x_0 of Eq. (13.3) (and, of

461

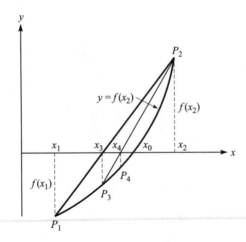

Figure 13.2.

course, also equation 13.4). Now setting $x = x_1$ on the right hand side of Eq. (13.4) we obtain the equation

$$g(x) = h(x_1),$$ (13.6)

which by hypothesis we can solve. If the solution is x_2, we set $x = x_2$ on the right hand side of Eq. (13.4) and obtain

$$g(x) = h(x_2).$$ (13.7)

By repeating this process, we obtain the nth approximation

$$g(x) = h(x_{n-1}).$$ (13.8)

From geometric considerations or interpretation of this procedure, we can see that the sequence $x_1, x_2, \ldots, x_n$ converges to the root $x = 0$ if, in the interval $2|x_1 - x_0|$ centered at x_0, the following conditions are met:

$$\left.\begin{array}{l} \text{(1) } |g'(x)| > |h'(x)|, \text{ and} \\ \text{(2) The derivatives are bounded.} \end{array}\right\}$$ (13.9)

Example 13.1
Find the approximate values of the real roots of the transcendental equation

$$e^x - 4x = 0.$$

Solution: Let $g(x) = x$ and $h(x) = e^x/4$, so the original equation can be rewritten as

$$x = e^x/4.$$

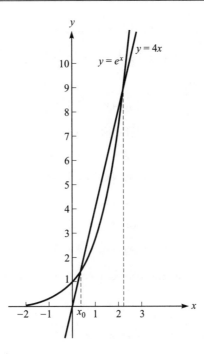

Figure 13.3.

According to Eq. (13.8) we have

$$x_{n+1} = e^{x_n}/4, \quad n = 1, 2, 3, \ldots. \tag{13.10}$$

There are two roots (see Fig. 13.3), with one around $x = 0.3$. If we take it as x_1, then we have, from Eq. (13.10)

$$x_2 = e^{x_1}/4 = 0.3374,$$

$$x_3 = e^{x_2}/4 = 0.3503,$$

$$x_4 = e^{x_3}/4 = 0.3540,$$

$$x_5 = e^{x_4}/4 = 0.3565,$$

$$x_6 = e^{x_5}/4 = 0.3571,$$

$$x_7 = e^{x_6}/4 = 0.3573.$$

The computations can be terminated at this point if only three-decimal-place accuracy is required.

The second root lies between 2 and 3. As the slope of $y = 4x$ is less than that of $y = e^x$, the first condition of Eq. (13.9) cannot be met, so we rewrite the original equation in the form

$$e^x = 4x, \quad \text{or} \quad x = \log 4x$$

463

and take $g(x) = x$, $h(x) = \log 4x$. We now have

$$x_{n+1} = \log 4x_n, \quad n = 1, 2, \ldots.$$

If we take $x_1 = 2.1$, then

$$x_2 = \log 4x_1 = 2.12823,$$
$$x_3 = \log 4x_2 = 2.14158,$$
$$x_4 = \log 4x_3 = 2.14783,$$
$$x_5 = \log 4x_4 = 2.15075,$$
$$x_6 = \log 4x_5 = 2.15211,$$
$$x_7 = \log 4x_6 = 2.15303,$$
$$x_8 = \log 4x_7 = 2.15316,$$

and we see that the value of the root correct to three decimal places is 2.153.

Newton's method

In Newton's method, the successive terms in the sequence of approximate values $x_1, x_2, \ldots, x_n$ that converges to the root is obtained by the intersection with the x-axis of the tangent line to the curve $y = f(x)$. Fig. 13.4 shows a portion of the graph of $f(x)$ close to one of its roots, x_0. We start with x_1, an initial guess of the value of the root x_0. Now the equation of the tangent line to $y = f(x)$ at P_1 is

$$y - f(x_1)f'(x_1)(x - x_1). \tag{13.11}$$

This tangent line intersects the x-axis at x_2 that is a better approximation to the root than x_1. To find x_2, we set $y = 0$ in Eq. (13.11) and find

$$x_2 = x_1 - f(x_1)/f'(x_1)$$

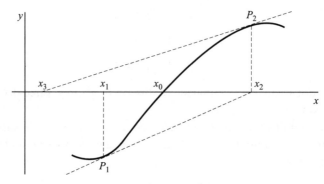

Figure 13.4.

provided $f'(x_1) \neq 0$. The equation of the tangent line at P_2 is

$$y - f(x_2) = f'(x_2)(x - x_2)$$

and it intersects the x-axis at x_3:

$$x_3 = x_2 - f(x_2)/f'(x_2).$$

This process is continued until we reach the desired level of accuracy. Thus, in general

$$x_{n+1} = x_n - \frac{f(x_n)}{f'(x_n)}, \quad n = 1, 2, \ldots. \tag{13.12}$$

Newton's method may fail if the function has a point of inflection, or other bad behavior, near the root. To illustrate Newton's method, let us consider the following trivial example.

Example 13.2
Solve, by Newton's method, $x^3 - 2 = 0$.

Solution: Here we have $y = x^3 - 2$. If we take $x_1 = 1.5$ (note that $1 < 2^{1/3} < 3$), then Eq. (13.12) gives

$$x_2 = 1.296296296,$$
$$x_3 = 1.260932225,$$
$$x_4 = 1.259921861,$$
$$x_5 = 1.25992105$$
$$x_6 = 1.25992105$$

$\left. \right\}$ repetition.

Thus, to eight-decimal-place accuracy, $2^{1/3} = 1.25992105$.

When applying Newton's method, it is often convenient to replace $f'(x_n)$ by

$$\frac{f(x_n + \delta) - f(x_n)}{\delta},$$

with δ small. Usually $\delta = 0.001$ will give good accuracy. Eq. (13.12) then reads

$$x_{n+1} = x_n - \frac{\delta f(x_n)}{f(x_n + \delta) - f(x_n)}, \quad n = 1, 2, \ldots. \tag{13.13}$$

Example 13.3
Solve the equation $x^2 - 2 = 0$.

Solution: Here $f(x) = x^2 - 2$. Take $x_1 = 1$ and $\delta = 0.001$, then Eq. (13.13) gives

$$x_2 = 1.499750125,$$

$$x_3 = 1.416680519,$$

$$x_4 = 1.414216580,$$

$$x_5 = 1.414213563,$$

$$\left.\begin{array}{l} x_6 = 1.414113562 \\ x_7 = 1.414113562 \end{array}\right\} x_6 = x_7.$$

Numerical integration

Very often definite integrations cannot be done in closed form. When this happens we need some simple and useful techniques for approximating definite integrals. In this section we discuss three such simple and useful methods.

The rectangular rule

The reader is familiar with the interpretation of the definite integral $\int_a^b f(x)dx$ as the area under the curve $y = f(x)$ between the limits $x = a$ and $x = b$:

$$\int_a^b f(x)dx = \sum_{i=1}^n f(\alpha_i)(x_i - x_{i-1}),$$

where $x_{i-1} \le \alpha_i \le x_i$, $a = x_0 < x_1 < x_2 < \cdots < x_n = b$. We can obtain a good approximation to this definite integral by simply evaluating such an area under the curve $y = f(x)$. We can divide the interval $a \le x \le b$ into n subintervals of length $h = (b - a)/n$, and in each subinterval, the function $f(\alpha_i)$ is replaced by a

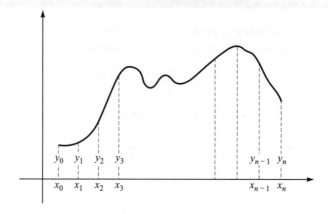

Figure 13.5.

straight line connecting the values at each head or end of the subinterval (or at the center point of the interval), as shown in Fig. 13.5. If we choose the head, $\alpha_i = x_{i-1}$, then we have

$$\int_a^b f(x)dx \approx h(y_0 + y_1 + \cdots + y_{n-1}),\qquad(13.14)$$

where $y_0 = f(x_0)$, $y_1 = f(x_1), \ldots, y_{n-1} = f(x_{n-1})$. This method is called the rectangular rule.

It will be shown later that the error decreases as n^2. Thus, as n increases, the error decreases rapidly.

The trapezoidal rule

The trapezoidal rule evaluates the small area of a subinterval slightly differently. The area of a trapezoid as shown in Fig. 13.6 is given by

$$\tfrac{1}{2}h(Y_1 + Y_2).$$

Thus, applied to Fig. 13.5, we have the approximation

$$\int_a^b f(x)dx \approx \frac{(b-a)}{n}(\tfrac{1}{2}y_0 + y_1 + y_2 + \cdots + y_{n-1} + \tfrac{1}{2}y_n).\qquad(13.15)$$

What are the upper and lower limits on the error of this method? Let us first calculate the error for a single subinterval of length $h(= (b-a)/n)$. Writing $x_i + h = z$ and $\varepsilon_i(z)$ for the error, we have

$$\int_{x_i}^z f(x)dx = \frac{h}{2}[y_i + y_z] + \varepsilon_i(z),$$

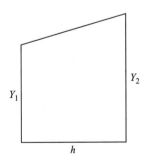

Y_2

Y_1

h

Figure 13.6.

467

where $y_i = f(x_i)$, $yz = f(z)$. Or

$$\varepsilon_i(z) = \int_{x_i}^{z} f(x)dx - \frac{h}{2}[f(x_i) - f(z)]$$

$$= \int_{x_i}^{z} f(x)dx - \frac{z - x_i}{2}[f(x_i) - f(z)].$$

Differentiating with respect to z:

$$\varepsilon_i'(z) = f(z) - [f(x_i) + f(z)]/2 - (z - x_i)f'(z)/2.$$

Differentiating once again,

$$\varepsilon_i''(z) = -(z - x_i)f''(z)/2.$$

If m_i and M_i are, respectively, the minimum and the maximum values of $f''(z)$ in the subinterval $[x_i, z]$, we can write

$$\frac{z - x_i}{2}m_i \le -\varepsilon_i''(z) \le \frac{z - x_i}{2}M_i.$$

Anti-differentiation gives

$$\frac{(z - x_i)^2}{4}m_i \le -\varepsilon_i'(z) \le \frac{(z - x_i)^2}{4}M_i$$

Anti-differentiation once more gives

$$\frac{(z - x_i)^3}{12}m_i \le -\varepsilon_i(z) \le \frac{(z - x_i)^3}{12}M_i.$$

or, since $z - x_i = h$,

$$\frac{h^3}{12}m_i \le -\varepsilon_i \le \frac{h^3}{12}M_i.$$

If m and M are, respectively, the minimum and the maximum of $f''(z)$ in the interval $[a, b]$ then

$$\frac{h^3}{12}m \le -\varepsilon_i \le \frac{h^3}{12}M \qquad \text{for all } i.$$

Adding the errors for all subintervals, we obtain

$$\frac{h^3}{12}nm \le -\varepsilon \le \frac{h^3}{12}nM$$

or, since $h = (b - a)/n$,

$$\frac{(b - a)^3}{12n^2}m \le -\varepsilon \le \frac{(b - a)^3}{12n^2}M. \qquad (13.16)$$

468

Thus, the error decreases rapidly as n increases, at least for twice-differentiable functions.

Simpson's rule

Simpson's rule provides a more accurate and useful formula for approximating a definite integral. The interval $a \leq x \leq b$ is subdivided into an even number of subintervals. A parabola is fitted to points a, $a + h$, $a + 2h$; another to $a + 2h$, $a + 3h$, $a + 4h$; and so on. The area under a parabola, as shown in Fig. 13.7, is (Problem 13.8)

$$\frac{h}{3}(y_1 + 4y_2 + y_3).$$

Thus, applied to Fig. 13.5, we have the approximation

$$\int_a^b f(x)dx \approx \frac{h}{3}(y_0 + 4y_1 + 2y_2 + 4y_3 + 2y_4 + \cdots + 2y_{n-2} + 4y_{n-1} + y_n), \quad (13.17)$$

with n even and $h = (b - a)/n$.

The analysis of errors for Simpson's rule is fairly involved. It has been shown that the error is proportional to h^4 (or inversely proportional to n^4).

There are other methods of approximating integrals, but they are not so simple as the above three. The method called Gaussian quadrature is very fast but more involved to implement. Many textbooks on numerical analysis cover this method.

Numerical solutions of differential equations

We noted in Chapter 2 that the methods available for the exact solution of differential equations apply only to a few, principally linear, types of differential equations. Many equations which arise in physical science and in engineering are not solvable by such methods and we are therefore forced to find ways of obtaining approximate solutions of these differential equations. The basic idea of approximate solutions is to specify a small increment h and to obtain approximate values of a solution $y = y(x)$ at x_0, $x_0 + h$, $x_0 + 2h$, ...

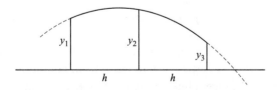

Figure 13.7.

469

The first-order ordinary differential equation

$$\frac{dy}{dx} = f(x, y), \tag{13.18}$$

with the initial condition $y = y_0$ when $x = x_0$, has the solution

$$y - y_0 = \int_{x_0}^{x} f(t, y(t))dt. \tag{13.19}$$

This integral equation cannot be evaluated because the value of y under the integral sign is unknown. We now consider three simple methods of obtaining approximate solutions: Euler's method, Taylor series method, and the Runge–Kutta method.

Euler's method

Euler proposed the following crude approach to finding the approximate solution. He began at the initial point (x_0, y_0) and extended the solution to the right to the point $x_1 = x_0 + h$, where h is a small quantity. In order to use Eq. (13.19) to obtain the approximation to $y(x_1)$, he had to choose an approximation to f on the interval $[x_0, x_1]$. The simplest of all approximations is to use $f(t, y(t)) = f(x_0, y_0)$. With this choice, Eq. (13.19) gives

$$y(x_1) = y_0 + \int_{x_0}^{x_1} f(x_0, y_0)dt = y_0 + f(x_0, y_0)(x_1 - x_0).$$

Letting $y_1 = y(x_1)$, we have

$$y_1 = y_0 + f(x_0, y_0)(x_1 - x_0). \tag{13.20}$$

From y_1, $y_1' = f(x_1, y_1)$ can be computed. To extend the approximate solution further to the right to the point $x_2 = x_1 + h$, we use the approximation: $f(t, y(t)) = y_1' = f(x_1, y_1)$. Then we obtain

$$y_2 = y(x_2) = y_1 + \int_{x_1}^{x_2} f(x_1, y_1)dt = y_1 + f(x_1, y_1)(x_2 - x_1).$$

Continuing in this way, we approximate y_3, y_4, and so on.

There is a simple geometrical interpretation of Euler's method. We first note that $f(x_0, y_0) = y'(x_0)$, and that the equation of the tangent line at the point (x_0, y_0) to the actual solution curve (or the integral curve) $y = y(x)$ is

$$y - y_0 = \int_{x_0}^{x} f(t, y(t))dt = f(x_0, y_0)(x - x_0).$$

Comparing this with Eq. (13.20), we see that (x_1, y_1) lies on the tangent line to the actual solution curve at (x_0, y_0). Thus, to move from point (x_0, y_0) to point (x_1, y_1) we proceed along this tangent line. Similarly, to move to point (x_2, y_2) we proceed parallel to the tangent line to the solution curve at (x_1, y_1), as shown in Fig. 13.8.

Table 13.1.

x	y (Euler)	y (actual)
1.0	3	3
1.1	3.4	3.43137
1.2	3.861	3.93122
1.3	4.3911	4.50887
1.4	4.99921	5.1745
1.5	5.69513	5.93977
1.6	6.48964	6.81695
1.7	7.39461	7.82002
1.8	8.42307	8.96433
1.9	9.58938	10.2668
2.0	10.9093	11.7463

The merit of Euler's method is its simplicity, but the successive use of the tangent line at the approximate values $y_1, y_2, \ldots$ can accumulate errors. The accuracy of the approximate vale can be quite poor, as shown by the following simple example.

Example 13.4
Use Euler's method to approximate solution to

$$y' = x^2 + y, \quad y(1) = 3 \text{ on interval } [1, 2].$$

Solution: Using $h = 0.1$, we obtain Table 13.1. Note that the use of a smaller step-size h will improve the accuracy.

Euler's method can be improved upon by taking the gradient of the integral curve as the means of obtaining the slopes at x_0 and $x_0 + h$, that is, by using the

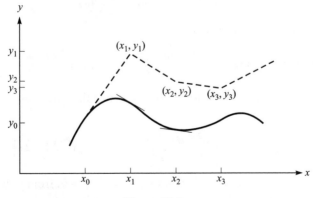

Figure 13.8.

471

approximate value obtained for y_1, we obtain an improved value, denoted by $(y_1)_1$:

$$(y_1)_1 = y_0 + \tfrac{1}{2}\{f(x_0, y_0) + f(x_0 + h, y_1)\}. \tag{13.21}$$

This process can be repeated until there is agreement to a required degree of accuracy between successive approximations.

The three-term Taylor series method

The rationale for this method lies in the three-term Taylor expansion. Let y be the solution of the first-order ordinary equation (13.18) for the initial condition $y = y_0$ when $x = x_0$ and suppose that it can be expanded as a Taylor series in the neighborhood of x_0. If $y = y_1$ when $x = x_0 + h$, then, for sufficiently small values of h, we have

$$y_1 = y_0 + h\left(\frac{dy}{dx}\right)_0 + \frac{h^2}{2!}\left(\frac{d^2y}{dx^2}\right)_0 + \frac{h^3}{3!}\left(\frac{d^3y}{dx^3}\right)_0 + \cdots. \tag{13.22}$$

Now

$$\frac{dy}{dx} = f(x, y),$$

$$\frac{d^2y}{dx^2} = \frac{\partial f}{\partial x} + \frac{dy}{dx}\frac{\partial f}{\partial y} = \frac{\partial f}{\partial x} + f\frac{\partial f}{\partial y},$$

and

$$\frac{d^3y}{dx^3} = \left(\frac{\partial}{\partial x} + f\frac{\partial}{\partial y}\right)\left(\frac{\partial f}{\partial x} + f\frac{\partial f}{\partial y}\right)$$

$$= \frac{\partial^2 f}{\partial x^2} + \frac{\partial f}{\partial y}\frac{\partial f}{\partial y} + 2f\frac{\partial^2 f}{\partial x \partial y} + f\left(\frac{\partial f}{\partial y}\right)^2 + f^2\frac{\partial^2 f}{\partial y^2}.$$

Equation (13.22) can be rewritten as

$$y_1 = y_0 + hf(x_0, y_0) + \frac{h^2}{2}\left[\frac{\partial f(x_0, y_0)}{\partial x} + f(x_0, y_0)\frac{\partial f(x_0, y_0)}{\partial y}\right],$$

where we have dropped the h^3 term. We now use this equation as an iterative equation:

$$y_{n+1} = y_n + hf(x_n, y_n) + \frac{h^2}{2}\left[\frac{\partial f(x_n, y_n)}{\partial x} + f(x_n, y_n)\frac{\partial f(x_n, y_n)}{\partial y}\right]. \tag{13.23}$$

That is, we compute $y_1 = y(x_0 + h)$ from y_0, $y_2 = y(x_1 + h)$ from y_1 by replacing x by x_1, and so on. The error in this method is proportional h^3. A good approxima-

Table 13.2.

n	x_n	y_n	y_{n+1}
0	1.0	−2.0	−2.1
1	1.1	−2.1	−2.2
2	1.2	−2.2	−2.3
3	1.3	−2.3	−2.4
4	1.4	−2.4	−2.5
5	1.5	−2.5	−2.6

tion can be obtained for y_n by summing a number of terms of the Taylor's expansion. To illustrate this method, let us consider a very simple example.

Example 13.5

Find the approximate values of y_1 through y_{10} for the differential equation $y' = x + y$, with the initial condition $x_0 = 1.0$ and $y = -2.0$.

Solution: Now $f(x, y) = x + y$, $\partial f/\partial x = \partial f/\partial y = 1$ and Eq. (13.23) reduces to

$$y_{n+1} = y_n + h(x_n + y_n) + \frac{h^2}{2}(1 + x_n + y_n).$$

Using this simple formula with $h = 0.1$ we obtain the results shown in Table 13.2.

The Runge–Kutta method

In practice, the Taylor series converges slowly and the accuracy involved is not very high. Thus we often resort to other methods of solution such as the Runge–Kutta method, which replaces the Taylor series, Eq. (13.23), with the following formula:

$$y_{n+1} = y_n + \frac{h}{6}(k_1 + 4k_2 + k_3), \tag{13.24}$$

where

$$k_1 = f(x_n, y_n), \tag{13.24a}$$

$$k_2 = f(x_n + h/2, \ y_n + hk_1/2), \tag{13.24b}$$

$$k_3 = f(x_n + h, y_0 + 2hk_2 - hk_1). \tag{13.24c}$$

This approximation is equivalent to Simpson's rule for the approximate integration of $f(x, y)$, and it has an error proportional to h^4. A beauty of the

473

Runge–Kutta method is that we do not need to compute partial derivatives, but it becomes rather complicated if pursued for more than two or three steps.

The accuracy of the Runge–Kutta method can be improved with the following formula:

$$y_{n+1} = y_n + \frac{h}{6}(k_1 + 2k_2 + 2k_3 + k_4), \tag{13.25}$$

where

$$k_1 = f(x_n, y_n), \tag{13.25a}$$

$$k_2 = f(x_n + h/2, y_n + hk_1/2), \tag{13.25b}$$

$$k_3 = f(x_n + h, y_0 + hk_2/2), \tag{13.25c}$$

$$k_4 = f(x_n + h, y_n + hk_3). \tag{13.25d}$$

With this formula the error in y_{n+1} is of order h^5.

You may wonder how these formulas are established. To this end, let us go back to Eq. (13.22), the three-term Taylor series, and rewrite it in the form

$$y_1 = y_0 + hf_0 + (1/2)h^2(A_0 + f_0B_0) + (1/6)h^3(C_0 + 2f_0D_0 + f_0^2E_0)$$
$$+ A_0B_0 + f_0B_0^2) + O(h^4), \tag{13.26}$$

where

$$A = \frac{\partial f}{\partial x}, \quad B = \frac{\partial f}{\partial y}, \quad C = \frac{\partial^2 f}{\partial x^2}, \quad D = \frac{\partial^2 f}{\partial x \partial y}, \quad E = \frac{\partial^2 f}{\partial y^2}$$

and the subscript 0 denotes the values of these quantities at (x_0, y_0).

Now let us expand k_1, k_2, and k_3 in the Runge–Kutta formula (13.24) in powers of h in a similar manner:

$$k_1 = hf(x_0, y_0),$$

$$k_2 = f(x_0 + h/2, y_0 + k_1h/2),$$

$$= f_0 + \frac{1}{2}h(A_0 + f_0B_0) + \frac{1}{8}h^2(C_0 + 2f_0D_0 + f_0^2E_0) + O(h^3).$$

Thus

$$2k_2 - k_1 = f_0 + h(A_0 + f_0B_0) + \cdots$$

and

$$\left(\frac{d}{dh}(2k_2 - k_1)\right)_{h=0} = f_0, \quad \left(\frac{d^2}{dh^2}(2k_2 - k_1)\right)_{h=0} = 2(A_0 + f_0B_0).$$

Then

$$k_3 = f(x_0 + h, y_0 + 2hk_2 - hk_1)$$
$$= f_0 + h(A_0 + f_0B_0) + (1/2)h^2\{C_0 + 2f_0D_0 + f_0^2E_0 + 2B_0(A_0 + f_0B_0)\}$$
$$+ O(h^3).$$

and

$$(1/6)(k_1 + 4k_2 + k_3) = hf_0 + (1/2)h^2(A_0 + f_0B_0)$$
$$+ (1/6)h^3(C_0 + 2f_0D_0 + f_0^2E_0 + A_0B_0 + f_0B_0^2) + O(h^4).$$

Comparing this with Eq. (13.26), we see that it agrees with the Taylor series expansion (up to the term in h^3) and the formula is established. Formula (13.25) can be established in a similar manner by taking one more term of the Taylor series.

Example 13.6
Using the Runge–Kutta method and $h = 0.1$, solve

$$y' = x - y^2/10; \quad x_0 = 0, \quad y_0 = 1.$$

Solution: With $h = 0.1$, $h^4 = 0.0001$ and we may use the Runge–Kutta third-order approximation.

First step: $x_0 = 0$, $y_0 = 0$, $f_0 = -0.1$,
    $k_1 = -0.1$, $y_0 + hk_1/2 = 0.995$,
    $k_2 = -0.049$, $2k_2 - k_1 = 0.002$, $k_3 = 0$,
    $y_1 = y_0 + \dfrac{h}{6}(k_1 + 4k_2 + k_1) = 0.9951.$

Second step: $x_1 = x_0 + h = 0.1$, $y_1 = 0.9951$, $f_1 = 0.001$,
    $k_1 = 0.001$, $y_1 + hk_1/2 = 0.9952$,
    $k_2 = 0.051$, $2k_2 - k_1 = 0.101$, $k_3 = 0.099$,
    $y_2 = y_1 + \dfrac{h}{6}(k_1 + 4k_2 + k_1) = 1.0002.$

Third step: $x_2 = x_1 + h = 0.2$, $y_2 = 1.0002$, $f_2 = 0.1$,
    $k_1 = 0.1$, $y_2 + hk_1/2 = 1.0052$,
    $k_2 = 0.149$, $2k_2 - k_1 = 0.198$, $k_3 = 0.196$,
    $y_3 = y_2 + \dfrac{h}{6}(k_1 + 4k_2 + k_1) = 1.0151.$

Equations of higher order. System of equations

The methods in the previous sections can be extended to obtain numerical solutions of equations of higher order. An nth-order differential equation is equivalent to n first-order differential equations in $n+1$ variables. Thus, for instance, the second-order equation

$$y'' = f(x, y, y'),\qquad(13.27)$$

with initial conditions

$$y(x_0) = y_0, \quad y'(x_0) = y_0',\qquad(13.28)$$

can be written as a system of two equations of first order by setting

$$y' = u,\qquad(13.29)$$

then Eqs. (13.27) and (13.28) become

$$u' = f(x, y, u),\qquad(13.30)$$

$$y(x_0) = y_0, \quad u(x_0) = u_0.\qquad(13.31)$$

The two first-order equations (13.29) and (13.30) with the initial conditions (13.31) are completely equivalent to the original second-order equation (13.27) with the initial conditions (13.28). And the methods in the previous sections for determining approximate solutions can be extended to solve this system of two first-order equations. For example, the equation

$$y'' - y = 2,$$

with initial conditions

$$y(0) = -1, \quad y'(0) = 1,$$

is equivalent to the system

$$y' = x + u, \quad u' = 1 + y,$$

with

$$y(0) = -1, \quad u(0) = 1.$$

These two first-order equations can be solved with Taylor's method (Problem 13.12).

The simple methods outlined above all have the disadvantage that the error in approximating to values of y is to a certain extent cumulative and may become large unless some form of checking process is included. For this reason, methods of solution involving finite difference are devised, most of them being variations of the Adams–Bashforth method that contains a self-checking process. This method is quite involved and because of limited space we shall not cover it here, but it is discussed in any standard textbook on numerical analysis.

Least-squares fit

We now look at the problem of fitting of experimental data. In some experimental situations there may be underlying theory that suggests the kind of function to be used in fitting the data. Often there may be no theory on which to rely in selecting a function to represent the data. In such circumstances a polynomial is often used. We saw earlier that the $m + 1$ coefficients in the polynomial

$$y = a_0 + a_1 x + \cdots + a_m x^m$$

can always be determined so that a given set of $m + 1$ points (x_i, y_i), where the xs may be unequal, lies on the curve described by the polynomial. However, when the number of points is large, the degree m of the polynomial is high, and an attempt to fit the data by using a polynomial is very laborious. Furthermore, the experimental data may contain experimental errors, and so it may be more sensible to represent the data approximately by some function $y = f(x)$ that contains a few unknown parameters. These parameters can then be determined so that the curve $y = f(x)$ fits the data. How do we determine these unknown parameters?

Let us represent a set of experimental data (x_i, y_i), where $i = 1, 2, \ldots, n$, by some function $y = f(x)$ that contains r parameters $a_1, a_2, \ldots, a_r$. We then take the deviations (or residuals)

$$d_i = f(x_i) - y_i \tag{13.32}$$

and form the weighted sum of squares of the deviations

$$S = \sum_{i=1}^{n} w_i (d_i)^2 = \sum_{i=1}^{n} w_i [f(x_i) - y_i]^2, \tag{13.33}$$

where the weights w_i express our confidence in the accuracy of the experimental data. If the points are equally weighted, the ws can all be set to 1.

It is clear that the quantity S is a function of as: $S = S(a_1, a_2, \ldots, a_r)$. We can now determine these parameters so that S is a minimum:

$$\frac{\partial S}{\partial a_1} = 0, \quad \frac{\partial S}{\partial a_2} = 0, \ldots, \quad \frac{\partial S}{\partial a_r} = 0. \tag{13.34}$$

The set of r equations (13.34) is called the normal equations and serves to determine the r unknown as in $y = f(x)$. This particular method of determining the unknown as is known as the method of least squares.

We now illustrate the construction of the normal equations with the simplest case: $y = f(x)$ is a linear function:

$$y = a_1 + a_2 x. \tag{13.35}$$

The deviations d_i are given by

$$d_i = (a_1 + a_2 x) - y_i$$

and so, assuming $w_i = 1$

$$S = \sum_{i=1}^n d_i^2 = (a_1 + a_2 x_1 - y_1)^2 + (a_1 + a_2 x_2 - y_2)^2 + \cdots + (a_1 + a_2 x_r - y_r)^2.$$

We now find the partial derivatives of S with respect to a_1 and a_2 and set these to zero:

$$\partial S/\partial a_1 = 2(a_1 + a_2 x_1 - y_1) + 2(a_1 + a_2 x_2 - y_2) + \cdots + 2(a_1 + a_2 x_n - y_n) = 0,$$

$$\partial S/\partial a_2 = 2x_1(a_1 + a_2 x_1 - y_1) + 2x_2(a_1 + a_2 x_2 - y_2) + \cdots + 2x_n(a_1 + a_2 x_n - y_n)$$
$$= 0.$$

Dividing out the factor 2 and collecting the coefficients of a_1 and a_2, we obtain

$$na_1 + \left(\sum_{i=1}^n x_i\right) a_2 = \sum_{i=1}^n y_1, \tag{13.36}$$

$$\left(\sum_{i=1}^n x_i\right) a_1 + \left(\sum_{i=1}^n x_i^2\right) a_2 = \sum_{i=1}^n x_i y_i. \tag{13.37}$$

These equations can be solved for a_1 and a_2.

Problems

13.1. Given six points $(-1,0)$, $(-0.8, 2)$, $(-0.6, 1)$, $(-0.4, -1)$, $(-0.2, 0)$, and $(0, -4)$, determine a smooth function $y = f(x)$ such that $y_i = f(x_i)$.

13.2. Find an approximate value of the real root of

$$x - \tan x = 0$$

near $x = 3\pi/2$.

13.3. Find the angle subtended at the center of a circle by an arc whose length is double the length of the chord.

13.4. Use Newton's method to solve

$$e^{x^2} - x^3 + 3x - 4 = 0,$$

with $x_0 = 0$ and $h = 0.001$.

13.5. Use Newton's method to find a solution of
$$\sin(x^3 + 2) = 1/x,$$
with $x_0 = 1$ and $h = 0.001$.

13.6. Approximate the following integrals using the rectangular rule, the trapezoidal rule, and Simpson's rule, with $n = 2, 4, 10, 20, 50$:

(a) $\displaystyle\int_0^{\pi/2} e^{-x^2} \sin(x^2 + 1)\,dx;$

(b) $\displaystyle\int_0^{\sqrt{2}} \frac{\sin(x^2) + 3x - 2}{x + 4}\,dx;$

(c) $\displaystyle\int_0^1 \frac{dx}{\sqrt{2 - \sin^2 x}}.$

13.7 Show that the area under a parabola, as shown in Fig. 13.7, is given by
$$A = \frac{h}{3}(y_1 + 4y_2 + y_3).$$

13.8. Using the improved Euler's method, find the value of y when $x = 0.2$ on the integral curve of the equation $y' = x^2 - 2y$ through the point $x = 0$, $y = 1$.

13.9. Using Taylor's method, find correct to four places of decimals values of y corresponding to $x = 0.2$ and $x = -0.2$ for the solution of the differential equation
$$dy/dx = x - y^2/10,$$
with the initial condition $y = 1$ when $x = 0$.

13.10. Using the Runge–Kutta method and $h = 0.1$, solve
$$y' = x^2 - \sin(y^2), \quad x_0 = 1 \text{ and } y_0 = 4.7.$$

13.11. Using the Runge–Kutta method and $h = 0.1$, solve
$$y' = ye^{-x^2}, \quad x_0 = 1 \text{ and } y_0 = 3.$$

13.12. Using Taylor's method, obtain the solution of the system
$$y' = x + u, \quad u' = 1 + y$$
with
$$y(0) = -1, \quad u(0) = 1.$$

13.13. Find to four places of decimals the solution between $x = 0$ and $x = 0.5$ of the equations
$$y' = \tfrac{1}{2}(y + u), \quad u' = \tfrac{1}{2}(y^2 - u^2),$$
with $y = u = 1$ when $x = 0$.

479

13.14. Find to three places of decimals a solution of the equation

$$y'' + 2xy' - 4y = 0,$$

with $y = y' = 1$ when $x = 0$.

13.15. Use Eqs. (13.36) and (13.37) to calculate the coefficients in $y = a_1 + a_2x$ to fit the following data: $(x, y) = (1, 1.7), (2, 1.8), (3, 2.3), (4, 3.2)$.

14

Introduction to probability theory

The theory of probability is so useful that it is required in almost every branch of science. In physics, it is of basic importance in quantum mechanics, kinetic theory, and thermal and statistical physics to name just a few topics. In this chapter the reader is introduced to some of the fundamental ideas that make probability theory so useful. We begin with a review of the definitions of probability, a brief discussion of the fundamental laws of probability, and methods of counting (some facts about permutations and combinations), probability distributions are then treated.

A notion that will be used very often in our discussion is 'equally likely'. This cannot be defined in terms of anything simpler, but can be explained and illustrated with simple examples. For example, heads and tails are equally likely results in a spin of a fair coin; the ace of spades and the ace of hearts are equally likely to be drawn from a shuffled deck of 52 cards. Many more examples can be given to illustrate the concept of 'equally likely'.

A definition of probability

Now a question that arises naturally is that of how shall we measure the probability that a particular case (or outcome) in an experiment (such as the throw of dice or the draw of cards) out of many equally likely cases that will occur. Let us flip a coin twice, and ask the question: what is the probability of it coming down heads at least once. There are four equally likely results in flipping a coin twice: *HH, HT, TH, TT*, where *H* stands for head and *T* for tail. Three of the four results are favorable to at least one head showing, so the probability of getting one head is 3/4. In the example of drawn cards, what is the probability of drawing the ace of spades? Obviously there is one chance out of 52, and the probability, accordingly, is 1/52. On the other hand, the probability of drawing an ace is four times as great $-4/52$, for there are four aces, equally likely. Reasoning in this way, we are led to give the notion of probability the following definition:

> If there are N mutually exclusive, collective exhaustive, and equally likely outcomes of an experiment, and n of these are favorable to an event A, then the probability $p(A)$ of an event A is n/N: $p = n/N$, or

$$p(A) = \frac{\text{number of outcomes favorable to } A}{\text{total number of results}}. \tag{14.1}$$

We have made no attempt to predict the result, just to measure it. The definition of probability given here is often called *a posteriori* probability.

The terms exclusive and exhaustive need some attention. Two events are said to be mutually exclusive if they cannot both occur together in a single trial; and the term collective exhaustive means that all possible outcomes or results are enumerated in the N outcomes.

If an event is certain not to occur its probability is zero, and if an event is certain to occur, then its probability is 1. Now if p is the probability that an event will occur, then the probability that it will fail to occur is $1 - p$, and we denote it by q:

$$q = 1 - p. \tag{14.2}$$

If p is the probability that an event will occur in an experiment, and if the experiment is repeated M times, then the expected number of times the event will occur is Mp. For sufficiently large M, Mp is expected to be close to the actual number of times the event will occur. For example, the probability of a head appearing when tossing a coin is $1/2$, the expected number of times heads appear is $4 \times 1/2$ or 2. Actually, heads will not always appear twice when a coin is tossed four times. But if it is tossed 50 times, the number of heads that appear will, on the average, be close to 25 ($50 \times 1/2 = 25$). Note that closeness is computed on a percentage basis: 20 is 20% of 25 away from 25 while 1 is 50% of 2 away from 2.

Sample space

The equally likely cases associated with an experiment represent the possible outcomes. For example, the 36 equally likely cases associated with the throw of a pair of dice are the 36 ways the dice may fall, and if 3 coins are tossed, there are 8 equally likely cases corresponding to the 8 possible outcomes. A list or set that consists of all possible outcomes of an experiment is called a sample space and each individual outcome is called a sample point (a point of the sample space). The outcomes composing the sample space are required to be mutually exclusive. As an example, when tossing a die the outcomes 'an even number shows' and 'number 4 shows' cannot be in the same sample space. Often there will be more than one sample space that can describe the outcome of an experiment but there is usually only one that will provide the most information. In a throw of a fair die, one sample space is the set of all possible outcomes $\{1, 2, 3, 4, 5, 6\}$, and another could be $\{even\}$ or $\{odd\}$.

A finite sample space is one that has only a finite number of points. The points of the sample space are weighted according to their probabilities. To see this, let the points have the probabilities

$$p_1, p_2, \ldots, p_N$$

with

$$p_1 + p_2 + \cdots + p_N = 1.$$

Suppose the first n sample points are favorable to another event A. Then the probability of A is defined to be

$$p(A) = p_1 + p_2 + \cdots + p_n.$$

Thus the points of the sample space are weighted according to their probabilities. If each point has the sample probability $1/n$, then $p(A)$ becomes

$$p(A) = \frac{1}{N} + \frac{1}{N} + \cdots + \frac{1}{N} = \frac{n}{N}$$

and this definition is consistent with that given by Eq. (14.1).

A sample space with constant probability is called uniform. Non-uniform sample spaces are more common. As an example, let us toss four coins and count the number of heads. An appropriate sample space is composed of the outcomes

0 heads, 1 head, 2 heads, 3 heads, 4 heads,

with respective probabilities, or weights

$1/16, \quad 4/16, \quad 6/16, \quad 4/16, \quad 1/16.$

The four coins can fall in $2 \times 2 \times 2 \times 2 = 2^4$, or 16 ways. They give no heads (all land tails) in only one outcome, and hence the required probability is 1/16. There are four ways to obtain 1 head: a head on the first coin or on the second coin, and so on. This gives 4/16. Similarly we can obtain the probabilities for the other cases.

We can also use this simple example to illustrate the use of sample space. What is the probability of getting at least two heads? Note that the last three sample points are favorable to this event, hence the required probability is given by

$$\frac{6}{16} + \frac{4}{16} + \frac{1}{16} = \frac{11}{16}.$$

Methods of counting

In many applications the total number of elements in a sample space or in an event needs to be counted. A fundamental principle of counting is this: if one thing can be done in n different ways and another thing can be done in m different ways, then both things can be done together or in succession in mn different ways. As an example, in the example of throwing a pair of dice cited above, there are 36 equally like outcomes: the first die can fall in six ways, and for each of these the second die can also fall in six ways. The total number of ways is

$$6 + 6 + 6 + 6 + 6 + 6 = 6 \times 6 = 36$$

and these are equally likely.

Enumeration of outcomes can become a lengthy process, or it can become a practical impossibility. For example, the throw of four dice generates a sample space with $6^4 = 1296$ elements. Some systematic methods for the counting are desirable. Permutation and combination formulas are often very useful.

Permutations

A permutation is a particular ordered selection. Suppose there are n objects and r of these objects are arranged into r numbered spaces. Since there are n ways of choosing the first object, and after this is done there are $n - 1$ ways of choosing the second object, ..., and finally $n - (r - 1)$ ways of choosing the rth object, it follows by the fundamental principle of counting that the number of different arrangements or permutations is given by

$$_nP_r = n(n-1)(n-2)\cdots(n-r+1). \tag{14.3}$$

where the product on the right-hand side has r factors. We call $_nP_r$ the number of permutations of n objects taken r at a time. When $r = n$, we have

$$_nP_n = n(n-1)(n-2)\cdots 1 = n!.$$

We can rewrite $_nP_r$ in terms of factorials:

$$_nP_r = n(n-1)(n-2)\cdots(n-r+1)$$

$$= n(n-1)(n-2)\cdots(n-r+1)\frac{(n-r)\cdots 2 \times 1}{(n-r)\cdots 2 \times 1}$$

$$= \frac{n!}{(n-r)!}.$$

When $r = n$, we have $_nP_n = n!/(n-n)! = n!/0!$. This reduces to $n!$ if we have $0! = 1$ and mathematicians actually take this as the definition of $0!$.

Suppose the n objects are not all different. Instead, there are n_1 objects of one kind (that is, indistinguishable from each other), n_2 that is of a second kind, ..., n_k

of a kth kind so that $n_1 + n_2 + \cdots + n_k = n$. A natural question is that of how many distinguishable arrangements are there of these n objects. Assuming that there are N different arrangements, and each distinguishable arrangement appears $n_1!$, $n_2!$, ... times, where $n_1!$ is the number of ways of arranging the n_1 objects, similarly for $n_2!$, ..., $n_k!$. Then multiplying N by $n_1! n_2!$, ..., $n_k!$ we obtain the number of ways of arranging the n objects if they were all distinguishable, that is, $_nP_n = n!$:

$$N n_1! n_2! \cdots n_k! = n! \qquad \text{or} \qquad N = n!/(n_1! n_2! \ldots n_k!).$$

N is often written as $_nP_{n_1 n_2 \ldots n_k}$, and then we have

$$_nP_{n_1 n_2 \ldots n_k} = \frac{n!}{n_1! n_2! \cdots n_k!}. \tag{14.4}$$

For example, given six coins: one penny, two nickels and three dimes, the number of permutations of these six coins is

$$_6P_{123} = 6!/1!2!3! = 60.$$

Combinations

A permutation is a particular ordered selection. Thus 123 is a different permutation from 231. In many problems we are interested only in selecting objects without regard to order. Such selections are called combinations. Thus 123 and 231 are now the same combination. The notation for a combination is $_nC_r$ which means the number of ways in which r objects can be selected from n objects without regard to order (also called the combination of n objects taken r at a time). Among the $_nP_r$ permutations there are $r!$ that give the same combination. Thus, the total number of permutations of n different objects selected r at a time is

$$r! \, _nC_r = \, _nP_r = \frac{n!}{(n-r)!}.$$

Hence, it follows that

$$_nC_r = \frac{n!}{r!(n-r)!}. \tag{14.5}$$

It is straightforward to show that

$$_nC_r = \frac{n!}{r!(n-r)!} = \frac{n!}{[n-(n-r)]!(n-r)!} = \, _nC_{n-r}.$$

$_nC_r$ is often written as

$$_nC_r = \binom{n}{r}.$$

The numbers (14.5) are often called binomial coefficients because they arise in the binomial expansion

$$(x+y)^n = x^n + \binom{n}{1}x^{n-1}y + \binom{n}{2}x^{n-2}y^2 + \cdots + \binom{n}{n}y^n.$$

When n is very large a direct evaluation of $n!$ is impractical. In such cases we use Stirling's approximate formula

$$n! \approx \sqrt{2\pi n}\, n^n e^{-n}.$$

The ratio of the left hand side to the right hand side approaches 1 as $n \to \infty$. For this reason the right hand side is often called an asymptotic expansion of the left hand side.

Fundamental probability theorems

So far we have calculated probabilities by directly making use of the definitions; it is doable but it is not always easy. Some important properties of probabilities will help us to cut short our computation works. These important properties are often described in the form of theorems. To present these important theorems, let us consider an experiment, involving two events A and B, with N equally likely outcomes and let

$n_1 = $ number of outcomes in which A occurs, but not B,

$n_2 = $ number of outcomes in which B occurs, but not A,

$n_3 = $ number of outcomes in which both A and B occur,

$n_4 = $ number of outcomes in which neither A nor B occurs.

This covers all possibilities, hence $n_1 + n_2 + n_3 + n_4 = N$.

The probabilities of A and B occurring are respectively given by

$$P(A) = \frac{n_1 + n_3}{N}, \qquad P(B) = \frac{n_2 + n_3}{N}, \tag{14.6}$$

the probability of either A or B (or both) occurring is

$$P(A + B) = \frac{n_1 + n_2 + n_3}{N}, \tag{14.7}$$

and the probability of both A and B occurring successively is

$$P(AB) = \frac{n_3}{N}. \tag{14.8}$$

Let us rewrite $P(AB)$ as

$$P(AB) = \frac{n_3}{N} = \frac{n_1 + n_3}{N} \frac{n_3}{n_1 + n_3}.$$

Now $(n_1 + n_3)/N$ is $P(A)$ by definition. After A has occurred, the only possible cases are the $(n_1 + n_3)$ cases favorable to A. Of these, there are n_3 cases favorable to B, the quotient $n_3/(n_1 + n_3)$ represents the probability of B when it is known that A occurred, $PA(B)$. Thus we have

$$P(AB) = P(A)P_A(B). \tag{14.9}$$

This is often known as the theorem of joint (or compound) probability. In words, the joint probability (or the compound probability) of A and B is the product of the probability that A will occur times the probability that B will occur if A does. $P_A(B)$ is called the conditional probability of B given A (that is, given that A has occurred).

To illustrate the theorem of joint probability (14.9), we consider the probability of drawing two kings in succession from a shuffled deck of 52 playing cards. The probability of drawing a king on the first draw is 4/52. After the first king has been drawn, the probability of drawing another king from the remaining 51 cards is 3/51, so that the probability of two kings is

$$\frac{4}{52} \times \frac{3}{51} = \frac{1}{221}.$$

If the events A and B are independent, that is, the information that A has occurred does not influence the probability of B, then $P_A(B) = P(B)$ and the joint probability takes the form

$$P(AB) = P(A)P(B), \text{ for independent events.} \tag{14.10}$$

As a simple example, let us toss a coin and a die, and let A be the event 'head shows' and B is the event '4 shows.' These events are independent, and hence the probability that 4 and a head both show is

$$P(AB) = P(A)P(B) = (1/2)(1/6) = 1/12.$$

Theorem (14.10) can be easily extended to any number of independent events $A, B, C, \ldots$.

Besides the theorem of joint probability, there is a second fundamental relationship, known as the theorem of total probability. To present this theorem, let us go back to Eq. (14.4) and rewrite it in a slightly different form

$$P(A + B) = \frac{n_1 + n_2 + n_3}{N}$$

$$= \frac{n_1 + n_2 + 2n_3 - n_3}{N} = \frac{(n_1 + n_3) + (n_2 + n_3) - n_3}{N}$$

$$= \frac{n_1 + n_3}{N} + \frac{n_2 + n_3}{N} - \frac{n_3}{N} = P(A) + P(B) - P(AB),$$

$$P(A + B) = P(A) + P(B) - P(AB). \tag{14.11}$$

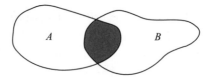

Figure 14.1.

This theorem can be represented diagrammatically by the intersecting points sets A and B shown in Fig. 14.1. To illustrate this theorem, consider the simple example of tossing two dice and find the probability that at least one die gives 2. The probability that both give 2 is 1/36. The probability that the first die gives 2 is 1/6, and similarly for the second die. So the probability that at least one gives 2 is

$$P(A + B) = 1/6 + 1/6 - 1/36 = 11/36.$$

For mutually exclusive events, that is, for events A, B which cannot both occur, $P(AB) = 0$ and the theorem of total probability becomes

$$P(A + B) = P(A) + P(B), \qquad \text{for mutually exclusive events.} \qquad (4.12)$$

For example, in the toss of a die, '4 shows' (event A) and '5 shows' (event B) are mutually exclusive, the probability of getting either 4 or 5 is

$$P(A + B) = P(A) + P(B) = 1/6 + 1/6 = 1/3.$$

The theorems of total and joint probability for uniform sample spaces established above are also valid for arbitrary sample spaces. Let us consider a finite sample space, its events E_i are so numbered that $E_1, E_2, \ldots, E_j$ are favorable to A, $E_{j+1}, \ldots, E_k$ are favorable to both A and B, and $E_{k+1}, \ldots, E_m$ are favorable to B only. If the associated probabilities are p_i, then Eq. (14.11) is equivalent to the identity

$$p_1 + \cdots + p_m = (p_1 + \cdots + p_j + p_{j+1} + \cdots + p_k)$$
$$+ (p_{j+1} + \cdots + p_k + p_{k+1} + \cdots + p_m) - (p_{j+1} + \cdots + p_m).$$

The sums within the three parentheses on the right hand side represent, respectively, $P(A)P(B)$, and $P(AB)$ by definition. Similarly, we have

$$P(AB) = p_{j+1} + \cdots + p_k$$

$$= (p_1 + \cdots + p_k) \left(\frac{p_{j+1}}{p_1 + \cdots + p_k} + \cdots + \frac{p_k}{p_1 + \cdots + p_k} \right)$$

$$= P(A)P_A(B),$$

which is Eq. (14.9).

Random variables and probability distributions

As demonstrated above, simple probabilities can be computed from elementary considerations. We need more efficient ways to deal with probabilities of whole classes of events. For this purpose we now introduce the concepts of random variables and a probability distribution.

Random variables

A process such as spinning a coin or tossing a die is called random since it is impossible to predict the final outcome from the initial state. The outcomes of a random process are certain numerically valued variables that are often called random variables. For example, suppose that three dimes are tossed at the same time and we ask how many heads appear. The answer will be 0, 1, 2, or 3 heads, and the sample space S has 8 elements:

$$S = \{TTT, HTT, THT, TTH, HHT, HTH, THH, HHH\}.$$

The random variable X in this case is the number of heads obtained and it assumes the values

$$0, 1, 1, 1, 2, 2, 2, 3.$$

For instance, $X = 1$ corresponds to each of the three outcomes: HTT, THT, TTH. That is, the random variable X can be thought of as a function of the number of heads appear.

A random variable that takes on a finite or countable infinite number of values (that is it has as many values as the natural numbers $1, 2, 3, \ldots$) is called a discrete random variable while one that takes on a non-countable infinite number of values is called a non-discrete or continuous random variable.

Probability distributions

A random variable, as illustrated by the simple example of tossing three dimes at the same time, is a numerical-valued function defined on a sample space. In symbols,

$$X(s_i) = x_i \qquad i = 1, 2, \ldots, n, \tag{14.13}$$

where s_i are the elements of the sample space and x_i are the values of the random variable X. The set of numbers x_i can be finite or infinite.

In terms of a random variable we will write $P(X = x_i)$ as the probability that the random variable X takes the value x_i, and $P(X < x_i)$ as the probability that the random variable takes values less than x_i, and so on. For simplicity, we often write $P(X = x_i)$ as p_i. The pairs (x_i, p_i) for $i = 1, 2, 3, \ldots$ define the probability distribution or probability function for the random variable X. Evidently any probability distribution p_i for a discrete random variable must satisfy the following conditions:

(i) $0 \le p_i \le 1$;
(ii) the sum of all the probabilities must be unity (certainty), $\sum_i p_i = 1$.

Expectation and variance

The expectation or expected value or mean of a random variable is defined in terms of a weighted average of outcomes, where the weighting is equal to the probability p_i with which x_i occurs. That is, if X is a random variable that can take the values $x_1, x_2, \ldots$, with probabilities $p_1, p_2, \ldots$, then the expectation or expected value $E(X)$ is defined by

$$E(X) = p_1 x_1 + p_2 x_2 + \cdots = \sum_i p_i x_i. \tag{14.14}$$

Some authors prefer to use the symbol μ for the expectation value $E(X)$. For the three dimes tossed at the same time, we have

$$x_i = 0 \quad 1 \quad 2 \quad 3$$
$$p_i = 1/8 \quad 3/8 \quad 3/8 \quad 1/8$$

and

$$E(X) = \frac{1}{8} \times 0 + \frac{3}{8} \times 1 + \frac{3}{8} \times 2 + \frac{1}{8} \times 3 = \frac{3}{2}.$$

We often want to know how much the individual outcomes are scattered away from the mean. A quantity measure of the spread is the difference $X - E(X)$ and this is called the deviation or residual. But the expectation value of the deviations is always zero:

$$E(X - E(X)) = \sum_i (x_i - E(X)) p_i = \sum_i x_i p_i - E(X) \sum_i p_i$$
$$= E(X) - E(X) \cdot 1 = 0.$$

This should not be particularly surprising; some of the deviations are positive, and some are negative, and so the mean of the deviations is zero. This means that the mean of the deviations is not very useful as a measure of spread. We get around the problem of handling the negative deviations by squaring each deviation, thereby obtaining a quantity that is always positive. Its expectation value is called the variance of the set of observations and is denoted by σ^2

$$\sigma^2 = E[(X - E(X))^2] = E[(X - \mu)^2]. \tag{14.15}$$

The square root of the variance, σ, is known as the standard deviation, and it is always positive.

We now state some basic rules for expected values. The proofs can be found in any standard textbook on probability and statistics. In the following c is a constant, X and Y are random variables, and $h(X)$ is a function of X:

(1) $E(cX) = cE(X)$;
(2) $E(X + Y) = E(X) + E(Y)$;
(3) $E(XY) = E(X)E(Y)$ (provided X and Y are independent);
(4) $E(h(X)) = \sum_i h(x_i)p_i$ (for a finite distribution).

Special probability distributions

We now consider some special probability distributions in which we will use all the things we have learned so far about probability.

The binomial distribution

Before we discuss the binomial distribution, let us introduce a term, the Bernoulli trials. Consider an experiment such as spinning a coin or throw a die repeatedly. Each spin or toss is called a trial. In any single trial there will be a probability p associated with a particular event (or outcome). If p is constant throughout (that is, does not change from one trial to the next), such trials are then said to be independent and are known as Bernoulli trials.

Now suppose that we have n independent events of some kind (such as tossing a coin or die), each of which has a probability p of success and probability of $q = (1 - p)$ of failure. What is the probability that exactly m of the events will succeed? If we select m events from n, the probability that these m will succeed and all the rest $(n - m)$ will fail is $p^m q^{n-m}$. We have considered only one particular group or combination of m events. How many combinations of m events can be chosen from n? It is the number of combinations of n things taken m at a time: $_nC_m$. Thus the probability that exactly m events will succeed from a group of n is

$$f(m) = P(X = m) = {}_nC_m p^m q^{(n-m)}$$

$$= \frac{n!}{m!(n - m)!} p^m q^{(n-m)}. \tag{14.16}$$

This discrete probability function (14.16) is called the binomial distribution for X, the random variable of the number of successes in the n trials. It gives the probability of exactly m successes in n independent trials with constant probability p. Since many statistical studies involve repeated trials, the binomial distribution has great practical importance.

Why is the discrete probability function (14.16) called the binomial distribution? Since for $m = 0, 1, 2, \ldots, n$ it corresponds to successive terms in the binomial expansion

$$(q + p)^n = q^n + {}_nC_1 q^{n-1} p + {}_nC_2 q^{n-2} p^2 + \cdots + p^n = \sum_{m=0}^{n} {}_nC_m p^m q^{n-m}.$$

To illustrate the use of the binomial distribution (14.16), let us find the probability that a one will appear exactly 4 times if a die is thrown 10 times. Here $n = 10$, $m = 4$, $p = 1/6$, and $q = (1 - p) = 5/6$. Hence the probability is

$$f(4) = P(X = 4) = \frac{10!}{4!6!} \left(\frac{1}{6}\right)^4 \left(\frac{5}{6}\right)^6 = 0.0543.$$

A few examples of binomial distributions, computed from Eq. (14.16), are shown in Figs. 14.2, and 14.3 by means of histograms.

One of the key requirements for a probability distribution is that

$$\sum_{m=0}^{n} f(m) = \sum_{m=0}^{n} {}_nC_m p^m q^{n-m} = 1. \tag{14.17}$$

To show that this is in fact the case, we note that

$$\sum_{m=0}^{n} {}_nC_m p^m q^{n-m}$$

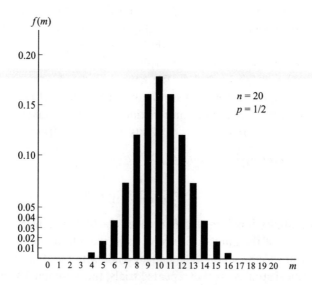

Figure 14.2.   The distribution is symmetric about $m = 10$.

492

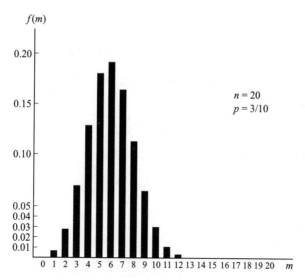

Figure 14.3. The distribution favors smaller value of m.

is exactly equal to the binomial expansion of $(q+p)^n$. But here $q+p=1$, so $(q+p)^n=1$ and our proof is established.

The mean (or average) number of successes, $\bar{m}$, is given by

$$\bar{m} = \sum_{m=0}^{n} m_n C_m p^m (1-p)^{n-m}. \tag{14.18}$$

The sum ranges from $m=0$ to n because in every one of the sets of trials the same number of successes between 0 and n must occur. It is similar to Eq. (14.17); the difference is that the sum in Eq. (14.18) contains an extra factor n. But we can convert it into the form of the sum in Eq. (14.17). Differentiating both sides of Eq. (14.17) with respect to p, which is legitimate as the equation is true for all p between 0 and 1, gives

$$\sum_n C_m [m p^{m-1}(1-p)^{n-m} - (n-m)p^m(1-p)^{n-m-1}] = 0,$$

where we have dropped the limits on the sum, remembering that m ranges from 0 to n. The last equation can be rewritten as

$$\sum m_n C_m p^{m-1}(1-p)^{n-m} = \sum (n-m)_n C_m p^m (1-p)^{n-m-1}$$
$$= n \sum {}_n C_m p^m (1-p)^{n-m-1} - \sum m_n C_m p^m (1-p)^{n-m-1}$$

or

$$\sum m_n C_m [p^{m-1}(1-p)^{n-m} + p^m(1-p)^{n-m-1}] = n \sum {}_n C_m p^m (1-p)^{n-m-1}.$$

493

Now multiplying both sides by $p(1-p)$ we get

$$\sum m_n C_m[(1-p)p^m(1-p)^{n-m} + p^{m+1}(1-p)^{n-m}] = np \sum {}_nC_m p^m(1-p)^{n-m}.$$

Combining the two terms on the left hand side, and using Eq. (14.17) in the right hand side we have

$$\sum m_n C_m p^m(1-p)^{n-m} = \sum mf(m) = np. \tag{14.19}$$

Note that the left hand side is just our original expression for $\bar{m}$, Eq. (14.18). Thus we conclude that

$$\bar{m} = np \tag{14.20}$$

for the binomial distribution.

The variance σ^2 is given by

$$\sigma^2 = \sum (m - \bar{m})^2 f(m) = \sum (m - np)^2 f(m); \tag{14.21}$$

here we again drop the summation limits for convenience. To evaluate this sum we first rewrite Eq. (14.21) as

$$\sigma^2 = \sum (m^2 - 2mnp + n^2p^2)f(m)$$
$$= \sum m^2 f(m) - 2np \sum mf(m) + n^2p^2 \sum f(m).$$

This reduces to, with the help of Eqs. (14.17) and (14.19),

$$\sigma^2 = \sum m^2 f(m) - (np)^2. \tag{14.22}$$

To evaluate the first term on the right hand side, we first differentiate Eq. (14.19):

$$\sum m_n C_m[mp^{m-1}(1-p)^{n-m} - (n-m)p^m(1-p)^{n-m-1}] = p,$$

then multiplying by $p(1-p)$ and rearranging terms as before

$$\sum m^2 {}_nC_m p^m(1-p)^{n-m} - np \sum m_n C_m p^m(1-p)^{n-m} = np(1-p).$$

By using Eq. (14.19) we can simplify the second term on the left hand side and obtain

$$\sum m^2 {}_nC_m p^m(1-p)^{n-m} = (np)^2 + np(1-p)$$

or

$$\sum m^2 f(m) = np(1-p+np).$$

Inserting this result back into Eq. (14.22), we obtain

$$\sigma^2 = np(1-p+np) - (np)^2 = np(1-p) = npq, \tag{14.23}$$

and the standard deviation σ;

$$\sigma = \sqrt{npq}. \tag{14.24}$$

Two different limits of the binomial distribution for large n are of practical importance: (1) $n \to \infty$ and $p \to 0$ in such a way that the product $np = \lambda$ remains constant; (2) both n and pn are large. The first case will result a new distribution, the Poisson distribution, and the second cases gives us the Gaussian (or Laplace) distribution.

The Poisson distribution

Now $np = \lambda$, so $p = \lambda/n$. The binomial distribution (14.16) then becomes

$$
\begin{aligned}
f(m) = P(X = m) &= \frac{n!}{m!(n-m)!} \left(\frac{\lambda}{n}\right)^m \left(1 - \frac{\lambda}{n}\right)^{n-m} \\
&= \frac{n(n-1)(n-2)\cdots(n-m+1)}{m!n^m} \lambda^m \left(1 - \frac{\lambda}{n}\right)^{n-m} \\
&= \left(1 - \frac{1}{n}\right)\left(1 - \frac{2}{n}\right)\cdots\left(1 - \frac{m-1}{n}\right)\frac{\lambda^m}{m!}\left(1 - \frac{\lambda}{n}\right)^{n-m}. \tag{14.25}
\end{aligned}
$$

Now as $n \to \infty$,

$$\left(1 - \frac{1}{n}\right)\left(1 - \frac{2}{n}\right)\cdots\left(1 - \frac{m-1}{n}\right) \to 1,$$

while

$$\left(1 - \frac{\lambda}{n}\right)^{n-m} = \left(1 - \frac{\lambda}{n}\right)^n \left(1 - \frac{\lambda}{n}\right)^{-m} \to \left(e^{-\lambda}\right)(1) = e^{-\lambda},$$

where we have made use of the result

$$\lim_{n \to \infty} \left(1 + \frac{\alpha}{n}\right)^n = e^\alpha.$$

It follows that Eq. (14.25) becomes

$$f(m) = P(X = m) = \frac{\lambda^m e^{-\lambda}}{m!}. \tag{14.26}$$

This is known as the Poisson distribution. Note that $\sum_{m=0}^{\infty} P(X = m) = 1$, as it should.

The Poisson distribution has the mean

$$E(X) = \sum_{m=0}^{\infty} \frac{m\lambda^m e^{-\lambda}}{m!} = \sum_{m=1}^{\infty} \frac{\lambda^m e^{-\lambda}}{(m-1)!} = \lambda \sum_{m=0}^{\infty} \frac{\lambda^m e^{-\lambda}}{m!}$$

$$= \lambda e^{-\lambda} \sum_{m=0}^{\infty} \frac{\lambda^m}{m!} = \lambda e^{-\lambda} e^{\lambda} = \lambda, \qquad (14.27)$$

where we have made use of the result

$$\sum_{m=0}^{\infty} \frac{\lambda^m}{m!} = e^{\lambda}.$$

The variance σ^2 of the Poisson distribution is

$$\sigma^2 = \text{Var}(X) = E[(X - E(X))^2] = E(X^2) - [E(X)]^2$$

$$= \sum_{m=0}^{\infty} \frac{m^2 \lambda^m e^{-\lambda}}{m!} - \lambda^2 = e^{-\lambda} \sum_{m=1}^{\infty} \frac{m\lambda^m}{(m-1)!} - \lambda^2$$

$$= e^{-\lambda} \lambda \frac{d}{d\lambda} \left(\lambda e^{\lambda} \right) - \lambda^2 = \lambda. \qquad (14.28)$$

To illustrate the use of the Poisson distribution, let us consider a simple example. Suppose the probability that an individual suffers a bad reaction from a flu injection is 0.001; what is the probability that out of 2000 individuals (a) exactly 3, (b) more than 2 individuals will suffer a bad reaction? Now X denotes the number of individuals who suffer a bad reaction and it is binomially distributed. However, we can use the Poisson approximation, because the bad reactions are assumed to be rare events. Thus

$$P(X = m) = \frac{\lambda^m e^{-\lambda}}{m!}, \quad \text{with } \lambda = mp = (2000)(0.001) = 2:$$

(a) $P(X = 3) = \dfrac{2^3 e^{-2}}{3!} = 0.18;$

(b) $P(X > 2) = 1 - [P(X = 0) + P(X = 1) + P(X = 2)]$

$$= 1 - \left[\frac{2^0 e^{-2}}{0!} + \frac{2^1 e^{-2}}{1!} + \frac{2^2 e^{-2}}{2!} \right]$$

$$= 1 - 5e^{-2} = 0.323.$$

An exact evaluation of the probabilities using the binomial distribution would require much more labor.

The Poisson distribution is very important in nuclear physics. Suppose that we have n radioactive nuclei and the probability for any one of these to decay in a given interval of time T is p, then the probability that m nuclei will decay in the interval T is given by the binomial distribution. However, n may be a very large number (such as 10^{23}), and p may be the order of 10^{-20}, and it is impractical to evaluate the binomial distribution with numbers of these magnitudes. Fortunately, the Poisson distribution can come to our rescue.

The Poisson distribution has its own significance beyond its connection with the binomial distribution and it can be derived mathematically from elementary considerations. In general, the Poisson distribution applies when a very large number of experiments is carried out, but the probability of success in each is very small, so that the expected number of successes is a finite number.

The Gaussian (or normal) distribution

The second limit of the binomial distribution that is of interest to us results when both n and pn are large. Clearly, we assume that m, n, and $n - m$ are large enough to permit the use of Stirling's formula ($n! \approx \sqrt{2\pi n} n^n e^{-n}$). Replacing $m!$, $n!$, and $(n - m)!$ by their approximations and after simplification, we obtain

$$P(X = m) \cong \left(\frac{np}{m}\right)^m \left(\frac{nq}{n - m}\right)^{n-m} \sqrt{\frac{n}{2\pi m(n - m)}}. \tag{14.29}$$

The binomial distribution has the mean value np (see Eq. (14.20)). Now let δ denote the deviation of m from np; that is, $\delta = m - np$. Then $n - m = nq - \delta$, and Eq. (14.29) becomes

$$P(X = m) = \frac{1}{\sqrt{2\pi npq(1 + \delta/np)(1 - \delta/np)}} \left(1 + \frac{\delta}{np}\right)^{-(np+\delta)} \left(1 - \frac{\delta}{nq}\right)^{-(nq=\delta)}$$

or

$$P(X = m)A = \left(1 + \frac{\delta}{np}\right)^{-(np+\delta)} \left(1 - \frac{\delta}{nq}\right)^{-(nq-\delta)},$$

where

$$A = \sqrt{2\pi npq\left(1 + \frac{\delta}{np}\right)\left(1 - \frac{\delta}{nq}\right)}.$$

Then

$$\log(P(X = m)A) \cong -(np + \delta)\log(1 + \delta/np) - (nq - \delta)\log(1 - \delta/nq).$$

Assuming $|\delta| < npq$, so that $|\delta/np| < 1$ and $|\delta/nq| < 1$, this permits us to write the two convergent series

$$\log\left(1 + \frac{\delta}{np}\right) = \frac{\delta}{np} - \frac{\delta^2}{2n^2p^2} + \frac{\delta^3}{3n^3p^3} - \cdots,$$

$$\log\left(1 - \frac{\delta}{nq}\right) = -\frac{\delta}{nq} - \frac{\delta^2}{2n^2q^2} - \frac{\delta^3}{3n^3q^3} - \cdots.$$

Hence

$$\log(P(X = m)A) \cong -\frac{\delta^2}{2npq} - \frac{\delta^3(p^2 - q^2)}{2 \times 3n^2p^2q^2} - \frac{\delta^4(p^3 + q^3)}{3 \times 4n^3p^3q^3} - \cdots.$$

Now, if $|\delta|$ is so small in comparison with npq that we ignore all but the first term on the right hand side of this expansion and A can be replaced by $(2\pi npq)^{1/2}$, then we get the approximation formula

$$P(X = m) = \frac{1}{\sqrt{2\pi npq}} e^{-\delta^2/2npq}. \tag{14.30}$$

When $\sigma = \sqrt{npq}$, Eq. (14.30) becomes

$$f(m) = P(X = m) = \frac{1}{\sqrt{2\pi}\sigma} e^{-\delta^2/2\sigma^2}. \tag{14.31}$$

This is called the Guassian, or normal, distribution. It is a very good approximation even for quite small values of n.

The Gaussian distribution is a symmetrical bell-shaped distribution about its mean μ, and σ is a measure of the width of the distribution. Fig. 14.4 gives a comparison of the binomial distribution and the Gaussian approximation.

The Gaussian distribution also has a significance far beyond its connection with the binomial distribution. It can be derived mathematically from elementary considerations, and is found to agree empirically with random errors that actually

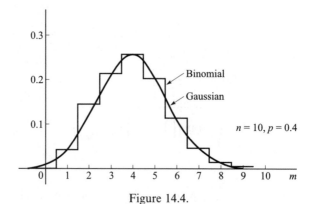

Figure 14.4.

occur in experiments. Everyone believes in the Gaussian distribution: mathematicians think that physicists have verified it experimentally and physicists think that mathematicians have proved it theoretically.

One of the main uses of the Gaussian distribution is to compute the probability

$$\sum_{m=m_1}^{m_2} f(m)$$

that the number of successes is between the given limits m_1 and m_2. Eq. (14.31) shows that the above sum may be approximated by a sum

$$\sum \frac{1}{\sqrt{2\pi}\sigma} e^{-\delta^2/2\sigma^2} \qquad (14.32)$$

over appropriate values of δ. Since $\delta = m - np$, the difference between successive values of δ is 1, and hence if we let $z = \delta/\sigma$, the difference between successive values of z is $\Delta z = 1/\sigma$. Thus Eq. (14.32) becomes the sum over z,

$$\sum \frac{1}{\sqrt{2\pi}} e^{-z^2/2} \Delta z. \qquad (14.33)$$

As $\Delta z \to 0$, the expression (14.33) approaches an integral, which may be evaluated in terms of the function

$$\Phi(z) = \int_0^z \frac{1}{\sqrt{2\pi}} e^{-z^2/2} dz = \frac{1}{\sqrt{2\pi}} \int_0^z e^{-z^2/2} dz. \qquad (14.34)$$

The function $(\Phi(z)$ is related to the extensively tabulated error function, erf(z):

$$\text{erf}(z) = \frac{2}{\sqrt{\pi}} \int_0^z e^{-z^2} dz, \quad \text{and} \quad \Phi(z) = \frac{1}{2}\text{erf}\left(\frac{z}{\sqrt{2}}\right).$$

These considerations lead to the following important theorem, which we state without proof: If m is the number of successes in n independent trials with constant probability p, the probability of the inequality

$$z_1 \leq \frac{m - np}{\sqrt{npq}} \leq z_2 \qquad (14.35)$$

approaches the limit

$$\frac{1}{\sqrt{2\pi}} \int_{z_1}^{z_2} e^{-z^2/2} dz = \Phi(z_2) - \Phi(z_1) \qquad (14.36)$$

as $n \to \infty$. This theorem is known as Laplace–de Moivre limit theorem.

To illustrate the use of the result (14.36), let us consider the simple example of a die tossed 600 times, and ask what the probability is that the number of ones will be between 80 and 110. Now $n = 600$, $p = 1/6$, $q = 1 - p = 5/6$, and m varies from 80 to 110. Hence

$$z_1 = \frac{80 - 100}{\sqrt{100(5/6)}} = -2.19 \quad \text{and} \quad z_1 = \frac{110 - 100}{\sqrt{100(5/6)}} = 1.09.$$

The tabulated error function gives

$$\Phi(z_2) = \Phi(1.09) = 0.362,$$

and

$$\Phi(z_1) = \Phi(-2.19) = -\Phi(2.19) = -0.486,$$

where we have made use of the fact that $\Phi(-z) = -\Phi(z)$, you can check this with Eq. (14.34). So the required probability is approximately given by

$$0.362 - (-0.486) = 0.848.$$

Continuous distributions

So far we have discussed several discrete probability distributions: since measurements are generally made only to a certain number of significant figures, the variables that arise as the result of an experiment are discrete. However, discrete variables can be approximated by continuous ones within the experimental error. Also, in some applications a discrete random variable is inappropriate. We now give a brief discussion of continuous variables that will be denoted by x. We shall see that continuous variables are easier to handle analytically.

Suppose we want to choose a point randomly on the interval $0 \le x \le 1$, how shall we measure the probabilities associated with that event? Let us divide this interval $(0, 1)$ into a number of subintervals, each of length $\Delta x = 0.1$ (Fig. 14.5), the point x is then equally likely to be in any of these subintervals. The probability that $0.3 < x < 0.6$, for example, is 0.3, as there are three favorable cases. The probability that $0.32 < x < 0.64$ is found to be $0.64 - 0.32 = 0.32$ when the interval is divided into 100 parts, and so on. From these we see that the probability for x to be in a given subinterval of (0, 1) is the length of that subinterval. Thus

$$P(a < x < b) = b - a, \qquad 0 \le a \le b \le 1. \tag{14.37}$$

Figure 14.5.

500

The variable x is said to be uniformly distributed on the interval $0 \leq x \leq 1$. Expression (14.37) can be rewritten as

$$P(a < x < b) = \int_a^b dx = \int_a^b 1dx.$$

For a continuous variable it is customary to speak of the probability density, which in the above case is unity. More generally, a variable may be distributed with an arbitrary density $f(x)$. Then the expression

$$f(z)dz$$

measures approximately the probability that x is on the interval

$$z < x < z + dz.$$

And the probability that x is on a given interval (a, b) is

$$P(a < x < b) = \int_a^b f(x)dx \qquad (14.38)$$

as shown in Fig. 14.6.

The function $f(x)$ is called the probability density function and has the properties:

(1) $f(x) \geq 0, \qquad (-\infty < x < \infty);$

(2) $\int_{-\infty}^{\infty} f(x)dx = 1$, a real-valued random variable must lie between $\pm\infty$.

The function

$$F(x) = P(X \leq x) = \int_{-\infty}^x f(u)du \qquad (14.39)$$

defines the probability that the continuous random variable X is in the interval $(-\infty, x)$, and is called the cumulative distributive function. If $f(x)$ is continuous, then Eq. (14.39) gives

$$F'(x) = f(x)$$

and we may speak of a probability differential $dF(x) = f(x)dx$.

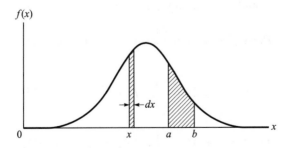

Figure 14.6.

By analogy with those for discrete random variables the expected value or mean and the variance of a continuous random variable X with probability density function $f(x)$ are defined, respectively, to be:

$$E(X) = \mu = \int_{-\infty}^{\infty} xf(x)dx; \tag{14.40}$$

$$\text{Var}(X) = \sigma^2 = E((X - \mu)^2) = \int_{-\infty}^{\infty} (x - \mu)^2 f(x)dx. \tag{14.41}$$

The Gaussian (or normal) distribution

One of the most important examples of a continuous probability distribution is the Gaussian (or normal) distribution. The density function for this distribution is given by

$$f(x) = \frac{1}{\sigma\sqrt{2\pi}} e^{-(x-\mu)^2/2\sigma^2}, \qquad -\infty < x < \infty, \tag{14.42}$$

where μ and σ are the mean and standard deviation, respectively. The corresponding distribution function is

$$F(x) = P(X \le x) = \frac{1}{\sigma\sqrt{2\pi}} \int_{-\infty}^{x} e^{-(u-\mu)^2/2\sigma^2} du. \tag{14.43}$$

The standard normal distribution has mean zero ($\mu = 0$) and standard deviation ($\sigma = 1$)

$$f(z) = \frac{1}{\sqrt{2\pi}} e^{-z^2/2}. \tag{14.44}$$

Any normal distribution can be 'standardized' by considering the substitution $z = (x - \mu)/\sigma$ in Eqs. (14.42) and (14.43). A graph of the density function (14.44), known as the standard normal curve, is shown in Fig. 14.7. We have also indicated the areas within 1, 2 and 3 standard deviations of the mean (that is between $z = -1$ and $+1$, -2 and $+2$, -3 and $+3$):

$$P(-1 \le Z \le 1) = \frac{1}{\sqrt{2\pi}} \int_{-1}^{1} e^{-z^2/2} dz = 0.6827;$$

$$P(-2 \le Z \le 2) = \frac{1}{\sqrt{2\pi}} \int_{-2}^{2} e^{-z^2/2} dz = 0.9545;$$

$$P(-3 \le Z \le 3) = \frac{1}{\sqrt{2\pi}} \int_{-3}^{3} e^{-z^2/2} dz = 0.9973.$$

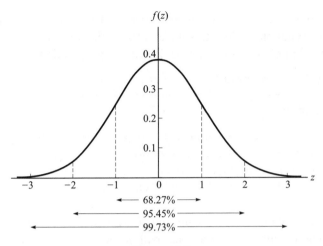

Figure 14.7.

The above three definite integrals can be evaluated by making numerical approximations. A short table of the values of the integral

$$F(x) = \frac{1}{\sqrt{2\pi}} \int_0^x e^{-t^2} dt \left(= \frac{1}{2} \frac{1}{\sqrt{2\pi}} \int_{-x}^x e^{-t^2} dt \right)$$

is included in Appendix 3. A more complete table can be found in *Tables of Normal Probability Functions*, National Bureau of Standards, Washington, DC, 1953.

The Maxwell–Boltzmann distribution

Another continuous distribution that is very important in physics is the Maxwell–Boltzmann distribution

$$f(x) = 4a\sqrt{\frac{a}{\pi}}x^2 e^{-ax^2}, \quad 0 \le x < \infty, \quad a > 0, \qquad (14.45)$$

where $a = m/2kT$, m is the mass, T is the temperature (K), k is the Boltzmann constant, and x is the speed of a gas molecule.

Problems

14.1 If a pair of dice is rolled what is the probability that a total of 8 shows?

14.2 Four coins are tossed, and we are interested in the number of heads. What is the probability that there is an odd number of heads? What is the probability that the third coin will land heads?

14.3 Two coins are tossed. A reliable witness tells us 'at least 1 coin showed heads.' What effect does this have on the uniform sample space?

14.4 The tossing of two coins can be described by the following sample space:

Event	no heads	one head	two head
Probability	1/4	1/2	1/4

What happens to this sample space if we know at least one coin showed heads but have no other specific information?

14.5 Two dice are rolled. What are the elements of the sample space? What is the probability that a total of 8 shows? What is the probability that at least one 5 shows?

14.6 A vessel contains 30 black balls and 20 white balls. Find the probability of drawing a white ball and a black ball in succession from the vessel.

14.7 Find the number of different arrangements or permutations consisting of three letters each which can be formed from the seven letters A, B, C, D, E, F, G.

14.8 It is required to sit five boys and four girls in a row so that the girls occupy the even seats. How many such arrangements are possible?

14.9 A balanced coin is tossed five times. What is the probability of obtaining three heads and two tails?

14.10 How many different five-card hands can be dealt from a shuffled deck of 52 cards? What is the probability that a hand dealt at random consists of five spades?

14.11 (a) Find the constant term in the expansion of $(x^2 + 1/x)^{12}$.
(b) Evaluate 50!.

14.12 A box contains six apples of which two are spoiled. Apples are selected at random without replacement until a spoiled one is found. Find the probability distribution of the number of apples drawn from the box, and present this distribution graphically.

14.13 A fair coin is tossed six times. What is the probability of getting exactly two heads?

14.14 Suppose three dice are rolled simultaneously. What is the probability that two 5s appear with the third face showing a different number?

14.15 Verify that $\sum_{m=0}^{\infty} P(X = m) = 1$ for the Poisson distribution.

14.16 Certain processors are known to have a failure rate of 1.2%. There are shipped in batches of 150. What is the probability that a batch has exactly one defective processor? What is the probability that it has two?

14.17 A Geiger counter is used to count the arrival of radioactive particles. Find:
(a) the probability that in time t no particles will be counted;
(b) the probability of exactly one count in time t.

14.18 Given the density function $f(x)$

$$f(x) = \begin{cases} kx^2 & 0 < x < 3 \\ 0 & \text{otherwise} \end{cases} :$$

(a) find the constant k;
(b) compute $P(1 < x < 2)$;
(c) find the distribution function and use it to find $P(1 < x \le (2))$.

Appendix 1

Preliminaries (review of fundamental concepts)

This appendix is for those readers who need a review; a number of fundamental concepts or theorem will be reviewed without giving proofs or attempting to achieve completeness.

We assume that the reader is already familiar with the classes of real numbers used in analysis. The set of *positive integers* (also known as natural numbers) 1, 2, ..., n admits the operations of addition without restriction, that is, they can be added (and therefore multiplied) together to give other positive integers. The set of *integers* 0, $\pm 1, \pm 2, \ldots, \pm n$ admits the operations of addition and subtraction among themselves. *Rational numbers* are numbers of the form p/q, where p and q are integers and $q \neq 0$. Examples of rational numbers are 2/3, $-10/7$. This set admits the further property of division among its members. The set of *irrational numbers* includes all numbers which cannot be expressed as the quotient of two integers. Examples of irrational numbers are $\sqrt{2}, \sqrt[3]{11}, \pi$ and any number of the form $\sqrt[n]{a/b}$, where a and b are integers which are perfect nth powers.

The set of real numbers contains all the rationals and irrationals. The important property of the set of real numbers $\{x\}$ is that it can be put into (1:1) correspondence with the set of points $\{P\}$ of a line as indicated in Fig. A.1.

The basic rules governing the combinations of real numbers are:

commutative law:	$a + b = b + a, \quad a \cdot b = b \cdot a;$
associative law:	$a + (b + c) = (a + b) + c, \quad a \cdot (b \cdot c) = (a \cdot b) \cdot c;$
distributive law:	$a \cdot (b + c) = a \cdot b + a \cdot c;$
index law	$a^m a^n = a^{m+n}, \quad a^m / a^n = a^{m-n} \quad (a \neq 0);$

where a, b, c, are algebraic symbols for the real numbers.

Problem A1.1
Prove that $\sqrt{2}$ is an irrational number.

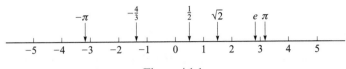

Figure A1.1.

(Hint: Assume the contrary, that is, assume that $\sqrt{2} = p/q$, where p and q are positive integers having no common integer factor.)

Inequalities

If x and y are real numbers, $x > y$ means that x is greater than y; and $x < y$ means that x is less than y. Similarly, $x \geq y$ implies that x is either greater than or equal to y. The following basic rules governing the operations with inequalities:

(1) Multiplication by a constant: If $x > y$, then $ax > ay$ if a is a positive number, and $ax < ay$ if a is a negative number.

(2) Addition of inequalities: If x, y, u, v are real numbers, and if $x > y$, and $u > v$, than $x + u > y + v$.

(3) Subtraction of inequalities: If $x > y$, and $u > v$, we cannot deduce that $(x - u) > (y - v)$. Why? It is evident that $(x - u) - (y - v) = (x - y) - (u - v)$ is not necessarily positive.

(4) Multiplication of inequalities: If $x > y$, and $u > v$, and x, y, u, v are *all* positive, then $xu > yv$. When some of the numbers are negative, then the result is not necessarily true.

(5) Division of inequalities: $x > y$ and $u > v$ do not imply $x/u > y/v$.

When we wish to consider the numerical value of the variable x without regard to its sign, we write $|x|$ and read this as 'absolute or mod x'. Thus the inequality $|x| \leq a$ is equivalent to $a \leq x \leq + a$.

Problem A1.2
Find the values of x which satisfy the following inequalities:

(a) $x^3 - 7x^2 + 21x - 27 > 0$,

(b) $|7 - 3x| < 2$,

(c) $\dfrac{5}{5x - 1} > \dfrac{2}{2x + 1}$. (Warning: cross multiplying is not permitted.)

Problem A1.3
If $a_1, a_2, \ldots, a_n$ and $b_1, b_2, \ldots, b_n$ are any real numbers, prove Schwarz's inequality:

$$(a_1 b_1 + a_2 b_2 + \cdots + a_n b_n)^2 \leq (a_1^2 + a_2^2 + \cdots + a_n^2)(b_1^2 + b_2^2 + \cdots + b_n^2).$$

507

Problem A1.4

Show that

$$\frac{1}{2}+\frac{1}{4}+\frac{1}{8}+\cdots+\frac{1}{2^{n-1}}\leq 1 \qquad \text{for all positive integers } n>1.$$

If $x_1, x_2, \ldots, x_n$ are n positive numbers, their arithmetic mean is defined by

$$A=\frac{1}{n}\sum_{k=1}^{n}x_k=\frac{x_1+x_2+\cdots+x_n}{n}$$

and their geometric mean by

$$G=\sqrt[n]{\prod_{k=1}^{n}x_k}=\sqrt[n]{x_1 x_2\cdots x_n},$$

where $\sum$ and $\prod$ are the summation and product signs. The harmonic mean H is sometimes useful and it is defined by

$$\frac{1}{H}=\frac{1}{n}\sum_{k=1}^{n}\frac{1}{x_k}=\frac{1}{n}\left(\frac{1}{x_1}+\frac{1}{x_2}+\cdots+\frac{1}{x_n}\right).$$

There is a basic inequality among the three means: $A\geq G\geq H$, the equality sign occurring when $x_1=x_2=\cdots=x_n$.

Problem A1.5

If x_1 and x_2 are two positive numbers, show that $A\geq G\geq H$.

Functions

We assume that the reader is familiar with the concept of functions and the process of graphing functions.

A polynomial of degree n is a function of the form

$$f(x)=p_n(x)=a_0 x^n+a_1 x^{n-1}+a_2 x^{n-2}+\cdots+a_n \qquad (a_j=\text{constant}, a_0\neq 0).$$

A polynomial can be differentiated and integrated. Although we have written $a_j=$ constant, they might still be functions of some other variable independent of x. For example,

$$t^{-3}x^3+\sin tx^2+\sqrt{t}x+t$$

is a polynomial function of x (of degree 3) and each of the *a*s is a function of a certain variable t: $a_0=t^{-3}, a_1=\sin t, a_2=t^{1/2}, a_3=t$.

The polynomial equation $f(x)=0$ has exactly n roots provided we count repetitions. For example, $x^3-3x^2+3x-1=0$ can be written $(x-1)^3=0$ so that the three roots are 1, 1, 1. Note that here we have used the binomial theorem

$$(a+x)^n=a^n+na^{n-1}x+\frac{n(n-1)}{2!}a^{n-2}x^2+\cdots+x^n.$$

A rational function is of the form $f(x) = p_n(x)/q_n(x)$, where $p_n(x)$ and $q_n(x)$ are polynomials.

A transcendental function is any function which is not algebraic, for example, the trigonometric functions sin x, cos x, etc., the exponential functions e^x, the logarithmic functions log x, and the hyperbolic functions sinh x, cosh x, etc.

The exponential functions obey the index law. The logarithmic functions are inverses of the exponential functions, that is, if $a^x = y$ then $x = \log_a y$, where a is called the base of the logarithm. If $a = e$, which is often called the natural base of logarithms, we denote $\log_e x$ by $\ln x$, called the natural logarithm of x. The fundamental rules obeyed by logarithms are

$$\ln(mn) = \ln m + \ln n, \quad \ln(m/n) = \ln m - \ln n, \quad \text{and} \quad \ln m^p = p \ln m.$$

The hyperbolic functions are defined in terms of exponential functions as follows

$$\sinh x = \frac{e^x - e^{-x}}{2}, \qquad \cosh x = \frac{e^x + e^{-x}}{2},$$

$$\tanh x = \frac{\sinh x}{\cosh x} = \frac{e^x - e^{-x}}{e^x + e^x}, \qquad \coth x = \frac{1}{\tanh x} = \frac{e^x + e^{-x}}{e^x - e^{-x}},$$

$$\operatorname{sech} x = \frac{1}{\cosh x} = \frac{2}{e^x + e^{-x}}, \qquad \operatorname{cosech} x = \frac{1}{\sinh x} = \frac{2}{e^x - e^{-x}}.$$

Rough graphs of these six functions are given in Fig. A1.2.

Some fundamental relationships among these functions are as follows:

$$\cosh^2 x - \sinh^2 x = 1, \quad \operatorname{sech}^2 x + \tanh^2 x = 1, \quad \coth^2 x - \operatorname{cosech}^2 x = 1,$$

$$\sinh(x \pm y) = \sinh x \cosh y \pm \cosh x \sinh y,$$

$$\cosh(x \pm y) = \cosh x \cosh y \pm \sinh x \sinh y,$$

$$\tanh(x \pm y) = \frac{\tanh x \pm \tanh y}{1 \pm \tanh x \tanh y}.$$

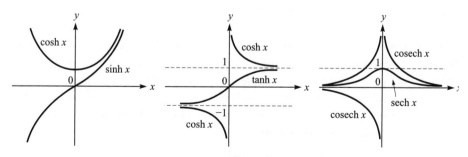

Figure A1.2. Hyperbolic functions.

Problem A1.6

Using the rules of exponents, prove that $\ln(mn) = \ln m + \ln n$.

Problem A1.7

Prove that: (a) $\sin^2 x = \frac{1}{2}(1 - \cos 2x)$, $\cos^2 x = \frac{1}{2}(1 + \cos 2x)$, and (b) $A\cos x + B\sin x = \sqrt{A^2 + B^2}\sin(x + \delta)$, where $\tan\delta = A/B$

Problem A1.8

Prove that: (a) $\cosh^2 x - \sinh^2 x = 1$, and (b) $^2x + \tanh^2 x = 1$.

Limits

We are sometimes required to find the limit of a function $f(x)$ as x approaches some particular value α:

$$\lim_{x \to \alpha} f(x) = l.$$

This means that if $|x - \alpha|$ is small enough, $|f(x) - l|$ can be made as small as we please. A more precise analytic description of $\lim_{x \to \alpha} f(x) = l$ is the following:

> For any $\varepsilon > 0$ (however small) we can always find a number η (which, in general, depends upon ε) such that $|f(x) - l| < \varepsilon$ whenever $|x - \alpha| < \eta$.

As an example, consider the limit of the simple function $f(x) = 2 - 1/(x - 1)$ as $x \to 2$. Then

$$\lim_{x \to 2} f(x) = 1$$

for if we are given a number, say $\varepsilon = 10^{-3}$, we can always find a number η which is such that

$$\left(2 - \frac{1}{x-1}\right) - 1 < 10^{-3} \tag{A1.1}$$

provided $|x - 2| < \eta$. In this case (A1.1) will be true if $1/(x - 1) > 1 - 10^{-3} = 0.999$. This requires $x - 1 < (0.999)^{-1}$, or $x - 2 < (0.999)^{-1} - 1$. Thus we need only take $\eta = (0.999)^{-1} - 1$.

The function $f(x)$ is said to be continuous at α if $\lim_{x \to \alpha} f(x) = l$. If $f(x)$ is continuous at each point of an interval such as $a \leq x \leq b$ or $a < x \leq b$, etc., it is said to be continuous in the interval (for example, a polynomial is continuous at all x).

The definition implies that $\lim_{x \to \alpha - 0} f(x) = \lim_{x \to \alpha + 0} f(x) = f(\alpha)$ at all points α of the interval (a, b), but this is clearly inapplicable at the endpoints a and b. At these points we define continuity by

$$\lim_{x \to a+0} f(x) = f(a) \quad \text{and} \quad \lim_{x \to b-0} f(x) = f(b).$$

A finite discontinuity may occur at $x = \alpha$. This will arise when $\lim_{x \to \alpha - 0} f(x) = l_1$, $\lim_{x \to \alpha - 0} f(x) = l_2$, and $l_1 \neq l_2$.

It is obvious that a continuous function will be bounded in any finite interval. This means that we can find numbers m and M independent of x and such that $m \leq f(x)M$ for $a \leq x \leq b$. Furthermore, we expect to find x_0, x_1 such that $f(x_0) = m$ and $f(x_1) = M$.

The order of magnitude of a function is indicated in terms of its variable. Thus, if x is very small, and if $f(x) = a_1 x + a_2 x^2 + a_3 x^3 + \cdots$ (a_k constant), its magnitude is governed by the term in x and we write $f(x) = O(x)$. When $a_1 = 0$, we write $f(x) = O(x^2)$, etc. When $f(x) = O(x^n)$, then $\lim_{x \to 0} \{f(x)/x^n\}$ is finite and/or $\lim_{x \to 0} \{f(x)/x^{n-1}\} = 0$.

A function $f(x)$ is said to be differentiable or to possess a derivative at the point x if $\lim_{h \to 0} [f(x + h) - f(x)]/h$ exists. We write this limit in various forms df/dx, f' or Df, where $D = d(\ )/dx$. Most of the functions in physics can be successively differentiated a number of times. These successive derivatives are written as $f'(x), f''(x), \ldots, f^n(x), \ldots$, or $Df, D^2 f, \ldots, D^n f, \ldots$.

Problem A1.9

If $f(x) = x^2$, prove that: (a) $\lim_{x \to 2} f(x) = 4$, and (b) $f(x)$ is continuous at $x = 2$.

Infinite series

Infinite series involve the notion of sequence in a simple way. For example, $\sqrt{2}$ is irrational and can only be expressed as a non-recurring decimal $1.414\ldots$. We can approximate to its value by a sequence of rationals, $1, 1.4, 1.41, 1.414, \ldots$ say $\{a_n\}$ which is a countable set limit of a_n whose values approach indefinitely close to $\sqrt{2}$. Because of this we say the limit of a_n as n tends to infinity exists and equals $\sqrt{2}$, and write $\lim_{n \to \infty} a_n = \sqrt{2}$.

In general, a sequence $u_1, u_2, \ldots, \{u_n\}$ is a function defined on the set of natural numbers. The sequence is said to have the limit l or to converge to l, if given any $\varepsilon > 0$ there exists a number $N > 0$ such that $|u_n - l| < \varepsilon$ for all $n > N$, and in such case we write $\lim_{n \to \infty} u_n = l$.

Consider now the sums of the sequence $\{u_n\}$

$$S_n = \sum_{r=1}^{n} u_r = u_1 + u_2 + u_3 + \cdots, \qquad (A.2)$$

where $u_r > 0$ for all r. If $n \to \infty$, then (A.2) is an infinite series of positive terms. We see that the behavior of this series is determined by the behavior of the sequence $\{u_n\}$ as it converges or diverges. If $\lim_{n \to \infty} S_n = s$ (finite) we say that (A.2) is convergent and has the sum s. When $s_n \to \infty$ as $n \to \infty$, we say that (A.2) is divergent.

Example A1.1.

Show that the series

$$\sum_{n=1}^{\infty} \frac{1}{2^n} = \frac{1}{2} + \frac{1}{2^2} + \frac{1}{2^3} + \cdots$$

is convergent and has sum $s = 1$.

Solution: Let

$$s_n = \frac{1}{2} + \frac{1}{2^2} + \frac{1}{2^3} + \cdots + \frac{1}{2^n},$$

then

$$\frac{1}{2} s_n = \frac{1}{2^2} + \frac{1}{2^3} + \cdots + \frac{1}{2^{n+1}}.$$

Subtraction gives

$$\left(1 - \frac{1}{2}\right) s_n = \frac{1}{2} - \frac{1}{2^{n+1}} = \frac{1}{2}\left(1 - \frac{1}{2^n}\right), \qquad \text{or} \qquad s_n = 1 - \frac{1}{2^n}.$$

Then since $\lim_{n\to\infty} s_n = \lim_{n\to\infty}(1 - 1/2^n) = 1$, the series is convergent and has the sum $s = 1$.

Example A1.2.

Show that the series $\sum_{n=1}^{\infty} (-1)^{n-1} = 1 - 1 + 1 - 1 + \cdots$ is divergent.

Solution: Here $s_n = 0$ or 1 according as n is even or odd. Hence $\lim_{n\to\infty} s_n$ does not exist and so the series is divergent.

Example A1.3.

Show that the geometric series $\sum_{n=1}^{\infty} ar^{n-1} = a + ar + ar^2 + \cdots$, where a and r are constants, (*a*) converges to $s = a/(1 - r)$ if $|r| < 1$, and (*b*) diverges if $|r| > 1$.

Solution: Let

$$s_n = a + ar + ar^2 + \cdots + ar^{n-1}.$$

Then

$$rs_n = \qquad ar + ar^2 + \cdots + ar^{n-1} + ar^n.$$

Subtraction gives

$$(1 - r)s_n = a - ar^n \qquad \text{or} \qquad s_n = \frac{a(1 - r^n)}{1 - r}.$$

(a) If $|r| < 1$,

$$\lim_{n\to\infty} s_n = \lim_{n\to\infty} \frac{a(1 - r^n)}{1 - r} = \frac{a}{1 - r}.$$

(b) If $|r| > 1$,

$$\lim_{n\to\infty} s_n = \lim_{n\to\infty} \frac{a(1 - r^n)}{1 - r}$$

does not exist.

Example A1.4.
Show that the p series $\sum_{n=1}^{\infty} 1/n^p$ converges if $p > 1$ and diverges if $p \le 1$.

Solution: Using $f(n) = 1/n^p$ we have $f(x) = 1/x^p$ so that if $p \ne 1$,

$$\int_1^\infty \frac{dx}{x^p} = \lim_{M\to\infty} \int_1^M x^{-p} dx = \lim_{M\to\infty} \left. \frac{x^{1-p}}{1 - p} \right|_1^M = \lim_{M\to\infty} \left[\frac{M^{1-p}}{1 - p} - \frac{1}{1 - p} \right].$$

Now if $p > 1$ this limit exists and the corresponding series converges. But if $p < 1$ the limit does not exist and the series diverges.
  If $p = 1$ then

$$\int_1^\infty \frac{dx}{x} = \lim_{M\to\infty} \int_1^M \frac{dx}{x} = \lim_{M\to\infty} \ln x \Big|_1^M = \lim_{M\to\infty} \ln M,$$

which does not exist and so the corresponding series for $p = 1$ diverges.
  This shows that $1 + \frac{1}{2} + \frac{1}{3} + \cdots$ diverges even though the nth term approaches zero.

Tests for convergence

There are several important tests for convergence of series of positive terms. Before using these simple tests, we can often weed out some very badly divergent series with the following preliminary test:

> If the terms of an infinite series do not tend to zero (that is, if $\lim_{n\to\infty} a_n \ne 0$), the series diverges. If $\lim_{n\to\infty} a_n = 0$, we must test further.

Four of the common tests are given below:

Comparison test
If $u_n \le v_n$ (all n), then $\sum_{n=1}^{\infty} u_n$ converges when $\sum_{n=1}^{\infty} v_n$ converges. If $u_n \ge v_n$ (all n), then $\sum_{n=1}^{\infty} u_n$ diverges when $\sum_{n=1}^{\infty} v_n$ diverges.

513

Since the behavior of $\sum_{n=1}^{\infty} u_n$ is unaffected by removing a finite number of terms from the series, this test is true if $u_n \leq v_n$ or $u_n \geq v_n$ for all $n > N$. Note that $n > N$ means from some term onward. Often, $N = 1$.

Example A1.5
(a) Since $1/(2^n + 1) \leq 1/2^n$ and $\sum 1/2^n$ converges, $\sum 1/(2^n + 1)$ also converges.
(b) Since $1/\ln n > 1/n$ and $\sum_{n=2}^{\infty} 1/n$ diverges, $\sum_{n=2}^{\infty} 1/\ln n$ also diverges.

Quotient test
If $u_{n+1}/u_n \leq v_{n+1}/v_n$ (all n), then $\sum_{n=1}^{\infty} u_n$ converges when $\sum_{n=1}^{\infty} v_n$ converges. And if $u_{n+1}/u_n \geq v_{n+1}/v_n$ (all n), then $\sum_{n=1}^{\infty} u_n$ diverges when $\sum_{n=1}^{\infty} v_n$ diverges.
 We can write

$$u_n = \frac{u_n}{u_{n-1}} \frac{u_{n-1}}{u_{n-2}} \cdots \frac{u_2}{u_1} u_1 \leq \frac{v_n}{v_{n-1}} \frac{v_{n-1}}{v_{n-2}} \cdots \frac{v_2}{v_1} \leq v_1$$

so that $u_n \leq v_n u_1$ which proves the quotient test by using the comparison test.
 A similar argument shows that if $u_{n+1}/u_n \geq v_{n+1}/v_n$ (all n), then $\sum_{n=1}^{\infty} u_n$ diverges when $\sum_{n=1}^{\infty} v_n$ diverges.

Example A1.6
Consider the series

$$\sum_{n=1}^{\infty} \frac{4n^2 - n + 3}{n^3 + 2n}.$$

For large n, $(4n^2 - n + 3)/(n^3 + 2n)$ is approximately $4/n$. Taking $u_n = (4n^2 - n + 3)/(n^3 + 2n)$ and $v_n = 1/n$, we have $\lim_{n \to \infty} u_n/v_n = 1$. Now since $\sum v_n = \sum 1/n$ diverges, $\sum u_n$ also diverges.

D'Alembert's ratio test:
$\sum_{n=1}^{\infty} u_n$ converges when $u_{n+1}/u_n < 1$ (all $n \geq N$) and diverges when $u_{n+1}/u_n > 1$.
 Write $v_n = x^{n-1}$ in the quotient test so that $\sum_{n=1}^{\infty} v_n$ is the geometric series with common ratio $v_{n+1}/v_n = x$. Then the quotient test proves that $\sum_{n=1}^{\infty} u_n$ converges when $x < 1$ and diverges when $x > 1$.
 Sometimes the ratio test is stated in the following form: if $\lim_{n \to \infty} u_{n+1}/u_n = \rho$, then $\sum_{n=1}^{\infty} u_n$ converges when $\rho < 1$ and diverges when $\rho > 1$.

Example A1.7
Consider the series

$$1 + \frac{1}{2!} + \frac{1}{3!} + \cdots + \frac{1}{n!} + \cdots.$$

Using the ratio test, we have

$$\frac{u_{n+1}}{u_n} = \frac{1}{(n+1)!} \div \frac{1}{n!} = \frac{n!}{(n+1)!} = \frac{1}{n+1} < 1,$$

so the series converges.

Integral test.
If $f(x)$ is positive, continuous and monotonic decreasing and is such that $f(n) = u_n$ for $n > N$, then $\sum u_n$ converges or diverges according as

$$\int_N^\infty f(x)dx = \lim_{M \to \infty} \int_N^M f(x)dx$$

converges or diverges. We often have $N = 1$ in practice.
   To prove this test, we will use the following property of definite integrals:

$$\text{If in } a \leq x \leq b, \ f(x) \leq g(x), \text{ then } \int_a^b f(x)dx \leq \int_a^b g(x)dx.$$

Now from the monotonicity of $f(x)$, we have

$$u_{n+1} = f(n+1) \leq f(x)f(n) = u_n, \qquad n = 1, 2, 3, \ldots.$$

Integrating from $x = n$ to $x = n+1$ and using the above quoted property of definite integrals we obtain

$$u_{n+1} \leq \int_n^{n+1} f(x)dx \leq u_n, \qquad n = 1, 2, 3, \ldots.$$

Summing from $n = 1$ to $M - 1$,

$$u_1 + u_2 + \cdots + u_M \leq \int_1^M f(x)dx \leq u_1 + u_2 + \cdots + u_{M-1}. \qquad \text{(A1.3)}$$

If $f(x)$ is strictly decreasing, the equality sign in (A1.3) can be omitted.
   If $\lim_{M \to \infty} \int_1^M f(x)dx$ exists and is equal to s, we see from the left hand inequality in (A1.3) that $u_1 + u_2 + \cdots + u_M$ is monotonically increasing and bounded above by s, so that $\sum u_n$ converges. If $\lim_{M \to \infty} \int_1^M f(x)dx$ is unbounded, we see from the right hand inequality in (A1.3) that $\sum u_n$ diverges.
   Geometrically, $u_1 + u_2 + \cdots + u_M$ is the total area of the rectangles shown shaded in Fig. A1.3, while $u_1 + u_2 + \cdots + u_{M-1}$ is the total area of the rectangles which are shaded and non-shaded. The area under the curve $y = f(x)$ from $x = 1$ to $x = M$ is intermediate in value between the two areas given above, thus illustrating the result (A1.3).

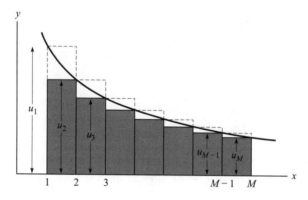

Figure A1.3.

Example A1.8

$\sum_{n=1}^{\infty} 1/n^2$ converges since $\lim_{M \to \infty} \int_1^M dx/x^2 = \lim_{M \to \infty}(1 - 1/M)$ exists.

Problem A1.10

Find the limit of the sequence $0.3, 0.33, 0.333, \ldots$, and justify your conclusion.

Alternating series test

An alternating series is one whose successive terms are alternately positive and negative $u_1 - u_2 + u_3 - u_4 + \cdots$. It converges if the following two conditions are satisfied:

(a) $|u_{n+1}| |u_n|$ for $n \geq 1$, $\qquad (b)$ $\lim_{n \to \infty} u_n = 0 \qquad \left(\text{or } \lim_{n \to \infty} |u_n| = 0 \right).$

The sum of the series to $2M$ is

$$S_{2M} = (u_1 - u_2) + (u_3 - u_4) + \cdots + (u_{2M-1} - u_{2M})$$

$$= u_1 - (u_2 - u_3) - (u_4 - u_5) - \cdots - (u_{2M-2} - u_{2M-1}) - u_{2M}.$$

Since the quantities in parentheses are non-negative, we have

$$S_{2M} \geq 0, \qquad S_2 \leq S_4 \leq S_6 \leq \cdots \leq S_{2M} \leq u_1.$$

Therefore $\{S_{2M}\}$ is a bounded monotonic increasing sequence and thus has the limit S.

Also $S_{2M+1} = S_{2M} + u_{2M+1}$. Since $\lim_{M \to \infty} S_{2M} = S$ and $\lim_{M \to \infty} u_{2M+1} = 0$ (for, by hypothesis, $\lim_{n \to \infty} u_n = 0$), it follows that $\lim_{M \to \infty} S_{2M+1} = \lim_{M \to \infty} S_{2M} + \lim_{M \to \infty} u_{2M+1} = S + 0 = S$. Thus the partial sums of the series approach the limit S and the series converges.

516

Problem A1.11

Show that the error made in stopping after *2M* terms is less than or equal to u_{2M+1}.

Example A1.9

For the series

$$1 - \frac{1}{2} + \frac{1}{3} - \frac{1}{4} + \cdots = \sum_{n=1}^{\infty} \frac{(-1)^{n-1}}{n},$$

we have $u_n = (-1)^{n+1}/n, |u_n| = 1/n, |u_{n+1}| = 1/(n+1)$. Then for $n \geq 1$, $|u_{n+1}| \leq |u_n|$. Also we have $\lim_{n \to \infty} |u_n| = 0$. Hence the series converges.

Absolute and conditional convergence

The series $\sum u_n$ is called absolutely convergent if $\sum |u_n|$ converges. If $\sum u_n$ converges but $\sum |u_n|$ diverges, then $\sum u_n$ is said to be conditionally convergent.

It is easy to show that if $\sum |u_n|$ converges, then $\sum u_n$ converges (in words, an absolutely convergent series is convergent). To this purpose, let

$$S_M = u_1 + u_2 + \cdots + u_M \qquad T_M = |u_1| + |u_2| + \cdots + |u_M|,$$

then

$$S_M + T_M = (u_1 + |u_1|) + (u_2 + |u_2|) + \cdots + (u_M + |u_M|)$$

$$\leq 2|u_1| + 2|u_2| + \cdots + 2|u_M|.$$

Since $\sum |u_n|$ converges and since $u_n + |u_n| \geq 0$, for $n = 1, 2, 3, \ldots$, it follows that $S_M + T_M$ is a bounded monotonic increasing sequence, and so $\lim_{M \to \infty} (S_M + T_M)$ exists. Also $\lim_{M \to \infty} T_M$ exists (since the series is absolutely convergent by hypothesis),

$$\lim_{M \to \infty} S_M = \lim_{M \to \infty} (S_M + T_M - T_M) = \lim_{M \to \infty} (S_M + T_M) - \lim_{M \to \infty} T_M$$

must also exist and so the series $\sum u_n$ converges.

The terms of an absolutely convergent series can be rearranged in any order, and all such rearranged series will converge to the same sum. We refer the reader to text-books on advanced calculus for proof.

Problem A1.12

Prove that the series

$$1 - \frac{1}{2^2} + \frac{1}{3^2} - \frac{1}{4^2} + \frac{1}{5^2} - \cdots$$

converges.

How do we test for absolute convergence? The simplest test is the ratio test, which we now review, along with three others – Raabe's test, the nth root test, and Gauss' test.

Ratio test

Let $\lim_{n\to\infty} |u_{n+1}/u_n| = L$. Then the series $\sum u_n$:

(a) converges (absolutely) if $L < 1$;
(b) diverges if $L > 1$;
(c) the test fails if $L = 1$.

Let us consider first the positive-term $\sum u_n$, that is, each term is positive. We must now prove that if $\lim_{n\to\infty} u_{n+1}/u_n = L < 1$, then necessarily $\sum u_n$ converges. By hypothesis, we can choose an integer N so large that for all $n \geq N, (u_{n+1}/u_n) < r$, where $L < r < 1$. Then

$$u_{N+1} < r u_N, \quad u_{N+2} < r u_{N+1} < r^2 u_N, \quad u_{N+3} < r u_{N+2} < r^3 u_N, \quad \text{etc.}$$

By addition

$$u_{N+1} + u_{N+2} + \cdots < u_N(r + r^2 + r^3 + \cdots)$$

and so the given series converges by the comparison test, since $0 < r < 1$.

When the series has terms with mixed signs, we consider $|u_1| + |u_2| + |u_3| + \cdots$, then by the above proof and because an absolutely convergent series is convergent, it follows that if $\lim_{n\to\infty} |u_{n+1}/u_n| = L < 1$, then $\sum u_n$ converges absolutely.

Similarly we can prove that if $\lim_{n\to\infty} |u_{n+1}/u_n| = L > 1$, the series $\sum u_n$ diverges.

Example A1.10

Consider the series $\sum_{n=1}^{\infty} (-1)^{n-1} 2^n/n^2$. Here $u_n = (-1)^{n-1} 2^n/n^2$. Then $\lim_{n\to\infty} |u_{n+1}/u_n| = \lim_{n\to\infty} 2n^2/(n+1)^2 = 2$. Since $L = 2 > 1$, the series diverges.

When the ratio test fails, the following three tests are often very helpful.

Raabe's test

Let $\lim_{n\to\infty} n(1 - |u_{n+1}/u_n|) = \ell$, then the series $\sum u_n$:

(a) converges absolutely if $\ell < 1$;
(b) diverges if $\ell > 1$.

The test fails if $\ell = 1$.

The nth root test

Let $\lim_{n\to\infty} \sqrt[n]{|u_n|} = R$, then the series $\sum u_n$:

(a) converges absolutely if $R < 1$;

(b) diverges if $R > 1$.

The test fails if $R = 1$.

Gauss' test

If

$$\left|\frac{u_{n+1}}{u_n}\right| = 1 - \frac{G}{n} + \frac{c_n}{n^2},$$

where $|c_n| < P$ for all $n > N$, then the series $\sum u_n$:

(a) converges (absolutely) if $G > 1$;

(b) diverges or converges conditionally if $G \leq 1$.

Example A1.11

Consider the series $1 + 2r + r^2 + 2r^3 + r^4 + 2r^5 + \cdots$. The ratio test gives

$$\left|\frac{u_{n+1}}{u_n}\right| = \begin{cases} 2|r|, & n \text{ odd} \\ |r|/2, & n \text{ even} \end{cases},$$

which indicates that the ratio test is not applicable. We now try the nth root test:

$$\sqrt[n]{|u_n|} = \begin{cases} \sqrt[n]{2|r^n|} = \sqrt[n]{2}|r|, & n \text{ odd} \\ \sqrt[n]{|r^n|} = |r|, & n \text{ even} \end{cases}$$

and so $\lim_{n\to\infty} \sqrt[n]{|u_n|} = |r|$. Thus if $|r| < 1$ the series converges, and if $|r| > 1$ the series diverges.

Example A1.12

Consider the series

$$\left(\frac{1}{3}\right)^2 + \left(\frac{1 \times 4}{3 \times 6}\right)^2 + \left(\frac{1 \times 4 \times 7}{3 \times 6 \times 9}\right)^2 + \cdots + \left(\frac{1 \times 4 \times 7 \cdots (3n-2)}{3 \times 6 \times 9 \cdots (3n)}\right) + \cdots.$$

The ratio test is not applicable, since

$$\lim_{n\to\infty} \left|\frac{u_{n+1}}{u_n}\right| = \lim_{n\to\infty} \left|\frac{(3n+1)}{(3n+3)}\right|^2 = 1.$$

But Raabe's test gives

$$\lim_{n\to\infty} n\left(1 - \left|\frac{u_{n+1}}{u_n}\right|\right) = \lim_{n\to\infty} n\left\{1 - \left(\frac{3n+1}{3n+3}\right)^2\right\} = \frac{4}{3} > 1,$$

and so the series converges.

Problem A1.13

Test for convergence the series

$$\left(\frac{1}{2}\right)^2 + \left(\frac{1 \times 3}{2 \times 4}\right)^2 + \left(\frac{1 \times 3 \times 5}{2 \times 4 \times 6}\right)^2 + \cdots + \left(\frac{1 \times 3 \times 5 \cdots (2n-1)}{1 \times 3 \times 5 \cdots (2n)}\right) + \cdots.$$

Hint: Neither the ratio test nor Raabe's test is applicable (show this). Try Gauss' test.

Series of functions and uniform convergence

The series considered so far had the feature that u_n depended just on n. Thus the series, if convergent, is represented by just a number. We now consider series whose terms are functions of x, $u_n = u_n(x)$. There are many such series of functions. The reader should be familiar with the power series in which the nth term is a constant times x^n:

$$S(x) = \sum_{n=0}^{\infty} a_n x^n. \tag{A1.4}$$

We can think of all previous cases as power series restricted to $x = 1$. In later sections we shall see Fourier series whose terms involve sines and cosines, and other series in which the terms may be polynomials or other functions. In this section we consider power series in x.

The convergence or divergence of a series of functions depends, in general, on the values of x. With x in place, the partial sum Eq. (A1.2) now becomes a function of the variable x:

$$s_n(x) = u_1 + u_2(x) + \cdots + u_n(x). \tag{A1.5}$$

as does the series sum. If we define $S(x)$ as the limit of the partial sum

$$S(x) = \lim_{n \to \infty} s_n(x) = \sum_{n=0}^{\infty} u_n(x), \tag{A1.6}$$

then the series is said to be convergent in the interval $[a, b]$ (that is, $a \leq x \leq b$), if for each $\varepsilon > 0$ and each x in $[a, b]$ we can find $N > 0$ such that

$$|S(x) - s_n(x)| < \varepsilon, \qquad \text{for all } n \geq N. \tag{A1.7}$$

If N depends only on ε and not on x, the series is called uniformly convergent in the interval $[a, b]$. This says that for our series to be uniformly convergent, it must be possible to find a finite N so that the remainder of the series after N terms, $\sum_{i=N+1}^{\infty} u_i(x)$, will be less than an arbitrarily small ε for all x in the given interval.

The domain of convergence (absolute or uniform) of a series is the set of values of x for which the series of functions converges (absolutely or uniformly).

We deal with power series in x exactly as before. For example, we can use the ratio test, which now depends on x, to investigate convergence or divergence of a series:

$$r(x) = \lim_{n\to\infty} \left|\frac{u_{n+1}}{u_n}\right| = \lim_{n\to\infty} \left|\frac{a_{n+1}x^{n+1}}{a_n x^n}\right| = |x| \lim_{n\to\infty} \left|\frac{a_{n+1}}{a_n}\right| = |x|r, \qquad r = \lim_{n\to\infty} \left|\frac{a_{n+1}}{a_n}\right|,$$

thus the series converges (absolutely) if $|x|r < 1$ or

$$|x| < R = \frac{1}{r} = \lim_{n\to\infty} \left|\frac{a_n}{a_{n+1}}\right|$$

and the domain of convergence is given by $R : -R < x < R$. Of course, we need to modify the above discussion somewhat if the power series does not contain every power of x.

Example A1.13
For what value of x does the series $\sum_{n=1}^{\infty} x^{n-1}/n \times 3^n$ converge?

Solution: Now $u_n = x^{n-1}/n \cdot 3^n$, and $x \neq 0$ (if $x = 0$ the series converges). We have

$$\lim_{n\to\infty} \left|\frac{u_{n+1}}{u_n}\right| = \lim_{n\to\infty} \frac{n}{3(n+1)}|x| = \frac{1}{3}|x|.$$

Then the series converges if $|x| < 3$, and diverges if $|x| > 3$. If $|x| = 3$, that is, $x = \pm 3$, the test fails.

If $x = 3$, the series becomes $\sum_{n=1}^{\infty} 1/3n$ which diverges. If $x = -3$, the series becomes $\sum_{n=1}^{\infty} (-1)^{n-1}/3n$ which converges. Then the interval of convergence is $-3 \leq x < 3$. The series diverges outside this interval. Furthermore, the series converges absolutely for $-3 < x < 3$ and converges conditionally at $x = -3$.

As for uniform convergence, the most commonly encountered test is the Weierstrass M test:

Weierstrass M test

If a sequence of positive constants $M_1, M_2, M_3, \ldots$, can be found such that: (a) $M_n \geq |u_n(x)|$ for all x in some interval $[a, b]$, and (b) $\sum M_n$ converges, then $\sum u_n(x)$ is uniformly and absolutely convergent in $[a, b]$.

The proof of this common test is direct and simple. Since $\sum M_n$ converges, some number N exists such that for $n \geq N$,

$$\sum_{i=N+1}^{\infty} M_i < \varepsilon.$$

This follows from the definition of convergence. Then, with $M_n \geq |u_n(x)|$ for all x in $[a, b]$,

$$\sum_{i=N+1}^{\infty} |u_i(x)| < \varepsilon.$$

Hence

$$|S(x) - s_n(x)| = \left| \sum_{i=N+1}^{\infty} u_i(x) \right| < \varepsilon, \qquad \text{for all } n \geq N$$

and by definition $\sum u_n(x)$ is uniformly convergent in $[a, b]$. Furthermore, since we have specified absolute values in the statement of the Weierstrass M test, the series $\sum u_n(x)$ is also seen to be absolutely convergent.

It should be noted that the Weierstrass M test only provides a sufficient condition for uniform convergence. A series may be uniformly convergent even when the M test is not applicable. The Weierstrass M test might mislead the reader to believe that a uniformly convergent series must be also absolutely convergent, and conversely. In fact, the uniform convergence and absolute convergence are independent properties. Neither implies the other.

A somewhat more delicate test for uniform convergence that is especially useful in analyzing power series is Abel's test. We now state it without proof.

Abel's test

If (a) $u_n(x) = a_n f_n(x)$, and $\sum a_n = A$, convergent, and (b) the functions $f_n(x)$ are monotonic $[f_{n+1}(x) \leq f_n(x)]$ and bounded, $0 \leq f_n(x) \leq M$ for all x in $[a, b]$, then $\sum u_n(x)$ converges uniformly in $[a, b]$.

Example A1.14

Use the Weierstrass M test to investigate the uniform convergence of

$$(a) \sum_{n=1}^{\infty} \frac{\cos nx}{n^4}, \qquad (b) \sum_{n=1}^{\infty} \frac{x^n}{n^{3/2}}, \qquad (c) \sum_{n=1}^{\infty} \frac{\sin nx}{n}.$$

Solution:

(a) $|\cos(nx)/n^4| \leq 1/n^4 = M_n$. Then since $\sum M_n$ converges (p series with $p = 4 > 1$), the series is uniformly and absolutely convergent for all x by the M test.

(b) By the ratio test, the series converges in the interval $-1 \leq x \leq 1$ (or $|x| \leq 1$). For all x in $|x| \leq 1$, $\left| x^n/n^{3/2} \right| = |x|^n/n^{3/2} \leq 1/n^{3/2}$. Choosing $M_n = 1/n^{3/2}$, we see that $\sum M_n$ converges. So the given series converges uniformly for $|x| \leq 1$ by the M test.

(c) $|\sin(nx)/n/n| \leq 1/n = M_n$. However, $\sum M_n$ does not converge. The M test cannot be used in this case and we cannot conclude anything about the uniform convergence by this test.

A uniformly convergent infinite series of functions has many of the properties possessed by the sum of finite series of functions. The following three are particularly useful. We state them without proofs.

(1) If the individual terms $u_n(x)$ are continuous in $[a, b]$ and if $\sum u_n(x)$ converges uniformly to the sum $S(x)$ in $[a, b]$, then $S(x)$ is continuous in $[a, b]$. Briefly, this states that a uniformly convergent series of continuous functions is a continuous function.

(2) If the individual terms $u_n(x)$ are continuous in $[a, b]$ and if $\sum u_n(x)$ converges uniformly to the sum $S(x)$ in $[a, b]$, then

$$\int_a^b S(x)dx = \sum_{n=1}^{\infty} \int_a^b u_n(x)dx$$

or

$$\int_a^b \sum_{n=1}^{\infty} u_n(x)dx = \sum_{n=1}^{\infty} \int_a^b u_n(x)dx.$$

Briefly, a uniform convergent series of continuous functions can be integrated term by term.

(3) If the individual terms $u_n(x)$ are continuous and have continuous derivatives in $[a, b]$ and if $\sum u_n(x)$ converges uniformly to the sum $S(x)$ while $\sum du_n(x)/dx$ is uniformly convergent in $[a, b]$, then the derivative of the series sum $S(x)$ equals the sum of the individual term derivatives,

$$\frac{d}{dx}S(x) = \sum_{n=1}^{\infty} \frac{d}{dx}u_n(x) \quad \text{or} \quad \frac{d}{dx}\left\{\sum_{n=1}^{\infty} u_n(x)\right\} = \sum_{n=1}^{\infty} \frac{d}{dx}u_n(x).$$

Term-by-term integration of a uniformly convergent series requires only continuity of the individual terms. This condition is almost always met in physical applications. Term-by-term integration may also be valid in the absence of uniform convergence. On the other hand term-by-term differentiation of a series is often not valid because more restrictive conditions must be satisfied.

Problem A1.14

Show that the series

$$\frac{\sin x}{1^3} + \frac{\sin 2x}{2^3} + \cdots + \frac{\sin nx}{n^3} + \cdots$$

is uniformly convergent for $-\pi \leq x \leq \pi$.

Theorems on power series

When we are working with power series and the functions they represent, it is very useful to know the following theorems which we will state without proof. We will see that, within their interval of convergence, power series can be handled much like polynomials.

(1) A power series converges uniformly and absolutely in any interval which lies entirely within its interval of convergence.

(2) A power series can be differentiated or integrated term by term over any interval lying entirely within the interval of convergence. Also, the sum of a convergent power series is continuous in any interval lying entirely within its interval of convergence.

(3) Two power series can be added or subtracted term by term for each value of x common to their intervals of convergence.

(4) Two power series, for example, $\sum_{n=0}^{\infty} a_n x^n$ and $\sum_{n=0}^{\infty} b_n x^n$, can be multiplied to obtain $\sum_{n=0}^{\infty} c_n x^n$, where $c_n = a_0 b_n + a_1 b_{n-1} + a_2 b_{n-2} + \cdots + a_n b_0$, the result being, valid for each x within the common interval of convergence.

(5) If the power series $\sum_{n=0}^{\infty} a_n x^n$ is divided by the power series $\sum_{n=0}^{\infty} b_n x^n$, where $b_0 \neq 0$, the quotient can be written as a power series which converges for sufficiently small values of x.

Taylor's expansion

It is very useful in most applied work to find power series that represent the given functions. We now review one method of obtaining such series, the Taylor expansion. We assume that our function $f(x)$ has a continuous nth derivative in the interval $[a, b]$ and that there is a Taylor series for $f(x)$ of the form

$$f(x) = a_0 + a_1(x - \alpha) + a_2(x - \alpha)^2 + a_3(x - \alpha)^3 + \cdots + a_n(x - \alpha)^n + \cdots,$$

$$(A1.8)$$

where α lies in the interval $[a, b]$. Differentiating, we have

$$f'(x) = a_1 + 2a_2(x - \alpha) + 3a_3(x - \alpha)^2 + \cdots + na_n(x - \alpha)^{n-1} + \cdots,$$

$$f''(x) = 2a_2 + 3 \cdot 2a_3(x - \alpha) + 4 \cdot 3a_4(x - a)^2 + \cdots + n(n - 1)a_n(x - \alpha)^{n-2} + \cdots,$$

$$\vdots$$

$$f^{(n)}(x) = n(n - 1)(n - 2) \cdots 1 \cdot a_n + \text{terms containing powers of } (x - \alpha).$$

We now put $x = \alpha$ in each of the above derivatives and obtain

$$f(\alpha) = a_0, \quad f'(\alpha) = a_1, \quad f''(\alpha) = 2a_2, \quad f'''(\alpha) = 3!a_3, \cdots, f^{(n)}(\alpha) = n!a_n,$$

where $f'(\alpha)$ means that $f(x)$ has been differentiated and then we have put $x = \alpha$; and by $f''(\alpha)$ we mean that we have found $f''(x)$ and then put $x = \alpha$, and so on. Substituting these into (A1.8) we obtain

$$f(x) = f(\alpha) + f'(\alpha)(x - \alpha) + \frac{1}{2!} f''(\alpha)(x - \alpha)^2 + \cdots + \frac{1}{n!} f^{(n)}(\alpha)(x - \alpha)^n + \cdots.$$

(A1.9)

This is the Taylor series for $f(x)$ about $x = \alpha$. The Maclaurin series for $f(x)$ is the Taylor series about the origin. Putting $\alpha = 0$ in (A1.9), we obtain the Maclaurin series for $f(x)$:

$$f(x) = f(0) + f'(0)x + \frac{1}{2!} f''(0)x^2 + \frac{1}{3!} f'''(0)x^3 + \cdots + \frac{1}{n!} f^{(n)}(0)x^n + \cdots.$$

(A1.10)

Example A1.15

Find the Maclaurin series expansion of the exponential function e^x.

Solution: Here $f(x) = e^x$. Differentiating, we have $f^{(n)}(0) = 1$ for all $n, n = 1, 2, 3 \ldots$. Then, by Eq. (A1.10), we have

$$e^x = 1 + x + \frac{1}{2!}x^2 + \frac{1}{3!}x^3 + \cdots = \sum_{n=0}^{\infty} \frac{x^n}{n!}, \qquad -\infty < x < \infty.$$

The following series are frequently employed in practice:

(1) $\sin x = x - \dfrac{x^3}{3!} + \dfrac{x^5}{5!} - \dfrac{x^7}{7!} + \cdots (-1)^{n-1} \dfrac{x^{2n-1}}{(2n-1)!} + \cdots,$ $\qquad -\infty < x < \infty.$

(2) $\cos x = 1 - \dfrac{x^2}{2!} + \dfrac{x^4}{4!} - \dfrac{x^6}{6!} + \cdots (-1)^{n-1} \dfrac{x^{2n-2}}{(2n-2)!} + \cdots,$ $\qquad -\infty < x < \infty.$

(3) $e^x = 1 + x + \dfrac{x^2}{2!} + \dfrac{x^3}{3!} + \cdots + \dfrac{x^{n-1}}{(n-1)!} + \cdots,$ $\qquad -\infty < x < \infty.$

(4) $\ln|(1 + x| = x - \dfrac{x^2}{2} + \dfrac{x^3}{3} - \dfrac{x^4}{4} + \cdots (-1)^{n-1} \dfrac{x^n}{n} + \cdots,$ $\qquad -1 < x \le 1.$

(5) $\dfrac{1}{2} \ln\left|\dfrac{1 + x}{1 - x}\right| = x + \dfrac{x^3}{3} + \dfrac{x^5}{5} + \dfrac{x^7}{7} + \cdots + \dfrac{x^{2n-1}}{2n - 1} + \cdots,$ $\qquad -1 < x < 1.$

(6) $\tan^{-1} x = x - \dfrac{x^3}{3} + \dfrac{x^5}{5} - \dfrac{x^7}{7} + \cdots (-1)^{n-1} \dfrac{x^{2n-1}}{2n - 1} + \cdots,$ $\qquad -1 \le x \le 1.$

(7) $(1 + x)^p = 1 + px + \dfrac{p(p - 1)}{2!}x^2 + \cdots + \dfrac{p(p - 1)\cdots(p - n + 1)}{n!}x^n + \cdots.$

This is the binomial series: (a) If p is a positive integer or zero, the series terminates. (b) If $p > 0$ but is not an integer, the series converges absolutely for $-1 \le x \le 1$. (c) If $-1 < p < 0$, the series converges for $-1 < x \le 1$. (d) If $p \le -1$, the series converges for $-1 < x < 1$.

Problem A1.16
Obtain the Maclaurin series for $\sin x$ (the Taylor series for $\sin x$ about $x = 0$).

Problem A1.17
Use series methods to obtain the approximate value of $\int_0^1 (1 - e^{-x})/x\,dx$.

We can find the power series of functions other than the most common ones listed above by the successive differentiation process given by Eq. (A1.9). There are simpler ways to obtain series expansions. We give several useful methods here.

(a) For example to find the series for $(x + 1)\sin x$, we can multiply the series for $\sin x$ by $(x + 1)$ and collect terms:

$$(x + 1)\sin x = (x + 1)\left(x - \frac{x^3}{3!} + \frac{x^5}{5!} - \cdots\right) = x + x^2 - \frac{x^3}{3!} - \frac{x^4}{3!} + \cdots.$$

To find the expansion for $e^x \cos x$, we can multiply the series for e^x by the series for $\cos x$:

$$e^x \cos x = \left(1 + x + \frac{x^2}{2!} + \frac{x^3}{3!} + \cdots\right)\left(1 - \frac{x^2}{2!} + \frac{x^4}{4!} + \cdots\right)$$

$$= 1 + x + \frac{x^2}{2!} + \frac{x^3}{3!} + \frac{x^4}{4!} \cdots$$

$$\quad - \frac{x^2}{2!} - \frac{x^3}{3!} - \frac{x^4}{2!2!} \cdots$$

$$\quad + \frac{x^4}{4!} \cdots$$

$$= 1 + x - \frac{x^3}{3} - \frac{x^4}{6} \cdots.$$

Note that in the first example we obtained the desired series by multiplication of a known series by a polynomial; and in the second example we obtained the desired series by multiplication of two series.

(b) In some cases, we can find the series by division of two series. For example, to find the series for $\tan x$, we can divide the series for $\sin x$ by the series for $\cos x$:

$$\tan x = \frac{\sin x}{\cos x} = \left(1 - \frac{x^2}{2} + \frac{x^4}{4!} \cdots\right) \Big/ \left(x - \frac{x^3}{3!} + \frac{x^5}{5!} \cdots\right) = x + \frac{1}{3}x^3 + \frac{2}{15}x^5 \cdots.$$

The last step is by long division

$$
\begin{array}{r}
x + \dfrac{1}{3}x^3 + \dfrac{2}{15}x^5 \cdots \\[2ex]
1 - \dfrac{x^2}{2!} + \dfrac{x^4}{4!} \cdots \overline{\smash{\big)}\, x - \dfrac{x^3}{3!} + \dfrac{x^5}{5!} \cdots} \\[2ex]
x - \dfrac{x^3}{2!} + \dfrac{x^5}{4!} \cdots \\[2ex]
\dfrac{x^3}{3} - \dfrac{x^5}{30} \cdots \\[2ex]
\dfrac{x^3}{3} - \dfrac{x^5}{6} \cdots \\[2ex]
\dfrac{2x^5}{15} \cdots, \quad \text{etc.}
\end{array}
$$

Problem A1.18
Find the series expansion for $1/(1+x)$ by long division. Note that the series can be found by using the binomial series: $1/(1+x) = (1+x)^{-1}$.

(c) In some cases, we can obtain a series expansion by substitution of a polynomial or a series for the variable in another series. As an example, let us find the series for e^{-x^2}. We can replace x in the series for e^x by $-x^2$ and obtain

$$e^{-x^2} = 1 - x^2 + \frac{(-x^2)^2}{2!} + \frac{(-x)^3}{3!} \cdots = 1 - x^2 + \frac{x^4}{2!} - \frac{x^6}{3!} \cdots.$$

Similarly, to find the series for $\sin \sqrt{x}/\sqrt{x}$ we replace x in the series for $\sin x$ by $\sqrt{x}$ and obtain

$$\frac{\sin \sqrt{x}}{\sqrt{x}} = 1 - \frac{x}{3!} + \frac{x^2}{5!} \cdots, \qquad x > 0.$$

Problem A1.19
Find the series expansion for $e^{\tan x}$.

Problem A1.20
Assuming the power series for e^x holds for complex numbers, show that

$$e^{ix} = \cos x + i \sin x.$$

(d) Find the series for $\tan^{-1} x$ (arc tan x). We can find the series by the successive differentiation process. But it is very tedious to find successive derivatives of $\tan^{-1} x$. We can take advantage of the following integration

$$\int_0^x \frac{dt}{1+t^2} = \tan^{-1} t \Big|_0^x = \tan^{-1} x.$$

We now first write out $(1+t^2)^{-1}$ as a binomial series and then integrate term by term:

$$\int_0^x \frac{dt}{1+t^2} = \int_0^x (1 - t^2 + t^4 - t^6 + \cdots) dt = t - \frac{t^3}{3} + \frac{t^5}{5} - \frac{t^7}{7} + \cdots \Big|_0^x.$$

Thus, we have

$$\tan^{-1} x = x - \frac{x^3}{3} + \frac{x^5}{5} - \frac{x^7}{7} + \cdots .$$

(e) Find the series for $\ln x$ about $x = 1$. We want a series of powers $(x - 1)$ rather than powers of x. We first write

$$\ln x = \ln[1 + (x - 1)]$$

and then use the series $\ln(1 + x)$ with x replaced by $(x - 1)$:

$$\ln x = \ln[1 + (x-1)] = (x - 1) - \frac{1}{2}(x - 1)^2 + \frac{1}{3}(x - 1)^3 - \frac{1}{4}(x - 1)^4 \cdots .$$

Problem A1.21

Expand $\cos x$ about $x = 3\pi/2$.

Higher derivatives and Leibnitz's formula for nth derivative of a product

Higher derivatives of a function $y = f(x)$ with respect to x are written as

$$\frac{d^2 y}{dx^2} = \frac{d}{dx}\left(\frac{dy}{dx}\right), \quad \frac{d^3 y}{dx^3} = \frac{d}{dx}\left(\frac{d^2 y}{dx^2}\right), \quad \cdots, \quad \frac{d^n y}{dx^n} = \frac{d}{dy}\left(\frac{d^{n-1} y}{dx^{n-1}}\right).$$

These are sometimes abbreviated to either

$$f''(x), f'''(x), \ldots, f^{(n)}(x) \quad \text{or} \quad D^2 y, D^3 y, \ldots, D^n y$$

where $D = d/dx$.

When higher derivatives of a product of two functions $f(x)$ and $g(x)$ are required, we can proceed as follows:

$$D(fg) = fDg + gDf$$

and

$$D^2(fg) = D(fDg + gDf) = fD^2g + 2Df \cdot Dg + D^2g.$$

Similarly we obtain

$$D^3(fg) = fD^3g + 3Df \cdot D^2g + 3D^2f \cdot Dg + gD^3f,$$

$$D^4(fg) = fD^4g + 4Df \cdot D^3g + 6D^2f \cdot D^2g + 4D^3 \cdot Dg + gD^4g,$$

and so on. By inspection of these results the following formula (due to Leibnitz) may be written down for nth derivative of the product fg:

$$D^n(fg) = f(D^ng) + n(Df)(D^{n-1}g) + \frac{n(n-1)}{2!}(D^2f)(D^{n-2}g) + \cdots$$

$$+ \frac{n!}{k!(n-k)!}(D^kf)(D^{n-k}g) + \cdots + (D^nf)g.$$

Example A1.16

If $f = 1 - x^2, g = D^2y$, where y is a function of x, say $u(x)$, then

$$D^n\{(1-x^2)D^2y\} = (1-x^2)D^{n+2}y - 2nxD^{n+1}y - n(n-1)D^ny.$$

Leibnitz's formula may also be applied to a differential equation. For example, y satisfies the differential equation

$$D^2y + x^2y = \sin x.$$

Then differentiating each term n times we obtain

$$D^{n+2}y + (x^2D^ny + 2nxD^{n-1}y + n(n-1)D^{n-2}y) = \sin\left(\frac{n\pi}{2} + x\right),$$

where we have used Leibnitz's formula for the product term x^2y.

Problem A1.22

Using Leibnitz's formula, show that

$$D^n(x^2 \sin x) = \{x^2 - n(n-1)\}\sin(x + n\pi/2) - 2nx\cos(x + n\pi/2).$$

Some important properties of definite integrals

Integration is an operation inverse to that of differentiation; and it is a device for calculating the 'area under a curve'. The latter method regards the integral as the limit of a sum and is due to Riemann. We now list some useful properties of definite integrals.

(1) If in $a \le x \le b, m \le f(x) \le M$, where m and M are constants, then

$$m(b-a) \le \int_a^b f(x)d \le M(b-a).$$

Divide the interval $[a, b]$ into n subintervals by means of the points $x_1, x_2, \ldots, x_{n-1}$ chosen arbitrarily. Let η_k be any point in the subinterval $x_{k-1} \leq \eta_k \leq x_k$, then we have

$$m\Delta x_k \leq f(\eta_k)\Delta x_k \leq M\Delta x_k, \quad k = 1, 2, \ldots, n,$$

where $\Delta x_k = x_k - x_{k-1}$. Summing from $k = 1$ to n and using the fact that

$$\sum_{k=1}^{n} \Delta x_k = (x_1 - a) + (x_2 - x_1) + \cdots + (b - x_{n-1}) = b - a,$$

it follows that

$$m(b - a) \leq \sum_{rk=1}^{n} f(\eta_k)\Delta x_k \leq M(b - a).$$

Taking the limit as $n \to \infty$ and each $\Delta x_k \to 0$ we have the required result.
(2) If in $a \leq x \leq b, f(x) \leq g(x)$, then

$$\int_a^b f(x)dx \leq \int_a^b g(x)dx.$$

(3) $\left| \int_a^b f(x)dx \right| \leq \int_a^b |f(x)|dx \quad$ if $a < b$.

From the inequality

$$|a + b + c + \cdots| \leq |a| + |b| + |c| + \cdots,$$

where $|a|$ is the absolute value of a real number a, we have

$$\left| \sum_{k=1}^{n} f(\eta_k)\Delta x_k \right| \leq \sum_{k=1}^{n} |f(\eta_k)\Delta x_k| = \sum_{k=1}^{n} |f(\eta_k)|\Delta x_k.$$

Taking the limit as $n \to \infty$ and each $\Delta x_k \to 0$ we have the required result.
(4) The mean value theorem: If $f(x)$ is continuous in $[a, b]$, we can find a point η in (a, b) such that

$$\int_a^b f(x)dx = (b - a)f(\eta).$$

Since $f(x)$ is continuous in $[a, b]$, we can find constants m and M such that $m \leq f(x) \leq M$. Then by (1) we have

$$m \leq \frac{1}{b - a} \int_a^b f(x)dx \leq M.$$

Since $f(x)$ is continuous it takes on all values between m and M; in particular there must be a value η such that

$$f(\eta) = \int_a^b f(x)dx/(b - a), \quad a < \eta < b.$$

The required result follows on multiplying by $b - a$.

Some useful methods of integration

(1) Changing variables: We use a simple example to illustrate this common procedure. Consider the integral

$$I = \int_0^\infty e^{-ax^2} dx,$$

which is equal to $(\pi/a)^{1/2}/2$. To show this let us write

$$I = \int_0^\infty e^{-ax^2} dx = \int_0^\infty e^{-ay^2} dy.$$

Then

$$I^2 = \int_0^\infty e^{-ax^2} dx \int_0^\infty e^{-ay^2} dy = \int_0^\infty \int_0^\infty e^{-a(x^2+y^2)} dxdy.$$

We now rewrite the integral in plane polar coordinates (r,θ): $x^2 + y^2 = r^2, dxdy = rdrd\theta$. Then

$$I^2 = \int_0^\infty \int_0^{\pi/2} e^{-ar^2} rd\theta dr = \frac{\pi}{2} \int_0^\infty e^{-ar^2} rdr = \frac{\pi}{2}\left(-\frac{e^{-ar^2}}{2a}\right)\Big|_0^\infty = \frac{\pi}{4a}$$

and

$$I = \int_0^\infty e^{-ax^2} dx = (\pi/a)^{1/2}/2.$$

(2) Integration by parts: Since

$$\frac{d}{dx}(uv) = u\frac{dv}{dx} + v\frac{du}{dx},$$

where $u = f(x)$ and $v = g(x)$, it follows that

$$\int u\left(\frac{dv}{dx}\right) dx = uv - \int v\left(\frac{du}{dx}\right) dx.$$

This can be a useful formula in evaluating integrals.

Example A1.17
Evaluate $I = \int \tan^{-1} xdx$

Solution: Since $\tan^{-1} x$ can be easily differentiated, we write $I = \int \tan^{-1} xdx = \int 1 \cdot \tan^{-1} xdx$ and let $u = \tan^{-1} x, dv/dx = 1$. Then

$$I = x\tan^{-1} x - \int \frac{xdx}{1+x^2} = x\tan^{-1} x - \frac{1}{2}\log(1 + x^2) + c.$$

Example A1.18

Show that

$$\int_{-\infty}^{\infty} x^2 e^{-ax^2} dx = \frac{\pi^{1/2}}{2a^{3/2}}.$$

Solution: Let us first consider the integral

$$I = \int_0^{\infty} e^{-ax^2} dx$$

'Integration-by-parts' gives

$$I = \int_b^c e^{-ax^2} dx = e^{-ax^2} x \Big|_b^c + 2 \int_b^c ax^2 e^{-ax^2} dx,$$

from which we obtain

$$\int_b^c x^2 e^{-ax^2} dx = \frac{1}{2a} \left(\int_b^c e^{-ax^2} dx - e^{-ax^2} x \Big|_b^c \right).$$

We let limits b and c become $-\infty$ and $+\infty$, and thus obtain the desired result.

Problem A1.23

Evaluate $I = \int xe^{\lambda x} dx$ (λ constant).

(3) Partial fractions: Any rational function $P(x)/Q(x)$, where $P(x)$ and $Q(x)$ are polynomials, with the degree of $P(x)$ less than that of $Q(x)$, can be written as the sum of rational functions having the form $A/(ax+b)^k$, $(Ax+B)/(ax^2+bx+c)^k$, where $k = 1, 2, 3, \ldots$ which can be integrated in terms of elementary functions.

Example A1.19

$$\frac{3x - 2}{(4x - 3)(2x + 5)^3} = \frac{A}{4x - 3} + \frac{B}{(2x + 5)^3} + \frac{C}{(2x + 5)^2} + \frac{D}{2x + 5},$$

$$\frac{5x^2 - x + 2}{(x^2 + 2x + 4)^2(x - 1)} = \frac{Ax + B}{(x^2 + 2x + 4)^2} + \frac{Cx + D}{x^2 + 2x + 4} + \frac{E}{x - 1}.$$

Solution: The coefficients A, B, C etc., can be determined by clearing the fractions and equating coefficients of like powers of x on both sides of the equation.

Problem A1.24

Evaluate

$$I = \int \frac{6 - x}{(x - 3)(2x + 5)} dx.$$

(4) Rational functions of sin x and cos x can always be integrated in terms of elementary functions by substitution $\tan(x/2) = u$, as shown in the following example.

Example A1.20

Evaluate

$$I = \int \frac{dx}{5 + 3\cos x}.$$

Solution: Let $\tan(x/2) = u$, then

$$\sin(x/2) = u/\sqrt{1 + u^2}, \qquad \cos(x/2) = 1/\sqrt{1 + u^2}$$

and

$$\cos x = \cos^2(x/2) - \sin^2(x/2) = \frac{1 - u^2}{1 + u^2};$$

also

$$du = \tfrac{1}{2}\sec^2(x/2)dx \quad \text{or} \quad dx = 2\cos^2(x/2) = 2du/(1 + u^2).$$

Thus

$$I = \int \frac{du}{u^2 + 4} = \tfrac{1}{2}\tan^{-1}(u/2) + c = \tfrac{1}{2}\tan^{-1}[\tfrac{1}{2}\tan x/2)] + c.$$

Reduction formulas

Consider an integral of the form $\int x^n e^{-x} dx$. Since this depends upon n let us call it I_n. Then using integration by parts we have

$$I_n = -x^n e^{-x} + n\int x^{n-1} e^{-x} dx = x^n e^{-x} + nI_{n-1}.$$

The above equation gives I_n in terms of $I_{n-1}(I_{n-2}, I_{n-3}$, etc.) and is therefore called a reduction formula.

Problem A1.25

Evaluate $I_n = \int_0^{\pi/2} \sin^n x dx = \int_0^{\pi/2} \sin x \sin^{n-1} x dx.$

Differentiation of integrals

(1) Indefinite integrals: We first consider differentiation of indefinite integrals. If $f(x, \alpha)$ is an integrable function of x and α is a variable parameter, and if

$$\int f(x, \alpha)dx = G(x, \alpha), \tag{A1.11}$$

then we have

$$\partial G(x, \alpha)/\partial x = f(x, \alpha). \tag{A1.12}$$

Furthermore, if $f(x, \alpha)$ is such that

$$\frac{\partial^2 G(x, \alpha)}{\partial x \partial \alpha} = \frac{\partial^2 G(x, \alpha)}{\partial \alpha \partial x},$$

then we obtain

$$\frac{\partial}{\partial x}\left[\frac{\partial G(x, \alpha)}{\partial \alpha}\right] = \frac{\partial}{\partial \alpha}\left[\frac{\partial G(x, \alpha)}{\partial x}\right] = \frac{\partial f(x, \alpha)}{\partial \alpha}$$

and integrating gives

$$\int \frac{\partial f(x, \alpha)}{\partial \alpha}dx = \frac{\partial G(x, \alpha)}{\partial \alpha}, \tag{A1.13}$$

which is valid provided $\partial f(x, \alpha)/\partial \alpha$ is continuous in x as well as α.

(2) Definite integrals: We now extend the above procedure to definite integrals:

$$I(\alpha) = \int_a^b f(x, \alpha)dx, \tag{A1.14}$$

where $f(x, \alpha)$ is an integrable function of x in the interval $a \leq x \leq b$, and a and b are in general continuous and differentiable (at least once) functions of α. We now have a relation similar to Eq. (A1.11):

$$I(\alpha) = \int_a^b f(x, \alpha)dx = G(b, \alpha) - G(a, \alpha) \tag{A1.15}$$

and, from Eq. (A1.13),

$$\int_a^b \frac{\partial f(x, \alpha)}{\partial \alpha}dx = \frac{\partial G(b, \alpha)}{\partial \alpha} - \frac{\partial G(a, \alpha)}{\partial \alpha}. \tag{A1.16}$$

Differentiating (A1.15) totally

$$\frac{dI(\alpha)}{d\alpha} = \frac{\partial G(b, \alpha)}{\partial b}\frac{db}{d\alpha} + \frac{\partial G(b, \alpha)}{\partial \alpha} - \frac{\partial G(a, \alpha)}{\partial a}\frac{da}{d\alpha} - \frac{\partial G(a, \alpha)}{\partial \alpha}:$$

which becomes, with the help of Eqs. (A1.12) and (A1.16),

$$\frac{dI(\alpha)}{d\alpha} = \int_a^b \frac{\partial f(x,\alpha)}{\partial \alpha}\,dx + f(b,\alpha)\frac{db}{d\alpha} - f(a,\alpha)\frac{da}{d\alpha}, \tag{A1.17}$$

which is known as Leibnitz's rule for differentiating a definite integral. If a and b, the limits of integration, do not depend on α, then Eq. (A1.17) reduces to

$$\frac{dI(\alpha)}{d\alpha} = \frac{d}{d\alpha}\int_a^b f(x,\alpha)\,dx = \int_a^b \frac{\partial f(x,\alpha)}{\partial \alpha}\,dx.$$

Problem A1.26

If $I(\alpha) = \displaystyle\int_0^{\alpha^2} \sin(\alpha x)/x\,dx$, find $dI/d\alpha$.

Homogeneous functions

A homogeneous function $f(x_1, x_2, \ldots, x_n)$ of the kth degree is defined by the relation

$$f(\lambda x_1, \lambda x_2, \ldots, \lambda x_n) = \lambda^k f(x_1, x_2, \ldots, x_n).$$

For example, $x^3 + 3x^2y - y^3$ is homogeneous of the third degree in the variables x and y.

If $f(x_1, x_2, \ldots, x_n)$ is homogeneous of degree k then it is straightforward to show that

$$\sum_{j=1}^n x_j \frac{\partial f}{\partial x_j} = kf.$$

This is known as Euler's theorem on homogeneous functions.

Problem A1.27
Show that Euler's theorem on homogeneous functions is true.

Taylor series for functions of two independent variables

The ideas involved in Taylor series for functions of one variable can be generalized. For example, consider a function of two variables (x, y). If all the nth partial derivatives of $f(x, y)$ are continuous in a closed region and if the $(n + 1)$st partial derivatives exist in the open region, then we can expand the function $f(x, y)$ about $x = x_0, y = y_0$ in the form

$$f(x_0 + h, y_0 + k) = f(x_0, y_0) + \left(h\frac{\partial}{\partial x} + k\frac{\partial}{\partial y} \right) f(x_0, y_0)$$

$$+ \frac{1}{2!} \left(h\frac{\partial}{\partial x} + k\frac{\partial}{\partial y} \right)^2 f(x_0, y_0)$$

$$+ \cdots + \frac{1}{n!} \left(h\frac{\partial}{\partial x} + k\frac{\partial}{\partial y} \right)^n f(x_0, y_0) + R_n,$$

where $h = \Delta x = x - x_0, k = \Delta y = y - y_0, R_n$, the remainder after n terms, is given by

$$R_n = \frac{1}{(n+1)!} \left(h\frac{\partial}{\partial x} + k\frac{\partial}{\partial y} \right)^{n+1} f(x_0 + \theta h, y_0 + \theta k), \quad 0 < \theta < 1,$$

and where we use the operator notation

$$\left(h\frac{\partial}{\partial x} + k\frac{\partial}{\partial y} \right) f(x_0, y_0) = hf_x(x_0, y_0) + kf_y(x_0, y_0),$$

$$\left(h\frac{\partial}{\partial x} + k\frac{\partial}{\partial y} \right)^2 f(x_0, y_0) = \left(h^2\frac{\partial^2}{\partial x^2} + 2hk\frac{\partial^2}{\partial x \partial y} + k^2\frac{\partial^2}{\partial y^2} \right) f(x_0, y_0),$$

etc., when we expand

$$\left(h\frac{\partial}{\partial x} + k\frac{\partial}{\partial y} \right)^n$$

formally by the binomial theorem.

When $\lim_{n \to \infty} R_n = 0$ for all (x, y) in a region, the infinite series expansion is called a Taylor series in two variables. Extensions can be made to three or more variables.

Lagrange multiplier

For functions of one variable such as $f(x)$ to have a stationary value (maximum or minimum) at $x = a$, we have $f'(a) = 0$. If $f''(a) < 0$ it is a relative maximum while if $f(a) > 0$ it is a relative minimum.

Similarly $f(x, y)$ has a relative maximum or minimum at $x = a, y = b$ if $f_x(a, b) = 0, f_y(a, b) = 0$. Thus possible points at which $f(x, y)$ has a relative maximum or minimum are obtained by solving simultaneously the equations

$$\partial f/\partial x = 0, \quad \partial f/\partial y = 0.$$

Sometimes we wish to find the relative maxima or minima of $f(x, y) = 0$ subject to some constraint condition $\phi(x, y) = 0$. To do this we first form the function $g(x, y) = f(x, y) + f(x, y)$ and then set

$$\partial g/\partial x = 0, \quad \partial g/\partial y = 0.$$

The constant λ is called a Lagrange multiplier and the method is known as the method of undetermined multipliers.

Appendix 2

Determinants

The determinant is a tool used in many branches of mathematics, science, and engineering. The reader is assumed to be familiar with this subject. However, for those who are in need of review, we prepared this appendix, in which the determinant is defined and its properties developed. In Chapters 1 and 3, the reader will see the determinant's use in proving certain properties of vector and matrix operations.

The concept of a determinant is already familiar to us from elementary algebra, where, in solving systems of simultaneous linear equation, we find it convenient to use determinants. For example, consider the system of two simultaneous linear equations

$$\left.\begin{array}{l} a_{11}x_1 + a_{12}x_2 = b_1, \\ a_{21}x_1 + a_{22}x_2 = b_2, \end{array}\right\} \tag{A2.1}$$

in two unknowns x_1, x_2 where a_{ij} $(i, j = 1, 2)$ are constants. These two equations represent two lines in the $x_1 x_2$ plane. To solve the system (A2.1), multiplying the first equation by a_{22}, the second by $-a_{12}$ and then adding, we find

$$x_1 = \frac{b_1 a_{22} - b_2 a_{12}}{a_{11} a_{22} - a_{21} a_{12}}. \tag{A2.2a}$$

Next, by multiplying the first equation by $-a_{21}$, the second by a_{11} and adding, we find

$$x_2 = \frac{b_2 a_{11} - b_1 a_{21}}{a_{11} a_{22} - a_{21} a_{12}}. \tag{A2.2b}$$

We may write the solutions (A2.2) of the system (A2.1) in the determinant form

$$x_1 = \frac{D_1}{D}, \quad x_2 = \frac{D_2}{D}, \tag{A2.3}$$

where

$$D_1 = \begin{vmatrix} b_1 & a_{12} \\ b_2 & a_{22} \end{vmatrix}, \qquad D_2 = \begin{vmatrix} a_{11} & b_1 \\ a_{21} & b_2 \end{vmatrix}, \qquad D = \begin{vmatrix} a_{11} & a_{12} \\ a_{21} & a_{22} \end{vmatrix} \qquad \text{(A2.4)}$$

are called determinants of second order or order 2. The numbers enclosed between vertical bars are called the elements of the determinant. The elements in a horizontal line form a row and the elements in a vertical line form a column of the determinant. It is obvious that in Eq. (A2.3) $D \neq 0$.

Note that the elements of determinant D are arranged in the same order as they occur as coefficients in Eqs. (A1.1). The numerator D_1 for x_1 is constructed from D by replacing its first column with the coefficients b_1 and b_2 on the right-hand side of (A2.1). Similarly, the numerator for x_2 is formed by replacing the second column of D by b_1, b_2. This procedure is often called Cramer's rule.

Comparing Eqs. (A2.3) and (A2.4) with Eq. (A2.2), we see that the determinant is computed by summing the products on the rightward arrows and subtracting the products on the leftward arrows:

$$\begin{vmatrix} a_{11} & a_{12} \\ a_{21} & a_{22} \end{vmatrix} = a_{11}a_{22} - a_{12}a_{21}, \quad \text{etc.}$$

$$(-) \qquad (+)$$

This idea is easily extended. For example, consider the system of three linear equations

$$\left. \begin{aligned} a_{11}x_1 + a_{12}x_2 + a_{13}x_3 &= b_1, \\ a_{21}x_1 + a_{22}x_2 + a_{23}x_3 &= b_2, \\ a_{31}x_1 + a_{32}x_2 + a_{33}x_3 &= b_3, \end{aligned} \right\} \qquad \text{(A2.5)}$$

in three unknowns x_1, x_2, x_3. To solve for x_1, we multiply the equations by

$$a_{22}a_{33} - a_{32}a_{23}, \qquad -(a_{12}a_{33} - a_{32}a_{13}), \qquad a_{12}a_{23} - a_{22}a_{13},$$

respectively, and then add, finding

$$x_1 = \frac{b_1 a_{22}a_{33} - b_1 a_{23}a_{32} + b_2 a_{13}a_{32} - b_2 a_{12}a_{33} + b_3 a_{12}a_{23} - b_3 a_{13}a_{22}}{a_{11}a_{22}a_{33} - a_{11}a_{32}a_{23} + a_{21}a_{32}a_{13} - a_{21}a_{12}a_{33} + a_{31}a_{12}a_{23} - a_{31}a_{22}a_{13}},$$

which can be written in determinant form

$$x_1 = D_1/D, \qquad \text{(A2.6)}$$

where

$$D = \begin{vmatrix} a_{11} & a_{12} & a_{13} \\ a_{21} & a_{22} & a_{23} \\ a_{31} & a_{32} & a_{33} \end{vmatrix}, \quad D_1 = \begin{vmatrix} b_1 & a_{12} & a_{13} \\ b_2 & a_{22} & a_{23} \\ b_3 & a_{32} & a_{33} \end{vmatrix}. \tag{A2.7}$$

Again, the elements of D are arranged in the same order as they appear as coefficients in Eqs. (A2.5), and D_1 is obtained by Cramer's rule. In the same manner we can find solutions for x_2, x_3. Moreover, the expansion of a determinant of third order can be obtained by diagonal multiplication by repeating on the right the first two columns of the determinant and adding the signed products of the elements on the various diagonals in the resulting array:

$$\begin{vmatrix} a_{11} & a_{12} & a_{13} \\ a_{21} & a_{22} & a_{23} \\ a_{31} & a_{32} & a_{33} \end{vmatrix} \begin{matrix} a_{11} & a_{12} \\ a_{21} & a_{22} \\ a_{31} & a_{32} \end{matrix}$$

$$(-) \ (-) \ (-) \quad (+) \ (+) \ (+)$$

This method of writing out determinants is correct only for second- and third-order determinants.

Problem A2.1
Solve the following system of three linear equations using Cramer's rule:

$$2x_1 - x_2 + 2x_3 = 2,$$
$$x_1 + 10x_2 - 3x_3 = 5,$$
$$-x_1 + x_2 + x_3 = -3.$$

Problem A2.2
Evaluate the following determinants

$$(a) \ \begin{vmatrix} 1 & 2 \\ 4 & 3 \end{vmatrix}, \quad (b) \ \begin{vmatrix} 5 & 1 & 8 \\ 15 & 3 & 6 \\ 10 & 4 & 2 \end{vmatrix}, \quad (c) \ \begin{vmatrix} \cos\theta & -\sin\theta \\ \sin\theta & \cos\theta \end{vmatrix}.$$

Determinants, minors, and cofactors

We are now in a position to define an nth-order determinant. A determinant of order n is a square array of n^2 quantities enclosed between vertical bars,

$$D = \begin{vmatrix} a_{11} & a_{12} & \cdots & a_{1n} \\ a_{21} & a_{22} & \cdots & a_{2n} \\ \vdots & \vdots & & \vdots \\ a_{n1} & a_{n2} & \cdots & a_{nn} \end{vmatrix}.$$

(A2.8)

By deleting the ith row and the kth column from the determinant D we obtain an $(n-1)$st order determinant (a square array of $n-1$ rows and $n-1$ columns between vertical bars), which is called the minor of the element a_{ik} (which belongs to the deleted row and column) and is denoted by M_{ik}. The minor M_{ik} multiplied by $(-)^{i+k}$ is called the cofactor of a_{ik} and is denoted by C_{ik}:

$$C_{ik} = (-1)^{i+k} M_{ik}.$$

(A2.9)

For example, in the determinant

$$\begin{vmatrix} a_{11} & a_{12} & a_{13} \\ a_{21} & a_{22} & a_{23} \\ a_{31} & a_{32} & a_{33} \end{vmatrix},$$

we have

$$C_{11} = (-1)^{1+1} M_{11} = \begin{vmatrix} a_{22} & a_{23} \\ a_{32} & a_{33} \end{vmatrix}, \qquad C_{32} = (-1)^{3+2} M_{32} = - \begin{vmatrix} a_{11} & a_{13} \\ a_{21} & a_{23} \end{vmatrix}, \qquad \text{etc.}$$

It is very convenient to get the proper sign (plus or minus) for the cofactor $(-1)^{i+k}$ by thinking of a checkerboard of plus and minus signs like this

$$\begin{vmatrix} + & - & + & - & \\ - & + & - & + & \\ + & - & + & - & \text{etc.} \\ - & + & - & + & \\ \text{etc.} & & & & \ddots \\ & & & & + & - \\ & & & & - & + \end{vmatrix}$$

thus, for the element a_{23} we can see that the checkerboard sign is minus.

Expansion of determinants

Now we can see how to find the value of a determinant: multiply each of one row (or one column) by its cofactor and then add the results, that is,

$$D = a_{i1}C_{i1} + a_{i2}C_{i2} + \cdots + a_{in}C_{in}$$

$$= \sum_{k=1}^{n} a_{ik}C_{ik} \qquad (i = 1, 2, \ldots, \text{or } n) \qquad \text{(A2.10a)}$$

(cofactor expansion along the ith row)

or

$$D = a_{1k}C_{1k} + a_{2k}C_{2k} + \cdots + a_{nk}C_{nk}$$

$$= \sum_{i=1}^{n} a_{ik}C_{ik} \qquad (k = 1, 2, \ldots, \text{or } n). \qquad \text{(A2.10b)}$$

(cofactor expansion along the kth column)

We see that D is defined in terms of n determinants of order $n - 1$, each of which, in turn, is defined in terms of $n - 1$ determinants of order $n - 2$, and so on; we finally arrive at second-order determinants, in which the cofactors of the elements are single elements of D. The method of evaluating a determinant just described is one form of Laplace's development of a determinant.

Problem A2.3
For a second-order determinant

$$D = \begin{vmatrix} a_{11} & a_{12} \\ a_{21} & a_{22} \end{vmatrix}$$

show that the Laplace's development yields the same value of D no matter which row or column we choose.

Problem A2.4
Let

$$D = \begin{vmatrix} 1 & 3 & 0 \\ 2 & 6 & 4 \\ -1 & 0 & 2 \end{vmatrix}.$$

Evaluate D, first by the first-row expansion, then by the first-column expansion. Do you get the same value of D?

Properties of determinants

In this section we develop some of the fundamental properties of the determinant function. In most cases, the proofs are brief.

(1) If all elements of a row (or a column) of a determinant are zero, the value of the determinant is zero.

Proof: Let the elements of the kth row of the determinant D be zero. If we expand D in terms of the ith row, then

$$D = a_{i1}C_{i1} + a_{i2}C_{i2} + \cdots + a_{in}C_{in}.$$

Since the elements $a_{i1}, a_{i2}, \ldots, a_{in}$ are zero, $D = 0$. Similarly, if all the elements in one column are zero, expanding in terms of that column shows that the determinant is zero.

(2) If all the elements of one row (or one column) of a determinant are multiplied by the same factor k, the value of the new determinant is k times the value of the original determinant. That is, if a determinant B is obtained from determinant D by multiplying the elements of a row (or a column) of D by the same factor k, then $B = kD$.

Proof: Suppose B is obtained from D by multiplying its ith row by k. Hence the ith row of B is ka_{ij}, where $j = 1, 2, \ldots, n$, and all other elements of B are the same as the corresponding elements of A. Now expand B in terms of the ith row:

$$B = ka_{i1}C_{i1} + ka_{i2}C_{i2} + \cdots + ka_{in}C_{in}$$

$$= k(a_{i1}C_{i1} + a_{i2}C_{i2} + \cdots + a_{in}C_{in})$$

$$= kD.$$

The proof for columns is similar.

Note that property (1) can be considered as a special case of property (2) with $k = 0$.

Example A2.1

If

$$D = \begin{vmatrix} 1 & 2 & 3 \\ 0 & 1 & 1 \\ 4 & -1 & 0 \end{vmatrix} \quad \text{and} \quad B = \begin{vmatrix} 1 & 6 & 3 \\ 0 & 3 & 1 \\ 4 & -3 & 0 \end{vmatrix},$$

then we see that the second column of B is three times the second column of D. Evaluating the determinants, we find that the value of D is -3, and the value of B is -9 which is three times the value of D, illustrating property (2).

Property (2) can be used for simplifying a given determinant, as shown in the following example.

Example A2.2

$$\begin{vmatrix} 1 & 3 & 0 \\ 2 & 6 & 4 \\ -1 & 0 & 2 \end{vmatrix} = 2 \begin{vmatrix} 1 & 3 & 0 \\ 1 & 3 & 2 \\ -1 & 0 & 2 \end{vmatrix} = 2 \times 3 \begin{vmatrix} 1 & 1 & 0 \\ 1 & 1 & 2 \\ -1 & 0 & 2 \end{vmatrix} = 2 \times 3 \times 2 \begin{vmatrix} 1 & 1 & 0 \\ 1 & 1 & 1 \\ -1 & 0 & 1 \end{vmatrix} = -12.$$

(3) The value of a determinant is not altered if its rows are written as columns, in the same order.

Proof: Since the same value is obtained whether we expand a determinant by any row or any column, thus we have property (3). The following example will illustrate this property.

Example A2.3

$$D = \begin{vmatrix} 1 & 0 & 2 \\ -1 & 1 & 0 \\ 2 & -1 & 3 \end{vmatrix} = 1 \times \begin{vmatrix} 1 & 0 \\ -1 & 3 \end{vmatrix} - 0 \times \begin{vmatrix} -1 & 0 \\ 2 & 3 \end{vmatrix} + 2 \times \begin{vmatrix} -1 & 1 \\ 2 & -1 \end{vmatrix} = 1.$$

Now interchanging the rows and the columns, then evaluating the value of the resulting determinant, we find

$$\begin{vmatrix} 1 & -1 & 2 \\ 0 & 1 & -1 \\ 2 & 0 & 3 \end{vmatrix} = 1 \times \begin{vmatrix} 1 & -1 \\ 0 & 3 \end{vmatrix} - (-1) \times \begin{vmatrix} 0 & -1 \\ 2 & 3 \end{vmatrix} + 2 \times \begin{vmatrix} 0 & 1 \\ 2 & 0 \end{vmatrix} = 1,$$

illustrating property (3).

(4) If any two rows (or two columns) of a determinant are interchanged, the resulting determinant is the negative of the original determinant.

Proof: The proof is by induction. It is easy to see that it holds for 2×2 determinants. Assuming the result holds for $n \times n$ determinants, we shall show that it also holds for $(n+1) \times (n+1)$ determinants, thereby proving by induction that it holds in general.

Let B be an $(n+1) \times (n+1)$ determinant obtained from D by interchanging two rows. Expanding B in terms of a row that is not one of those interchanged, such as the kth row, we have

$$B = \sum_{j=1}^{n} (-1)^{j+k} b_{kj} M'_{kj},$$

where M'_{kj} is the minor of b_{kj}. Each b_{kj} is identical to the corresponding a_{kj} (the elements of D). Each M'_{kj} is obtained from the corresponding M_{kj} (of a_{kj}) by interchanging two rows. Thus $b_{kj} = a_{kj}$, and $M'_{kj} = -M_{kj}$. Hence

$$B = -\sum_{j=1}^{n} (-1)^{j+k} b_{kj} M_{kj} = -D.$$

The proof for columns is similar.

Example A2.4
Consider

$$D = \begin{vmatrix} 1 & 0 & 2 \\ -1 & 1 & 0 \\ 2 & -1 & 3 \end{vmatrix} = 1.$$

Now interchanging the first two rows, we have

$$B = \begin{vmatrix} -1 & 1 & 0 \\ 1 & 0 & 2 \\ 2 & -1 & 3 \end{vmatrix} = -1$$

illustrating property (4).

(5) If corresponding elements of two rows (or two columns) of a determinant are proportional, the value of the determinant is zero.

Proof: Let the elements of the ith and jth rows of D be proportional, say, $a_{ik} = c a_{jk}, k = 1, 2, \ldots, n$. If $c = 0$, then $D = 0$. For $c \neq 0$, then by property (2), $D = cB$, where the ith and jth rows of B are identical. Interchanging these two rows, B goes over to $-B$ (by property (4)). But the rows are identical, the new determinant is still B. Thus $B = -B, B = 0$, and $D = 0$.

Example A2.5

$$B = \begin{vmatrix} 1 & 1 & 2 \\ -1 & -1 & 0 \\ 2 & 2 & 8 \end{vmatrix} = 0, \quad D = \begin{vmatrix} 3 & 6 & -4 \\ 1 & -1 & 3 \\ -6 & -12 & 8 \end{vmatrix} = 0.$$

In B the first and second columns are identical, and in D the first and the third rows are proportional.

(6) If each element of a row of a determinant is a binomial, then the determinant can be written as the sum of two determinants, for example,

$$\begin{vmatrix} 4x+2 & 3 & 2 \\ x & 4 & 3 \\ 3x-1 & 2 & 1 \end{vmatrix} = \begin{vmatrix} 4x & 3 & 2 \\ x & 4 & 3 \\ 3x & 2 & 1 \end{vmatrix} + \begin{vmatrix} 2 & 3 & 2 \\ 0 & 4 & 3 \\ -1 & 2 & 1 \end{vmatrix}.$$

Proof: Expanding the determinant by the row whose terms are binomials, we will see property (6) immediately.

(7) If we add to the elements of a row (or column) any constant multiple of the corresponding elements in any other row (or column), the value of the determinant is unaltered.

Proof: Applying property (6) to the determinant that results from the given addition, we obtain a sum of two determinants: one is the original determinant and the other contains two proportional rows. Then by property (4), the second determinant is zero, and the proof is complete.

It is advisable to simplify a determinant before evaluating it. This may be done with the help of properties (7) and (2), as shown in the following example.

Example A2.6

Evaluate

$$
D = \begin{vmatrix} 1 & 24 & 21 & 93 \\ 2 & -37 & -1 & 194 \\ -2 & 35 & 0 & -171 \\ -3 & 177 & 63 & 234 \end{vmatrix}.
$$

To simplify this, we want the first elements of the second, third and last rows all to be zero. To achieve this, add the second row to the third, and add three times the first to the last, subtract twice the first row from the second; then develop the resulting determinant by the first column:

$$
D = \begin{vmatrix} 1 & 24 & 21 & 93 \\ 0 & -85 & -43 & 8 \\ 0 & -2 & -1 & 23 \\ 0 & 249 & 126 & 513 \end{vmatrix} = \begin{vmatrix} -85 & -43 & 8 \\ -2 & -1 & 23 \\ 249 & 126 & 513 \end{vmatrix}.
$$

We can simplify the resulting determinant further. Add three times the first row to the last row:

$$
D = \begin{vmatrix} -85 & -43 & 8 \\ -2 & -1 & 23 \\ -6 & -3 & 537 \end{vmatrix}.
$$

Subtract twice the second column from the first, and then develop the resulting determinant by the first column:

$$
D = \begin{vmatrix} 1 & -43 & 8 \\ 0 & -1 & 23 \\ 0 & -3 & 537 \end{vmatrix} = \begin{vmatrix} -1 & 23 \\ -3 & 537 \end{vmatrix} = -537 - 23 \times (-3) = -468.
$$

By applying the product rule of differentiation we obtain the following theorem.

Derivative of a determinant

If the elements of a determinant are differentiable functions of a variable, then the derivative of the determinant may be written as a sum of individual determinants, for example,

$$\frac{d}{dx}\begin{vmatrix} a & b & c \\ e & f & g \\ h & m & n \end{vmatrix} = \begin{vmatrix} a' & b' & c' \\ e & f & g \\ h & m & n \end{vmatrix} + \begin{vmatrix} a & b & c \\ e' & f' & g' \\ h & m & n \end{vmatrix} + \begin{vmatrix} a & b & c \\ e & f & g \\ h' & m' & n' \end{vmatrix},$$

where $a, b, \ldots, m, n$ are differentiable functions of x, and the primes denote derivatives with respect to x.

Problem A2.5
Show, without computation, that the following determinants are equal to zero:

$$\begin{vmatrix} 0 & a & -b \\ -a & 0 & c \\ b & -c & 0 \end{vmatrix}, \quad \begin{vmatrix} 0 & 2 & -3 \\ -2 & 0 & 4 \\ 3 & -4 & 0 \end{vmatrix}.$$

Problem A2.6
Find the equation of a plane which passes through the three points $(0, 0, 0)$, $(1, 2, 5)$, and $(2, -1, 0)$.

Appendix 3

$$\text{Table of}^* \; F(x) = \frac{1}{\sqrt{2\pi}} \int_0^x e^{-t^2/2} dt.$$

x	0.0	0.01	0.02	0.03	0.04	0.05	0.06	0.07	0.08	0.09
0.0	0.0000	0.0040	0.0080	0.0120	0.0160	0.0199	0.0239	0.0279	0.0319	0.0359
0.1	0.0398	0.0438	0.0478	0.0517	0.0557	0.0596	0.0636	0.0675	0.0714	0.0753
0.2	0.0793	0.0832	0.0871	0.0910	0.0948	0.0987	0.1026	0.1064	0.1103	0.1141
0.3	0.1179	0.1217	0.1255	0.1293	0.1331	0.1368	0.1406	0.1443	0.1480	0.1517
0.4	0.1554	0.1591	0.1628	0.1664	0.1700	0.1736	0.1772	0.1808	0.1844	0.1879
0.5	0.1915	0.1950	0.1985	0.2019	0.2054	0.2088	0.2123	0.2157	0.2190	0.2224
0.6	0.2257	0.2291	0.2324	0.2357	0.2389	0.2422	0.2454	0.2486	0.2517	0.2549
0.7	0.2580	0.2611	0.2642	0.2673	0.2704	0.2734	0.2764	0.2794	0.2823	0.2852
0.8	0.2881	0.2910	0.2939	0.2967	0.2995	0.3023	0.3051	0.3078	0.3106	0.3133
0.9	0.3159	0.3186	0.3212	0.3238	0.3264	0.3289	0.3315	0.3340	0.3365	0.3389
1.0	0.3413	0.3438	0.3461	0.3485	0.3508	0.3531	0.3554	0.3577	0.3599	0.3621
1.1	0.3643	0.3665	0.3686	0.3708	0.3729	0.3749	0.3770	0.3790	0.3810	0.3830
1.2	0.3849	0.3869	0.3888	0.3907	0.3925	0.3944	0.3962	0.3980	0.3997	0.4015
1.3	0.4032	0.4049	0.4066	0.4082	0.4099	0.4115	0.4131	0.4147	0.4162	0.4177
1.4	0.4192	0.4207	0.4222	0.4236	0.4251	0.4265	0.4279	0.4292	0.4306	0.4319
1.5	0.4332	0.4345	0.4357	0.4370	0.4382	0.4394	0.4406	0.4418	0.4429	0.4441
1.6	0.4452	0.4463	0.4474	0.4484	0.4495	0.4505	0.4515	0.4525	0.4535	0.4545
1.7	0.4554	0.4564	0.4573	0.4582	0.4591	0.4599	0.4608	0.4616	0.4625	0.4633
1.8	0.4641	0.4649	0.4656	0.4664	0.4671	0.4678	0.4686	0.4693	0.4699	0.4706
1.9	0.4713	0.4719	0.4726	0.4732	0.4738	0.4744	0.4750	0.4756	0.4761	0.4767
2.0	0.4472	0.4778	0.4783	0.4788	0.04793	0.4798	0.4803	0.4808	0.4812	0.4817
2.1	0.4821	0.4826	0.4830	0.4834	0.4838	0.4842	0.4846	0.4850	0.4854	0.4857
2.2	0.4861	0.4864	0.4868	0.4871	0.4875	0.4878	0.4881	0.4884	0.4887	0.4890
2.3	0.4893	0.4896	0.4898	0.4901	0.4904	0.4906	0.4909	0.4911	0.4913	0.4916
2.4	0.4918	0.4920	0.4922	0.4925	0.4927	0.4929	0.4931	0.4932	0.4934	0.4936
2.5	0.4938	0.4940	0.4941	0.4943	0.4945	0.4946	0.4948	0.4949	0.4951	0.4952
2.6	0.4953	0.4955	0.4956	0.4957	0.4959	0.4960	0.4961	0.4962	0.4963	0.4964
2.7	0.4965	0.4966	0.4967	0.4968	0.4969	0.4970	0.4971	0.4972	0.4973	0.4974
2.8	0.4974	0.4975	0.4976	0.4977	0.4977	0.4978	0.4979	0.4979	0.4980	0.4981
2.9	0.4981	0.4982	0.4982	0.4983	0.4984	0.4984	0.4985	0.4986	0.4986	0.4986
3.0	0.4987	0.4987	0.4987	0.4988	0.4988	0.4989	0.4989	0.4989	0.4990	0.4990

x	0.0	0.2	0.4	0.6	0.8
1.0	0.3413447	0.3849303	0.4192433	0.4452007	0.4640697
2.0	0.4772499	0.4860966	0.4918025	0.4953388	0.4974449
3.0	0.4986501	0.4993129	0.4998409	0.4999277	0.4999277
4.0	0.4999683	0.4999867	0.4999946	0.4999979	0.4999992

Further reading

Anton, Howard, *Elementary Linear Algebra*, 3rd ed., John Wiley, New York, 1982.

Arfken, G. B., Weber, H. J., *Mathematical Methods for Physicists*, 4th ed., Academic Press, New York, 1995.

Boas, Mary L., *Mathematical Methods in the Physical Sciences*, 2nd ed., John Wiley, New York, 1983.

Butkov, Eugene, *Mathematical Physics*, Addison-Wesley, Reading (MA), 1968.

Byon, F. W., Fuller, R. W., *Mathematics of Classical and Quantum Physics*, Addison-Wesley, Reading (MA), 1968.

Churchill, R. V., Brown, J. W., Verhey, R. F., *Complex Variables & Applications*, 3rd ed., McGraw-Hill, New York, 1976.

Harper, Charles, *Introduction to Mathematical Physics*, Prentice Hall, Englewood Cliffs, NJ, 1976.

Kreyszig, E., *Advanced Engineering Mathematics*, 3rd ed., John Wiley, New York, 1972.

Joshi, A. W., *Matrices and Tensor in Physics*, John Wiley, New York, 1975.

Joshi, A. W., *Elements of Group Theory for Physicists*, John Wiley, New York, 1982.

Lass, Harry, *Vector and Tensor Analysis*, McGraw-Hill, New York, 1950.

Margenus, Henry, Murphy, George M., *The Mathematics of Physics and Chemistry*, D. Van Nostrand, New York, 1956.

Mathews, Fon, Walker, R. L., *Mathematical Methods of Physics*, W. A. Benjamin, New York, 1965.

Spiegel, M. R., *Advanced Mathematics for Engineers and Scientists*, Schaum's Outline Series, McGraw-Hill, New York, 1971.

Spiegel, M. R., *Theory and Problems of Vector Analysis*, Schaum's Outline Series, McGraw-Hill, New York, 1959.

Wallace, P. R., *Mathematical Analysis of Physical Problems*, Dover, New York, 1984.

Wong, Chun Wa, *Introduction to Mathematical Physics, Methods and Concepts*, Oxford, New York, 1991.

Wylie, C., *Advanced Engineering Mathematics*, 2nd ed., McGraw-Hill, New York, 1960.

Index